Structural Integrity

Volume 21

The *Structural Integrity* book series is a high level academic and professional series publishing research on all areas of Structural Integrity. It promotes and expedites the dissemination of new research results and tutorial views in the structural integrity field.

The Series publishes research monographs, professional books, handbooks, edited volumes and textbooks with worldwide distribution to engineers, researchers, educators, professionals and libraries.

Topics of interested include but are not limited to:

- Structural integrity
- Structural durability
- Degradation and conservation of materials and structures
- Dynamic and seismic structural analysis
- Fatigue and fracture of materials and structures
- Risk analysis and safety of materials and structural mechanics
- Fracture Mechanics
- Damage mechanics
- Analytical and numerical simulation of materials and structures
- Computational mechanics
- Structural design methodology
- Experimental methods applied to structural integrity
- Multiaxial fatigue and complex loading effects of materials and structures
- Fatigue corrosion analysis
- Scale effects in the fatigue analysis of materials and structures
- Fatigue structural integrity
- Structural integrity in railway and highway systems
- Sustainable structural design
- Structural loads characterization
- Structural health monitoring
- Adhesives connections integrity
- Rock and soil structural integrity.

**** Indexing: The books of this series are submitted to Web of Science, Scopus, Google Scholar and Springerlink ****

This series is managed by team members of the ESIS/TC12 technical committee.

Springer and the Series Editors welcome book ideas from authors. Potential authors who wish to submit a book proposal should contact Dr. Mayra Castro, Senior Editor, Springer (Heidelberg), e-mail: mayra.castro@springer.com

More information about this series at http://www.springer.com/series/15775

Alexandre Cury · Diogo Ribeiro · Filippo Ubertini ·
Michael D. Todd
Editors

Structural Health Monitoring Based on Data Science Techniques

Editors
Alexandre Cury
Department of Applied and Computational Mechanics
Federal University of Juiz de Fora
Juiz de Fora, Brazil

Diogo Ribeiro
Department of Civil Engineering, School of Engineering
Polytechnic Institute of Porto
Porto, Portugal

Filippo Ubertini
Department of Civil and Environmental Engineering
University of Perugia
Perugia, Italy

Michael D. Todd
Department of Structural Engineering
University of California San Diego
La Jolla, CA, USA

ISSN 2522-560X ISSN 2522-5618 (electronic)
Structural Integrity
ISBN 978-3-030-81718-3 ISBN 978-3-030-81716-9 (eBook)
https://doi.org/10.1007/978-3-030-81716-9

This Springer imprint is published by the registered company Springer Nature Switzerland AG
The registered company address is: Gewerbestrasse 11, 6330 Cham, Switzerland

Foreword

One can speculate that there has been an interest in detecting damage in engineered systems since man has used tools. Over time, early ad hoc qualitative damage detection procedures, many of which were vibration-based, evolved into more refined approaches that became what we know today as non-destructive evaluation (NDE) methods. One drawback of most NDE methods is that the system being inspected must be taken out of service and often disassembled before such methods can be used. Structural health monitoring (SHM) attempts to address this shortcoming by developing more continuous, automated in situ damage detection capabilities that strives to minimize the human-in-loop aspect of the assessment process.

The term structural health monitoring begins to appear regularly in the technical literature around the late 1980s and early 1990s. These early studies focused primarily on deterministic, inverse physics-based modeling approaches that identified the presence, location, and extent of damage. When researchers and practitioners attempted to apply such methods to in situ structures, various limitations were identified including difficulties handling the mismatch between measured and analytical degrees of freedom, the almost exclusive use of linear models when simulating both the undamaged and damaged system response, and the inability to handle the operational and environmental variability that all real-world systems experience. The latter limitation associated with the fact that operational and environmental variability will cause changes in the SHM system sensor readings and these changes must be distinguished from changes in sensor reading caused by damage has proven to be one of the most significant challenges associated with transitioning SHM research to practice.

In the late 1990s and early 2000s, various research groups started to recognize that SHM is not a deterministic problem. Instead, they proposed to address SHM through more data-driven approaches based on general statistical-pattern-recognition-based methodologies. Although many variations of this statistical pattern recognition approach have been proposed in different SHM studies; almost all encompass three common components: 1. A deployed sensing system typically monitoring kinematic response quantities; 2. the extraction of damage-sensitive features from the raw sensor data; and 3. the statistical classification of those features into damage and undamaged categories. A common misconception with these approaches is that they preclude the use of physics-based models when, in fact, the pattern recognition will always be improved when it is based on knowledge of the physics governing the system response in both its undamaged and damaged states.

This paradigm shift from inverse deterministic modeling to statistical-pattern-recognition-based SHM began the process of adopting many data-driven algorithms from disparate fields such as radar and sonar detection, machine learning, speech-pattern recognition, statistical decision theory and econometrics to the SHM problem. In aggregate, these fields represent components of the more general field referred to as data science. This focus on applying elements of data science to SHM has mostly replaced the earlier deterministic inverse modeling approaches. Furthermore, data science offers approaches that can better address the random and systematic changes in sensor measurements caused by operational and environmental variability and can produce a quantified probability of detection measure. Both attributes are essential for the adaptation of SHM by asset owners and regulatory agencies.

Currently, all scientific and engineering fields are benefitting from the rapid advances in data science and the associated availability of general software tools for implementing these algorithms. The field of structural health monitoring is one such beneficiary. However, as in other technical fields, innovation and application-specific knowledge are required to effectively adapt these general tools to domain-specific problems. This book provides numerous examples from aerospace, civil, and mechanical engineering applications that demonstrate how SHM researchers have taken the tools of data science and creatively adapted them to address many problems that have been limiting the more widespread adaptation of SHM by industry. The chapters in this book show the breadth of data science methodologies that can be applied to SHM. Furthermore, these chapters demonstrate that advances in data science can impact every aspect of a SHM process. As such, this book will provide experienced researchers new to the data science field an overview of how such tools can be used in a damage detection context. Additionally, this book will provide those just beginning their technical careers with ideas for new research and application directions to pursue and the associated technologies they will need to learn that will be the foundation for making future advances in SHM.

Dr. Charles Farrar
Los Alamos National Laboratory
Los Alamos, New Mexico, USA

Preface

Structural health monitoring (SHM) may be defined as the general process of making an assessment, based on appropriate analyses of in situ measured data, about the current ability of a structural component or system to perform its intended design function(s) successfully. A successful SHM strategy may enable significant ownership cost reduction in a life cycle perspective through maintenance optimization, performance maximization during operation, unscheduled downtime minimization, and/or enable significant life safety advantage through catastrophic failure mitigation. Broadly speaking, SHM strategies for most applications necessarily integrate real-time data acquisition, feature extraction from the acquired data, statistical modeling of the features, and classification of the features to make informed decisions; the ultimate global goal of SHM systems is to direct economically efficient and/or safety-maximized structural health decision making for the general purpose of long-term effective life cycle management.

An explosion of approaches that address some or part of this overall SHM strategy has occurred in recent years, across many different structural applications ranging from civil to aerospace to industrial/mechanical systems. A significant fraction of this growth has been fueled by ubiquitous "Internet of Things (IoT)" data streams from diverse sources, advances in computing such as cloud computing, and the adoption and development of advanced analytics techniques drawn from machine learning and data science. This domain of advancement in SHM can address some of the paramount challenges in long-term monitoring of civil structures such as, but not limited to, (i) structural complexity, (ii) operational and environmental variability (e.g., loading conditions, operating environment), (iii) complex, interconnected degradation and failure modes, (iv) challenges in monitoring very large-scale structures with potentially localized failure modes (e.g., pitting corrosion), and (v) data reliability and security, including long-term functionalities of sensor networks.

Damage identification as well as continuous condition monitoring are among the most important aspects related to proper operation of structural systems to ensure their integrity, safety, and desirable operational properties. In recent years, an exponential development of damage identification methods as well as condition monitoring has been observed. The degradation process of structural systems is usually due to a combination of reasons, such as materials aging, ineffective maintenance, design or constructive issues, unexpected loading events, natural hazards (e.g., earthquakes), and more.

Most damage identification strategies are developed primarily based on the signals monitored over time, often seeking for an effective fusion between heterogeneous sensor data, such as, history of structural accelerations, displacements, strains, time series of environmental parameters, and more. However, with the evolution of computational and information technologies, remarkable improvements are being observed in data acquisition systems, which, in turn, demand further development of structural monitoring tools and techniques to deal with large volumes of data. Hence, analyses that were earlier performed incipiently with a reduced number of variables, i.e., by means of modal and/or probability/statistical analyses, now are being automatically carried out with the aid of powerful machine learning methods, such as artificial neural networks and support vector machines.

One observes, however, that some key aspects still play major roles on the performance of damage identification algorithms applied to large-scale structural systems: (i) the high dimensionality of the parameters monitored; (ii) environmental/operational factors, such as temperature, humidity, and traffic; (iii) structural complexity; (iv) reliability of the measured data; (v) low sensitivity of global structural response to local damage; and (vi) the need to integrate physical/engineering knowledge in machine learning algorithms enabling an effective SHM data to decisions process.

This book has 22 chapters and contains a representative collection of actual uses of data science in SHM, ranging from civil to mechanical/aerospace system applications. Chapters 1–3 cover different Bayesian-based strategies for structural damage detection. Chapters 4–8 address the use of data-driven techniques and their aptness for real-time structural condition assessment, especially considering raw vibration measurements as inputs. Chapters 9 and 10 continue this discussion by bringing physics-based and reduced order modeling aspects into the SHM paradigm. Chapter 11 discusses how deep learning can assist image processing for increasing safety in construction sites. Chapters 12–15 consider the influence of both environmental and operational effects and present strategies to circumvent them when it comes to structural damage detection. Chapters 16–20 explore some recent concepts regarding explainable artificial intelligence, physics-informed and interpretable machine learning, as well as novel developments involving population-based SHM. Chapters 21 and 22 conclude this book with an overview of structural damage detection via remotely sensed data and with a discussion about new designs for SHM systems.

In summary, this book is addressed to scientists, engineers, designers, technicians, stakeholders, and contractors who seek an up-to-date view of the recent advances in the field of data science applied to SHM.

Juiz de Fora, Brazil — Alexandre Cury
Porto, Portugal — Diogo Ribeiro
Perugia, Italy — Filippo Ubertini
La Jolla, USA — Michael D. Todd

Acknowledgements

The editors would like to express their sincere gratitude to several professionals and their organizations that directly or indirectly contribute for the realization of this book; to the Springer Structural Integrity Series Editors, Dr. José A. F. O. Correia and Prof. Abílio M. P. De Jesus for all the motivation and encouragement to submit a book proposal in a new and challenging thematic area for structural health monitoring; to the Senior Editor from Springer, Dr. Mayra Castro, as well as the Coordinators of the Springer Production team, Vijay Kumar and Augustus Vinoth, for the prompt support in addressing all questions during the book preparation; and to all authors of the book chapters for their extraordinary commitment with the objectives of the publication and their enthusiasm to be part of this project. The authors come from several companies/institutions from both academia and industry around the world.

Editor Alexandre Cury would like to thank his family, colleagues, and undergraduate and graduate students which were and are part of the Graduate Programs in Civil Engineering at the University of Juiz de Fora (PEC/UFJF) and University of Ouro Preto (PROPEC/UFOP). Their constant support and encouragement were essential to all the academic accomplishments that drove him to be a part of this project. Special thanks to colleague and friend Prof. Flávio Barbosa (UFJF) for presenting the initial concepts of SHM and structural damage detection; to Dr. Christian Cremona (Bouygues Construction) for being a deeply personal and scientific inspiration; and last but not least, to the financial support agencies FAPEMIG, CNPq and CAPES for their pivotal role in funding research and innovation in Brazil during these challenging economic times.

Editor Diogo Ribeiro would like to express his gratitude to some colleagues and institutions that provided the necessary support and motivation to conclude this editorial project; to the School of Engineering of Polytechnic of Porto (ISEP) and CONSTRUCT R&D unit for providing all the conditions to conduct research in the fields of SHM and damage identification; to the friends and professors from ISEP, Maria da Fátima Portela, Maria do Rosário Oliveira, Ricardo Santos and Teresa Neto, for the permanent companionship and mutual help, particularly in assuming several background activities that provide me the necessary time for the book project; to my Ph.D. student and friend Andreia Meixedo for all the fruitful discussions and mutual learning during the last years in the thematic of damage identification; and to Prof. Rui Calçada from Faculty of Engineering of University of Porto (FEUP) for the constant scientific inspiration as well as for the encouraging words at the beginning of this editorial project.

Editor Filippo Ubertini would like to express his gratitude to all colleagues, post-doctoral scholars, Ph.D. students, and technicians currently working within his "SHM Lab" group at University of Perugia, as well as to the formerly graduated Ph.D. students, Master students, and international visiting students, some of which are now building their own careers in academia at an international level. Special thanks also go to worldwide distinguished collaborators and particularly to Prof. Simon Laflamme from Iowa State University, with whom a quite fruitful scientific collaboration in SHM and a sincere friendship have started more than 10 years ago.

Michael D. Todd would like to thank first and foremost all of the students and postdoctoral scholars whom he has had the good fortune to mentor throughout his career. Working with them and helping them launch and develop their careers are the most gratifying part of being an academic. He also thanks his many collaborators worldwide for a very satisfying and productive career, but he would like to express special gratitude to Chuck Farrar at the Los Alamos National Laboratory, who introduced him to structural health monitoring more than 20 years ago. He has truly been a mentor and a friend.

Finally, the editors would like to thank all the readers of this book for their interest and passion in shaping the future of data science applied to structural health monitoring.

Contents

Vibration-Based Structural Damage Detection Using Sparse Bayesian Learning Techniques

Rongrong Hou, Xiaoyou Wang, and Yong Xia

Abstract Vibration-based structural damage detection constantly involves uncertainties, including measurement noise, methodology, and modeling errors. Bayesian inference provides a rigorous framework to consider uncertainties and obtain probabilistic solutions. In recent decades, sparse Bayesian learning (SBL) and the closely related automatic relevance determination model have been extensively used, resulting in sparse solution. Given that damage typically occurs in limited sections or members, particularly at the early stage of structural failure, the SBL method is developed for structural damage detection using vibration data. However, analytical posterior probability density function is unavailable owing to the high-dimensional integral in the evidence and nonlinear relationship between the measured modal and structural parameters. Therefore, a range of techniques are utilized to obtain solutions based on analytical approximations or numerical sampling, including the expectation–maximization, Laplace approximation, variational Bayesian inference, and delayed rejection adaptive metropolis techniques. Numerical and experimental examples demonstrate that the proposed SBL method can accurately locate and quantify sparse damage. In addition, the mechanisms, advantages, and limitations of different analytical and numerical techniques are described and compared, and the corresponding suggestions for their applications are proposed.

Keywords Vibration-based structural damage detection · Uncertainties · Sparse Bayesian learning · Analytical approximations · Numerical sampling

R. Hou · X. Wang · Y. Xia (✉)
The Hong Kong Polytechnic University, Kowloon, Hong Kong, China
e-mail: ceyxia@polyu.edu.hk

R. Hou
e-mail: rongrong.hou@connect.polyu.hk

X. Wang
e-mail: xiaoyou.wang@connect.polyu.hk

R. Hou
Harbin Institute of Technology, Harbin, China

A. Cury et al. (eds.), *Structural Health Monitoring Based on Data Science Techniques*, Structural Integrity 21, https://doi.org/10.1007/978-3-030-81716-9_1

1 Sparse Bayesian Learning for Structural Damage Detection

Structural damage detection constantly involves uncertainties, which may be categorized as modeling errors, methodology errors, and measurement noise [1–3]. Moreover, operational and environmental variations cause significant changes in the identified modal parameters [4, 5]. Numerous studies have proposed probabilistic approaches to address the uncertainties in structural damage detection. Representative approaches include perturbation techniques [3, 6], Monte Carlo simulation [7], statistical pattern recognition [8], and Bayesian methods [9–13]. Among these methods, Bayesian inference has attracted considerable attention since the 1990s [9].

In practice, structural damage commonly appears in a few sections or members only, particularly at the early stage. Therefore, structural damage possesses sparsity compared with the numerous elements of the entire structure. Sparse Bayesian learning (SBL) is effective in promoting sparsity in the inferred predictors, which treats the sparse feature as the prior [14]. SBL is widely applied to sparse signal reconstruction and compressed sensing [15, 16] and has been developed for structural damage detection in recent years [17–22]. However, the application of the SBL framework to the damage detection of real structures suffers from the common limitation. For the majority of problems, an analytical posterior probability density function (PDF) is not available owing to the computationally prohibited integration in the evidence, which is the result of the high-dimension and nonlinear relationship between the measured modal data and structural parameters [23]. Analytical and numerical techniques have been developed to circumvent this difficulty.

The analytical methods include hierarchical modelling and asymptotic techniques. Huang et al. [17, 18] proposed a hierarchical Bayesian model (HBM) by expanding the nonlinear eigenvalue problem as the product of a series of coupled hierarchical linear PDFs, thereby simplifying the calculation and making the posterior PDF analytically tractable. However, HBM is of low efficiency given the need to estimate a series of additional parameters introduced by hyper-priors. Moreover, the successful application of HBM requires proper assumptions of all hyper-priors, which is not guaranteed owing to the demand for sufficient information on individual hyper-parameter. Given that the uncertainties of the posterior are directly related to the variance of the prior [24], the inappropriate assumptions of hyper-priors tend to increase the uncertainties of the prior, further reduce the robustness of the model, and lead to a decrease in the sensitivity of the posterior PDF to the prior information. This phenomenon worsens with the increase of hierarchical layers.

The numerical techniques include the expectation–maximization (EM) technique and sampling techniques, such as Markov chain Monte Carlo (MCMC) simulation. With MCMC, samples consistent with the posterior probability distribution can be generated [25, 26]. Thereafter, the asymptotic solution of the posterior PDF is obtained from these samples. One advantage of the MCMC approach is that it can provide a full characterization of the posterior PDF instead of merely focusing on the most probable values (MPVs) [25]. However, the basic MCMC technique is extraordinarily time-consuming for models containing numerous parameters or involving partial differential equations [26]. To improve computational efficiency, the Metropolis–Hastings algorithm and Gibbs sampling (GS) technique, as special cases of MCMC, have been developed to deal with multi-dimensional problems. Huang et al. [19] applied the GS technique to provide a full treatment of the posterior PDFs of uncertain parameters for structural damage assessment. However, GS is only applicable to standard distributions and converges slowly for numerous hierarchical models [24].

In this regard, this chapter will introduce three techniques based on analytical approximations or numerical sampling to improve the efficiency, accuracy, and feasibility of the aforementioned algorithms.

2 Bayesian Probabilistic Framework

2.1 Structural Model Class

The structural model class $\mathcal{M}$ is based on a set of linear structural models, in which each model has a known mass matrix $\mathbf{M}$ and uncertain stiffness matrix $\mathbf{K}$ parameterized by the stiffness parameters as follows:

$$\mathbf{K} = \sum_{i=1}^{n} \alpha_i \mathbf{K}_i, \tag{1}$$

where $\mathbf{K}_i$ is the ith element stiffness matrix that can be obtained from the finite element model of a structure, α_i is the ith element stiffness parameter to be updated according to the observed data, and n is the number of structural elements. The change in mass is assumed to be negligible when damage occurs.

Suppose that the element stiffness parameter reduces to $\overline{\alpha}_i$, the stiffness reduction factor (SRF) is defined as follows [27]:

$$\theta_i = \frac{\overline{\alpha}_i - \alpha_i}{\alpha_i}. \tag{2}$$

SRF, as the damage index, indicates damage location and damage severity. Given that structural damage typically occurs at several sections or members only, $\boldsymbol{\theta}$ is a sparse vector with only several nonzero components at the damaged locations.

The rth structural eigenvalue and the corresponding mode shape are governed by the following eigenvalue equation:

$$(\mathbf{K} - \boldsymbol{\lambda}_r \mathbf{M})\phi_r = \mathbf{0}. \tag{3}$$

Suppose that N_{m} modes of vibration have been identified from modal testing, the identified eigenvalues and mode shapes can be expressed as follows:

$$\hat{\boldsymbol{\lambda}} = \left\{\hat{\lambda}_1, \hat{\lambda}_2, \ldots, \hat{\lambda}_{N_{\mathrm{m}}}\right\}, \tag{4}$$

$$\hat{\boldsymbol{\psi}} = \left[\hat{\boldsymbol{\phi}}_1, \hat{\boldsymbol{\phi}}_2, \ldots, \hat{\boldsymbol{\phi}}_{N_{\mathrm{m}}}\right], \tag{5}$$

where $\hat{\boldsymbol{\phi}}_r \in R^{N_{\mathrm{p}}}$ denotes the identified mode shape of the rth mode at the N_{p} measurement points. A set of modal data is expressed as follows:

$$\mathcal{D} = \left[\hat{\boldsymbol{\lambda}}, \hat{\boldsymbol{\psi}}\right]. \tag{6}$$

2.2 Bayesian Model Updating Framework

Bayes' theorem is used to develop a PDF for the damage indexes $\boldsymbol{\theta}$, conditional on the measured modal data and chosen class of models [28]:

$$p(\boldsymbol{\theta}|\mathcal{D}, \mathcal{M}) = c^{-1} p(\mathcal{D}|\boldsymbol{\theta}, \mathcal{M}) p(\boldsymbol{\theta}|\mathcal{M}), \tag{7}$$

where $p(\boldsymbol{\theta}|\mathcal{D}, \mathcal{M})$ is the posterior PDF of the damage indexes given the modal data $\mathcal{D}$ and model class $\mathcal{M}$; $c = p(\mathcal{D}|\mathcal{M})$ is a normalizing constant referred to as evidence; $p(\mathcal{D}|\boldsymbol{\theta}, \mathcal{M})$ is the likelihood function representing the PDF of modal data given the damage indexes; and $p(\boldsymbol{\theta}|\mathcal{M})$ is the prior PDF of the damage indexes. To simplify the notation, the dependence of the PDF on $\mathcal{M}$ is dropped thereafter.

2.3 Likelihood Function for Structural Modal Parameters

According to the axioms of probability, PDF of the modal data $p(\mathcal{D}|\boldsymbol{\theta})$ in Eq. (7) can be expressed as follows [28]:

$$p(\mathcal{D}|\boldsymbol{\theta}) = \prod_{r=1}^{N_{\mathrm{m}}} p\left(\hat{\lambda}_r|\boldsymbol{\theta}\right) p\left(\hat{\boldsymbol{\phi}}_r|\boldsymbol{\theta}\right). \tag{8}$$

Given the damage index, the modal parameters are deemed independent mode by mode.

The measurement errors $\boldsymbol{\varepsilon}_r$ and $\boldsymbol{e}_r$ are assumed to follow the Gaussian distribution with a zero mean and diagonal variance matrix [29], which are, respectively, given as follows:

$$\boldsymbol{\varepsilon}_r = \frac{\hat{\lambda}_r - \lambda_r}{\hat{\lambda}_r} \sim N\left(0, \beta^{-1}\right), \tag{9}$$

$$\boldsymbol{e}_r = \hat{\boldsymbol{\phi}}_r - \boldsymbol{\phi}_r \sim N\left(0, \gamma^{-1}\boldsymbol{I}\right), \tag{10}$$

where the hyper-parameters β and γ equal the reciprocal of the variance, representing the precision of the measured eigenvalue and mode shapes, respectively.

The resulting likelihood function of $\boldsymbol{\theta}$ based on the measured eigenvalues $\hat{\lambda}$ and $\hat{\boldsymbol{\phi}}$ are given as follows:

$$p\left(\hat{\boldsymbol{\lambda}}|\boldsymbol{\theta}, \beta\right) = \left(\frac{\beta}{2\pi}\right)^{\frac{N_{\mathrm{m}}}{2}} \exp\left\{-\frac{\beta}{2}\sum_{r=1}^{N_{\mathrm{m}}}\left[\frac{\hat{\lambda}_r - \lambda_r(\boldsymbol{\theta})}{\hat{\lambda}_r}\right]^2\right\}, \tag{11}$$

$$p\left(\hat{\boldsymbol{\phi}}|\boldsymbol{\theta}, \gamma\right) = \left(\frac{\gamma}{2\pi}\right)^{\frac{N_{\mathrm{p}}\cdot N_{\mathrm{m}}}{2}} \exp\left\{-\frac{\gamma}{2}\sum_{r=1}^{N_{\mathrm{m}}}\left\|\hat{\boldsymbol{\phi}}_r - \boldsymbol{\phi}_r(\boldsymbol{\theta})\right\|_2^2\right\}. \tag{12}$$

2.4 Prior PDF of the Damage Indexes

In the Bayesian framework, the damage indexes $\boldsymbol{\theta}$ are assigned with a prior distribution according to engineering judgments and knowledge. Moreover, the suggestion is to adopt the conjugate prior to simplify the calculation. SBL proposes to use the automatic relevance determination (ARD) prior to incorporate a preference for sparser parameters [17, 18, 30]. ARD prior following the Gaussian distribution is conjugated to the likelihood function in this study and adopted as the prior, which is expressed as follows:

$$p(\boldsymbol{\theta}|\boldsymbol{\alpha}) = \prod_{i=1}^{n} p(\theta_i|\alpha_i) = \left(\frac{1}{2\pi}\right)^{\frac{n}{2}} \prod_{i=1}^{n}\left[\alpha_i^{\frac{1}{2}} \exp\left\{-\frac{1}{2}\alpha_i\theta_i^2\right\}\right], \tag{13}$$

where the individual hyper-parameter α_i represents the precision of the associated damage index θ_i.

2.5 Posterior PDF of the Damage Indexes

From Eq. (7), the posterior PDF of the damage indexes $\boldsymbol{\theta}$ is expressed as follows:

$$
\begin{aligned}
p\left(\theta|\hat{\lambda},\hat{\psi},\alpha,\beta,\gamma\right) &= c^{-1}p\left(\hat{\lambda},\hat{\psi}|\theta,\beta,\gamma\right)p(\theta|\alpha) \\
&= c^{-1}p\left(\hat{\lambda}|\theta,\beta\right)p\left(\hat{\psi}|\theta,\gamma\right)p(\theta|\alpha) \\
&= c^{-1}\left(\frac{\beta}{2\pi}\right)^{\frac{N_m}{2}}\left(\frac{\gamma}{2\pi}\right)^{\frac{N_p\cdot N_m}{2}}\left(\frac{1}{2\pi}\right)^{\frac{n}{2}}\left(\prod_{i=1}^{n}\alpha_i^{\frac{1}{2}}\right) \\
&\exp\left\{-\frac{\beta}{2}\sum_{r=1}^{N_m}\left[\frac{\hat{\lambda}_r-\lambda_r(\theta)}{\hat{\lambda}_r}\right]^2-\frac{\gamma}{2}\sum_{r=1}^{N_m}\left\|\hat{\phi}_r-\phi_r(\theta)\right\|_2^2-\frac{1}{2}\sum_{i=1}^{n}\left(\alpha_i\theta_i^2\right)\right\},
\end{aligned} \tag{14}
$$

with the distributions on the right-hand side as defined by Eqs. (11), (12), and (13). The evidence is calculated by integrating $p\left(\hat{\lambda},\hat{\psi},\theta|\alpha,\beta,\gamma\right)$ with respect to θ:

$$
c = p\left(\hat{\lambda},\hat{\psi}|\alpha,\beta,\gamma\right) = \int p\left(\hat{\lambda},\hat{\psi},\theta|\alpha,\beta,\gamma\right)d\theta. \tag{15}
$$

The maximum posterior (MAP) estimate of the damage indexes $\tilde{\theta}$ in Eq. (14) can be calculated by maximizing $\ln p\left(\theta|\hat{\lambda},\hat{\psi},\alpha,\beta,\gamma\right)$, or equivalently minimizing the following objective function:

$$
J(\theta) = \beta\sum_{r=1}^{N_m}\left[\frac{\hat{\lambda}_r-\lambda_r(\theta)}{\hat{\lambda}_r}\right]^2+\gamma\sum_{r=1}^{N_m}\left\|\hat{\phi}_r-\phi_r(\theta)\right\|_2^2+\sum_{i=1}^{n}\left(\alpha_i\theta_i^2\right). \tag{16}
$$

The items unrelated with θ are omitted in the preceding equation. Note that Eq. (16) is similar to the regularization technique, in which the first two terms are equivalent to the data-fitting terms with different weights and the third term to the regularization term with α as the regularization parameter. We will demonstrate that Eq. (16) closely resembles the l_0 regularization (Refer to Eq. 20).

The minimization of Eq. (16) cannot be directly obtained because parameters $\{\alpha,\beta,\gamma\}$ are unknown. MPVs of the hyper-parameters $\{\alpha,\beta,\gamma\}$ can be calculated by maximizing $p\left(\alpha,\beta,\gamma|\hat{\lambda},\hat{\psi}\right)$. In particular,

$$
p\left(\alpha,\beta,\gamma|\hat{\lambda},\hat{\psi}\right) = \frac{p\left(\hat{\lambda},\hat{\psi}|\alpha,\beta,\gamma\right)p(\alpha,\beta,\gamma)}{p\left(\hat{\lambda},\hat{\psi}\right)} \propto p\left(\hat{\lambda},\hat{\psi}|\alpha,\beta,\gamma\right), \tag{17}
$$

where we assume that $p(\alpha,\beta,\gamma)$ is uniformly distributed as hyper-prior information is unknown. Consequently, the hyper-parameters $\{\alpha,\beta,\gamma\}$ can be estimated by maximizing $p\left(\hat{\lambda},\hat{\psi}|\alpha,\beta,\gamma\right)$ (i.e., evidence defined in Eq. 15). However, the computation of the integral in Eq. (15) is intractable because the frequencies and mode shapes are in nonlinear relations with $\boldsymbol{\theta}$. Therefore, analytical and numerical techniques are required. In the following sections, EM, Laplace's approximation and variational Bayesian inference (VBI), and delayed rejection adaptive metropolis (DRAM) algorithm will be introduced to solve this problem.

3 Bayesian Inference Using the EM Algorithm

The EM algorithm is proposed to maximize the natural log evidence $\ln p\left(\hat{\boldsymbol{\lambda}}, \hat{\boldsymbol{\psi}}|\boldsymbol{\alpha}, \beta, \gamma\right)$ iteratively. This algorithm alternates between performing an expectation (E) step and maximization (M) step. In particular, $\boldsymbol{\theta}$ is regarded as a latent variable, and $\left\{\boldsymbol{\theta}, \hat{\boldsymbol{\lambda}}, \hat{\boldsymbol{\psi}}\right\}$ is referred to as the complete data set. The complete-data natural log likelihood function is expressed as follows:

$$\ln p\left(\boldsymbol{\theta}, \hat{\boldsymbol{\lambda}}, \hat{\boldsymbol{\psi}}|\boldsymbol{\alpha}, \beta, \gamma\right) = \ln p\left(\hat{\boldsymbol{\lambda}}|\boldsymbol{\theta}, \beta\right) + \ln p\left(\hat{\boldsymbol{\psi}}|\boldsymbol{\theta}, \gamma\right) + \ln p(\boldsymbol{\theta}|\boldsymbol{\alpha}). \tag{18}$$

Given the difficulty of direct maximization of $\ln p\left(\hat{\boldsymbol{\lambda}}, \hat{\boldsymbol{\psi}}|\boldsymbol{\alpha}, \beta, \gamma\right)$ with respect to $\{\boldsymbol{\alpha}, \beta, \gamma\}$, the EM algorithm proposes to maximize the expectation of the complete-data $E\left[\ln p\left(\boldsymbol{\theta}, \hat{\boldsymbol{\lambda}}, \hat{\boldsymbol{\psi}}|\boldsymbol{\alpha}, \beta, \gamma\right)\right]$ instead [31, 32], such that:

$$\begin{aligned} E\left[\ln p\left(\boldsymbol{\theta}, \hat{\boldsymbol{\lambda}}, \hat{\boldsymbol{\psi}}|\boldsymbol{\alpha}, \beta, \gamma\right)\right] &= \frac{N_{\mathrm{m}}}{2}\ln\left(\frac{\beta}{2\pi}\right) - \frac{\beta}{2}E\left\{\sum_{r=1}^{N_{\mathrm{m}}}\left[\frac{\hat{\lambda}_r - \lambda_r(\boldsymbol{\theta})}{\hat{\lambda}_r}\right]^2\right\} \\ &+ \frac{N_{\mathrm{p}}N_{\mathrm{m}}}{2}\ln\left(\frac{\gamma}{2\pi}\right) - \frac{\gamma}{2}E\left[\sum_{r=1}^{N_{\mathrm{m}}}\left\|\hat{\boldsymbol{\phi}}_r - \boldsymbol{\phi}_r(\boldsymbol{\theta})\right\|_2^2\right] \\ &+ \frac{n}{2}\ln\left(\frac{1}{2\pi}\right) + \frac{1}{2}\sum_{i=1}^{n}\ln\alpha_i - \frac{1}{2}\sum_{i=1}^{n}\alpha_i E\left(\theta_i^2\right). \end{aligned} \tag{19}$$

In practice, the complete data set is not available, and the latent variable $\boldsymbol{\theta}$ is given by the posterior distribution $p\left(\boldsymbol{\theta}|\hat{\boldsymbol{\lambda}}, \hat{\boldsymbol{\psi}}, \boldsymbol{\alpha}, \beta, \gamma\right)$. In the E step, given the current values of the hyper-parameters $\{\boldsymbol{\alpha}, \beta, \gamma\}^{\mathrm{old}}$, the posterior distribution of $\boldsymbol{\theta}$ given by $p\left(\boldsymbol{\theta}|\hat{\boldsymbol{\lambda}}, \hat{\boldsymbol{\psi}}, \boldsymbol{\alpha}^{\mathrm{old}}, \beta^{\mathrm{old}}, \gamma^{\mathrm{old}}\right)$ is used to determine the expectation of the complete-data $E\left[\ln p\left(\boldsymbol{\theta}, \hat{\boldsymbol{\lambda}}, \hat{\boldsymbol{\psi}}|\boldsymbol{\alpha}, \beta, \gamma\right)\right]$. In the subsequent M step, the new estimate $\{\boldsymbol{\alpha}, \beta, \gamma\}^{\mathrm{new}}$ is obtained by maximizing the expectation with respect to $\{\boldsymbol{\alpha}, \beta, \gamma\}$. By differentiating Eq. (19) with respect to $\boldsymbol{\alpha}$, β, and γ, and setting these derivatives to zero thereafter, the hyper-parameters are solved as follows:

$$\alpha_i = \frac{1}{E_{\boldsymbol{\theta}}\left(\theta_i^2\right)}, \tag{20}$$

$$\beta = \frac{N_{\mathrm{m}}}{E_{\boldsymbol{\theta}}\left\{\sum_{r=1}^{N_{\mathrm{m}}}\left[\frac{\hat{\lambda}_r - \lambda_r(\boldsymbol{\theta})}{\hat{\lambda}_r}\right]^2\right\}}, \tag{21}$$

$$\gamma = \frac{N_{\mathrm{p}} \cdot N_{\mathrm{m}}}{E_{\boldsymbol{\theta}}\left[\sum_{r=1}^{N_{\mathrm{m}}}\left\|\hat{\boldsymbol{\phi}}_r - \boldsymbol{\phi}_r(\boldsymbol{\theta})\right\|_2^2\right]}, \tag{22}$$

where $E_{\boldsymbol{\theta}}$ denotes an expectation with respect to the posterior distribution of $\boldsymbol{\theta}$ using the current estimates of the hyper-parameters $\{\boldsymbol{\alpha}, \beta, \gamma\}^{\mathrm{old}}$.

3.1 Posterior Sampling

Posterior sampling is conducted to approximate the expectations in Eqs. (20), (21), and (22). We first approximate the conditional posterior PDF $p\left(\boldsymbol{\theta}|\hat{\boldsymbol{\lambda}}, \hat{\boldsymbol{\psi}}, \boldsymbol{\alpha}, \beta, \gamma\right)$ for stiffness parameter $\boldsymbol{\theta}$ in (14) by a multivariate Gaussian distribution using Laplace approximation [9]. The mean of the Gaussian distribution is the MAP estimate $\tilde{\boldsymbol{\theta}}$, which is calculated by minimizing the objective function in Eq. (16). The covariance matrix Σ_θ of the Gaussian distribution is equal to the inverse of the Hessian matrix calculated at $\tilde{\boldsymbol{\theta}}$, where the (i, j) components of the Hessian matrix is given as follows:

$$\mathbf{H}_{ij}\left(\tilde{\boldsymbol{\theta}}\right) = \left.\frac{\partial^2 J(\boldsymbol{\theta})}{\partial\boldsymbol{\theta}_i \partial\boldsymbol{\theta}_j}\right|_{\boldsymbol{\theta}=\tilde{\boldsymbol{\theta}}}. \tag{23}$$

We calculate the variance for each damage index θ_i independently [28]. Thereafter, we generate samples from the posterior PDF $p\left(\boldsymbol{\theta}|\hat{\boldsymbol{\lambda}}, \hat{\boldsymbol{\psi}}, \boldsymbol{\alpha}, \beta, \gamma\right)$, and the probabilistic information encapsulated in $p\left(\boldsymbol{\theta}|\hat{\boldsymbol{\lambda}}, \hat{\boldsymbol{\psi}}, \boldsymbol{\alpha}, \beta, \gamma\right)$ is characterized by the posterior samples $\boldsymbol{\theta}^{(k)}$, $k = 1, \ldots, K$. The expectations in Eqs. (20), (21), and (22) are eventually approximated by Eqs. (24), (25), and (26), respectively:

$$E_{\boldsymbol{\theta}}\left(\theta_i^2\right) = \int \theta_i^2 p\left(\theta_i|\hat{\boldsymbol{\lambda}}, \hat{\boldsymbol{\psi}}, \boldsymbol{\alpha}, \beta, \gamma\right) \mathrm{d}\theta_i \approx \frac{1}{K}\sum_{k=1}^{K}\left((\theta_i)^{(k)}\right)^2, \tag{24}$$

$$E_{\boldsymbol{\theta}}\left\{\sum_{r=1}^{N_{\mathrm{m}}}\left[\frac{\hat{\lambda}_r - \lambda_r(\boldsymbol{\theta})}{\hat{\lambda}_r}\right]^2\right\} = \int \sum_{r=1}^{N_{\mathrm{m}}}\left[\frac{\hat{\lambda}_r - \lambda_r(\boldsymbol{\theta})}{\hat{\lambda}_r}\right]^2 p\left(\theta_i|\hat{\boldsymbol{\lambda}}, \hat{\boldsymbol{\psi}}, \boldsymbol{\alpha}, \beta, \gamma\right) \mathrm{d}\boldsymbol{\theta}$$

$$\approx \frac{1}{K}\sum_{k=1}^{K}\sum_{r=1}^{N_{\mathrm{m}}}\left[\frac{\hat{\lambda}_r - \lambda_r\left(\boldsymbol{\theta}^{(k)}\right)}{\hat{\lambda}_r}\right]^2, \tag{25}$$

$$E_{\boldsymbol{\theta}}\left[\sum_{r=1}^{N_m}\left\|\hat{\boldsymbol{\phi}}_r - \boldsymbol{\phi}_r(\boldsymbol{\theta})\right\|_2^2\right] = \int\sum_{r=1}^{N_{\mathrm{m}}}\left\|\hat{\boldsymbol{\phi}}_r - \boldsymbol{\phi}_r(\boldsymbol{\theta})\right\|^2 p\left(\theta_i|\hat{\boldsymbol{\lambda}}, \hat{\boldsymbol{\psi}}, \boldsymbol{\alpha}, \beta, \gamma\right)\mathrm{d}\boldsymbol{\theta}$$
$$\approx \frac{1}{K}\sum_{r=1}^{N_{\mathrm{m}}}\left\|\hat{\boldsymbol{\phi}}_r - \boldsymbol{\phi}_r\left(\boldsymbol{\theta}^{(k)}\right)\right\|^2. \tag{26}$$

3.2 Likelihood Sampling

Given the complexity of the posterior PDF $p\left(\boldsymbol{\theta}|\hat{\boldsymbol{\lambda}}, \hat{\boldsymbol{\psi}}, \boldsymbol{\alpha}, \beta, \gamma\right)$, which may not be Gaussian, another sampling method is based on the likelihood function of $\boldsymbol{\theta}$. N_{s} sets of modal data $\mathcal{D}_j = \left[\hat{\boldsymbol{\lambda}}^{(j)}, \hat{\boldsymbol{\psi}}^{(j)}\right]$ $(j = 1, 2, \ldots, N_{\mathrm{s}})$ are generated according to the measured modal data following Gaussian distribution. The mean of the Gaussian distribution is equal to the measured modal data with assigned variance. For each data set, given the current estimates of the hyper-parameters $\{\boldsymbol{\alpha}, \beta, \gamma\}$, the MAP estimate $\tilde{\boldsymbol{\theta}}$ is calculated by minimizing the objective function in Eq. (16). Thereafter, the expectation is taken with respect to the MAP values of $\boldsymbol{\theta}$ as follows:

$$E_{\tilde{\boldsymbol{\theta}}}\left(\theta_i^2\right) = E\left(\tilde{\theta}_i^2\right), \tag{27}$$

$$E_{\tilde{\boldsymbol{\theta}}}\left\{\sum_{r=1}^{N_{\mathrm{m}}}\left[\frac{\hat{\lambda}_r - \lambda_r(\boldsymbol{\theta})}{\hat{\lambda}_r}\right]^2\right\} = E\left\{\sum_{r=1}^{N_{\mathrm{m}}}\left[\frac{\hat{\lambda}_r - \lambda_r\left(\tilde{\boldsymbol{\theta}}\right)}{\hat{\lambda}_r}\right]^2\right\}, \tag{28}$$

$$E_{\tilde{\boldsymbol{\theta}}}\left[\sum_{r=1}^{N_{\mathrm{m}}}\left\|\hat{\boldsymbol{\phi}}_r - \boldsymbol{\phi}_r(\boldsymbol{\theta})\right\|_2^2\right] = E\left(\sum_{r=1}^{N_{\mathrm{m}}}\left\|\hat{\boldsymbol{\phi}}_r - \boldsymbol{\phi}_r\left(\tilde{\boldsymbol{\theta}}\right)\right\|_2^2\right). \tag{29}$$

3.3 Summary

Each iteration uses the estimates of $\{\boldsymbol{\alpha}, \beta, \gamma\}$ to determine the posterior distribution of the latent variable $\boldsymbol{\theta}$. The current distribution of $\boldsymbol{\theta}$ is utilized to improve the estimates of $\{\boldsymbol{\alpha}, \beta, \gamma\}$. The proposed EM algorithm is implemented as follows:

1. Initialize the hyper-parameters $\{\boldsymbol{\alpha}, \beta, \gamma\}^{(0)}$ and latent variable $\boldsymbol{\theta}^{(0)}$.

2. At the jth iteration,
   E step: Compute the MAP estimates of $\tilde{\boldsymbol{\theta}}^{(j)}$ through minimizing $J(\boldsymbol{\theta})$ in Eq. (16) given hyper-parameters $\{\boldsymbol{\alpha}, \beta, \gamma\}^{(j-1)}$; calculate the expectations in Eqs. (20)–(22) using Eqs. (24)–(26) for the posterior sampling, and Eqs. (27)–(29) for the likelihood sampling;
   M step: Through maximization of $E\left[\ln p(\boldsymbol{\theta}, \hat{\boldsymbol{\lambda}}, \hat{\boldsymbol{\psi}}|\boldsymbol{\alpha}, \beta, \gamma)\right]$ with respect to $\boldsymbol{\alpha}$, β, and γ, update the hyper-parameters to $\{\boldsymbol{\alpha}, \beta, \gamma\}^{(j)}$ according to Eqs. (20), (21), and (22), given $\tilde{\boldsymbol{\theta}}^{(j)}$.
3. Repeat Step 2 for the $(i+1)$th iteration until the convergence criterion is met (i.e., $\|\tilde{\boldsymbol{\theta}}^{(j)} - \tilde{\boldsymbol{\theta}}^{(j-1)}\|_2 / \|\tilde{\boldsymbol{\theta}}^{(j)}\|_2 \leq \text{Tol}$).

Posterior sampling is conducted after E step once $\tilde{\boldsymbol{\theta}}^{(j)}$ is obtained at each iteration step. By contrast, likelihood sampling is conducted at the initialization stage only.

4 Bayesian Inference Based on the Laplace Approximation

The system is assumed to be globally identifiable based on the measurements. By employing the Laplace approximation, $p\left(\hat{\boldsymbol{\lambda}}, \hat{\boldsymbol{\psi}}, \boldsymbol{\theta}|\boldsymbol{\alpha}, \beta, \gamma\right)$, which is to be integrated in Eq. (15), can be approximated as a Gaussian distribution [31]:

$$p\left(\hat{\boldsymbol{\lambda}}, \hat{\boldsymbol{\psi}}, \boldsymbol{\theta}|\boldsymbol{\alpha}, \beta, \gamma\right) \cong p\left(\hat{\boldsymbol{\lambda}}, \hat{\boldsymbol{\psi}}, \tilde{\boldsymbol{\theta}}|\boldsymbol{\alpha}, \beta, \gamma\right) \exp\left\{-\frac{1}{2}\left(\boldsymbol{\theta} - \tilde{\boldsymbol{\theta}}\right)^{\mathrm{T}} \mathbf{A} \left(\boldsymbol{\theta} - \tilde{\boldsymbol{\theta}}\right)\right\}, \tag{30}$$

where $\tilde{\boldsymbol{\theta}}$ represents the mode of $p\left(\hat{\boldsymbol{\lambda}}, \hat{\boldsymbol{\psi}}, \boldsymbol{\theta}|\boldsymbol{\alpha}, \beta, \gamma\right)$ and is equivalent to the MAP estimate calculated using Eq. (16). Matrix $\mathbf{A}$ denotes the inverse of the variance, which equals the Hessian matrix at $\tilde{\boldsymbol{\theta}}$ and is defined as follows:

$$\mathbf{A} = -\nabla\nabla \ln p\left(\hat{\boldsymbol{\lambda}}, \hat{\boldsymbol{\psi}}, \boldsymbol{\theta}|\boldsymbol{\alpha}, \beta, \gamma\right)\Big|_{\theta=\tilde{\theta}} = \frac{\partial^2 J(\boldsymbol{\theta})}{\partial \boldsymbol{\theta}_i \partial \boldsymbol{\theta}_j}\Big|_{\theta=\tilde{\theta}} = \mathbf{W} + \beta \mathbf{H} + \gamma \mathbf{P}. \tag{31}$$

The Hessian matrix is decomposed into three items (i.e., $\mathbf{W}$ is a diagonal matrix with entries $\mathbf{W}_{ii} = \alpha_i$), and $\mathbf{H}$ and $\mathbf{P}$ are given as follows:

$$\begin{aligned}
\mathbf{H}_{ij} &= \frac{\partial}{\partial\theta_i \partial\theta_j} \frac{1}{2} \sum_{r=1}^{N_{\mathrm{m}}} \left[\frac{\hat{\lambda}_r - \lambda_r\left(\tilde{\boldsymbol{\theta}}\right)}{\hat{\lambda}_r}\right]^2 \\
&= \sum_{r=1}^{N_{\mathrm{m}}} \frac{1}{\hat{\lambda}_r^2} \cdot \frac{\partial \lambda_r\left(\tilde{\boldsymbol{\theta}}\right)}{\partial\theta_i} \cdot \frac{\partial \lambda_r\left(\tilde{\boldsymbol{\theta}}\right)}{\partial\theta_j} - \sum_{r=1}^{N_{\mathrm{m}}} \left[\frac{\hat{\lambda}_r - \lambda_r\left(\tilde{\boldsymbol{\theta}}\right)}{\hat{\lambda}_r}\right] \cdot \frac{1}{\hat{\lambda}_r} \cdot \frac{\partial \lambda_r\left(\tilde{\boldsymbol{\theta}}\right)}{\partial\theta_i \partial\theta_j},
\end{aligned} \tag{32}$$

$$\mathbf{P}_{ij} = \frac{\partial}{\partial\theta_i\partial\theta_j}\frac{1}{2}\sum_{r=1}^{N_\mathrm{m}}\sum_{k=1}^{N_\mathrm{p}}\left[\hat{\phi}_{k,r} - \phi_{k,r}\left(\tilde{\boldsymbol{\theta}}\right)\right]^2$$
$$= \sum_{r=1}^{N_\mathrm{m}}\sum_{k=1}^{N_\mathrm{p}}\frac{\partial\phi_{k,r}\left(\tilde{\boldsymbol{\theta}}\right)}{\partial\theta_i}\cdot\frac{\partial\phi_{k,r}\left(\tilde{\boldsymbol{\theta}}\right)}{\partial\theta_j} - \sum_{r=1}^{N_\mathrm{m}}\sum_{k=1}^{N_\mathrm{p}}\left[\hat{\phi}_{k,r} - \phi_{k,r}\left(\tilde{\boldsymbol{\theta}}\right)\right]\cdot\frac{\partial\phi_{k,r}\left(\tilde{\boldsymbol{\theta}}\right)}{\partial\theta_i\partial\theta_j}. \quad (33)$$

The derivatives of the eigenvalues and eigenvectors with respect to the stiffness parameter can be calculated using the Nelson's method [33] or the substructural approach [34].

Given that the integral of a normalized Gaussian distribution equals 1, the evidence $p\left(\hat{\boldsymbol{\lambda}}, \hat{\boldsymbol{\psi}}|\boldsymbol{\alpha}, \beta, \gamma\right)$ can be solved as follows:

$$p\left(\hat{\boldsymbol{\lambda}}, \hat{\boldsymbol{\psi}}|\boldsymbol{\alpha}, \beta, \gamma\right) = \int p\left(\hat{\boldsymbol{\lambda}}, \hat{\boldsymbol{\psi}}, \boldsymbol{\theta}|\boldsymbol{\alpha}, \beta, \gamma\right)\mathrm{d}\boldsymbol{\theta}$$
$$\cong p\left(\hat{\boldsymbol{\lambda}}, \hat{\boldsymbol{\psi}}, \tilde{\boldsymbol{\theta}}|\boldsymbol{\alpha}, \beta, \gamma\right)\cdot\int \exp\left\{-\frac{1}{2}\left(\boldsymbol{\theta} - \tilde{\boldsymbol{\theta}}\right)^\mathrm{T}\mathbf{A}\left(\boldsymbol{\theta} - \tilde{\boldsymbol{\theta}}\right)\right\}\mathrm{d}\boldsymbol{\theta}$$
$$= p\left(\hat{\boldsymbol{\lambda}}, \hat{\boldsymbol{\psi}}, \tilde{\boldsymbol{\theta}}|\boldsymbol{\alpha}, \beta, \gamma\right)\frac{(2\pi)^{n/2}}{|\mathbf{A}|^{1/2}}, \quad (34)$$

In this way, the asymptotic analytical expression of the evidence is obtained. For convenience of calculation, the logarithm form is used and shown as follows:

$$\ln p\left(\hat{\boldsymbol{\lambda}}, \hat{\boldsymbol{\psi}}|\boldsymbol{\alpha}, \beta, \gamma\right) = \ln p\left(\hat{\boldsymbol{\lambda}}, \hat{\boldsymbol{\psi}}, \tilde{\boldsymbol{\theta}}|\boldsymbol{\alpha}, \beta, \gamma\right) + \frac{n}{2}\ln(2\pi) - \frac{1}{2}\ln|\mathbf{A}|$$
$$= \ln p\left(\hat{\boldsymbol{\lambda}}|\tilde{\boldsymbol{\theta}}, \beta\right)\ln p\left(\hat{\boldsymbol{\lambda}}|\tilde{\boldsymbol{\theta}}, \beta\right) + \ln p\left(\hat{\boldsymbol{\psi}}|\tilde{\boldsymbol{\theta}}, \gamma\right)$$
$$+ \ln p\left(\tilde{\boldsymbol{\theta}}|\boldsymbol{\alpha}\right) + \frac{n}{2}\ln(2\pi) - \frac{1}{2}\ln|\mathbf{A}|$$
$$= \frac{N_\mathrm{m}}{2}\ln\left(\frac{\beta}{2\pi}\right) + \frac{N_\mathrm{p}N_\mathrm{m}}{2}\ln\left(\frac{\gamma}{2\pi}\right) + \frac{n}{2}\ln\left(\frac{1}{2\pi}\right)$$
$$+ \frac{1}{2}\sum_{i=1}^{n}\ln\alpha_i - \frac{1}{2}\sum_{i=1}^{n}\left(\alpha_i\tilde{\theta}_i^2\right) - \frac{\beta}{2}\sum_{r=1}^{N_\mathrm{m}}\left[\frac{\hat{\lambda}_r - \lambda_r\left(\tilde{\theta}\right)}{\hat{\lambda}_r}\right]^2$$
$$- \frac{\gamma}{2}\sum_{r=1}^{N_\mathrm{m}}\left\|\hat{\boldsymbol{\phi}}_r - \boldsymbol{\phi}_r\left(\tilde{\theta}\right)\right\|_2^2 + \frac{n}{2}\ln(2\pi) - \frac{1}{2}\ln|\mathbf{A}|. \quad (35)$$

MPVs of $\{\boldsymbol{\alpha}, \beta, \gamma\}$ can be estimated by maximizing $p\left(\hat{\boldsymbol{\lambda}}, \hat{\boldsymbol{\psi}}|\boldsymbol{\alpha}, \beta, \gamma\right)$. From Eq. (35), setting the derivative of $p\left(\hat{\boldsymbol{\lambda}}, \hat{\boldsymbol{\psi}}|\boldsymbol{\alpha}, \beta, \gamma\right)$ with respect to $\boldsymbol{\alpha}, \beta, \gamma$ equal to zero results in the following equations:

$$\alpha_i = \frac{1}{\tilde{\theta}_i^2 + \left[(\mathbf{W}+\beta\mathbf{H}+\gamma\mathbf{P})^{-1}\right]_{ii}}, \tag{36}$$

$$\beta = \frac{N_\mathrm{m}}{\sum_{r=1}^{N_\mathrm{m}}\left[\frac{\hat{\lambda}_r-\lambda_r\left(\tilde{\boldsymbol{\theta}}\right)}{\hat{\lambda}_r}\right]^2 + \sum_{i=1}^{n}\left[(\mathbf{W}+\beta\mathbf{H}+\gamma\mathbf{P})^{-1}\mathbf{H}\right]_{ii}}, \tag{37}$$

$$\gamma = \frac{N_\mathrm{m}N_\mathrm{p}}{\sum_{r=1}^{N_\mathrm{m}}\sum_{k=1}^{N_\mathrm{p}}\left[\hat{\phi}_{k,r} - \phi_{k,r}\left(\tilde{\boldsymbol{\theta}}\right)\right]^2 + \sum_{i=1}^{n}\left[(\mathbf{W}+\beta\mathbf{H}+\gamma\mathbf{P})^{-1}\mathbf{P}\right]_{ii}}. \tag{38}$$

Note that the solutions of $\tilde{\boldsymbol{\theta}}$ (Eq. 16) and $\{\boldsymbol{\alpha}, \beta, \gamma\}$ are coupled. Therefore, they can be determined from an iterative process of Eqs. (16) and (36)–(38).

The preceding formulations can be summarized as follows:

1. Initialize the hyper-parameters $\{\boldsymbol{\alpha}, \beta, \gamma\}^{(0)}$ and $\boldsymbol{\theta}^{(0)}$.
2. For $j = 1, 2, \ldots$, update $\tilde{\boldsymbol{\theta}}^{(j)}$ using Eq. (16) with $\boldsymbol{\alpha}^{(j-1)}, \beta^{(j-1)}, \gamma^{(j-1)}$.
3. Update $\alpha_i^{(j)}$ using Eq. (36) with $\tilde{\boldsymbol{\theta}}^{(j)}, \beta^{(j-1)}, \gamma^{(j-1)}$.
4. Update $\beta^{(j)}$ using Eq. (37) with $\tilde{\boldsymbol{\theta}}^{(j)}, \alpha_i^{(j)}, \gamma^{(j-1)}$.
5. Update $\gamma^{(j)}$ using Eq. (38) with $\tilde{\boldsymbol{\theta}}^{(j)}, \alpha_i^{(j)}, \beta^{(j)}$.
6. Let $j = j + 1$, repeat Steps 2–5 until the convergence criterion is satisfied (i.e., $\|\tilde{\boldsymbol{\theta}}^{(j)} - \tilde{\boldsymbol{\theta}}^{(j-1)}\|_2/\|\tilde{\boldsymbol{\theta}}^{(j)}\|_2 \le \mathrm{Tol}$).

5 Bayesian Inference Based on the VBI-DRAM Algorithm

5.1 *VBI*

The mechanism of VBI is to propose a tractable PDF to approximate the target PDF (i.e., posterior PDF $p\left(\boldsymbol{\theta}|\hat{\boldsymbol{\lambda}}, \hat{\boldsymbol{\psi}}, \boldsymbol{\alpha}, \beta, \gamma\right)$) [31].

The evidence can be calculated as follows:

$$c = p\left(\hat{\boldsymbol{\lambda}}, \hat{\boldsymbol{\psi}}|\boldsymbol{\alpha}, \beta, \gamma\right) = \frac{p\left(\hat{\boldsymbol{\lambda}}, \hat{\boldsymbol{\psi}}, \boldsymbol{\theta}|\boldsymbol{\alpha}, \beta, \gamma\right)}{p\left(\boldsymbol{\theta}|\hat{\boldsymbol{\lambda}}, \hat{\boldsymbol{\psi}}, \boldsymbol{\alpha}, \beta, \gamma\right)}. \tag{39}$$

Taking the logarithm of the two sides in Eq. (39), the formulation changes as follows:

$$\ln p\left(\hat{\boldsymbol{\lambda}}, \hat{\boldsymbol{\psi}}|\boldsymbol{\alpha}, \beta, \gamma\right) = \ln\frac{p\left(\hat{\boldsymbol{\lambda}}, \hat{\boldsymbol{\psi}}, \boldsymbol{\theta}|\boldsymbol{\alpha}, \beta, \gamma\right)}{Q} - \ln\frac{p\left(\boldsymbol{\theta}|\hat{\boldsymbol{\lambda}}, \hat{\boldsymbol{\psi}}, \boldsymbol{\alpha}, \beta, \gamma\right)}{Q}, \tag{40}$$

where $\boldsymbol{Q}(\boldsymbol{\theta}, \boldsymbol{\alpha}, \beta, \gamma)$ is the proposed PDF to approximate $p\left(\boldsymbol{\theta}|\hat{\boldsymbol{\lambda}}, \hat{\boldsymbol{\psi}}, \boldsymbol{\alpha}, \beta, \gamma\right)$, which is simplified as $\boldsymbol{Q}$.

Given that $\ln p\left(\hat{\boldsymbol{\lambda}}, \hat{\boldsymbol{\psi}}|\boldsymbol{\alpha}, \beta, \gamma\right)$ is irrelevant to $\boldsymbol{\theta}$, taking the expectation of the two sides in Eq. (40) with respect to $\boldsymbol{Q}$ yields the following equations:

$$\ln p\left(\hat{\boldsymbol{\lambda}}, \hat{\boldsymbol{\psi}}|\boldsymbol{\alpha}, \beta, \gamma\right) = \mathcal{L}(\boldsymbol{Q}) + D_{\mathrm{KL}}\left\{\boldsymbol{Q}||p\left(\boldsymbol{\theta}|\hat{\boldsymbol{\lambda}}, \hat{\boldsymbol{\psi}}, \boldsymbol{\alpha}, \beta, \gamma\right)\right\}, \tag{41}$$

$$\mathcal{L}(\boldsymbol{Q}) = E_{\boldsymbol{Q}}\left[\ln \frac{p\left(\hat{\boldsymbol{\lambda}}, \hat{\boldsymbol{\psi}}, \boldsymbol{\theta}|\boldsymbol{\alpha}, \beta, \gamma\right)}{\boldsymbol{Q}}\right], \tag{42}$$

$$D_{\mathrm{KL}}\left\{\boldsymbol{Q}||p\left(\boldsymbol{\theta}|\hat{\boldsymbol{\lambda}}, \hat{\boldsymbol{\psi}}, \boldsymbol{\alpha}, \beta, \gamma\right)\right\} = \int \boldsymbol{Q} \ln \frac{\boldsymbol{Q}}{p\left(\boldsymbol{\theta}|\hat{\boldsymbol{\lambda}}, \hat{\boldsymbol{\psi}}, \boldsymbol{\alpha}, \beta, \gamma\right)} \mathrm{d}\boldsymbol{\theta}, \tag{43}$$

where $\mathcal{L}(\boldsymbol{Q})$ represents the lower bound of $\boldsymbol{Q}$ and D_{KL} is the KL divergence [32] between $\boldsymbol{Q}$ and posterior PDF $p\left(\boldsymbol{\theta}|\hat{\boldsymbol{\lambda}}, \hat{\boldsymbol{\psi}}, \boldsymbol{\alpha}, \beta, \gamma\right)$.

Moreover, $D_{\mathrm{KL}} \geq 0$ has been proven [31]. $D_{\mathrm{KL}} = 0$ when $\boldsymbol{Q} = p\left(\boldsymbol{\theta}|\hat{\boldsymbol{\lambda}}, \hat{\boldsymbol{\psi}}, \boldsymbol{\alpha}, \beta, \gamma\right)$. Therefore, increasing the proximity of $\boldsymbol{Q}$ to the posterior PDF is equivalent to minimizing D_{KL}. Given that the posterior PDF $p\left(\boldsymbol{\theta}|\hat{\boldsymbol{\lambda}}, \hat{\boldsymbol{\psi}}, \boldsymbol{\alpha}, \beta, \gamma\right)$ is unknown, D_{KL} cannot be calculated directly. According to Eq. (41), minimizing D_{KL} is equivalent to maximize $\mathcal{L}(\boldsymbol{Q})$ [31]. To obtain the independent posterior PDF of the damage index and parameters, $\boldsymbol{Q}$ is factorized into two components based on mean field theory [35] as follows:

$$\boldsymbol{Q}(\boldsymbol{\theta}, \boldsymbol{\alpha}, \beta, \gamma) = q(\boldsymbol{\theta})q(\boldsymbol{\alpha}, \beta, \gamma). \tag{44}$$

The maximization of $\mathcal{L}(\boldsymbol{Q})$ can be achieved by optimizing each factor in turn through solving the expectation of the numerator $p\left(\hat{\boldsymbol{\lambda}}, \hat{\boldsymbol{\psi}}, \boldsymbol{\theta}|\boldsymbol{\alpha}, \beta, \gamma\right)$ in $\mathcal{L}(\boldsymbol{Q})$ with respect to other factors [31]. Therefore, factor $q(\boldsymbol{\alpha}, \beta, \gamma)$ can be derived by calculating the expectation of $\ln p\left(\hat{\boldsymbol{\lambda}}, \hat{\boldsymbol{\psi}}, \boldsymbol{\theta}|\boldsymbol{\alpha}, \beta, \gamma\right)$ with respect to $\boldsymbol{\theta}$.

$$\ln q(\boldsymbol{\alpha}, \beta, \gamma) = E_{\boldsymbol{\theta}}\left[\ln p\left(\hat{\boldsymbol{\lambda}}, \hat{\boldsymbol{\psi}}, \boldsymbol{\theta}|\boldsymbol{\alpha}, \beta, \gamma\right)\right] + \text{const}. \tag{45}$$

Parameters $\{\boldsymbol{\alpha}, \beta, \gamma\}$ are assumed to be independent from each other, leading to the following factorization:

$$q(\boldsymbol{\alpha}, \beta, \gamma) = q(\boldsymbol{\alpha})q(\beta)q(\gamma), \tag{46}$$

$$q(\boldsymbol{\alpha}) = \prod_{i=1}^{n} q(\alpha_i). \tag{47}$$

Therefore,

$$\ln q(\boldsymbol{\alpha}, \beta, \gamma) = \sum_{i=1}^{n} \ln q(\alpha_i) + \ln q(\beta) + \ln q(\gamma). \tag{48}$$

According to Eqs. (19) and (45), the logarithm of the posterior PDF of the individual parameter is obtained thereafter, and all parameters follow the gamma distribution:

$$q(\alpha_i) \propto (\alpha_i)^{\frac{1}{2}} \cdot \exp\left\{-\frac{\alpha_i}{2} E_{\boldsymbol{\theta}}\left(\theta_i^2\right)\right\}, \tag{49}$$

$$q(\beta) \propto (-\beta)^{\frac{N_m}{2}} \cdot \exp\left\{-\frac{\beta}{2} E_{\boldsymbol{\theta}}\left(\sum_{r=1}^{N_m}\left[\frac{\hat{\lambda}_r - \lambda_r(\boldsymbol{\theta})}{\hat{\lambda}_r}\right]^2\right)\right\}, \tag{50}$$

$$q(\gamma) \propto (\gamma)^{\frac{N_p N_m}{2}} \cdot \exp\left\{-\frac{\gamma}{2} E_{\boldsymbol{\theta}}\left[\sum_{r=1}^{N_m}\left\|\hat{\boldsymbol{\phi}}_r - \boldsymbol{\phi}_r(\boldsymbol{\theta})\right\|_2^2\right]\right\}. \tag{51}$$

Therefore, the mean and variance of each parameter are calculated as follows:

$$E(\alpha_i) = \frac{3}{E\left(\theta_i^2\right)}; \quad \mathrm{Var}(\alpha_i) = \frac{6}{\left[E\left(\theta_i^2\right)\right]^2}, \tag{52}$$

$$E(\beta) = \frac{N_m + 2}{E_{\boldsymbol{\theta}}\left(\sum_{r=1}^{N_m}\left[\frac{\hat{\lambda}_r - \lambda_r(\boldsymbol{\theta})}{\hat{\lambda}_r}\right]^2\right)}; \quad \mathrm{Var}(\beta) = \frac{2N_m + 4}{\left\{E_{\boldsymbol{\theta}}\left(\sum_{r=1}^{N_m}\left[\frac{\hat{\lambda}_r - \lambda_r(\boldsymbol{\theta})}{\hat{\lambda}_r}\right]^2\right)\right\}^2}, \tag{53}$$

$$E(\gamma) = \frac{N_p N_m + 2}{E_{\boldsymbol{\theta}}\left[\sum_{r=1}^{N_m}\left\|\hat{\boldsymbol{\phi}}_r - \boldsymbol{\phi}_r(\boldsymbol{\theta})\right\|_2^2\right]}; \quad \mathrm{Var}(\gamma) = \frac{2N_p N_m + 4}{\left\{E_{\boldsymbol{\theta}}\left[\sum_{r=1}^{N_m}\left\|\hat{\boldsymbol{\phi}}_r - \boldsymbol{\phi}_r(\boldsymbol{\theta})\right\|_2^2\right]\right\}^2}. \tag{54}$$

Factor $q(\boldsymbol{\theta})$ can be similarly derived by calculating the expectation of $\ln p\left(\hat{\boldsymbol{\lambda}}, \hat{\boldsymbol{\psi}}, \boldsymbol{\theta} | \boldsymbol{\alpha}, \beta, \gamma\right)$ with respect to $\{\boldsymbol{\alpha}, \beta, \gamma\}$:

$$\ln q(\boldsymbol{\theta}) = E_{\boldsymbol{\alpha},\beta,\gamma}\left[\ln p\left(\hat{\boldsymbol{\lambda}}, \hat{\boldsymbol{\psi}}, \boldsymbol{\theta} | \boldsymbol{\alpha}, \beta, \gamma\right)\right] + \mathrm{const}, \tag{55}$$

where the items independent with $\boldsymbol{\theta}$ are merged into the constant item. Therefore,

$$q(\boldsymbol{\theta}) \propto \exp\left\{-\frac{E(\beta)}{2}\sum_{r=1}^{N_\mathrm{m}}\left[\frac{\hat{\lambda}_r-\lambda_r(\boldsymbol{\theta})}{\hat{\lambda}_r}\right]^2 - \frac{E(\gamma)}{2}\sum_{r=1}^{N_\mathrm{m}}\left\|\hat{\boldsymbol{\phi}}_r-\boldsymbol{\phi}_r(\boldsymbol{\theta})\right\|_2^2 - \sum_{i=1}^{n}\left[\frac{E(\alpha_i)}{2}\cdot\theta_i^2\right]\right\}. \quad (56)$$

Equations (52)–(54) and (56) are coupled and should be iteratively calculated. The iterations are corresponding to the variational Bayesian EM step [31]. Given that VBI does not guarantee to converge to the global optimum, the parameters are initialized differently at the beginning of the iteration. The iterative process requires the calculation of three expectations, namely $E\left(\theta_i^2\right)$, $E_{\boldsymbol{\theta}}\left(\sum_{r=1}^{N_\mathrm{m}}\left[\frac{\hat{\lambda}_r-\lambda_r(\boldsymbol{\theta})}{\hat{\lambda}_r}\right]^2\right)$, and $E_{\boldsymbol{\theta}}\left[\sum_{r=1}^{N_\mathrm{m}}\left\|\hat{\boldsymbol{\phi}}_r-\boldsymbol{\phi}_r(\boldsymbol{\theta})\right\|_2^2\right]$, which is a full Bayesian analysis that considers posterior uncertainties of $\boldsymbol{\theta}$. However, the specific distribution of $\boldsymbol{\theta}$ cannot be directly recognized from Eq. (56) because of the nonlinear relationship between $\boldsymbol{\theta}$ and the modal parameters. In the next subsection, the numerical DRAM algorithm is used to obtain the statistical distribution of $\boldsymbol{\theta}$.

5.2 DRAM Algorithm

The DRAM algorithm is a combination of the DR and AM algorithms and is applicable to standard and nonstandard probabilistic distributions, provided that the probability proportional to the target PDF is available [36]. This algorithm is considerably efficient when applied to high-dimensional problems [37].

The DRAM algorithm is used to generate samples of $\boldsymbol{\theta}$ using Eq. (56). A two-layer DR is used for simplicity and efficiency. That is, if the secondary sample is rejected, then the new sample is set equal to the previous sample, and a third-layer sampling is no longer performed. A Gaussian distribution $\mathcal{N}(\mu, C)$ is adopted as the proposed PDF. To accelerate the convergence and shorten the burn-in period of the DRAM algorithm, the mean μ of the Gaussian distribution is initially set as $\ddot{\boldsymbol{\theta}}$, which is calculated by minimizing the objective function in Eq. (16). The mean μ and covariance matrix C are adjusted with the progress of sampling. The DRAM algorithm is described as follows.

Given the proposed sampling PDF $\mathcal{N}(\mu, C)$, sample $\boldsymbol{\theta}^{(1)}$, scale factor ρ to reduce the covariance ($\rho < 1$), bound N_t, trivial constant ε, and number of samples N_s,

1. Generate the candidate sample $x_1 \sim \mathcal{N}(\mu, C)$.
2. Calculate the acceptance ratio of the candidate sample, $\xi_{j1}\left(\boldsymbol{\theta}^{(j)}, x_1\right) = \min\left\{1, \frac{q(x_1)\mathcal{N}_{\boldsymbol{\theta}^{(j)}}(x_1, C)}{q\left(\boldsymbol{\theta}^{(j)}\right)\mathcal{N}_{x_1}\left(\boldsymbol{\theta}^{(j)}, C\right)}\right\}$.
3. Randomly generate $\boldsymbol{\mu}$ from uniform distribution $\mathcal{U}(0, 1)$.
4. If $\boldsymbol{\mu} < \xi_{j1}\left(\boldsymbol{\theta}^{(j)}, x_1\right)$, $\boldsymbol{\theta}^{(j+1)} = x_1$, then go to Step 8; otherwise, go to Step 5.

5. Generate the secondary candidate sample $x_2 \sim \mathcal{N}(\mu, \rho C)$.
6. Calculate the acceptance ratio of the secondary candidate sample,
$\xi_{21}(x_2, x_1) = \min\left\{1, \frac{q(x_1)\mathcal{N}_{x_2}(x_1, C_0)}{q(x_2)\mathcal{N}_{x_1}(x_2, C_0)}\right\}$.
$\xi_{j2}\left(\theta^{(j)}, x_2\right) = \min\left\{1, \frac{q(x_2)\mathcal{N}_{x_1}(x_2, C)\mathcal{N}_{\theta^{(j)}}(x_2, C)[1-\xi_{21}(x_2, x_1)]}{q(\theta^{(j)})\mathcal{N}_{x_1}(\theta^{(j)}, C)\mathcal{N}_{x_2}(\theta^{(j)}, C)\left[1-\xi_{j1}(\theta^{(j)}, x_1)\right]}\right\}$.
7. If $\boldsymbol{\mu} < \xi_{j2}\left(\boldsymbol{\theta}^{(j)}, x_2\right)$, then $\boldsymbol{\theta}^{(j+1)} = x_2$. Else, $\boldsymbol{\theta}^{(j+1)} = \boldsymbol{\theta}^{(j)}$.
8. Adjust the sampling covariance
$C = \begin{cases} C, \; j+1 < N_t \\ S_d \operatorname{cov}\left(\boldsymbol{\theta}^{(1)}, \boldsymbol{\theta}^{(2)}, \ldots, \boldsymbol{\theta}^{(j+1)}\right) + S_d \varepsilon \boldsymbol{I}_d, \; j+1 > N_t \end{cases}$.
9. Let $\mu = \boldsymbol{\theta}^{(j+1)}$.
10. Let $j = j+1$, repeat Steps 1–8 until $j = (n_s - 1)$.
11. The samples following the posterior PDF of $\boldsymbol{\theta}$ is obtained, $\left(\boldsymbol{\theta}^{(1)}, \boldsymbol{\theta}^{(2)}, \ldots, \boldsymbol{\theta}^{(N_s)}\right)$. The most probable value and uncertainty of $\boldsymbol{\theta}$ are calculated from the samples.

5.3 *Summary*

The damage index $\boldsymbol{\theta}$ is sampled using the DRAM sampling technique according to Eqs. (16) and (56), and MPVs $\boldsymbol{\theta}$ are calculated from the obtained samples. The parameters $\{\boldsymbol{\alpha}, \beta, \gamma\}$ are updated according to Eqs. (52)–(54). The proposed VBI-DRAM algorithm is implemented as follows:

1. Initialize the hyper-parameters $\{\boldsymbol{\alpha}, \beta, \gamma\}^{(0)}$ and latent variable $\boldsymbol{\theta}^{(0)}$.
2. At the jth iteration,
Solve $\tilde{\boldsymbol{\theta}}$ by minimizing $J(\boldsymbol{\theta})$ in Eq. (16);
Let $\mu_0 = \boldsymbol{\theta}^{(1)} = \tilde{\boldsymbol{\theta}}$, generate n_s samples of $\boldsymbol{\theta}\left(\boldsymbol{\theta}^{(1)} \sim \boldsymbol{\theta}^{(N_s)}\right)$ based on the DRAM algorithm according to Eq. (56), and identify $\boldsymbol{\theta}_{\text{MAP}}^{(j)}$;
Based on the generated samples, calculate $E(\alpha_i)^{(j)}$ and $\operatorname{Var}(\alpha_i)^{(j)}$ from Eq. (52);
Calculate $E(\beta)^{(j)}$ and $\operatorname{Var}(\beta)^{(j)}$ from Eq. (53);
Calculate $E(\gamma)^{(j)}$ and $\operatorname{Var}(\gamma)^{(j)}$ from Eq. (54).
3. Repeat Step 2 for the $(j+1)$th iteration until the convergence criterion is met (i.e., $\|\boldsymbol{\theta}_{\text{MAP}}^{(j)} - \boldsymbol{\theta}_{\text{MAP}}^{(j-1)}\|_2 / \|\boldsymbol{\theta}_{\text{MAP}}^{(j)}\|_2 \le \text{Tol}$).

6 Numerical Example

6.1 *Model Description*

As shown in Fig. 1, a cantilever beam is first utilized as a numerical preliminary study. The mass density and Young's modulus are 7.67×10^3 kg/m^3 and 7.0×10^{10} N/m^2, respectively. The beam is modeled with 45 equal Euler–Bernoulli beam elements (i.e., $n = 45$), 20-mm long each. Damage is simulated by the reduction of the bending

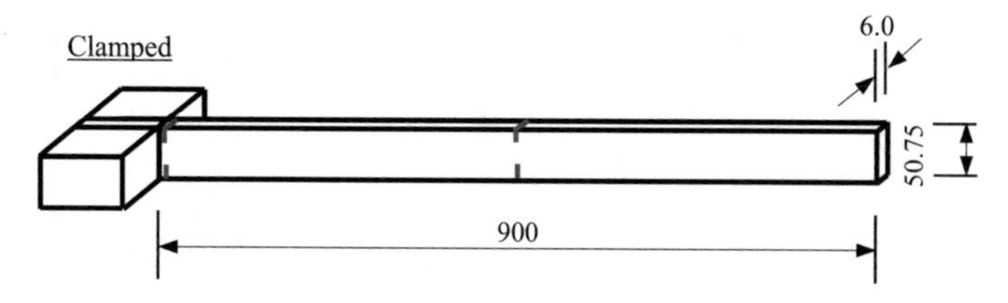

Fig. 1 Geometric configuration of the beam structure (unit: mm)

stiffness while mass remains unchanged. Element 1 at the clamped end and Element 23 at the mid-span (Fig. 1) are damaged by 50% (i.e., SRF(1) = SRF(23) = −50%).

6.2 Damage Identification

Natural frequencies are used for damage detection. Items associated with mode shapes are excluded, and the objective function in Eq. (16) is simplified as follows:

$$J_1(\boldsymbol{\theta}) = \beta \sum_{r=1}^{N_m} \left[\frac{\hat{\lambda}_r - \lambda_r(\boldsymbol{\theta})}{\hat{\lambda}_r} \right]^2 + \sum_{i=1}^{n} \left(\alpha_i \theta_i^2 \right). \tag{57}$$

The first six natural frequencies of the beam before and after damage are listed in Table 1.

Damage index and hyper-parameters should be initialized first. Modal experiment experiences have shown that natural frequencies may comprise 1% noise [38]. Therefore, a noise level of 1% is assigned to the frequencies (i.e., $\beta^{(0)} = 1/(1\%)^2 = 1 \times 10^4$). The uncertainty level of damage index is assumed as 10% of the exact damage index. In particular, the initial value $\alpha_i^{(0)} = 1/(10\%)^2 = 100\,(i = 1, 2, \ldots, 45)$. Initial damage indexes are set as $\boldsymbol{\theta}^{(0)} = \{0, \ldots, 0\}^{\mathrm{T}}$, indicating a lack of damage. The convergence criterion for the iteration process is set as $Tol = 0.01$. Identification error δ is defined as follows:

Table 1 Frequencies of the beam in the undamaged and damaged states

Mode No.	Undamaged freq. (Hz)	Damaged freq. (Hz)	Change ratio (%)
1	6.02	5.75	−4.56
2	37.75	35.67	−5.50
3	105.73	102.44	−3.11
4	207.25	197.69	−4.61
5	342.70	333.96	−2.55
6	512.07	492.45	−3.83
Average of frequency change (%)			−4.03

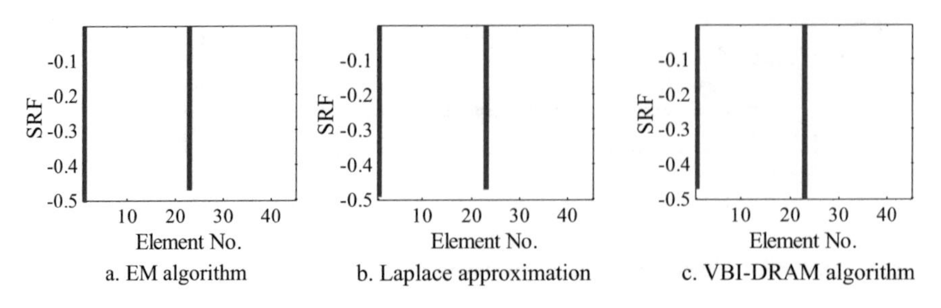

Fig. 2 Damage identification results of the beam

$$\delta = \sqrt{\frac{\left\|\tilde{\boldsymbol{\theta}} - \overline{\boldsymbol{\theta}}\right\|_2^2}{n}}, \tag{58}$$

where $\overline{\boldsymbol{\theta}}$ denotes the actual damage indexes.

For the EM algorithm using the likelihood sampling, 50 sets of natural frequencies are generated (i.e., $N_s = 50$) with Gaussian distribution having a zero mean and 1% standard deviation of the true values. Within EM, each set of sampled natural frequencies results in one set of MAP values of the damage indexes, from which the expectations are calculated according to Eqs. (27)–(29). Damage indexes converge after four iterations only. The mean of the MAP values of $\boldsymbol{\theta}$ in the final iteration is shown in Fig. 2a. The actual damaged elements are correctly located and quantified.

The proposed Laplace approximation algorithm in Sect. 4 is applied to detect damage. Random noise following the normal distribution with a zero mean and variance of 1% is directly added to the natural frequencies. After four iterations, the results converge and the two damaged elements are accurately detected (Fig. 2b).

The noisy natural frequencies used in the Laplace approximation algorithm were used in the VBI-DRAM algorithm for damage identification. In particular, $\tilde{\boldsymbol{\theta}}$ is first solved according to Eq. (57), and the DRAM algorithm is used thereafter to sample $\boldsymbol{\theta}$ according to the target PDF Eq. (56). In the DRAM algorithm, the covariance matrix C in Step 1 is assumed as a diagonal matrix with the entry equal to 0.05 and subsequently adjusted according to Step 8. Scale factor ρ is set to 0.01, and $S_d = 2.4^2/d$ (d is the dimension of $\boldsymbol{\theta}$, 45 here). Bound N_t is set to 1000, and ε is set to 10^{-8} to ensure that the covariance is positive and semidefinite. The number of samples N_s is set to 5000. Convergence is achieved after four iterations. The actual damaged elements are correctly identified upon convergence, as shown in Fig. 2c.

The identification errors of the EM technique, Laplace approximation, and VBI-DRAM algorithm are 0.41%, 0.45%, and 0.42%, respectively. The corresponding computation times are 5082, 116, and 319 s, respectively, which are obtained using a PC with Intel Core I7-8700 CPU and 20 GB RAM. Laplace approximation costs the least computational time because no sampling is required. The EM technique requires more extensive computational time than that of the VBI-DRAM algorithm.

The reason is that the numerical sampling of the damage index in the VBI-DRAM algorithm is directly conducted on PDF (Eq. 56).

7 Experimental Study

7.1 Model Description

Hou et al. [39] tested a three-story steel frame (Fig. 3). Each story is 0.5 m high, and the span is 0.5 m. Beams and columns have the same cross-section dimension of 75.0 mm × 5.0 mm. The Young's modulus of the steel is 2.0×10^{11} N/m^2, and mass density is 7.92×10^3 kg/m^3. The frame was excited with an instrumented hammer. The structure was impacted eight times and each impact lasted for 30 s. The eight signals were averaged to improve the modal identification accuracy. Measurement points were chosen every 100 mm, resulting in a total of 39 measurement points. Bruel & Kjaer accelerometers with a magnetic base were firmly mounted on the frame to measure the vertical acceleration of the beam and horizontal acceleration of the column. Sampling frequency was set as 2000 Hz.

The frame is divided into 225 Euler–Bernoulli beam elements, with each measuring 20-mm long. Two cuts were introduced to the frame model. Cuts 1 and 2 were located at the column end and beam/column joint, corresponding to elements 1 and 176, respectively. Saw cuts have the same length of 20 mm and depth of 22.5 mm, leading to a reduction in the moment of inertia of the cut sections by 60%

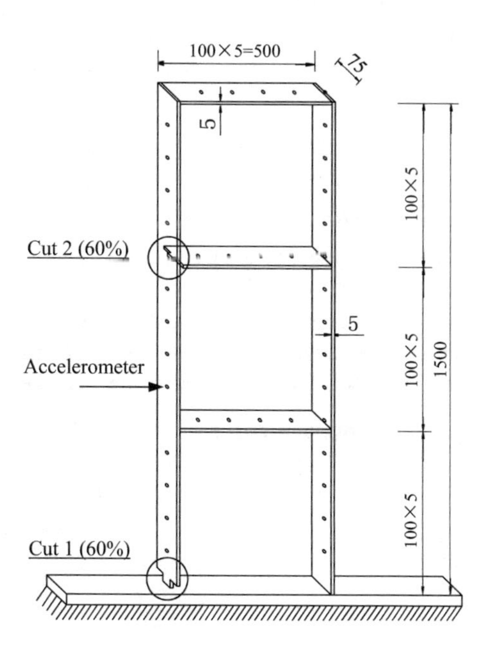

Fig. 3 Configuration of the experimental frame (unit: mm)

Table 2 Measured modal data of the frame in undamaged and damaged states

Mode No.	Undamaged freq. (Hz)	Damaged freq. (Hz)	MAC
1	4.23	4.08 (−3.53)	95.78
2	14.03	13.45 (−4.11)	97.49
3	25.45	25.13 (−1.23)	99.01
4	44.81	44.69 (−0.27)	97.59
5	58.12	57.28 (−1.44)	91.46
6	68.36	66.11 (−3.29)	88.14
7	72.27	71.42 (−1.18)	85.80
Average (%)		(−2.32)	93.61

Note Values in parentheses are the frequency change ratios (%) between the damaged and undamaged states. MAC = modal assurance criterion

(i.e., SRF(1) = SRF(176) = −60%). The measured first seven frequencies and MAC of the frame structure before and after damage are listed in Table 2.

7.2 *Damage Identification*

Natural frequencies and mode shapes are used for damage identification. Given that natural frequencies can be measured more accurately than mode shapes, their uncertainty levels are set as 1% and 5%, respectively. Thus, the hyper-parameters are initialized as $\beta^{(0)} = 1 \times 10^4$, $\gamma^{(0)} = 400$ and $\alpha_i^{(0)} = 100$ $(i = 1, 2, \ldots, 225)$. The initial damage indexes and convergence criterion are the same as those in the numerical example.

EM Algorithm. Given that only one set of measurement modal data is available, 50 sets of modal data $\mathcal{D}_j = [\hat{\lambda}^{(j)}, \hat{\psi}^{(j)}]\,(j = 1, 2, \ldots, 50)$ are generated through numerical simulation $\hat{\lambda}_r^{(j)} \sim N\left(\hat{\lambda}_r, (0.01\hat{\lambda}_r)^2\right)$ and $\hat{\phi}_r^{(j)} \sim N\left(\hat{\phi}_r, (0.05)^2 I\right)$. MPVs of the damage indexes are calculated by minimizing the objective function in Eq. (16) iteratively using the likelihood sampling. The process converges after five iterations, and only the results in the last iteration are shown in Fig. 4a for brevity. The two damaged elements are located successfully, while the severity of the damage at column end is larger than the true value (i.e., with 60% reduction).

Laplace Approximation. Following the iterative procedures summarized in Sect. 4, the proposed Laplace approximation algorithm is applied to detect damage of the frame. Convergence is achieved within five iterations, and the results are shown in Fig. 4b. Upon convergence, two damaged elements are accurately detected with no false identifications.

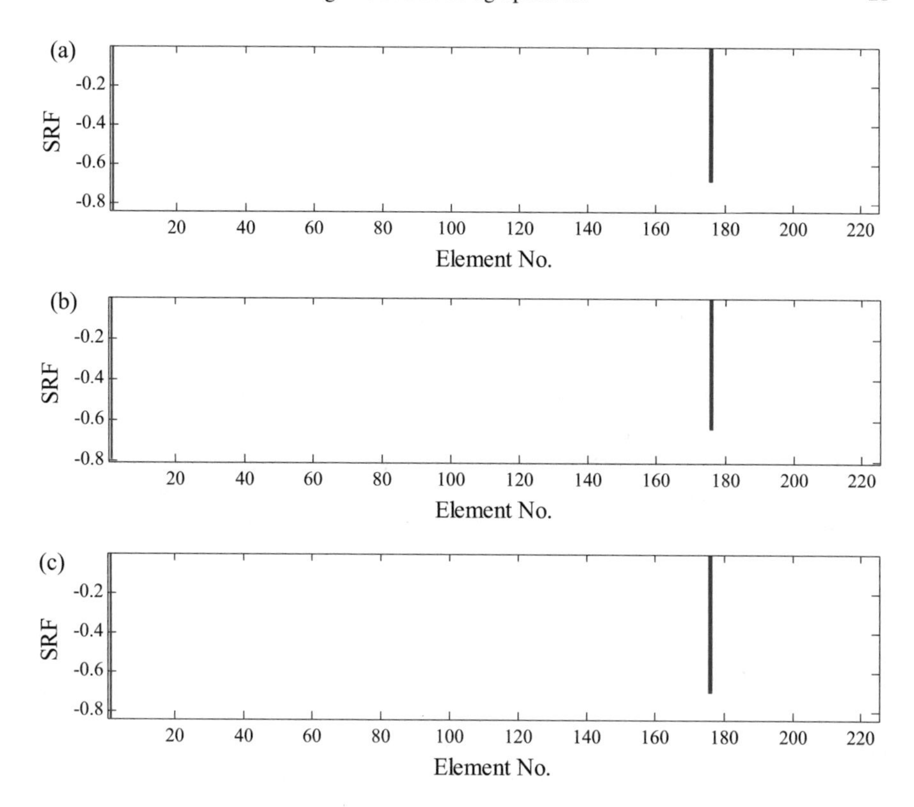

Fig. 4 Damage identification results of the frame, **a** EM algorithm, **b** Laplace approximation, **c** VBI-DRAM algorithm

VBI-DRAM Algorithm. In the DRAM algorithm, bound N_t is set to 1000 and ε is set to 10^{-8} to ensure that the covariance is positive and semidefinite. All other parameters are the same as the numerical example. Convergence is achieved after four iterations. The actual damaged elements are correctly identified, as shown in Fig. 4c.

During the iteration, the hyper-parameters also change continuously. Variations of α_1 and α_{100} for the Laplace approximation are shown in Fig. 5. Although α_i of all elements are initialized identically as 100, their values upon convergence vary remarkably. This sparse mechanism of the ARD model involves each variable being assigned with an individual hyper-parameter. In particular, α_{100} corresponding to the undamaged elements becomes significantly large. Therefore, the associated items are penalized considerably in the optimization (see Eq. 16), forcing the damage indexes θ_i to zero and realizing sparse damage detection.

The identification errors of the EM technique, Laplace approximation, and VBI-DRAM algorithm are 1.71%, 1.29%, and 1.73%, respectively. The corresponding computation times are 657, 246, and 73 min, respectively. The three techniques have good accuracy even though damage identification errors increase compared with

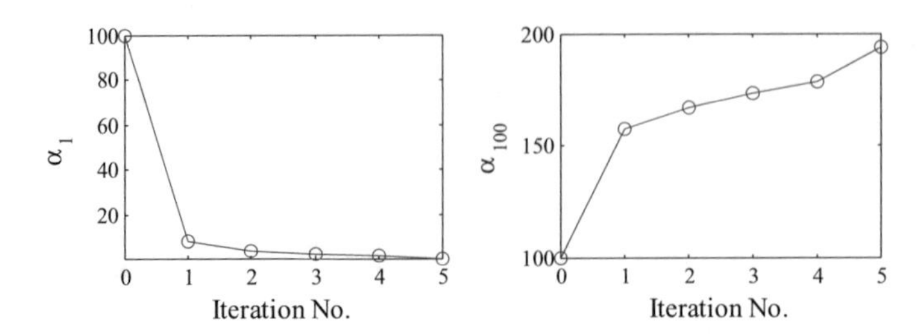

Fig. 5 Variation of hyper-parameters during the iterative process for the Laplace approximation

those in the numerical example. The possible reason is that the number of unknowns in the experimental example is substantially larger than that in the numerical one. Similarly, the EM technique requires the most computational time. However, the VBI-DRAM algorithm is more efficient than the Laplace approximation for this case because the calculation of the large-scale Hessian matrix is extremely time-consuming.

8 Comparison and Discussions

From the theoretical aspect, the posterior uncertainties of hyper-parameters are considered in the VBI-DRAM algorithm, while the other two techniques do not consider these uncertainties. Laplace approximation assumes that posterior PDF follows the standard distribution, which may not be realistic in practice. Moreover, the accuracy of the approximation depends on the dimension of the parameter vector. If the number of measurements is considerably fewer than the number of model parameters, then the corresponding results may be inaccurate.

From the computational perspective, the EM technique requires calculating the Hessian matrix in the posterior sampling or assuming the variance empirically in the likelihood sampling; both are extremely time-consuming. For the Laplace approximation, solutions of the hyper-parameters are expressed in a closed form. Thus, damage indexes and hyper-parameters are solved directly without sampling. In VBI, the calculation of the Hessian matrix is avoided. The numerical sampling of the damage index is directly conducted on the derived PDF, which is proportional to its posterior PDF. However, the VBI-DRAM algorithm requires the selection of numerous parameters to ensure the convergence and stability of the algorithm.

The numerical and experimental examples show that the damage detection results using the three techniques are accurate. The EM technique is the least efficient. Laplace approximation is markedly efficient for low-dimensional problems because no sampling is required. The VBI-DRAM algorithm is substantially efficient in dealing with high-dimensional problems. Therefore, the VBI-DRAM algorithm is recommended for large-scale structures with thousands of elements, in which case substructuring methods may be combined.

In summary, the main advantages of the EM algorithm are its simplicity and ease of implementation. Laplace approximation is the most efficient for low-dimensional problem. The VBI-DRAM algorithm can provide a full characterization of the posterior PDF instead of merely focusing on MPVs. The EM technique and VBI-DRAM algorithm are suitable for standard and nonstandard probability distributions. Thus, they are widely applicable.

9 Conclusions

This study proposes an SBL model for probabilistic structural damage detection using modal parameters. Sparsity of structural damage is exploited as an important prior information from the Bayesian perspective. Analytical approximation and numerical sampling techniques are introduced to deal with the computationally prohibited integration in the evidence.

A numerical cantilever beam and an experimental three-story frame are utilized to verify their effectiveness. The results show that damage location and severity can be accurately detected using the three techniques, even when the measurement data are considerably fewer than the damage indexes. Comparative studies indicate that each algorithm has its own advantages.

Acknowledgements The study was supported by RGC-GRF (Project No. 15201920).

References

1. Hemez FM (2005) Uncertainty quantification and the verification and validation of computational models. In: Inman DJ (ed) Damage prognosis: for aerospace, civil and mechanical systems. Wiley, West Sussex, pp 201–220
2. Hou RR, Xia Y (2021) Review on the new development of vibration-based damage identification for civil engineering structures: 2010–2019. J Sound Vib 491:115741
3. Xia Y, Hao H (2003) Statistical damage identification of structures with frequency changes. J Sound Vib 263(4):853–870
4. Sohn H, Dzwonczyk M, Straser E et al (1999) An experimental study of temperature effect on modal parameters of the Alamosa Canyon Bridge. Earthq Eng Struct D 28(8):878–897
5. Xia Y, Hao H, Zanardo G et al (2006) Long term vibration monitoring of an RC slab: temperature and humidity effect. Eng Struct 28(3):441–452
6. Hua XG, Ni YQ, Chen ZQ et al (2008) An improved perturbation method for stochastic finite element model updating. Int J Numer Meth Eng 73(13):1845–1864
7. Yeo I, Shin S, Lee HS et al (2000) Statistical damage assessment of framed structures from static responses. J Eng Mech 126(4):414–421
8. Gul M, Catbas FN (2009) Statistical pattern recognition for structural health monitoring using time series modeling: theory and experimental verifications. Mech Syst Signal Process 23(7):2192–2204
9. Beck JL, Katafygiotis LS (1998) Updating models and their uncertainties. I: Bayesian statistical framework. J Eng Mech 124(4):455–461

10. Ching J, Beck JL (2004) Bayesian analysis of the phase II IASC-ASCE structural health monitoring experimental benchmark data. J Eng Mech 130(10):1233–1244
11. Zhu YC, Au SK (2018) Bayesian operational modal analysis with asynchronous data, part I: most probable value. Mech Syst Signal Process 98:652–666
12. Yan WJ, Katafygiotis LS (2019) An analytical investigation into the propagation properties of uncertainty in a two-stage fast Bayesian spectral density approach for ambient modal analysis. Mech Syst Signal Process 118:503–533
13. Huang Y, Shao CS, Wu B et al (2019) State-of-the-art review on Bayesian inference in structural system identification and damage assessment. Adv Struct Eng 22(6):1329–1351
14. Tipping ME (2001) Sparse Bayesian learning and the relevance vector machine. J Mach Learn Res 1:211–244
15. Malioutov D, Cetin M, Willsky AS (2005) A sparse signal reconstruction perspective for source localization with sensor arrays. IEEE Trans Signal Process 53(8):3010–3022
16. Fang J, Shen YN, Li FW (2015) Support knowledge-aided sparse Bayesian learning for compressed sensing. In: Proceedings of IEEE international conference on acoustics. IEEE, Brisbane
17. Huang Y, Beck JL (2015) Hierarchical sparse Bayesian learning for structural health monitoring with incomplete modal data. Int J Uncertain Quantification 5(2):139–169
18. Huang Y, Beck JL et al (2017) Hierarchical sparse Bayesian learning for structural damage detection: theory, computation and application. Struct Saf 64:37–53
19. Huang Y, Beck JL, Li H (2017) Bayesian system identification based on hierarchical sparse Bayesian learning and Gibbs sampling with application to structural damage assessment. Comput Meth Appl Mech Eng 318:382–411
20. Hou RR, Xia Y, Zhou XQ et al (2019) Sparse Bayesian learning for structural damage detection using expectation–maximization technique. Struct Control Health Monit 26(5):e2343
21. Wang XY, Hou RR, Xia Y et al (2020) Laplace approximation in sparse Bayesian learning for structural damage detection. Mech Syst Signal Process 140:106701
22. Wang XY, Hou RR, Xia Y et al (2020) Structural damage detection based on variational Bayesian inference and delayed rejection adaptive Metropolis algorithm. Struct Health Monit. https://doi.org/10.1177/1475921720921256
23. Beal MJ (2003) Variational algorithms for approximate Bayesian inference. Dissertation, University of London
24. Rouder JN, Lu J (2005) An introduction to Bayesian hierarchical models with an application in the theory of signal detection. Psychon Bull Rev 12:573–604
25. Lam HF, Yang JH, Au SK (2015) Bayesian model updating of a coupled-slab system using field test data utilizing an enhanced Markov chain Monte Carlo simulation algorithm. Eng Struct 102:144–155
26. Lam HF, Yang JH, Au SK (2018) Markov chain Monte Carlo-based Bayesian method for structural model updating and damage detection. Struct Control Health Monit 25(4):e2140
27. Zhou XQ, Xia Y, Weng S (2015) L_1 regularization approach to structural damage detection using frequency data. Struct Health Monit 14(6):571–582
28. Vanik MW, Beck JL, Au SK (2000) Bayesian probabilistic approach to structural health monitoring. J Eng Mech 126(7):738–745
29. Jaynes ET (2003) Probability theory: the logic of science. Cambridge University Press, Cambridge
30. Mackay DJC (1992) Bayesian methods for adaptive models. Dissertation, California Institute of Technology
31. Bishop CM (2006) Pattern recognition and machine learning. Springer, New York
32. Dempster AP, Laird N, Rubin D (1997) Maximum likelihood for incomplete data via the EM algorithm. J R Stat Soc B 39:1–38
33. Nelson RB (1976) Simplified calculation of eigenvector derivatives. AIAA J 14(9):1201–1205
34. Weng S, Zhu HP, Xia Y et al (2013) Substructuring approach to the calculation of higher-order eigensensitivity. Comput Struct 117:23–33
35. Parisi G (1988) Statistical field theory. Addison-Wesley, Boston

36. Haario H, Laine M, Mira A et al (2006) DRAM: efficient adaptive MCMC. Stat Comput 16:339–354
37. Wan HP, Ren WX (2016) Stochastic model updating utilizing Bayesian approach and Gaussian process model. Mech Syst Signal Process 70–71:245–268
38. Mottershead JE, Friswell MI (1993) Model updating in structural dynamics: a survey. J Sound Vib 167(2):347–375
39. Hou RR, Xia Y, Zhou XQ (2018) Structural damage detection based on l_1 regularization using natural frequencies and mode shapes. Struct Control Health Monit 25(3):e2017

Bayesian Deep Learning for Vibration-Based Bridge Damage Detection

Davíð Steinar Ásgrímsson, Ignacio González, Giampiero Salvi, and Raid Karoumi

Abstract A machine learning approach to damage detection is presented for a bridge structural health monitoring (SHM) system. The method is validated on the renowned Z24 bridge benchmark dataset where a sensor instrumented, three-span bridge was monitored for almost a year before being deliberately damaged in a realistic and controlled way. Several damage cases were successfully detected, making this a viable approach in a data-based bridge SHM system. The method addresses directly a critical issue in most data-based SHM systems, which is that the collected training data will not contain all natural weather events and load conditions. A SHM system that is trained on such limited data must be able to handle uncertainty in its predictions to prevent false damage detections. A Bayesian autoencoder neural network is trained to reconstruct raw sensor data sequences, with uncertainty bounds in prediction. The uncertainty-adjusted reconstruction error of an unseen sequence is compared to a healthy-state error distribution, and the sequence is accepted or rejected based on the fidelity of the reconstruction. If the proportion of rejected sequences goes over a predetermined threshold, the bridge is determined to be in a damaged state. This is a fully operational, machine learning-based bridge damage detection system that is learned directly from raw sensor data.

Keywords Structural health monitoring · Bridge damage detection · Machine learning · Bayesian deep learning · Autoencoders · Z24 bridge benchmark

D. S. Ásgrímsson (✉) · I. González · G. Salvi · R. Karoumi (✉)
KTH Royal Institute of Technology, 100 44 Stockholm, Sweden
e-mail: dsas@kth.se

R. Karoumi
e-mail: raidk@kth.se

I. González
e-mail: ignaciog@kth.se

G. Salvi
e-mail: giampiero.salvi@ntnu.no

G. Salvi
NTNU Norwegian University of Science and Technology, 7491 Trondheim, Norway

A. Cury et al. (eds.), *Structural Health Monitoring Based on Data Science Techniques*, Structural Integrity 21, https://doi.org/10.1007/978-3-030-81716-9_2

1 Background

Damaged bridges are not a thing of the past. On August 1, 2007, the I-35W bridge in Minneapolis collapsed, killing 13 and injuring 145. On August 14, 2018, a major bridge in Genoa, Italy, partially collapsed, killing 43 people. One can wonder if a modern, data-based structural health monitoring system alongside strict maintenance monitoring could have prevented these events.

This chapter presents an end-to-end method of detecting damage in a real-life bridge using only sensor measurements, based on the previous work of [1–3]. Even though the results are a promising step into a more modern structural health monitoring of bridges, we believe that we are far from being able to replace standard monitoring methods of bridges. Using the presented method as a replacement for standard bridge inspection could be dangerous and possibly life-threatening. What this chapter presents is a method that could complement standard procedure and possibly improve the speed and accuracy of damage detection. If the method confidently predicts a damaged state, the prediction will have to be confirmed manually.

1.1 Structural Health Monitoring

There are two main approaches for structural health monitoring (SHM) of bridges: a model-based approach and a data-based approach. In both cases, sensors of different kinds are attached to the structure to measure its response to external stimulation. In a model-based approach, a finite element analysis is done specifically for each bridge which is calibrated with the sensor data. This analysis produces a baseline for values of displacement, strain, and vibration at each point. The real sensor measurements are then compared to the calibrated finite element model, and damage is assessed. In a data-based approach, only the sensor data is used for damage assessment. One aspect that data-based approaches to SHM have in common is a comparison of a current state to a baseline state. Sensor data is collected over a certain period, considered to be the structurally healthy baseline, and deviations from the baseline are assessed as structural damage. One of the biggest challenges of bridge SHM is the natural environmental response. Significant changes in a bridge's natural frequency and stiffness can occur due to temperature changes, making a modal approach to SHM impractical [4, 5]. This variability calls for more complex modeling. Vibration-based methods address some weaknesses of traditional modal methods and can be more robust to environmental variability and more sensitive to damage [6–8]. The use of artificial intelligence for SHM has recently shown great promise in many different applications [9], and machine learning methods are becoming established as an efficient alternative approach to classical modeling techniques in structural engineering [10].

Gonzalez and Karoumi [2] use bridge vibrational data and a specialized bridge weigh in motion system that records the bridges' load position, magnitude, and speed.

A neural network was trained to predict the bridges' vibrations at each time step. A Gaussian process was then used to characterize the prediction errors. Neves et al. [3] extend this idea from [2] without the specialized weigh in motion system.

There is a prominent need for addressing uncertainty in a data-based approach to SHM [11]. The value of addressing uncertainty has been shown as being more robust than plain neural networks for damage detection in a composite airfoil structure [10]. Uncertainty is therefore a central theme of this chapter.

1.2 Machine Learning

Here, some fundamental concepts are reviewed from the field of machine learning that will be useful to understand the method section.

Linear regression is the most basic building block of a neural network, a linear combination of $D + 1$-dimensional input variables $\mathbf{X} = x_0, \ldots, x_D$ and weight parameters $\mathbf{w} = w_0, \ldots, w_D$. A bias term x_0 is added to the D-dimensional input, which makes the input into the model $D + 1$-dimensional. The output prediction $\hat{y}$ is on the form of a matrix dot product, Eq. (1).

$$\hat{y} = \mathbf{w}^T \mathbf{X} \tag{1}$$

Given N observations of $(\mathbf{X}, y)$ pairs, there exists a closed form solution to the $\mathbf{w}$ that minimizes the error function. For linear regression, the error function is usually the mean squared error function which corresponds to a maximum likelihood solution if the error is assumed to be normally distributed with zero mean.

Logistic regression is very similar in form, except that the output prediction is a binary classification. A nonlinear activation function $\sigma(x)$ outputs a value between 0 and 1 that can be interpreted as the probability of a class being 1, Eq. (2).

$$\hat{y} = \sigma\left(\mathbf{w}^T \mathbf{X}\right) \tag{2}$$

A feed-forward neural network is a composite of many logistic regression blocks. The blocks are often grouped together in layers, and some of the layers are not observed and therefore often called hidden (Fig. 1). Optimizing the model parameters in this case requires more sophisticated methods. The most common methods are based on back-propagation [12–14] using variations of stochastic gradient descent [15, 16].

A neural network with many hidden layers is often called a deep neural network. A neural network can theoretically approximate any continuous function [17, 18]. In practice, there is a balance between the width and the depth of the hidden layers. By increasing the width and depth, the neural network can approximate functions of higher complexity. Recently, focusing on increasing the depth has been shown to be effective [19–21]. The cascading effect of a deep neural network presents a way to represent multiple levels of abstraction [22] as the model output $\hat{y}$ can be interpreted

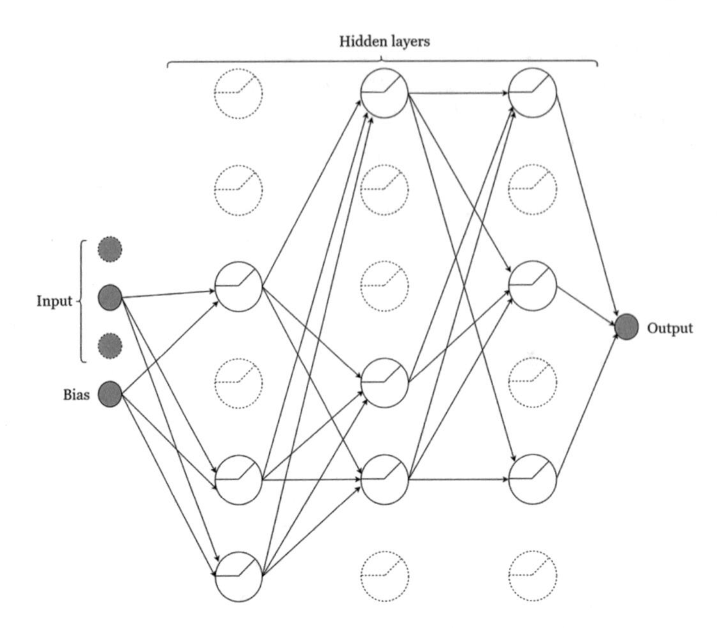

Fig. 1 A single forward pass of a neural network with ReLU activation function and dropout 0.5, where a node connection is randomly dropped with 50% probability. Dropped nodes are depicted as dotted, and observable nodes are shaded gray

as a composite of simpler functions $g_i()$.

$$\hat{y} = g_4(g_3(g_2(g_1(\mathbf{X})))) \tag{3}$$

An autoencoder (Fig. 2) is a neural network that is trained to reconstruct a given input, through the constraints of a lower-dimensional bottleneck layer. This can be seen as an unsupervised method since there is no requirement of labeled data. The main features of an autoencoder are the encoder, the code, and the decoder. The encoder maps the input to a lower-dimensional code representation [23]. The decoder then reconstructs the original input from the low-dimensional code.

For many problems, the ability to represent uncertainty is crucial. Neural networks have been successful in certain areas, but they do not output model uncertainty by default. Bayesian neural networks [24, 25], however, take uncertainty into account in a principled manner.

Gal and Ghahramani [26] presented an interpretation of dropout [27] as approximate Bayesian inference in deep Gaussian processes [28]. This method allows to obtain model uncertainty from any standard neural networks using dropout [27]. Usually, dropout is used during training of neural networks to reduce overfitting, but for prediction no dropout is used. Each unit is dropped with probability p, making the final network more robust; see Fig. 1. When this method is applied to a deep feed-forward neural network, it approximates a deep Gaussian process [28], and the model is both deep and can handle uncertainty in prediction. Gal and Ghahramani [26] make

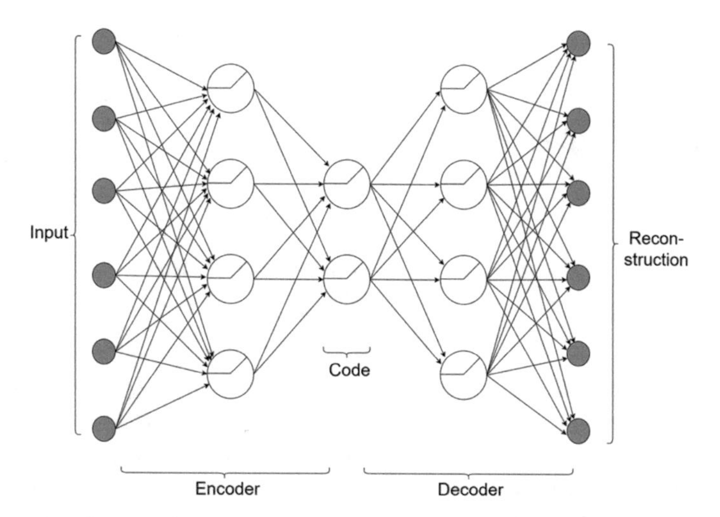

Fig. 2 An example of an autoencoder network that encodes the six-dimensional input to a two-dimensional code. This compact representation is then decoded to reconstruct the original six-dimensional input. This example has three hidden layers, and ReLU activation function and the encoder and decoder have a mirrored structure

a theoretically grounded observation of averaging many stochastic forward passes with dropout, to achieve a prediction with uncertainty. Training a neural network with dropout can be seen as training a collection of clipped networks with extensive weight sharing. Averaging many dropout forward passes is an approximation of a deep Gaussian process posterior [29], referred to as *MC dropout* [26]. The mean of the predictions is simply the predictive mean, and the variance of the predictions is the uncertainty. The method is theoretically grounded, easy to implement, and computationally efficient. One of the challenges of MC dropout is to achieve well-calibrated uncertainty estimates, and some extensions have been made [30, 31] to address this issue.

2 Method

The need for uncertainty in SHM is explained in Fig. 3. The collected training data for bridge structural health monitoring cannot be truly complete in a reasonable time frame, due to potentially large variations in regular environmental variables. Such a complete recording could possibly take many years. Therefore, a predictive model is nearly always going to be trained on a limited healthy-state dataset. It is important to note that a healthy state can still contain old damages even though this base state is considered as undamaged. It is of course not possible to detect damages induced in the bridge before the sensors were installed.

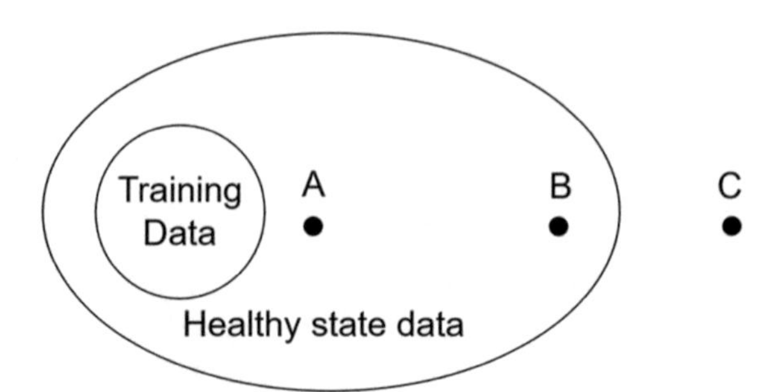

Fig. 3 A central theme of this chapter. The presented method addresses the problem of differentiating between data points **B** and **C**

A crucial observation is what happens when the model is faced with unfamiliar input data, depicted as points **B** and **C** in Fig. 3. The predictive error for test point **A** close to the training dataset should be lower than for a point **B** far away. Using only high predictive error for damage detection would likely falsely flag point **B** as damaged. By including model uncertainty, the reconstruction of point **B** will have large error, but the model should be able to explain the large error by having large uncertainty and therefore not label the large error as anomalous. If the true values fall outside the predictive uncertainty bounds of the model as in point **C**, the model cannot explain the large difference between real and predicted values and therefore correctly reject test point **C**.

The method to detect change in structural condition presented in this chapter builds upon the work of [1–3]. A Bayesian deep autoencoder is trained on the collected healthy bridge vibrational and environmental sensor data to reconstruct the given healthy-state input data. The autoencoder neural network reconstructs the input data with an uncertainty interval by using MC dropout. The reconstruction error of the unseen healthy-state data is then quantified, and a 95th percentile rejection threshold is determined. If the total sequence reconstruction error of a sequence in an unknown state falls outside the determined rejection threshold, then the sequence is rejected. If enough sequences are rejected, the bridge is determined to be in a damaged state.

2.1 Assumptions

A model is a set of assumptions about the data, and the following are the most important assumptions of the modeling. Some of them are limiting, and some make the problem more feasible. These assumptions are general enough so the method can be applied to a similar bridge in structure, and no assumption was created specifically to match the particular Z24 bridge structure.

Vibration sensor data is correlated.

Since the sensors are all connected to the same continuous bridge structure, it is reasonable to assume that the sensor recordings are correlated. When one sensor is detecting high vibration, it is assumed that the other sensors are detecting high vibrations too. The assumption that the vibration sensor data is correlated makes the prediction problem easier than if they were uncorrelated, the model will learn the interplay of the sensors and how the sensors react to stimulation in unison in a healthy state. If the bridge is damaged near a particular sensor, the interplay between the sensors will change, with higher error rates, and therefore makes it easier to detect damage. Due to this assumption, it is important that the input to the model contains data of all sensors simultaneously on the bridge. This type of input is referred to as a sliding window input in time. The sliding window input needs to contain the simultaneous sensor measurements of all the sensors. If each sensor measurement would be reconstructed separately, this critical information of how the sensors react in unison to stimulation would be lost.

Enough information is contained within a short time frame.

The chosen neural network model architecture can only take a fixed, limited size input. A reasonable input to the model would be a sliding window of sensor data in the order of seconds. This is a limitation to the model since the true behavior could possibly not be captured in such a short time frame. To address this issue, multiple consecutive sliding window inputs are reconstructed. This is referred to as a *sequence*, and the total summed reconstruction error is used to either accept or reject a sequence. The sequence size is in the order of minutes. Note that more complex recurrent neural network architectures exist that can take the full sequence as input by using internal state to model the temporal dynamic behavior.

Vibrations are dependent on environmental variables.

Environmental variables such as air temperature, soil temperature, humidity, wind speed, as well as vehicle passing have a big impact on the response of a bridge. It is therefore very important to include these variables as an input to the predictive model.

The vibration data is hierarchical.

The nature of a deep neural network is in the cascading composition of simpler elements, as described above. The vibrations of a bridge are the result of many interplaying variables such as weather, winds, and loads. It is therefore assumed that the data is inherently compositional, and a deep neural network would be a good fit for this problem.

A vibration sensor measures local condition.

One of the pillars of bridge structural health monitoring is the localization of damage. To address this, it is assumed that if the real sensor values for a given sensor are far away from the model's predictive uncertainty of that sensor, the bridge is damaged in the region of the physical location of the sensor. This assumption holds in many damage scenarios, but for a damage in a main longitudinal load carrying beam can also be seen in sensors located far from the damage location compared to a damage of a secondary load carrying member such as a crossbeam which will only influence local measurements. This is out of the reach of the suggested approach.

Damage determination is not instant.

A damaged bridge generally behaves in a more complex manner than an undamaged bridge. Damage results in more vibration, higher damping, and more complex nonlinear behavior. It is therefore assumed that after damage is present, it will continuously affect the behavior of the bridge. Note that in real life this is not always the case. Due to this assumption, multiple sequences are needed to determine bridge state. It is not reasonable that the model can determine if the bridge is in a damaged state by only looking at reconstruction error from a sensor input of a few seconds. A more reasonable approach is to look at the total reconstruction error from multiple long, continuous sequences.

2.2 The Damage Detection Algorithm

For the damage detection, three algorithms are presented. Note that all three algorithms use the exact same trained model. This makes it possible to deploy all three in a real-time system without significant increase in computational load. All three methods reconstruct a given 10-min-long sensor sequence, in a succession of non-overlapping 2 s sliding window inputs, with the autoencoder neural network trained on the healthy-state data. The reconstruction error for the whole sequence is summed and compared to the distribution of total sequence error for the healthy-state test data. If the total reconstruction error of the given sequence is over the 95% cutoff rejection threshold, the sequence is rejected. If more than 5% of sequences are rejected, the bridge is determined to be in a damaged state.

The **standard method** reconstructs the input in one single forward pass of the neural network. The method does not take into consideration any notion of uncertainty. This is a standard way of prediction in a neural network.

The **mean method** uses *MC Dropout* [26] to stochastically reconstruct the input 100 times and to use the mean of the 100 reconstructions as the final prediction. This method gives a Maximum A Posteriori estimation.

The **uncertainty method** uses *MC Dropout* [26] to stochastically reconstruct the input 100 times. True values that fall within 2 standard deviations from the mean of the predictions are then considered to have 0 error. Only values that fall outside the uncertainty range are taken into the reconstruction error sum. The reasoning behind this approach is to be able to disregard reconstruction error that falls within the range of expected values. This method therefore only reports errors that are considerably different from the true values.

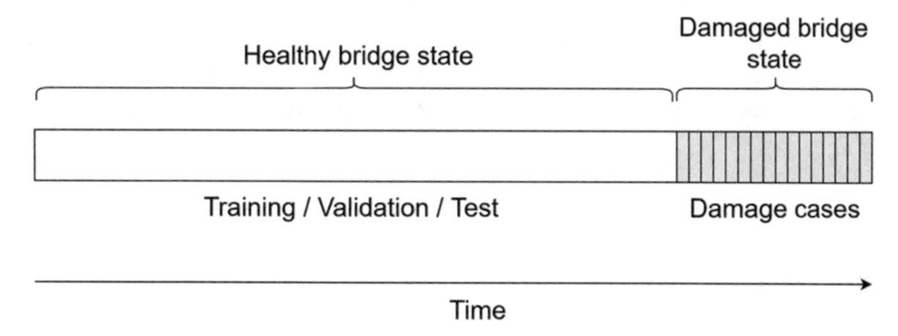

Fig. 4 Dataset visualized. Only the healthy bridge state data is used to train the models. The best model is picked by performance on the validation data. The errors are quantified on the unseen test data. The true damage detection performance is then assessed

2.3 *Validation*

To prevent a biased model being picked, a model hyperparameter search is performed on the training set and validated on the validation set. Multiple models are trained on the training set using random search for hyperparameter optimization [32]. The model with the lowest mean error on the validation set is then picked as the final model used for damage detection. Note that the damage cases are never seen or validated during the training, so all model parameters are determined from a healthy-state data. See Fig. 4 for a visual representation of the dataset split.

2.4 *Determining the Sequence Error Threshold*

The central idea in this method is to reject a small part of the healthy-state sequences to obtain a rejection criterion. The sequences with the highest reconstruction error are used as a benchmark for comparison sequences of a bridge in an unknown state. This threshold is chosen to be the nth percentile value, where n is chosen to be one of 95th, 99th, or 99.9th. The chosen percentile is not optimized in this study to find the n with the best damage detection performance, to reflect how this method would be used in a real-life SHM scenario where you do not have any damaged state data.

If significantly more sequences than the chosen percentile value in an unknown state are rejected, it is assumed that the state of the bridge has changed and therefore it is determined to be damaged. The higher the chosen percentile value, the more sequences are needed to make a statistically significant determination of the bridge state. Due to limitations of the chosen dataset of the Z24 bridge, namely the small number of sequences in each damaged state, we chose to use the 95th percentile as our rejection threshold to be able to confidently assess each bridge state.

Sequences with reconstruction error above the 95th percentile are rejected in the healthy-state test set. This threshold gives a baseline of comparison of summed sequence error of a bridge in an unknown state. By doing so, we know that 5% of never seen, healthy-state sequences are rejected. If a significantly higher portion than 5% of sequences of a bridge in an unknown state are rejected, the bridge is considered to be damaged. An example of the determined threshold can be seen in the Results chapter (Fig. 6).

3 Experiments

3.1 Z24 Benchmark

One of the challenges of a machine learning-based bridge SHM is that there are very few datasets that contain data of a bridge in a healthy state and in a damaged state. The Z24 bridge benchmark [33, 34] has sensor recordings of a real full-sized bridge that was monitored for almost a year before being deliberately damaged in a realistic and controlled way. The damage was introduced in 15 steps, each step damaging the bridge further in a realistic manner. This is useful to assess the model's sensitivity, to see at what stage the model can determine a damage state. The Z24 bridge was built in 1963—a typical posttensioned box girder highway bridge (Fig. 5) with a main span of 30 m over an underlying road and two connecting side spans of 14 m [33]. Bridge sensor measurements were recorded such as bridge vibrations, air temperature, soil temperature, humidity, wind speed, and vehicle passing. Most of the environmental sensors measured temperature. Ten-minute samples with a sampling rate of 100 Hz were recorded by seven accelerometers.

The Z24 benchmark contains data when the temperature drops below freezing, which has a big effect on the behavior of the bridge [4]. It is a challenge to assess environmental effects versus damage events, as bridges can present drastic seasonal changes [35]. In this method, events such as freezing temperatures receive no special treatment. These events are included in the training and test datasets, and therefore included in the 95% rejection threshold. The dataset presents an opportunity to determine a model's ability able to assess such events and better assess environmental changes versus damage cases. The Z24 bridge data is split into two parts, long-term continuous sensor monitoring for a period of one year and a short-term progressive

Fig. 5 Side view of a three-span highway beam bridge, where the main span goes over a road. For a more detailed view of the Z24 bridge, refer to [33]

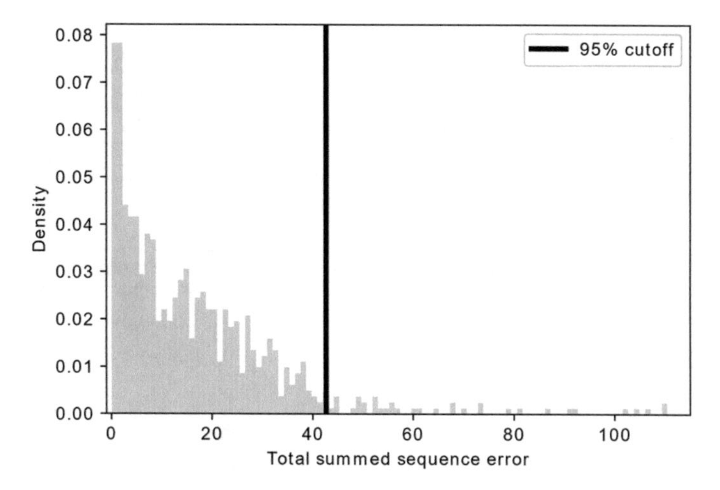

Fig. 6 Distribution of healthy-state test set summed sequence errors, showing the determined 95% rejection threshold for the uncertainty method model

damage with much higher spatial resolution instrumentation. The presented method only uses data from the long-term sensor measurements.

3.2 *Training*

The healthy bridge state data is cut up into 10-min-long sequences. The sequences are randomly sampled without substitution so that 70% belong to the training set, 15% to the validation set, and 15% to the test set. See Fig. 4 for a visualization of the dataset split.

A feed-forward autoencoder neural network with ReLU activation function (Fig. 2) is trained using dropout [27] on the healthy-state test dataset to reconstruct the healthy-state bridge sensor data. The autoencoder reconstructs vibration sensor data from all sensors at once. Environmental sensor data is also included in the input. The model is trained on non-overlapping sliding time window inputs corresponding to a 2 s time interval. The Adam optimizer [16] is used to train the model. The model was trained on a cloud computer with 16 CPUs, 50 GB memory, and a 16 GB GPU NVIDIA Tesla P100. The full training data was around 50 GB. The final parameters of the trained model after hyperparameter search can be seen in Table 1.

Table 1 Final parameters of the trained autoencoder neural network

Parameter	Value
Input size	1458
Encode layer width	256
Z code layer width	128
Decode layer width	256
Dropout	0.1
Learning rate	0.0003

4 Results

The main results can be seen in Table 2. Each row in the table represents a damage case. Gray cells mark when over 5% of the 10-min sequences are rejected and therefore the bridge is determined to be in a damage state. Since there are varying number of sequences available for each damage case, the results must be interpreted carefully. For example, damage case 2 had over 100 recorded sequences, and damage case 13 had only 10.

The presented method can confidently detect damage cases 5–7, 9–11, and 13–14, where all three methods determine the state to be damaged. Note that the three different methods of detecting damage are all using the same trained model. The sequence in Fig. 7 shows a comparison of the three presented reconstruction methods for a healthy and damaged state.

Table 2 Percentage of rejected sequences of the different damage cases

Damage Case	Description	Sequences [#]	Uncertainty [%]	Mean [%]	Standard [%]
0	Healthy state (randomly sampled)	103	1.9	1.0	1.0
1	Lowering of pier, 20 mm	35	5.7	14.3	51.4
2	Lowering of pier, 40 mm	110	0.9	6.4	24.5
3	Lowering of pier, 80 mm	24	4.2	4.2	29.2
4	Lowering of pier, 95 mm	20	5.0	5.0	45.0
5	Lifting of pier, tilt of foundation	22	13.6	18.2	50.0
6	Spalling of concrete at soffit, 12 m²	24	17.4	17.4	30.4
7	Spalling of concrete at soffit, 24 m²	20	20.0	25.0	35.0
8	Landslide of 1m at abutment	88	1.1	3.4	6.8
9	Failure of concrete hinge	44	6.8	9.1	15.9
10	Failure of 2 anchor heads	24	20.8	29.2	37.5
11	Failure of 4 anchor heads	73	8.2	11.0	12.3
12	Rupture of 2/16 tendons	15	0.0	0.0	6.7
13	Rupture of 4/16 tendons	10	10.0	10.0	10.0
14	Rupture of 6/16 tendons	27	11.1	14.8	22.2

Grayed cells mark where the bridge state is determined to be damaged, since over 5% of sequences are classified as damaged

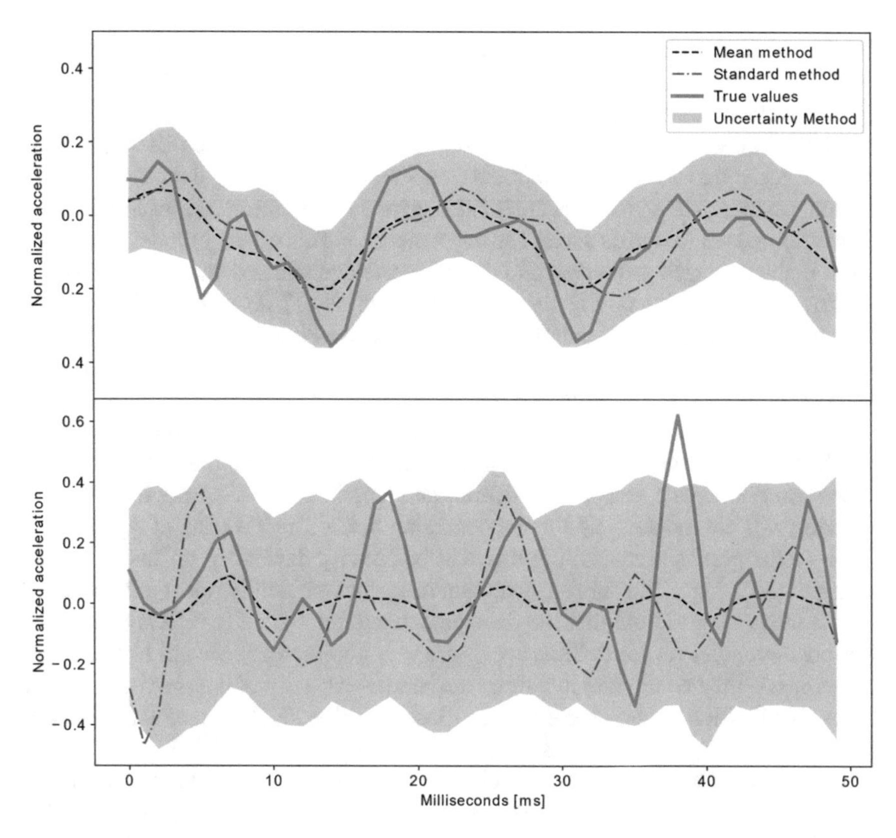

Fig. 7 All three model methods compared for a given input. The plotted data only shows one sensor and a 50-ms window for clarity. Damage state 7 below, healthy state above. Take note of the increasing uncertainty in the damage scenario. The true values in the damage scenario fall outside the 2 standard deviation ranges of the predictions at around 20 and 40 ms

5 Discussion

The method of using the same model in several ways is a novel approach to structural health monitoring, utilizing the trained model to a fuller extent. The standard method is more performant in detecting damage, while the uncertainty method is much more cautious in the damage assessment. The standard method is much better at detecting damage, but it does not take uncertainty into account in any way. Therefore, in an extreme weather event that is not present in the training data, this model would more likely wrongly determine the bridge to be in a damage state. This is a known issue in SHM, false damage detection. When the uncertainty method determines the bridge to be in a damaged state, it is a more certain determination. This is important in real-life usage of a structural health monitoring system to reduce false damage classification.

An interesting result is that being better at reconstruction does not equal being better at detecting damage. The mean method achieves lower reconstruction error than the standard method in all cases, healthy or damaged. This does not mean that the mean method is better at detecting damage; in fact, it is the opposite that is true. The standard method reconstructs healthy sequences adequately, but reconstructs damaged sequences poorly and is therefore better at detecting the difference. This does not mean that the standard method overfitted on the training data, since it does not achieve remarkably lower reconstruction error on the training dataset compared to the testing and validation datasets. The mean model and uncertainty model generalize better to the damaged sequences and therefore achieve worse damage detection.

Why is the method failing for certain damage cases? As seen in Table 2, the damage detection did not confidently reject cases 1–4, 8, and 12. Note that the standard method rejected all damage cases, but the uncertainty was too great for the uncertainty method to reject these cases.

The lowering of the pier is a rare occurrence in a real-life situation, but not unheard of. The severity of such an event depends on the specific design of the structure. In a three-span bridge with reasonable stiffness and length, as the Z24 bridge, a ~40 mm lowering will not induce a significant damaging force. The lowering of a pier could still be a dangerous scenario, but this will be heavily dependent on the structural design of the bridge. The damage detection does not confidently reject this scenario.

It is interesting to note that the damage detection confidently rejects cases 6–7, spalling of concrete at soffit. Such events are very common and are not immediately dangerous as they do not directly affect the bearing capacity of the concrete deck. It is possible that the sensors were located closely to the affected area, and therefore the damage detection was more sensitive to the changes.

The method does not confidently reject damage case 8, "landslide of 1 m at abutment." Such an event will have a specific effect on the bridge response in particular cases since the landfill supports the bridge only in a certain load scenario. During the damage cases, the bridge was closed to passing traffic. This could explain that the damage was not detected, since the agitation of the bridge was only coming from traffic passing under the bridge which is not as substantial.

Damage case 12, "rupture of 2/16 tendons," was not rejected confidently. The rupture of tendons is rare and possibly dangerous but is very hard to detect from standard visual inspection. Therefore, it might be more common than we think. The main reason that this damage scenario was not detected is that there were very few sequences, only 10—corresponding to only 120 min. It is therefore reasonable to say that the damage detection could have been more accurate given a greater number of sequences.

5.1 Damage Sensitivity

The presented method faces a well-known challenge in bridge structural health monitoring that damage is best assessed when the bridge is actively vibrating. This is clearly shown in Fig. 8, where the reconstruction error drops during the night when the bridge is in low use.

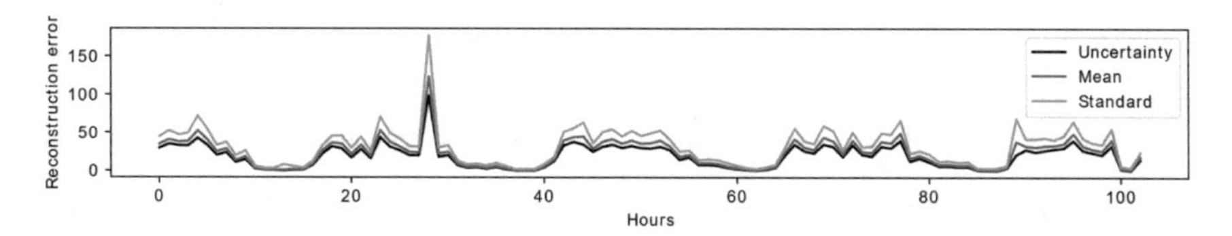

Fig. 8 Reconstruction error of a healthy-state data. Here, it is clearly shown that the reconstruction error is dependent on activity, as the error drops notably down overnight

The healthy-state training data was captured, while the bridge was in full use. Vehicles could drive over the bridge and under. When the bridge was damaged, it was closed for traffic. For every damage case, the bridge was only excited by underneath traffic which results in much lower vibration levels. Therefore, the damage detection is even more challenging and requires a more sensitive damage detection.

The timescale of damage detection is more nuanced since the damage detection is based on evaluating multiple sequences. The damage sensitivity is also dependent on bridge activity. In the case of considerable damage, the model would start rejecting the 10-min sequences in succession. For the model to assess damage with more certainty, it would need multiple 10-min sequences. Therefore, the timescale of reliable damage detection is in the order of hours, possibly days.

6 Conclusion

The results are promising for the field of bridge structural health monitoring. The autoencoder neural network model trained on the complete sensor data was able to detect several realistic damage cases on the Z24 bridge benchmark [34], which has been used in research for over 20 years. Due to the size of the full data and computational limitations, few researchers have presented methods of structural health monitoring on the full data. To our knowledge, this is the first time a model has been trained on the full healthy-state data and successfully detected damage cases while taking no special care of freezing temperature variations of the healthy-state data.

The method is performant and is a prominent step toward modern infrastructure. The methods' ability to detect damage in realistic damage scenarios of a real bridge has many economic, social, and ethical aspects. Such a system could possibly detect damage before a traditional bridge inspector could and prevent a costly repair. The model can reconstruct a full 10-min sequence and classify it in the order of seconds, on a standard modern personal computer. This means that it is computationally very fast and can be used in real time. The training of the model on the other hand requires considerable computing power.

Acknowledgements We thank Dr. Edwin Reynders at KU Leuven for providing the Z24 benchmark dataset; without the contribution, this work would not have been possible.

References

1. Ásgrímsson DS (2019) Quantifying uncertainty in structural condition with Bayesian deep learning—a study on the Z-24 bridge benchmark. KTH
2. Gonzalez I, Karoumi R (2015) BWIM aided damage detection in bridges using machine learning. J Civ Struct Health Monit 5(5):715–725
3. Neves AC, González I, Leander J, Karoumi R (2017) Structural health monitoring of bridges: a model-free ANN-based approach to damage detection. J Civ Struct Health Monit 7(5):689–702
4. Peeters B, De Roeck G (2001) One-year monitoring of the Z 24-bridge: environmental effects versus damage events. Earthquake Eng Struct Dyn 30(2):149–171
5. Gonzales I, Ülker-Kaustell M, Karoumi R (2013) Seasonal effects on the stiffness properties of a ballasted railway bridge. Eng Struct 57:63–72
6. Casas JR, Moughty JJ (2017) Bridge damage detection based on vibration data: past and new developments. Front Built Environ 3:4
7. Das S, Saha P, Patro SK (2016) Vibration-based damage detection techniques used for health monitoring of structures: a review. J Civ Struct Health Monit 6(3):477–507
8. Neves AC, González I, Karoumi R, Leander J (2020) The influence of frequency content on the performance of artificial neural network-based damage detection systems tested on numerical and experimental bridge data. Struct Health Monit 1475921720924320
9. Gomes GF, Mendéz YAD, Alexandrino PdaSL, da Cunha SS, Ancelotti AC (2018) The use of intelligent computational tools for damage detection and identification with an emphasis on composites—a review. Compos Struct
10. Teimouri H, Milani AS, Loeppky J, Seethaler R (2017) A Gaussian process-based approach to cope with uncertainty in structural health monitoring. Struct Health Monit 16(2):174–184
11. Simoen E, De Roeck G, Lombaert G (2015) Dealing with uncertainty in model updating for damage assessment: a review. Mech Syst Signal Process 56:123–149
12. Kelley HJ (1960) Gradient theory of optimal flight paths. ARS J 30(10):947–954
13. Bryson AE, Denham WF (1962) A steepest-ascent method for solving optimum programming problems. J Appl Mech 29(2):247–257
14. Rumelhart DE, Hinton GE, Williams RJ (1986) Learning representations by back-propagating errors. Nature 323(6088):533
15. Sutskever I, Martens J, Dahl G, Hinton G (2013) On the importance of initialization and momentum in deep learning. In: International conference on machine learning, pp 1139–1147
16. Kingma DP, Ba J (2014) Adam: a method for stochastic optimization. arXiv Preprint arXiv1412.6980
17. Leshno M, Lin VY, Pinkus A, Schocken S (1993) Multilayer feedforward networks with a nonpolynomial activation function can approximate any function. Neural Netw 6(6):861–867
18. Cybenko G (1989) Approximation by superpositions of a sigmoidal function. Math Control Signals Syst 2(4):303–314
19. Krizhevsky A, Sutskever I, Hinton GE (2012) Imagenet classification with deep convolutional neural networks. In: Advances in neural information processing systems, pp 1097–1105
20. Simonyan K, Zisserman A (2014) Very deep convolutional networks for large-scale image recognition. arXiv Preprint arXiv1409.1556
21. He K, Zhang X, Ren S, Sun J (2016) Deep residual learning for image recognition. In: Proceedings of the IEEE conference on computer vision and pattern recognition, pp 770–778
22. LeCun Y, Bengio Y, Hinton G (2015) Deep learning. Nature 521(7553):436
23. Hinton GE, Salakhutdinov RR (2006) Reducing the dimensionality of data with neural networks. Science (80-) 313(5786):504–507
24. Neal RM (1996) Bayesian learning for neural networks. Springer-Verlag New York Inc., New York
25. MacKay DJC (1992) A practical Bayesian framework for backpropagation networks. Neural Comput 4(3):448–472
26. Gal Y, Ghahramani Z (2016) Dropout as a Bayesian approximation: representing model uncertainty in deep learning. In: International conference on machine learning, pp 1050–1059

27. Srivastava N, Hinton G, Krizhevsky A, Sutskever I, Salakhutdinov R (2014) Dropout: a simple way to prevent neural networks from overfitting. J Mach Learn Res 15(1):1929–1958
28. Damianou A, Lawrence N (2013) Deep Gaussian processes. In: Artificial intelligence and statistics, pp 207–215
29. Gal Y (2016) Uncertainty in deep learning. University of Cambridge
30. Li Y, Gal Y (2017) Dropout inference in Bayesian neural networks with alpha-divergences. arXiv Preprint arXiv1703.02914
31. Gal Y, Hron J, Kendall A (2017) Concrete dropout. In: Advances in neural information processing systems, pp 3584–3593
32. Bergstra J, Bengio Y (2012) Random search for hyper-parameter optimization. J Mach Learn Res 13:281–305
33. Reynders E, De Roeck G (2009) Continuous vibration monitoring and progressive damage testing on the Z24 bridge. In: Encyclopedia of structural health monitoring
34. Reynders E, De Roeck G (2014) Vibration-based damage identification: the Z24 bridge benchmark. In: Encyclopedia of earthquake engineering, pp 1–8
35. Gonzalez I, Karoumi R (2014) Analysis of the annual variations in the dynamic behavior of a ballasted railway bridge using Hilbert transform. Eng Struct 60:126–132

Diagnosis, Prognosis, and Maintenance Decision Making for Civil Infrastructure: Bayesian Data Analytics and Machine Learning

Manuel A. Vega, Zhen Hu, Yichao Yang, Mayank Chadha, and Michael D. Todd

Abstract Due to the aging of civil infrastructure and the associated economic impact, there is an increasing need to continuously monitor structural and non-structural components for system life-cycle management, including maintenance prioritization. For complex infrastructure, this monitoring process involves different types of data sources collected at different timescales and resolutions, including but not limited to abstracted rating data from human inspections, historical failure record data, uncertain cost data, high-fidelity physics-based simulation data, and online high-resolution structural health monitoring (SHM) data. The heterogeneity of the data sources poses challenges to implementing a diagnostic/prognostic framework for decision making for life-cycle actions such as maintenance. Using quoin blocks components of a miter gate as an example, this chapter presents a holistic Bayesian data analytics and machine learning (ML) framework to demonstrate how to integrate various data sources using Bayesian and ML methods for effective SHM, and Prognostics and Health Management (PHM). In particular, this chapter discusses how Bayesian data analytics and ML methods can be applied to (1) diagnosis of bearing loss-of-contact degradation in quoin blocks; (2) optimized sensor placement for SHM on the gate; (3) fusion of various data sources for effective PHM; and (4) deciding maintenance strategies by considering the behavioral aspect of human decision making under uncertainty.

M. A. Vega · Y. Yang · M. Chadha · M. D. Todd (✉)
University of California San Diego, La Jolla, CA 92093, USA
e-mail: mdtodd@eng.ucsd.edu

M. A. Vega
e-mail: mvegaloo@eng.ucsd.edu

Y. Yang
e-mail: yiy018@eng.ucsd.edu

M. Chadha
e-mail: machadha@eng.ucsd.edu

Z. Hu
University of Michigan-Dearborn, Dearborn, MI 48128, USA
e-mail: zhennhu@umich.edu

A. Cury et al. (eds.), *Structural Health Monitoring Based on Data Science Techniques*, Structural Integrity 21, https://doi.org/10.1007/978-3-030-81716-9_3

Keywords Bayesian methods · Uncertainty quantification · FE model · Surrogate model · Damage estimation · Remaining useful life prediction · Decision theory

1 Introduction

Advances in sensing technologies, accelerated by the "Internet of things," have allowed collection of large amounts of data about our civil infrastructure, which includes complex transportation networks both over land and through our inland waterway navigation corridors. Among the most important reasons for this data collection are damage/state diagnostics and predictions of future state performance. Such assessments can lead to improved life-cycle management of civil infrastructure systems, which is critical to keep these systems continuously operational under increasingly constrained budgets. Figure 1 shows an overview of the association between diagnosis, prognosis, and maintenance decision making for civil infrastructure; more detail can be found in [1]. In a general sense, Fig. 1 shows the fundamental workflow of a "digital twin" for structural asset life-cycle diagnosis and prognosis.

For damage diagnosis, engineers can rely on supervised learning algorithms when sufficient life-cycle data is available [2–4]. On the other hand, when life-cycle data is limited, engineers typically rely on physics-based modeling (such as finite element (FE) models) and model updating techniques to estimate the unknown parameters required to infer the current state of the system as shown in Fig. 1.

Prognostics and Health Management (PHM) is the notion of augmenting current structural state diagnostic information gleaned by inspections or SHM to make predictions of the future state and reliability of the system based on degradation models or historic degradation/failure data [5]. When such prior data is available

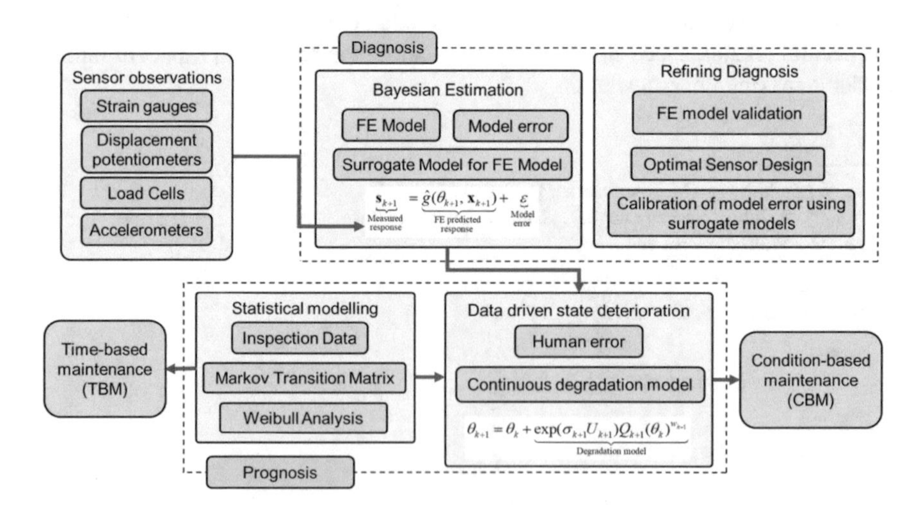

Fig. 1 Diagnosis, prognosis, and maintenance decision-making framework for civil infrastructure

(not very common in the civil infrastructure domain), a data-driven degradation model is possible, but more commonly physics-based approaches [6, 7] or empirical approaches to build the model [8–10] are required.

Additionally, PHM uses its prediction capabilities to inform life-cycle management, which targets optimization of a desired system performance criterion (e.g., cost, availability, reliability, etc.). For life-cycle management, maintenance approaches can be roughly classified into two categories, namely time-based maintenance (TBM) and condition-based maintenance (CBM). This term is closely related to condition monitoring (CM), which usually refers to implementation of state diagnostics applied to rotating machinery [11]. When applied to civil and aerospace systems, CM is referred to as SHM, so these terms are used interchangeably in a general sense. When information from an SHM (equivalent, a CM) process is used to trigger maintenance decisions, a CBM decision policy arises. TBM and CBM approaches have been benefited by advances in various fields such as data analytics, machine learning, computational mechanics, Bayesian statistics, and reliability engineering.

As stated before, diagnostic and prognostic approaches are either physics-based (e.g., FE model updating) [12–14] or data-driven [15–18]. For some engineering systems, hybrid approaches that combine the physics-based approach with data-driven approach to improve the CBM predictive capabilities are useful. However, the study of hybrid approaches [19, 20] has been very limited, and even more limited for large civil infrastructure. Other limitations that occur, for example, are that the monitoring process sometimes involves different types of data sources collected at different timescales and resolutions, such as abstracted rating data from human inspections, historical failure record data, uncertain cost data, high-fidelity physics-based simulation data, and online high-resolution structural health monitoring (SHM) data. The heterogeneity of the data sources poses challenges to the diagnostic/prognostic implementation of decision making for maintenance.

This chapter presents a holistic framework for diagnosis, prognosis, and maintenance decision making for civil infrastructure using Bayesian data analytics and machine learning methods. It combines a physics-based approach for diagnosis with data-driven approaches using various data sources for prognostics. In summary, this chapter discusses how to: (1) fuse various data sources using Bayesian methods; (2) perform damage diagnostics and prognosis using Bayesian data analysis and machine learning; (3) optimize maintenance strategies; and (4) apply these concepts to a real-world problem using a miter gate example, drawn from a navigation lock system used in the inland waterways navigation corridor.

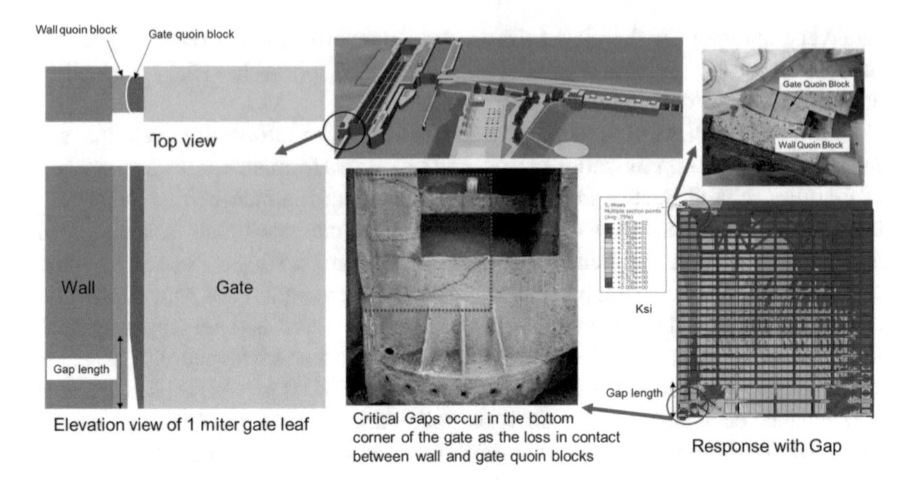

Fig. 2 FEM modeling of a leaf of a miter gate including gap length deterioration

2 Summary of Data Sources

2.1 Physics-Based Simulation Data

For civil systems, the approach is usually carried out by using a physics-based model (e.g., finite element (FE) model) of the structure [2–4]. It is fundamentally an inverse problem because the system parameters are estimated from measured response quantities. Generally, a physics-based model is a "forward" problem where the system responses (e.g., FE output response) are predicted as a function of the (known) system parameters (e.g., FE inputs). Synthetic system parameters may be used to obtain the FE system response, whose responses can be compared to the "true" system as shown in Fig. 2.

2.2 Inspection Data

Regular condition assessments are conducted in critical structures such as bridges and offshore structures. These assessments are obtained from an inspection process, which can be part of a periodic or non-periodic inspection policy. One may reasonably hypothesize that these inspections should reflect a deterioration state. These states can be in a continuous or discrete form, e.g., an inspection assignment of A, B, C, D, F, CF, such as is utilized for hydraulic structures owned by the US Army Corps of Engineers (USACE). Inspections can be performed by inspectors, drones [21], or even robots [22]. The resulting data format is usually highly abstracted in the format of ratings as mentioned above.

In [23], the failure condition in critical components is estimated using a transition matrix built from the discrete inspection ratings. The following is an example of a transition matrix built from the reported six damage ratings:

$$\mathbf{P}_{\text{Report}} = P\left(\mathbf{X}_{t+1}^{R} | \mathbf{X}_{t}^{R}\right) = \begin{bmatrix} P\left(A_{t+1}^{R} | A_{t}^{R}\right) & \cdots & P\left(\text{CF}_{t+1}^{R} | A_{t}^{R}\right) \\ \vdots & \ddots & \vdots \\ P\left(A_{t+1}^{R} | \text{CF}_{t}^{R}\right) & \cdots & P\left(\text{CF}_{t+1}^{R} | \text{CF}_{t}^{R}\right) \end{bmatrix}, \quad \forall i, j = 1, \ldots, 6, \tag{1}$$

where $P(\cdot)$ is a probability operator, $\mathbf{X}_{t}^{R} = \left[A_{t}^{R}, B_{t}^{R}, C_{t}^{R}, D_{t}^{R}, F_{t}^{R}, \text{CF}_{t}^{R}\right]$ are the damage ratings at time step t, and "|" is a conditional operator.

The above transition matrix obtained from inspection data can be used to estimate the possible condition (and failure condition) after n time steps as follows:

$$P\left(\mathbf{X}_{n+k}^{R}\right) = P\left(\mathbf{X}_{k}^{R}\right) \cdot \mathbf{P}_{\text{Report}}^{n}, \tag{2}$$

where $P\left(\mathbf{X}_{k}^{R}\right) = \left[P\left(X_{1,k}^{R}\right), P\left(X_{2,k}^{R}\right), P\left(X_{3,k}^{R}\right), P\left(X_{4,k}^{R}\right), P\left(X_{5,k}^{R}\right), P\left(X_{6,k}^{R}\right)\right]$ are the condition probabilities at the current time or initial time (i.e., $k = 0$) where $X_{1,k}^{R}$ and $X_{6,k}^{R}$ represents A_{k}^{R} and CF_{k}^{R}, respectively. The condition CF represents the complete failure condition. $\mathbf{P}_{\text{Report}}$ is the transition matrix built from the discrete ratings, and $P\left(\mathbf{X}_{n+k}^{R}\right)$ are the condition probabilities at n time steps in the future. The current time $P\left(\mathbf{X}_{k}^{R}\right)$ may be either obtained from current inspections or using SHM data.

2.3 Human Errors

Human error can greatly affect the reliability of the inspection assessment. For example, human psychology influences inspectors to make conservative or non-conservative assessment that can greatly influence maintenance decisions. Benchmark data may be available to account for the accuracy of the assessment given the inspector qualification, training, and certification [24]. A human observation error matrix can be obtained/estimated as follows to probabilistically measure the human error

$$\mathbf{P}_{\text{human}} = \begin{bmatrix} P_{11}^{h} & P_{12}^{h} & \cdots & P_{16}^{h} \\ P_{21}^{h} & P_{22}^{h} & \cdots & P_{26}^{h} \\ \vdots & \vdots & \ddots & \vdots \\ P_{61}^{h} & P_{62}^{h} & \cdots & P_{66}^{h} \end{bmatrix}$$

$$= \begin{bmatrix} P(A_t^R|A_t^{\text{tr}}) & P(A_t^R|B_t^{\text{tr}}) & \cdots & P(A_t^R|\text{CF}_t^{\text{tr}}) \\ P(B_t^R|A_t^{\text{tr}}) & P(B_t^R|B_t^{\text{tr}}) & \cdots & P(B_t^R|\text{CF}_t^{\text{tr}}) \\ \vdots & \vdots & \ddots & \vdots \\ P(\text{CF}_t^R|A_t^{\text{tr}}) & P(\text{CF}_t^R|B_t^{\text{tr}}) & \cdots & P(\text{CF}_t^R|\text{CF}_t^{\text{tr}}) \end{bmatrix}, \tag{3}$$

in which $P_{ik}^h = \Pr\{X_{i,t}^{\text{obs}}|X_{j,t}^{\text{tr}}\}$ is the probability that the reported OCA rating is k given that the true OCA rating is i.

An example of a conservative human error matrix is given as below

$$\mathbf{P}_{\text{human}} = \begin{bmatrix} 1 & 0 & 0 & 0 & 0 & 0 \\ 0.04 & 0.96 & 0 & 0 & 0 & 0 \\ 0 & 0.40 & 0.60 & 0 & 0 & 0 \\ 0 & 0.03 & 0.17 & 0.80 & 0 & 0 \\ 0 & 0 & 0 & 0.03 & 0.97 & 0 \\ 0 & 0 & 0 & 0 & 0.03 & 0.97 \end{bmatrix}. \tag{4}$$

The above $\mathbf{P}_{\text{human}}$ models the behavior of an inspector that regularly tends to assess a component to be in a better condition than reality. Accounting for $\mathbf{P}_{\text{human}}$ allows to estimate the $\mathbf{P}_{\text{true}}$ and $P(\mathbf{X}_k^{\text{tr}})$ from $\mathbf{P}_{\text{Report}}$ as will be discussed in Sect. 4.2. The terms $\mathbf{P}_{\text{true}}$ and $P(\mathbf{X}_k^{\text{tr}})$ are the true transition matrix and the true condition probabilities, respectively, at the current time.

2.4 SHM Data

SHM data involves periodically sampled response measurements from spatially distributed sensors, extraction of damage-sensitive features from these measurements and damage diagnosis using these features with either an inverse-problem approach or a data-driven approach.

As stated earlier, SHM diagnostic capabilities can inform the current state of the structure. However, SHM data inevitably will contain noise due to a variety of stochastic influences, not the least of which result from environmental and operational variability. In this chapter, the sensor monitoring data (i.e., strain measurement data for the miter gate example) at time step t_i is defined as $\mathbf{s}_i = [s_{i1}, s_{i2}, \ldots, s_{iN_S}]$, where N_S is the number of sensors. Also, $\mathbf{s}_{1:n} \triangleq \{\mathbf{s}_1, \mathbf{s}_2, \ldots, \mathbf{s}_n\}$ defines the sensor measurements collected up to t_n.

Next, the following section will discuss how to utilize the above data sources for the damage diagnostics, prognostics, and maintenance planning using Bayesian data analytics and machine learning.

3 Damage Diagnostics Using Bayesian Data Analysis and Machine Learning

In this section, a summary of how to perform damage diagnostics using simulation data and SHM data based on machine learning and recursive Bayesian updating is presented.

3.1 Surrogate Modeling for Physics-Based Models Using Machine Learning

For damage diagnosis with limited SHM data, inverse modeling via physics-based models such as finite element (FE) models have been used with data from SHM systems to estimate parameters that infer some form of damage state. A fast, efficient simulation of complex FE models is essential for appropriately fast damage diagnosis. Depending on the dimension space of the inputs and outputs of interest, different machine learning techniques can be chosen to build "cheap" yet accurate surrogate models of the physics-based FE model. Two of the commonly used techniques, artificial neural networks and Gaussian process regression, are briefly summarized as below.

Artificial Neural Network (ANN). ANNs are an attractive option for surrogate emulation of FE models. This type of supervised learning model works well with classification (for discrete classes) and regression (for continuous processes) problems. However, ANNs are effective when a large amount of data is available, and they are built to create point estimates rather than probabilistic estimates. Some researchers [25–28] have used Bayesian inference to estimate the ANN's weight and model parameters, which has been referred to as Bayesian neural networks (BNN). BNNs are good for high-dimensional spaces and better handle the issue of limited data availability.

Gaussian Process Regression (GP). GP (or Kriging) models are an attractive option for surrogate architectures because they are built to quantify the uncertainty in the estimations rather than simply point-based estimates, as most other supervised learning models (e.g., ANNs, support vector machines) do. Several researchers use GP regression to build Bayesian prediction models for civil engineering structures [23, 29]. A GP surrogate model, $\hat{\gamma}_j = \hat{g}_j(\mathbf{x})$ is defined as

$$\hat{\gamma}_j = \hat{g}_j(\mathbf{x}) = \mathbf{f}(\mathbf{x})^{\mathrm{T}}\boldsymbol{\alpha} + Z(\mathbf{x}), \tag{5}$$

where $\boldsymbol{\alpha}$, $\mathbf{f}(\mathbf{x})^{\mathrm{T}}$, and $Z(\mathbf{x}) \sim N\big(0, \sigma_{\mathrm{GP}}^2\rho(\cdot,\cdot)\big)$ are the coefficients of the trend function, the trend function, and a stationary Gaussian process, respectively.

The stationary Gaussian process uses a correlation function $\rho(\cdot,\cdot)$ to quantify the correlation between responses at any two points as below

$$\rho\left(\mathbf{x}, \mathbf{x}'\right) = \exp\left\{-\sum_{l=1}^{N_V} \omega_l (x_l - x_l')^2\right\}, \tag{6}$$

where N_V is the number of variables, and $\boldsymbol{\omega} = (\omega_1, \ldots, \omega_{N_V})^T$ is the vector of roughness parameters.

Furthermore, the aforementioned GP hyper-parameters $\boldsymbol{\upsilon} = \left(\boldsymbol{\alpha}, \sigma_{GP}^2, \boldsymbol{\omega}\right)$ are estimated using the maximum likelihood estimation method. After the estimation of the hyper-parameters $\boldsymbol{\upsilon}$ for any given inputs $\mathbf{x}$, the GP prediction is given by

$$\hat{\gamma}_j = \hat{g}_j(\mathbf{x}) \sim N\left(\mu_j(\mathbf{x}), \sigma_j^2(\mathbf{x})\right), \quad \forall j = 1, 2, \ldots, r, \tag{7}$$

where $\mu_j(\mathbf{x})$ and $\sigma_j^2(\mathbf{x})$ are the mean and variance of the prediction of γ_j, respectively, for the input $\mathbf{x}$.

In applications such as the miter gate presented in this chapter, the output space of the simulation or the sensors available (e.g., hundreds or thousands of nodes in the FE model) can be large. Also, it is known that sensors located close to each other may contain highly correlated information. Therefore, dimension reduction techniques are usually used in conjunction with GP models or ANNs to build surrogate models for the physics-based FE model. A commonly used technique is singular value decomposition (SVD). SVD is a linear algebra technique used to transform high-dimensional matrices into a reduced dimensional space preserving most of the original information. Figure 3 presents a generalize procedure of surrogate modeling based on a GP model and dimension reduction methods for a model with inputs $\mathbf{x}$ and

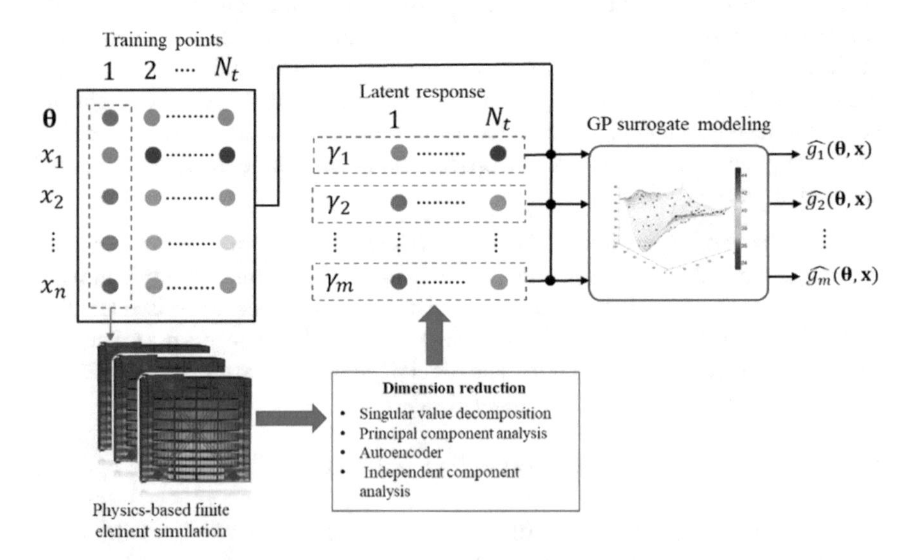

Fig. 3 Building surrogate model for FEM model using GP and dimension reduction

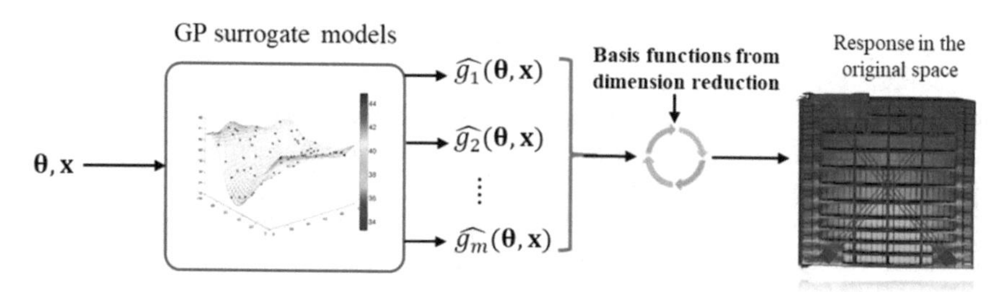

Fig. 4 Prediction using GP surrogate modeling and dimension reduction

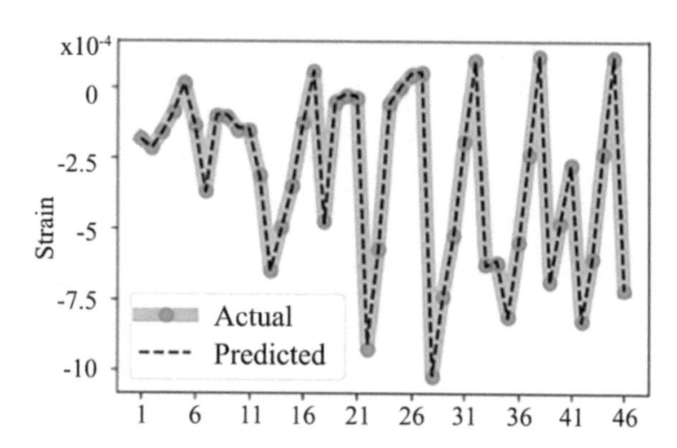

Fig. 5 Testing accuracy of Kriging model

$\boldsymbol{\theta}$. Following that, Fig. 4 shows the prediction process using the trained Kriging model for given FEM input parameters and features obtained from dimension reduction.

The proposed framework explained in this work is based on a surrogate model using a Kriging model combined with SVD to develop a fast emulator of the FEM. Figure 5 shows the testing accuracy obtained at 46 sensors installed in the Greenup miter gate at a particular point in time, which shows how close ML-based surrogate model emulates the original FE model.

3.2 *Damage Diagnostics Using Recursive Bayesian Updating*

Given the SHM data and a ML-based surrogate model of the physics-based FE model, damage diagnosis can be performed using Bayesian estimation of parameters, a_n, that directly relate to a damage mode (e.g., cracks, gap, thickness loss due to corrosion, etc. See Fig. 2 for examples). Bayesian estimation of this parameters can be performed based on the following state-space equation

$$\begin{aligned} &\text{State equation: } a_n = h(a_{n-1}) + \varepsilon_h, \\ &\text{Measurement equation: } \mathbf{s}_n = \hat{g}(a_n, \mathbf{x}_n) + \varepsilon_n, \end{aligned} \tag{8}$$

where $h(a_{n-1})$ is the state equation that describes the evolution of a_n over time, ε_h is the process noise, $\hat{g}(a_n, \mathbf{x}_n)$ is the ML surrogate model built in Sect. 3.1 where $\mathbf{x}_n$ represents the known/measurable input variables to the physics-based model (e.g., loads, geometry, material properties, etc.), and $\mathbf{s}_n$ is the observations from SHM system as discussed in Sect. 2.4.

Note that the state equation given in Eq. (8) may be unknown due to a lack of understanding of the damage evolution mechanism. In that situation, a random walk type equation can be used as the state equation with large process noise as discussed in [23]. Assume there are N_S strain sensors installed in a structure such as the miter gate and the damage parameter, a_n, to infer at time step t_n is the extent of the loss of bearing contacting (i.e., gap length). Then, the posterior probability density function of the gap length a_n at time step t_n conditioned on strain measurements $\mathbf{s}_{1:n}$ is estimated using Bayesian inference method recursively as follows

$$f(a_n|\mathbf{s}_{1:n}) = \frac{f(\mathbf{s}_n|a_n)f(a_n|\mathbf{s}_{1:n-1})}{\int f(\mathbf{s}_n|a_n)f(a_n|\mathbf{s}_{1:n-1})\mathrm{d}a_n} \propto f(\mathbf{s}_n|a_n)f(a_n|\mathbf{s}_{1:n-1}), \tag{9}$$

which $f(a_n|\mathbf{s}_{1:n-1})$ is defined as follows

$$f(a_n|\mathbf{s}_{1:n-1}) = \int f(a_n|a_{n-1})f(a_{n-1}|\mathbf{s}_{1:n-1})\mathrm{d}a_{n-1}, \tag{10}$$

in which $f(\mathbf{s}_n|a_n)$ is the likelihood function, obtained from the measurement equation, of observing $\mathbf{s}_n$ for given a_n at time step t_n, and the term $f(a_n|a_{n-1})$ represents the probability distribution of a_n for a given a_{n-1} obtained using the state equation, which describes the damage evolution over time.

The recursive Bayesian updating of Eqs. (9) and (10) is analytically intractable. In practical application, various filtering methods, such as particle filtering [30], extended Kalman filter [31], and unscented Kalman filter [32], have been developed to approximate recursive updating process.

Figure 6 presents an illustrative example of strain measurement data $\mathbf{s}_n$ of 10 sensors ($N_S = 10$) over a certain time period of interest. After that, Fig. 7 shows the damage diagnostics results of gap length of miter gate (i.e., the gap damage in the miter gate as shown in Fig. 2) over time using recursive Bayesian updating and the ML-based surrogate models. It shows that the recursive Bayesian inference method can effectively perform damage diagnostics by fusing the information from physics-based simulation model and SHM data. More details of the integration of ML-based surrogate model and SHM observation data using Bayesian recursive updating for damage diagnostics are available in [23].

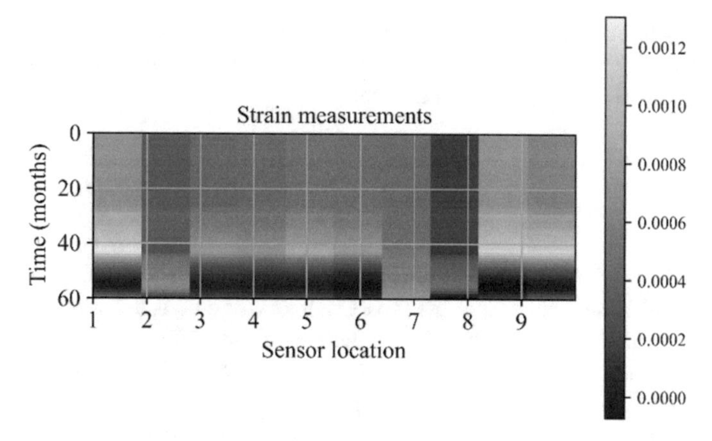

Fig. 6 Strain measurement from ten sensors over time

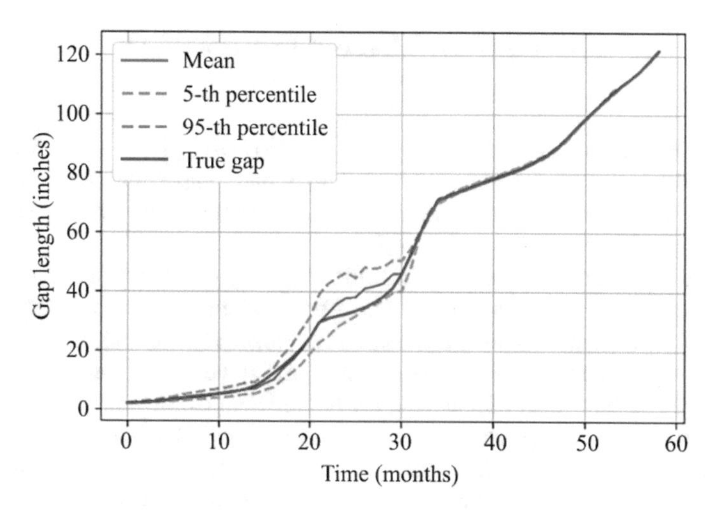

Fig. 7 Gap length diagnostics result using Bayesian method and ML surrogate model (note that gap here refers to the damage mode shown in Fig. 2)

3.3 Sensor Placement Optimization for Damage Diagnostics Using Machine Learning

Sensor placement optimization (SPO) plays a critical role in improving the effectiveness of the SHM system for damage diagnostics (see Eqs. 9 and 10). Designing an optimal sensor network is very challenging in practice since the observations are not available in the design stage. In that case, the physics-based simulation model needs to be employed to provide information on how to properly allocate the sensors.

The physics-based model, moreover, is computationally very demanding. Machine learning techniques play a vital role in overcoming the challenges in SPO.

A SPO model can be generalized as follows

$$\begin{aligned} \mathbf{d}^* &= \underset{\mathbf{d}\in\Omega_\mathrm{d}}{\mathrm{argmax}}\{\Psi(\mathbf{d})\}, \\ \text{s.t. } & C(\mathbf{d}) \leq C_\mathrm{e}, \end{aligned} \tag{11}$$

where $\mathbf{d}$ is a sensor network design, Ω_d is the sensor design domain, $\Psi(\mathbf{d})$ is a cost function, $C(\mathbf{d})$ is the total cost of the sensor network, and C_e is the allowable budget.

The cost function $\Psi(\mathbf{d})$ may be formulated from different perspectives. For example, probability of detection [33], Bayes risk [34], and information gain [35] have been used as cost functions in sensor placement design optimization. Taking the information gain measured by the Kullback–Leibler (KL) divergence as an example, $\Psi(\mathbf{d})$ is formulated as

$$\Psi(\mathbf{d}) = \int\limits_{\boldsymbol{\theta}}\int\limits_{\mathbf{s}|\boldsymbol{\theta}} f_{\boldsymbol{\theta}}(\boldsymbol{\theta}) f_{\mathbf{s}|\boldsymbol{\theta},\mathbf{d}}(\mathbf{s}|\boldsymbol{\theta},\mathbf{d}) D_{\mathrm{KL}}(\mathbf{d},\mathbf{s})\mathbf{d}\mathbf{s}\mathbf{d}\boldsymbol{\theta}, \tag{12}$$

where $D_{\mathrm{KL}}(\mathbf{d},\mathbf{s}) = \int_{\boldsymbol{\theta}} f_{\boldsymbol{\theta}|\mathbf{s},\mathbf{d}}(\boldsymbol{\theta}|\mathbf{s},\mathbf{d})\log\left[f_{\boldsymbol{\theta}|\mathbf{s},\mathbf{d}}(\boldsymbol{\theta}|\mathbf{s},\mathbf{d})/f_{\boldsymbol{\theta}}(\boldsymbol{\theta})\right]\mathbf{d}\boldsymbol{\theta}$ is the KL divergence (e.g., relative entropy) for given observations $\mathbf{s}$ and sensor placement design $\mathbf{d}$, which measures the difference (i.e., information gain) between the prior and posterior distributions of damage state variables $\boldsymbol{\theta}$.

Solving the sensor placement optimization model given in Eq. (11) is extremely challenging due to the high computational effort required in the repeated evaluations of Eq. (12). Bayesian data analytics and machine learning techniques are essential to overcome the challenge. First, ML models and Bayesian inference methods as discussed in Sects. 3.1 and 3.2 enable for the efficient estimation of the posterior distributions of the damage states variables. More importantly, ML-based optimization methods make it possible to solve the model given in Eq. (11) when a large number of sensors need to be allocated to a large civil infrastructure.

For example, when sensors need to be placed on the miter gate to detect the "gap" of the gate due to damage, the dimension of the design variables given in Eq. (11) will be very large. Assuming that 20 sensors need to be placed, the number of design variables will be 60, if the three-dimensional coordinates of each sensor are considered as design variables. In that case, directly solving the optimization model of Eq. (11) will be computationally prohibitive. Alternatively, a greedy-based framework can be employed to place the sensor one-by-one. In order to identify the optimal placement of the ith sensor, the optimization model given in Eq. (11) is re-formulated as

$$\begin{aligned} \mathbf{d}_i^* &= \underset{\mathbf{d}_i\in\Omega_\mathrm{d}}{\mathrm{argmax}}\left\{\Psi\left(\mathbf{d}_i,\mathbf{d}_{1:i-1}^*\right)\right\}, \\ \text{s.t. } & C(\mathbf{d}) \leq C_\mathrm{e} \text{ and } \mathbf{d} = \left\{\mathbf{d}_i \cup \mathbf{d}_{1:i-1}^*\right\}, \end{aligned} \tag{13}$$

where $\mathbf{d}^*_{1:i-1}$ are the coordinates of the previous $i - 1$ sensors and $\mathbf{d}_i$ are the coordinates of the ith sensor.

Through the formulation of the model in Eq. (13), the dimension of the design variables is reduced to 3 in each iteration of the greedy optimization scheme. Even for the three-dimensional optimization model, the global optimization of Eq. (13) is still computationally challenging. Bayesian optimization method, which is also known as the efficient global optimization method, can be employed to efficiently solve the optimization model by leveraging the prediction capability of the Gaussian process model [36, 37]. GP-based Bayesian optimization is a process of adaptively training a GP surrogate model for the objective function of an optimization model. In each iteration of the adaptive training of the GP, training data are identified as those design locations which have the highest probability of being the maximum/minimum design point. The key to the GP-based optimization is the definition of the expected improvement function (EIF)

$$\mathrm{EIF}(\mathbf{d}) = \left(\mu(\mathbf{d}) - \varphi^*\right)\Phi\left(\frac{\mu(\mathbf{d}) - \varphi^*}{\sigma(\mathbf{d})}\right) + \sigma(\mathbf{d})\phi\left(\frac{\mu(\mathbf{d}) - \varphi^*}{\sigma(\mathbf{d})}\right), \quad (14)$$

where $\phi(\cdot)$ and $\Phi(\cdot)$ are, respectively, the probability density function and cumulative distribution function of a standard normal random variable, φ^* is the current best values in the training dataset, $\mu(\mathbf{d})$ and $\sigma(\mathbf{d})$ are the mean and standard deviation of the GP surrogate model prediction.

By maximizing the EIF in Eq. (14), new training points may be identified to adaptively refine the GP surrogate model to approach the optimal value. Figure 8 shows an illustrative example of Bayesian optimization using GP.

As shown in Fig. 8, an initial GP surrogate model is trained first. Based on the trained GP surrogate model, the EIF values over the design space are computed as shown in Fig. 8a. By maximizing the EIF, a new training point is identified, and the GP model is retrained in the second iteration as shown in Fig. 8b. After a few of iterations, as shown in Fig. 9, the maximum point can be identified. These iterations show that the GP-based optimization needs very few evaluations of the objective function to identify the global optimization, which is much more efficient than the other global optimization algorithms, such as genetic algorithm and simulated annealing.

The ML-based optimization method allows us to effectively allocate the optimal sensors and thereby increases the effectiveness of damage diagnostics as discussed in Sect. 3.2. Note that the greedy algorithm-based sensor placement design is an approximation of the original model. It may not find the "true" globally optimized sensor network. This limitation can also be mitigated using other ML methods, such as the reinforcement learning method, which may be better suited for dynamic optimization problems. Figure 10 presents an example of sensor placement optimization results of miter gate obtained using the method presented in this section. More details are available in [38].

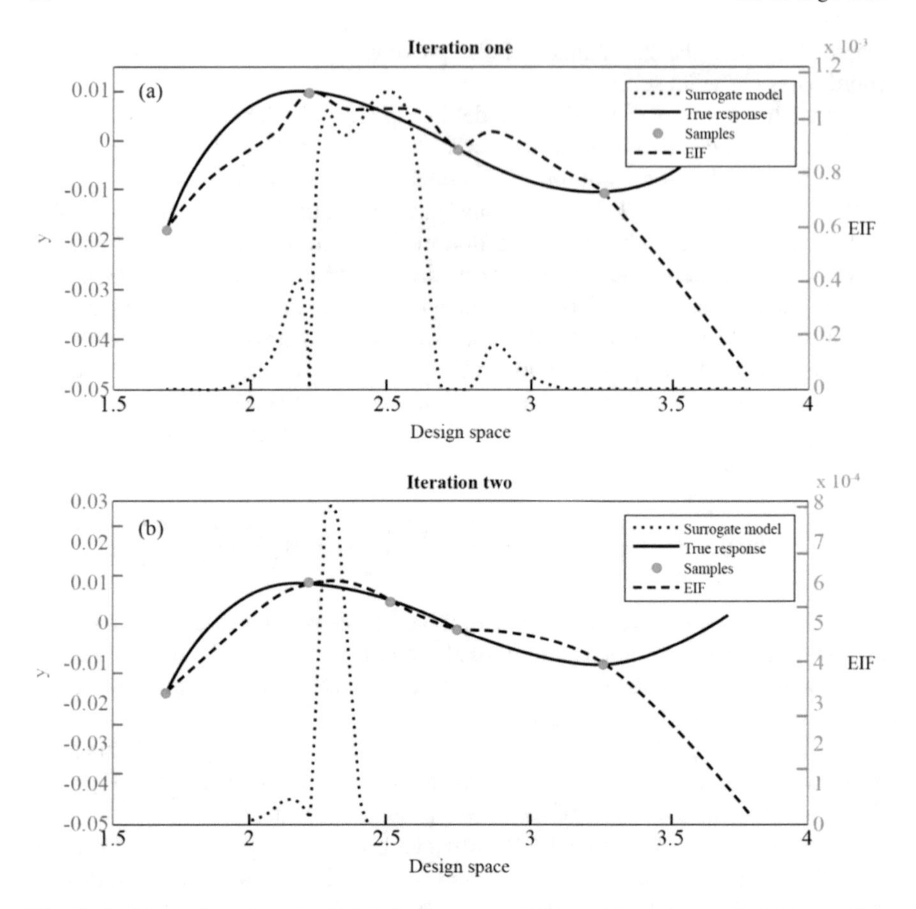

Fig. 8 An illustrative example of optimization using GP-based Bayesian optimization, **a** first iteration; **b** second iteration

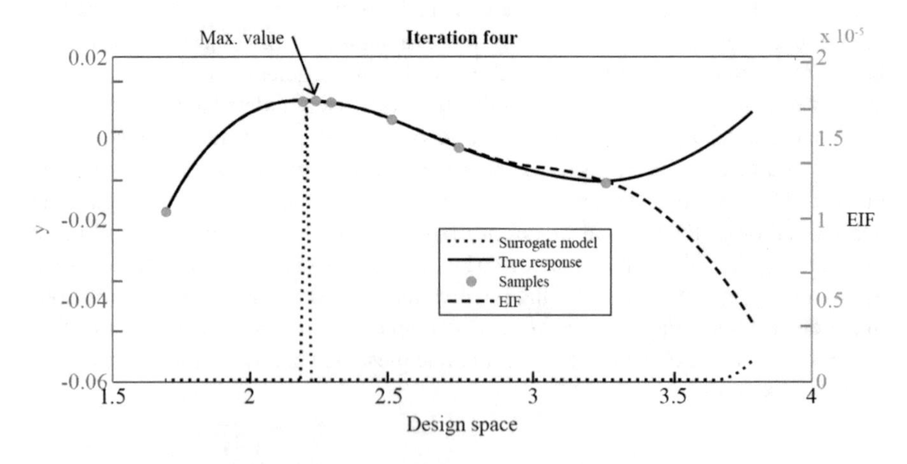

Fig. 9 An illustrative example of optimization using GP after converges

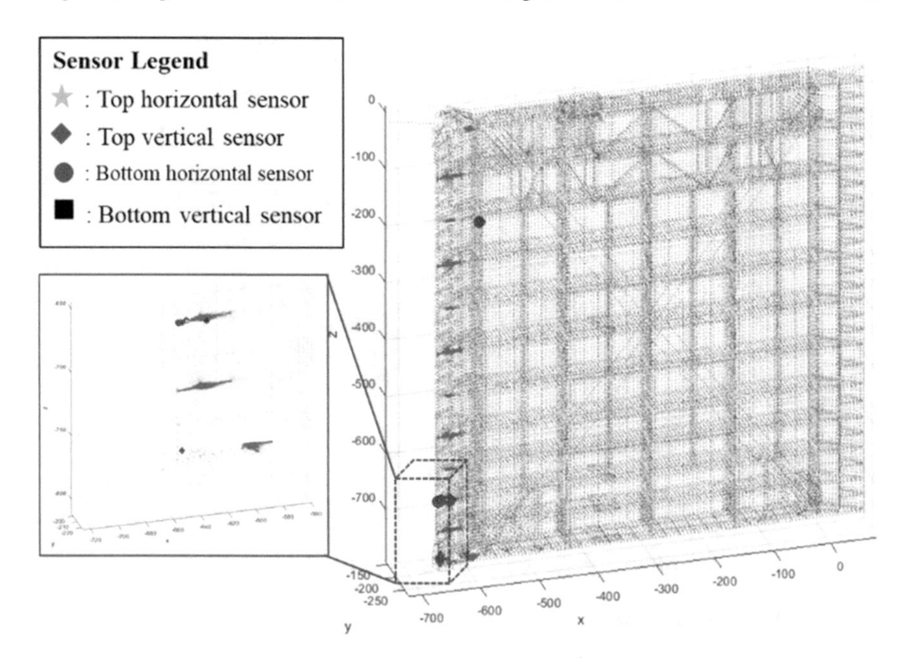

Fig. 10 An example of sensor placement optimization results using machine learning

4 Failure Prognostics Using Bayesian Data Analysis and Machine Learning

Failure prognostics is a process of predicting the end of life (EOL) of civil infrastructures to inform life-cycle management. Based on the state estimation from failure diagnostics as discussed in Sect. 3, the state of the system from the current time to a future time is obtained to predict the potential failure time or estimate the remaining useful life (RUL) of the system. Figure 11 shows a schematic of how to use the predictions to calculate the EOL and RUL distributions.

An essential part of the failure prognostics or RUL estimation is the degradation modeling since it is required to perform the projection of the state into future. To build such degradation models for prognosis purposes, researchers have tried to model the evolution/degradation of damage using physical degradation models such as applications in fatigue crack growth [39–41] and corrosion growth [6, 7]. These physical degradation models are developed based on the understanding of the physical behavior and are usually validated by experiments. On the other hand, empirical degradation models are used when the evolution/degradation of damage is not well understood either due to the limited understanding of the physical phenomenon or when the damage state cannot be measured continuously but rather occasionally. Approaches that combine physics-based approaches with data-driven methods have also been developed in recent years and lead to a group of hybrid approaches. A

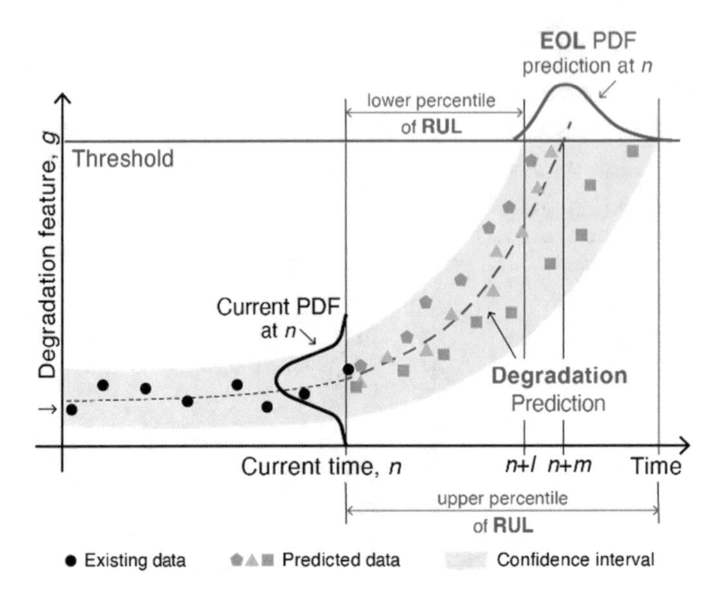

Fig. 11 RUL estimation based on failure prognostics

comprehensive review of prognostics approaches for rotating machine is available in [42].

For prognostics of civil infrastructure, however, the challenges come from the abstracted data sources and the lack of a degradation model. For example, the inspection data are highly abstracted ratings as discussed in Sect. 2.2 and the ratings may be polluted by human errors as presented in Sect. 2.3. This section presents how to perform failure prognostics in this situation through three different approaches, namely

- Failure prognostics based on inspection data
- Failure prognostics using a continuous degradation model mapped from inspection data
- Integrated failure diagnostics and prognostics using dynamic Bayesian networks (DBN).

In the following sections, the aforementioned three approaches are explained in detail.

4.1 Failure Prognostics Based on Inspection Data

As been discussed in Sect. 2.2, inspection data contains some degradation information of civil infrastructures even though they are highly abstracted. Since the damage state estimated from SHM is in continuous form, the continuous state can be converted into

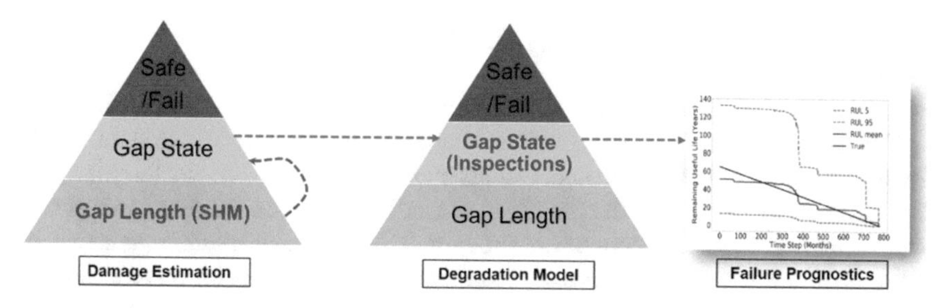

Fig. 12 Failure prognostics using inspection data

inspection data in discrete state through certain protocols. Taking the gap damage of a miter gate given in Fig. 2 as example, as shown in Fig. 12, the estimated gap length can be converted into gap state, which can then be used for failure prognostics using the gap state transition matrix given in Sect. 2.2.

More specifically, a certain protocol is usually needed to map the continuous damage state to the abstracted discrete state of inspection data. For instance, the gap length a_t of a miter gate (see Fig. 2) can be mapped into a gap state $X_{i,t}$ based on an engineering protocol as follows

$$R = h_{\text{OCA}}(a_t, \boldsymbol{\beta}) = \begin{cases} X_{1,t} = A, \ a_t \in [0, \beta_1] \\ X_{2,t} = B, \ a_t \in [\beta_1, \beta_2] \\ X_{3,t} = C, \ a_t \in [\beta_2, \beta_3] \\ X_{4,t} = D, \ a_t \in [\beta_3, \beta_4] \\ X_{5,t} = F, \ a_t \in [\beta_4, \beta_5] \\ X_{6,t} = \text{CF}, \ a_t \in [\beta_5, \infty) \end{cases}, \tag{15}$$

where $\boldsymbol{\beta} = [\beta_1, \ldots, \beta_5]$ are protocol parameters defined by the field engineers.

Based on the mapping defined in the above equation, SHM data $\mathbf{s}_{1:n}$ can be used to estimate the gap state at current time step (i.e., highly abstracted inspection data) as follows

$$\begin{aligned} P(X_{i,n}|\mathbf{s}_{1:n}) &= \Pr\{X_n = X_{i,n}|\mathbf{s}_{1:n}\} \\ &= \begin{cases} \int_{\beta_{i-1}}^{\beta_i} f(a_n|\mathbf{s}_{1:n})\mathrm{d}a_n, & \text{if } i \leq 5 \\ \int_{\beta_{i-1}}^{\infty} f(a_n|\mathbf{s}_{1:n})\mathrm{d}a_n, & \text{otherwise} \end{cases}, \quad \forall i = 1, \ldots, 6, \end{aligned} \tag{16}$$

where the PDF $f(a_n|\mathbf{s}_{1:n})$ of damage parameter a_n, is estimated using the Bayesian updating method presented in Sect. 3.2.

Once the gap state at current time step is estimated, the damage state of the failure m time steps into the future is obtained through the transition matrix given in Sect. 2.2 as

$$P(\mathbf{X}_{n+m}|\mathbf{s}_{1:n}) = P(\mathbf{X}_n|\mathbf{s}_{1:n}) \cdot \mathbf{P}^m_{\text{Report}}, \tag{17}$$

in which $\mathbf{P}_{\text{Report}}$ and $P(\mathbf{X}_n|\mathbf{s}_{1:n})$ are, respectively, given in Eqs. (1) and (16).

Using Eq. (17), the RUL for the system based on the current damage state and the future failure state can be estimated. Details on such predictions can be found in [23]. This approach assumes that the transition matrix $\mathbf{P}_{\text{Report}}$ can accurately represent the underlying degradation pattern of the structure. However, this assumption is usually not true since the abstracted ratings are often biased by human errors as discussed in Sect. 2.3. To tackle this issue, an approach has been proposed in [10] to map the reported transition matrix to a more useful transition matrix that has eliminated some of the effects of the human errors. In order to map the reported rating transition matrix $\mathbf{P}_{\text{Report}}$ to the underlying "true" transition matrix $\mathbf{P}_{\text{True}}$, the underlying true transition rating is defined at time t as X_t^{tr} and that at $t + 1$ as X_{t+1}^{tr}. Similarly, the reported ratings from field engineers are defined at time t as X_t^{obs} and that at time $t + 1$ as X_{t+1}^{obs}. Based on these definitions and using the human error matrix given in Eq. (3), $\mathbf{P}_{\text{Report}}$ can then be mapped into $\mathbf{P}_{\text{True}}$ by following the procedure shown in Fig. 13. More details of Fig. 13 are available in [1]. Once $\mathbf{P}_{\text{True}}$ is obtained, it can be used to substitute $\mathbf{P}_{\text{Report}}$ in Eq. (17) to get more accurate failure prognostics results.

In addition, a Bayesian method has also been developed in [23] to update the errors in the transition matrix using SHM data. The advantage of the approach presented here is that it requires minimal information for failure prognostics. The disadvantage is that there is very large uncertainty in the obtained RUL estimation results. Alternatively, a stochastic continuous degradation model can be built to improve the confidence of such predictions of the damage parameters which leads to better failure prognostics.

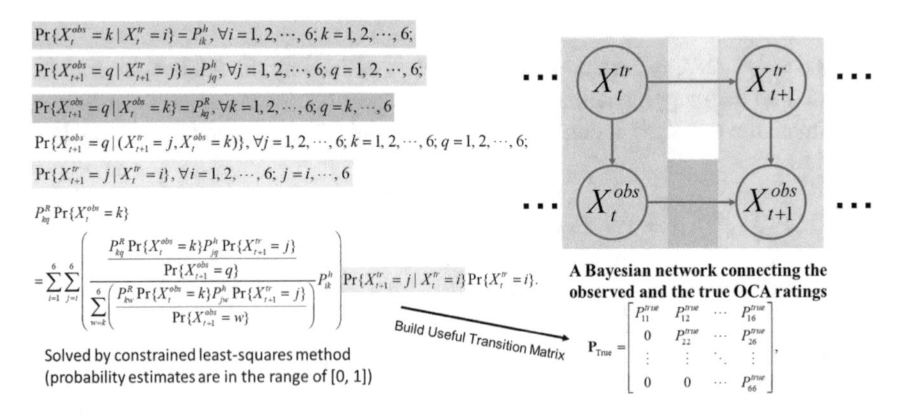

Fig. 13 Mapping between reported transition matrix to compensated/true transition matrix

4.2 *Mapping Inspection Data into Continuous Degradation Model for Failure Prognostics*

Using the gap growth of miter gate as an example, as shown in Fig. 14, an alternative approach to perform failure prognostics is to map the abstracted inspection data into a degradation model in continuous space and then perform failure prognostics in the continuous space. It is expected that prognostics in continuous space can increase the confidence of RUL estimation.

For a transition matrix $\mathbf{P}_{\text{Report}}$ given in Sect. 2.2 or $\mathbf{P}_{\text{True}}$ obtained in Sect. 4.1 after mitigating human error bias, a continuous degradation model can be obtained through Bayesian or optimization-based calibration. More details of the mapping from inspection data to a degradation model are available in [10]. Here, a brief summary of the major steps is presented. When optimization-based method is employed, the optimization model is given by

$$g_{\text{opt}}(\boldsymbol{\theta}; \mathbf{P}_{\text{True}}) = \left\| \hat{\mathbf{P}}(\boldsymbol{\theta}) - \mathbf{P}_{\text{True2}} \right\|, \tag{18}$$

where $\hat{\mathbf{P}}(\boldsymbol{\theta})$ is the simulated transition matrix for a given degradation model with parameters $\boldsymbol{\theta}$.

The most critical part is how to estimate $\boldsymbol{\theta}$ for given $\mathbf{P}_{\text{True}}$. To do that, $\hat{\mathbf{P}}(\boldsymbol{\theta})$ is calibrated for any given $\boldsymbol{\theta}$. The degradation model is modeled as a multi-stage degradation model as follows

$$\frac{\mathrm{d}a(t)}{\mathrm{d}t} = \exp\big(\sigma_j(t)U(t)\big)Q_j(t)(a(t))^{w_j(t)}, \quad \forall j = 1, \ldots, N_d, \tag{19}$$

where $a(t)$ is the gap length at time t, Q_i and w_i are stage-dependent degradation parameters, σ_i is a standard deviation variable of degradation stage i, and $U(t)$ is a stationary standard Gaussian process with autocorrelation function given by

$$\text{cov}(U(t_1), U(t_2)) = \exp(-\zeta |t_2 - t_1|), \tag{20}$$

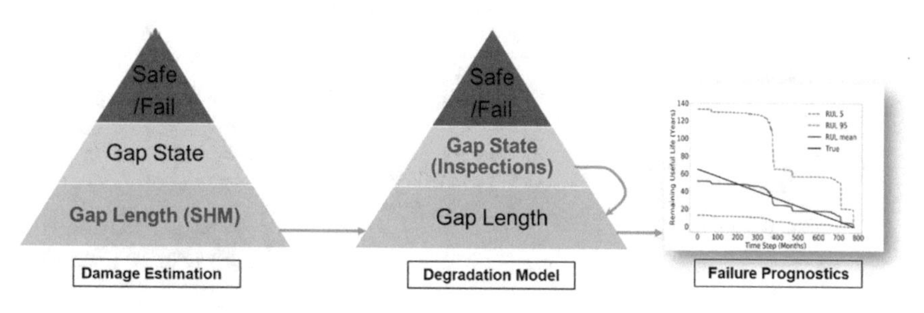

Fig. 14 Failure prognostics in continuous space by mapping inspection data into a degradation model

in which ζ is a correlation length parameter.

For the above degradation model and given degradation model parameters, a large number of realizations of the degradation curves is simulated using the Monte Carlo simulation method. The obtained realizations of the degradation curves in the continuous space can then be converted into ratings in the discrete states as follows

$$X_n = X_{\text{stage}}(a_n) = \begin{cases} X_{1,n}, & a_n \in [e_0,\ e_1) \\ X_{2,n}, & a_n \in [e_1,\ e_2) \\ X_{3,n}, & a_n \in [e_2,\ e_3) \\ X_{4,n}, & a_n \in [e_3,\ e_4) \\ X_{5,n}, & a_n \in [e_4,\ e_5) \\ X_{6,n}, & a_n \geq e_5 \end{cases}, \tag{21}$$

where e_j, $\forall j = 1, \ldots, 5$ are parameters that govern the transition between different stages of degradations in the continuous space.

After the simulated degradation curves are converted into degradation ratings using Eq. (21), the simulated transition matrix $\hat{\mathbf{P}}(\boldsymbol{\theta})$ can be obtained. For the above degradation model, the parameters of the degradation model, $\boldsymbol{\theta}$, can therefore be summarized as follows

$$\begin{aligned} &\boldsymbol{\theta} \triangleq (\boldsymbol{\theta}_1, \boldsymbol{\theta}_2, \boldsymbol{\theta}_3, \boldsymbol{\theta}_4, \boldsymbol{\theta}_5, e_1, e_2, e_3, e_4, \sigma), \quad \text{where} \\ &\boldsymbol{\theta}_{\mathbf{j}} \triangleq \{\sigma_j, \zeta_j, Q_j, w_j, \quad j = 1, 2, \ldots 5\}. \end{aligned} \tag{22}$$

Then, the degradation model can be estimated as follows using the optimization model given in Eq. (18). Figure 15 summarizes the overall procedure of estimating $\hat{\mathbf{P}}(\boldsymbol{\theta})$ for a given degradation model and model parameters $\boldsymbol{\theta}$.

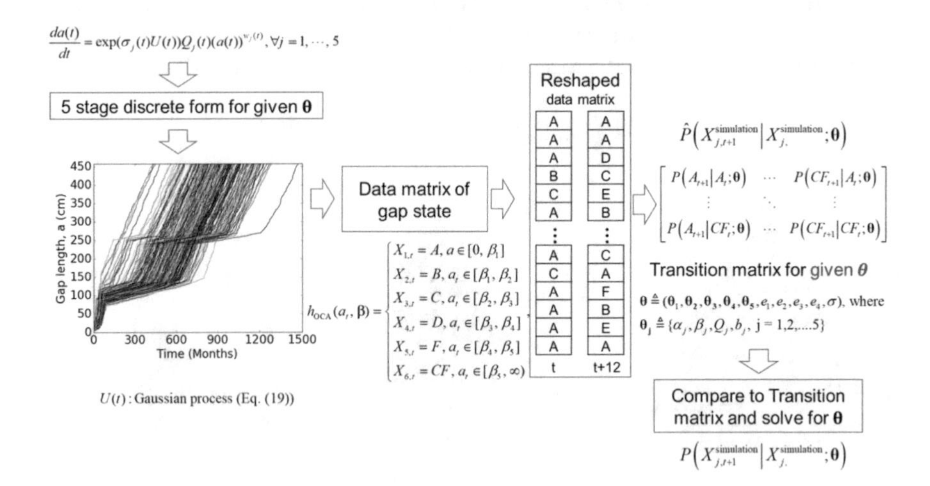

Fig. 15 Overview of obtaining simulated transition matrix for given $\boldsymbol{\theta}$

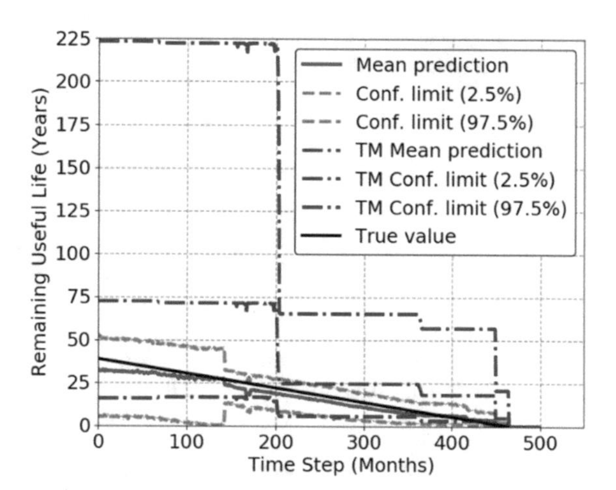

Fig. 16 Comparison of RUL estimates using discrete degradation model (Sect. 4.1) and piecewise continuous degradation model (Sect. 4.2)

Once the degradation model is available, it can be used for failure prognostics in the continuous space as illustrated in Fig. 11. This procedure, however, is not limited to the model given in Eq. (19). It is also applicable to other degradation models and can be integrated into a Bayesian framework. More detailed discussions of this approach can be found in [10].

Figure 16 shows a comparison of the RUL estimates obtained using the approaches presented in Sect. 4.1 (denoted as TM mean prediction, TM Conf. limit) and Sect. 4.2, respectively. It shows that mapping inspection data into a continuous degradation model can significantly increase the confidence of the failure prognostics results.

4.3 *Integrated Failure Diagnostics and Prognostics Using Dynamic Bayesian Networks*

Failure diagnostics and prognostics of civil infrastructure usually require the usage of multiple models including degradation model [39–41] and physics-based model [12–14], as discussed earlier. In addition to various analysis models, heterogeneous data and uncertainty sources are involved in the process of diagnostics and prognostics. A flexible tool that can be used to tackle the challenges in diagnostics and prognostics caused by the heterogeneity of model and data sources is Bayesian networks (BN).

A BN, which is also called probabilistic graphic model, is a directed acyclic graph that connects different variables in a probabilistic way. It allows for the flexible integration of multi-type of models and information sources in a systematic Bayesian framework, and thereby enable decision makers to update information and reduce uncertainty in a holistic manner [43]. Due to its capability of fusing information and data sources, BN plays a vital role in building digital twins for SHM in various assets, from aerospace engineering to civil and mechanical engineering [40, 44].

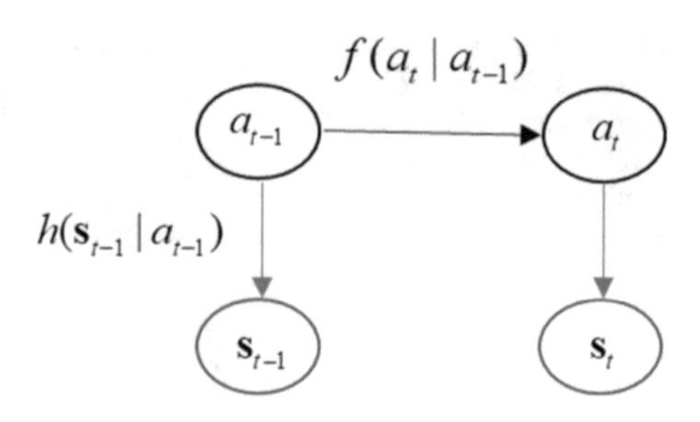

Fig. 17 An example of dynamic Bayesian network

For n random variables (nodes), $X_1, X_2, \ldots$, and X_n, BN represents the joint probability density function $p(\mathbf{X})$ as follows

$$p(\mathbf{X}) = p(X_1, X_2, \ldots, X_n) = \prod_{i=1}^{n} p(X_i|\pi_i), \tag{23}$$

where π_i is a set of parent nodes of X_i, $p(X_i|\pi_i)$ is the conditional probability mass (CPM) function or conditional probability density (CPD) function, and nodes without parent nodes are called root nodes. For root nodes, assume $p(X_i|\pi_i) = p(X_i)$.

A type of widely used BN in failure diagnostics and prognostics is the dynamic Bayesian network (DBN). Figure 17 shows a simple example of DBN. As shown in this figure, the DBN consists of state variable denoted by a_t and measurement variable represented by $\mathbf{s}_t$. The CPD function $f(a_t|a_{t-1})$ describes the transition of the state variable a_t over time and $h(\mathbf{s}_{t-1}|a_{t-1})$ models the probabilistic relationship between the state variable and measurement variable. When the state variables are variables related to the failure modes or degradation stages of the civil infrastructure, the measurements collected from measurement variable $\mathbf{s}_t$ can be used for failure diagnostics using Bayesian inference methods. Based on the failure diagnostics or estimation of damage related state variables, failure prognostics may then be performed according the transition of state variables over time, which is governed by the CPD function $f(a_t|a_{t-1})$.

In practical engineering applications, the node in the DBN can be a mathematical model, a finite element model, or a data-driven machine learning model. The DBN used for failure diagnostics and prognostics can be constructed using physics-based method [12–14], Bayesian network learning method [44], or a hybrid of physics and data-driven methods [19, 20].

Figure 18 shows a schematic dynamic Bayesian network at one time instant for the failure diagnostics and prognostics of a miter gate (see Fig. 2), which is used an example to explain the presented approaches in this chapter. Note that, for the sake of simplification, the transient BN is not depicted. As shown in this figure, the DBN connects variables of a degradation model with variables of a strain analysis model of the miter gate. For example, the parent nodes of node e_1 in Fig. 18 are nodes σ_{e} and μ_1. The CPDs in the DBN can be derived according to the approaches discussed in Sects. 3.2 and 4.2 of this chapter. Using the strain measurement collected from

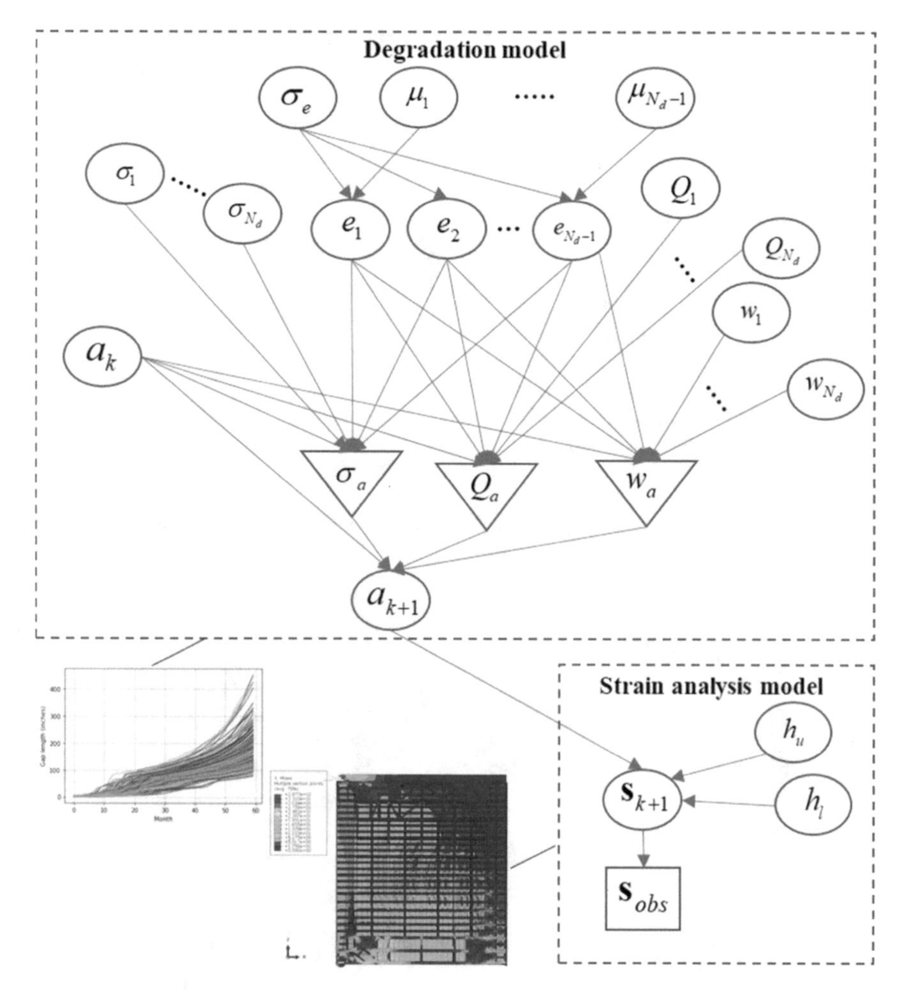

Fig. 18 A schematic Bayesian network of a miter gate at one time instant for diagnostics and prognostics

SHM system, the uncertain variables of the degradation model can be updated. The updated degradation model can then be used to estimate RUL of the gate dynamically over time.

Figure 19 shows an illustrative example of the posterior distribution updating of parameter w_1 over time using the strain measurements of the miter gate. Following that, Fig. 20 shows the RUL estimation of the miter gate over time. It shows that the confidence of RUL estimation increases over time as more and more measurements are collected. Additionally, it is worth mentioning that the degradation model and strain analysis model are updated in an integrated manner in the DBN-based framework. This example illustrates the flexibility of DBN in connecting multiple models

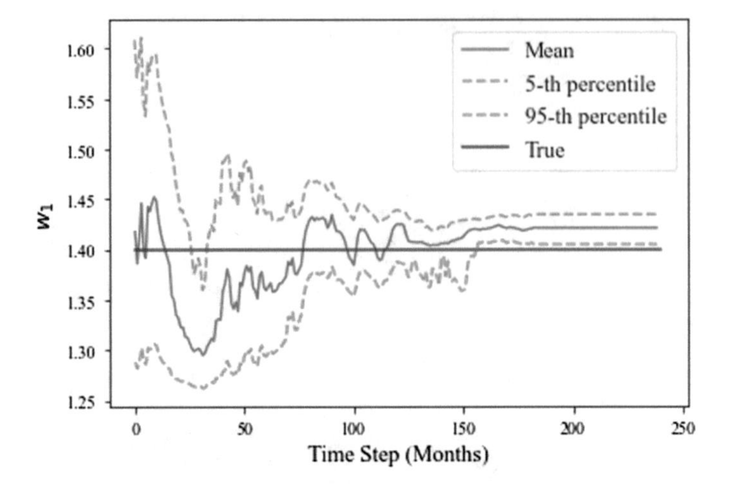

Fig. 19 An illustrative example of updating posterior distribution of a degradation model parameter using DBN over time

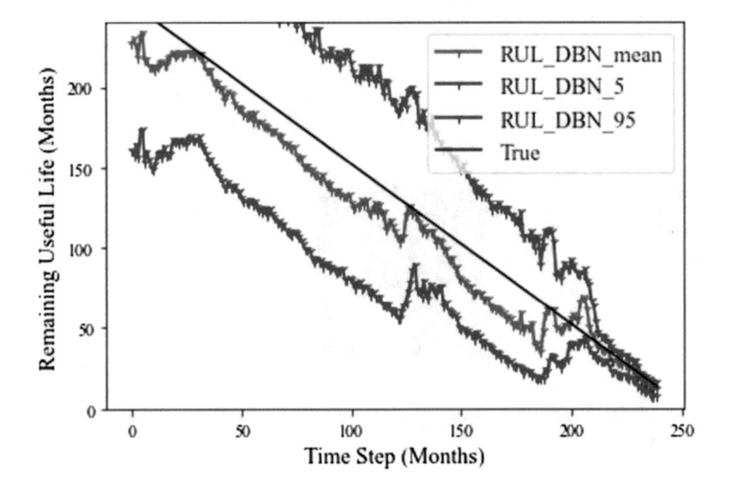

Fig. 20 RUL estimation over time of the miter gate

for failure diagnostics and prognostics. A good application of DBN for the failure diagnostics and prognostics is available in [45].

Moreover, the degradation model parameters obtained in Sect. 4.2 using optimization or Bayesian-based method can be used as prior information for the DBN-based method.

In summary, this section presents three different approaches for failure diagnostics using Bayesian data analytics and machine learning. The advantage of the approach presented in Sect. 4.1 requires minimal information for failure prognostics. The advantages of the approaches presented in Sects. 4.2 and 4.3 are that they can provide prediction results with less uncertainty. Next, the maintenance planning based on failure diagnostics and prognostics is briefly introduced.

5 Optimization of Maintenance Strategy

As mentioned in Sect. 1, there are two types of maintenance strategies, namely TBM and CBM. TBM (also known as periodic-based maintenance) assumes that the estimated failure behavior is statistically or experientially known [46]. Statistical modeling, such as Weibull analysis [47], is widely used in TBM to identify failure characteristics of a component or system. The goal of TBM models is to find the optimal policy that minimize a cost function. TBM approaches have been developed for both repairable or non-repairable systems [48]. The complexity of a TBM model depends on the targeted system such as single-system, multi-systems, parallel and series structure. A more extensive review of TBM applications can be found here [49]. For example, in miter gates, historical data is available in the form of discrete ratings.

Figure 21 shows a TBM approach using the transition matrix given in Sect. 2.2, whose goal is to estimate the optimal maintenance time based on a well-known cost function [48], which weight the probability of failure, $F(t)$, with the preventive, C_p, and unscheduled (or emergency), C_u, maintenance costs.

CBM is the most modern and popular maintenance technique among researchers and industry. CBM has gained increasing attention recently as a preferred approach to TBM. CBM is a maintenance approach that combines data-driven reliability models and information from a condition monitoring process. Based on the underlying degradation model, CBM models can be categorized into two subgroups: (1) models that assume discrete-state deterioration (such as in Sect. 4.1) and (2) models that assume

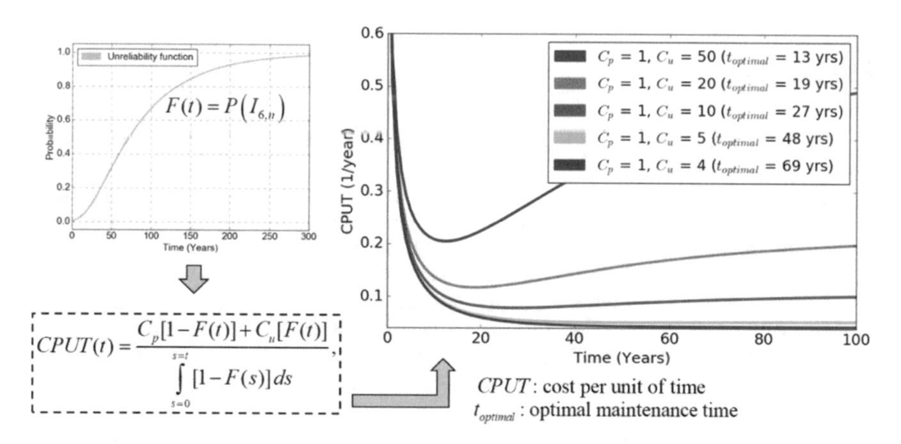

Fig. 21 TBM approach using discrete ratings

continuous state deterioration (such as in Sects. 4.2 and 4.3). A most extensive list of CBM application can be found in here [50–53]. Most of the CBM applications available in the literature are for mechanical systems, aerospace systems, or manufacturing systems. For large civil engineering infrastructure, most of the applications have been applied to bridge engineering [54–56]. In CBM, maintenance schedules are predicted based on the results from diagnosis and prognosis, as discussed in Sects. 3 and 4. For diagnosis and prognosis, CBM approaches can be classified into physics-based approach [12–14], data-driven approach [15–18], and hybrid approach [19, 20]. The approaches presented in this book chapter can be classified as hybrid approaches since they combine physics-based approach with data-driven approach to improve CBM predictive capabilities.

6 Conclusion

This chapter presents comprehensive failure diagnostics, prognostics, and maintenance planning approaches using machine learning, Bayesian data analysis, computational mechanics, and reliability engineering. The presented framework is aimed to allow real-time assessments of civil structures that have different forms of available data. The challenge of heterogeneity of the data sources has been successfully overcome to provide diagnostic/prognostic capabilities to the structure of interest. Additional steps have been discussed to improve these capabilities such as optimal sensor design and accounting for human error. Currently, the features used to update the models used in this work have been based on SHM data and human inspections. However, this framework can be adapted to work with more advanced data sources enhanced with supervised learning algorithms such images captured from drones or robots.

References

1. Vega MA (2020) Diagnosis, prognosis, and maintenance decision making for civil infrastructure. UC San Diego
2. Moaveni B, Conte JP, Hemez FM (2009) Uncertainty and sensitivity analysis of damage identification results obtained using finite element model updating. Comput Civ Infrastruct Eng 24:320–334. https://doi.org/10.1111/j.1467-8667.2008.00589.x
3. Jang S, Li J, Spencer BF (2013) Corrosion estimation of a historic truss bridge using model updating. J Bridge Eng 18:678–689. https://doi.org/10.1061/(ASCE)BE.1943-5592.0000403
4. Vega MA, Ramancha MR, Conte JP, Todd MD (2020) Efficient Bayesian inference of miter gates using high-fidelity models. In: Model validation and uncertainty quantification, volume 3: proceedings of the 38th IMAC, a conference and exposition on structural dynamics 2020, Houston, TX. Springer
5. Kim N-H, An D, Choi J-H (2017) Prognostics and health management of engineering systems. Springer International Publishing, Cham. https://doi.org/10.1007/978-3-319-44742-1

6. Guedes Soares C, Garbatov Y, Zayed A (2011) Effect of environmental factors on steel plate corrosion under marine immersion conditions. Corros Eng Sci Technol 46:524–541. https://doi.org/10.1179/147842209X12559428167841
7. Wang C, Elsayed EA (2020) Stochastic modeling of corrosion growth. Reliab Eng Syst Saf 204:107120. https://doi.org/10.1016/j.ress.2020.107120
8. Si X-S, Wang W, Hu C-H, Zhou D-H (2011) Remaining useful life estimation—a review on the statistical data driven approaches. Eur J Oper Res 213:1–14. https://doi.org/10.1016/j.ejor.2010.11.018
9. Wang Z, Shafieezadeh A (2020) Real-time high-fidelity reliability updating with equality information using adaptive Kriging. Reliab Eng Syst Saf 195:106735. https://doi.org/10.1016/j.ress.2019.106735
10. Vega MA, Hu Z, Fillmore TB, Smith MD, Todd MD (2021) A novel framework for integration of abstracted inspection data and structural health monitoring for damage prognosis of miter gates. Reliab Eng Syst Saf 211:107561
11. Mitchell JS (2007) From vibration measurements to condition based maintenance seventy years of continuous progress. Sound Vib 41:62–78
12. Orchard ME, Vachtsevanos GJ (2007) A particle filtering approach for on-line failure prognosis in a planetary carrier plate. Int J Fuzzy Log Intell Syst 7:221–227. https://doi.org/10.5391/IJFIS.2007.7.4.221
13. Daigle M, Goebel K (2011) Multiple damage progression paths in model-based prognostics. In: 2011 aerospace conference. IEEE, pp 1–10. https://doi.org/10.1109/AERO.2011.5747574
14. An D, Choi J-H, Schmitz TL, Kim NH (2011) In situ monitoring and prediction of progressive joint wear using Bayesian statistics. Wear 270:828–838. https://doi.org/10.1016/j.wear.2011.02.010
15. Zio E, Di Maio F (2010) A data-driven fuzzy approach for predicting the remaining useful life in dynamic failure scenarios of a nuclear system. Reliab Eng Syst Saf 95:49–57. https://doi.org/10.1016/j.ress.2009.08.001
16. Mohanty S, Das S, Chattopadhyay A, Peralta P (2009) Gaussian process time series model for life prognosis of metallic structures. J Intell Mater Syst Struct 20:887–896. https://doi.org/10.1177/1045389X08099602
17. Galar D, Kumar U, Lee J, Zhao W (2012) Remaining useful life estimation using time trajectory tracking and support vector machines. J Phys Conf Ser 364:012063. https://doi.org/10.1088/1742-6596/364/1/012063
18. Ye Z-S, Xie M (2015) Stochastic modelling and analysis of degradation for highly reliable products. Appl Stoch Model Bus Ind 31:16–32. https://doi.org/10.1002/asmb.2063
19. Xu J, Xu L (2011) Health management based on fusion prognostics for avionics systems. J Syst Eng Electron 22:428–436. https://doi.org/10.3969/j.issn.1004-4132.2011.03.010
20. Liao L, Kottig F (2014) Review of hybrid prognostics approaches for remaining useful life prediction of engineered systems, and an application to battery life prediction. IEEE Trans Reliab 63:191–207. https://doi.org/10.1109/TR.2014.2299152
21. Spencer BF, Hoskere V, Narazaki Y (2019) Advances in computer vision-based civil infrastructure inspection and monitoring. Engineering 5:199–222. https://doi.org/10.1016/j.eng.2018.11.030
22. Gibb S, La HM, Le T, Nguyen L, Schmid R, Pham H (2018) Nondestructive evaluation sensor fusion with autonomous robotic system for civil infrastructure inspection. J Field Robot 35:988–1004. https://doi.org/10.1002/rob.21791
23. Vega MA, Hu Z, Todd MD (2020) Optimal maintenance decisions for deteriorating quoin blocks in miter gates subject to uncertainty in the condition rating protocol. Reliab Eng Syst Saf 204:107147. https://doi.org/10.1016/j.ress.2020.107147
24. Campbell LE, Connor RJ, Whitehead JM, Washer GA (2020) Benchmark for evaluating performance in visual inspection of fatigue cracking in steel bridges. J Bridge Eng 25:04019128. https://doi.org/10.1061/(ASCE)BE.1943-5592.0001507
25. Yin T, Zhu H (2018) Probabilistic damage detection of a steel truss bridge model by optimally designed bayesian neural network. Sensors 18:3371. https://doi.org/10.3390/s18103371

26. Chua CG, Goh ATC (2005) Estimating wall deflections in deep excavations using Bayesian neural networks. Tunn Undergr Space Technol 20:400–409. https://doi.org/10.1016/j.tust.2005.02.001
27. Arangio S, Bontempi F (2015) Structural health monitoring of a cable-stayed bridge with Bayesian neural networks. Struct Infrastruct Eng 11:575–587. https://doi.org/10.1080/15732479.2014.951867
28. Vega MA, Todd MD (2020) A variational Bayesian neural network for structural health monitoring and cost-informed decision-making in miter gates. Struct Health Monit. https://doi.org/10.1177/1475921720904543
29. Parno M, O'Connor D, Smith M (2018) High dimensional inference for the structural health monitoring of lock gates, 1–29
30. Doucet A, de Freitas N, Gordon N (2001) Sequential Monte Carlo methods in practice. Springer New York, New York, NY. https://doi.org/10.1007/978-1-4757-3437-9
31. Mochnac J, Marchevsky S, Kocan P (2009) Bayesian filtering techniques: Kalman and extended Kalman filter basics. In: 2009 19th international conference radioelektronika. IEEE, pp 119–122. https://doi.org/10.1109/RADIOELEK.2009.5158765
32. Julier SJ, Uhlmann JK (2004) Unscented filtering and nonlinear estimation. Proc IEEE 92:401–422. https://doi.org/10.1109/JPROC.2003.823141
33. Tenney R, Sandell N (1981) Detection with distributed sensors. IEEE Trans Aerosp Electron Syst AES-17:501–510. https://doi.org/10.1109/TAES.1981.309178
34. Flynn EB, Todd MD (2010) A Bayesian approach to optimal sensor placement for structural health monitoring with application to active sensing. Mech Syst Signal Process 24:891–903. https://doi.org/10.1016/j.ymssp.2009.09.003
35. Nath P, Hu Z, Mahadevan S (2017) Sensor placement for calibration of spatially varying model parameters. J Comput Phys 343:150–169. https://doi.org/10.1016/j.jcp.2017.04.033
36. Jones DR, Schonlau M, Welch WJ (1998) Efficient global optimization of expensive black-box functions. J Glob Optim 13:455–492
37. Hu Z, Du X (2015) Mixed efficient global optimization for time-dependent reliability analysis. J Mech Des 137. https://doi.org/10.1115/1.4029520
38. Yang Y, Chadha M, Hu Z, Vega MA, Parno MD, Todd MD (2021) A probabilistic optimal sensor design approach for structural health monitoring using risk-weighted f-divergence. Mech Syst Signal Process 161:107920
39. An D, Choi J-H, Kim NH (2012) Identification of correlated damage parameters under noise and bias using Bayesian inference. Struct Health Monit 11:293–303. https://doi.org/10.1177/1475921711424520
40. Li C, Mahadevan S, Ling Y, Choze S, Wang L (2017) Dynamic Bayesian network for aircraft wing health monitoring digital twin. AIAA J 55:930–941. https://doi.org/10.2514/1.J055201
41. Leung MSH, Corcoran J, Cawley P, Todd MD (2019) Evaluating the use of rate-based monitoring for improved fatigue remnant life predictions. Int J Fatigue 120:162–174. https://doi.org/10.1016/j.ijfatigue.2018.11.012
42. Cubillo A, Perinpanayagam S, Esperon-Miguez M (2016) A review of physics-based models in prognostics: application to gears and bearings of rotating machinery. Adv Mech Eng 8:168781401666466. https://doi.org/10.1177/1687814016664660
43. Sankararaman S, Ling Y, Mahadevan S (2011) Uncertainty quantification and model validation of fatigue crack growth prediction. Eng Fract Mech 78:1487–1504. https://doi.org/10.1016/j.engfracmech.2011.02.017
44. Hu Z, Mahadevan S (2018) Bayesian network learning for data-driven design. ASCE-ASME J Risk Uncertainty Eng Syst Part B Mech Eng 4:1–12. https://doi.org/10.1115/1.4039149
45. Rafiq MI, Chryssanthopoulos MK, Sathananthan S (2015) Bridge condition modelling and prediction using dynamic Bayesian belief networks. Struct Infrastruct Eng 11:38–50. https://doi.org/10.1080/15732479.2013.879319
46. Yam RCM, Tse PW, Li L, Tu P (2001) Intelligent predictive decision support system for condition-based maintenance. Int J Adv Manuf Technol 17:383–391. https://doi.org/10.1007/s001700170173

47. Weibull W (1951) A statistical distribution function of wide applicability. J Appl Mech 103:293–297
48. Barlow R, Hunter L (1960) Optimum preventive maintenance policies. Oper Res 8:90–100. https://doi.org/10.1287/opre.8.1.90
49. Ahmad R, Kamaruddin S (2012) An overview of time-based and condition-based maintenance in industrial application. Comput Ind Eng 63:135–149. https://doi.org/10.1016/j.cie.2012.02.002
50. Zhu Y, Elsayed EA, Liao H, Chan LY (2010) Availability optimization of systems subject to competing risk. Eur J Oper Res 202:781–788. https://doi.org/10.1016/j.ejor.2009.06.008
51. Tian Z, Jin T, Wu B, Ding F (2011) Condition based maintenance optimization for wind power generation systems under continuous monitoring. Renew Energy 36:1502–1509. https://doi.org/10.1016/j.renene.2010.10.028
52. Tian Z, Liao H (2011) Condition based maintenance optimization for multi-component systems using proportional hazards model. Reliab Eng Syst Saf 96:581–589. https://doi.org/10.1016/j.ress.2010.12.023
53. Alaswad S, Xiang Y (2017) A review on condition-based maintenance optimization models for stochastically deteriorating system. Reliab Eng Syst Saf 157:54–63. https://doi.org/10.1016/j.ress.2016.08.009
54. Petcherdchoo A, Neves LA, Frangopol DM (2008) Optimizing lifetime condition and reliability of deteriorating structures with emphasis on bridges. J Struct Eng 134:544–552. https://doi.org/10.1061/(ASCE)0733-9445(2008)134:4(544)
55. Saydam D, Frangopol DM (2015) Risk-based maintenance optimization of deteriorating bridges. J Struct Eng 141:04014120. https://doi.org/10.1061/(ASCE)ST.1943-541X.0001038
56. Gong C, Frangopol DM (2020) Condition-based multiobjective maintenance decision making for highway bridges considering risk perceptions. J Struct Eng 146:04020051. https://doi.org/10.1061/(ASCE)ST.1943-541X.0002570

Real-Time Machine Learning for High-Rate Structural Health Monitoring

Simon Laflamme, Chao Hu, and Jacob Dodson

Abstract Advances in science and engineering are empowering high-rate dynamic systems, such as hypersonic vehicles, advanced weaponries, and active shock and blast mitigation strategies. The real-time estimation of the structural health of high-rate systems, termed high-rate structural health monitoring (HRSHM), is critical in designing decision mechanisms that can ensure structural integrity and performance. However, this is a difficult task, because three aspects uniquely characterize these systems: (1) large uncertainties in the external loads; (2) high levels of non-stationarities and heavy disturbances; and (3) unmodeled dynamics generated from changes in system configurations. In addition, because these systems are experiencing events of high amplitudes (often beyond 100 g) over short durations (under 100 ms), a successful feedback mechanism is one that can operate under 1 ms. A solution to the unique system characteristics and temporal constraint is the design and application of real-time learning algorithms. Here, we review and discuss a real-time learning algorithm for HRSHM applications. In particular, after introducing the HRSHM challenge, we explore fast real-time learning for time series prediction using conventional and deep neural networks and discuss a path to rapid real-time state estimation.

Keywords High-rate · Structural health monitoring · Neural network · Deep learning · Physics-informed machine learning

---

S. Laflamme (✉)
Department of Civil, Construction, and Environmental Engineering,
Iowa State University, Ames, IA 50021, USA
e-mail: laflamme@iastate.edu

C. Hu
Department of Mechanical Engineering, Iowa State University, Ames, IA 50021, USA

J. Dodson
Air Force Research Laboratory, Munitions Directorate, Eglin AFB, FL 32542, USA

A. Cury et al. (eds.), *Structural Health Monitoring Based on Data Science Techniques*, Structural Integrity 21, https://doi.org/10.1007/978-3-030-81716-9_4

1 Introduction

High-rate systems are engineering systems experiencing high-amplitude dynamic events (often beyond 100 g) over short durations (under 100 ms). Examples include structures exposed to blast, high-impact car crashes, hypersonic vehicles, and advanced weaponries. The development and implementation of feedback, decision, and reaction mechanisms in high-rate systems could dramatically improve their functionality and safety. For example, it could enable active blast mitigation strategies [23], smart airbag deployments [18], and damage-based decision for hypersonic systems [16]. Yet, the implementation of these feedback mechanisms is a difficult task due to the complexity of the dynamics and extreme time constraints under consideration.

In particular, the dynamics of high-rate systems has the following unique characteristics, defined in the introductory paper on high-rate state estimation [11]. The characteristics are:

1. large uncertainties in the external loads;
2. high levels of non-stationarities and heavy disturbances; and
3. unmodeled dynamics generated from changes in system configurations.

First, the external loads, for example, those caused by blasts and ballistic impacts, are largely uncertain and thus difficult to predict and estimate. Second, the dynamics, especially which follow the high-amplitude impact, results in high non-stationarities and large uncertainties in the system. Third, the impact will likely result in significant unmodeled dynamics generated from changes in system configurations, for instance changes in structure or boundary conditions. An example of signals of a high-rate system acquired from laboratory experimentations is illustrated in Fig. 1. The system (Fig. 1a) consists of an electronic unit housing circuit boards equipped with high-g accelerometers (Fig. 1a, right), securely held in a fixture (Fig. 1a, center), and impacted using an MTS-66 accelerated droptower (Fig. 1a, left). A section of the recorded time series signal from accelerometer "accel 1" following five consecutive tests is plotted in Fig. 1b. The following high-rate characteristics are: (1) the response is of high amplitude, in the thousands of g-force ($1\ \mathrm{kg_n} = 9810\ \mathrm{m/s^2} = 32{,}200\ \mathrm{ft/s^2}$); (2) the dynamics has high nonlinearities; (3) the response is altered after each test, which could be attributed to the whipping of cables, damage of the electronics assembly, and/or change of the internal boundary conditions of the electronics; and (4) the change in dynamics occurs in the sub-millisecond range.

It follows that these unique dynamic characteristics complicate physical modeling in the formulation of a feedback decision system. In addition, the rapid changes in dynamics exert important constraints on the close-loop time scales. The terminology for these short time scale are defined as [7]:

1. under 1 ms for high rate (HR)
2. under 100 μs for very high rate (VHR); and
3. under 1 μs for ultra high rate (UHR).

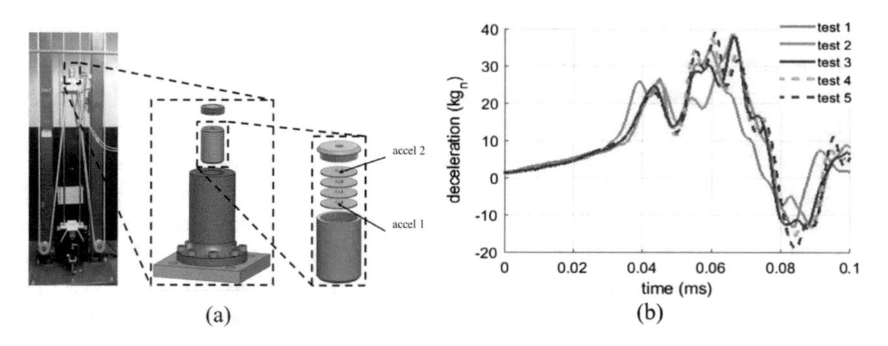

Fig. 1 Example of high-rate system: **a** experimental setup; and **b** time series signals from five consecutive tests

Of interest to this chapter are algorithms enabling high-rate feedback. Work in state estimation often uses convergence speed and computing time as key metrics to define performance of an algorithm. Convergence speed is the time taken for an estimation to stabilize within a given threshold of the true value. It usually influences the quality of decision, and increases with the increasing complexity of the dynamic system under consideration. Computing time is the time it takes for the algorithm to process data and produce an estimate. In real-time applications, as it is desirable for feedback systems, the computing time must remain under the feedback decision rate. Similar to the convergence time, the computing time is expected to increase with the complexity of the dynamics. Early work in 1995 on state estimation for induction motors using observers reported 73 ms computing time using backward difference [4]. With technological progress in computing, more recent work reported magnitude faster algorithms, with 19 μs and 86 μs using respectively a sliding mode observer and an extended Kalman filter [24], and 5 μs using a Luenberger or sliding mode observer [26].

While these computation times are impressive, a key limitation is in their limited applicability to the high-rate problem due to the required level of physical knowledge or linearity. It was argued in [11] that, given the unique dynamic characteristics of high-rate systems, the use of adaptive algorithms was critical in generating successful decision strategies. However, adaptive algorithms are known to require longer computing times, which also limits their applicability to high-rate mechanisms. Yet, being able to create fast adaptive algorithms with adequate convergence could open new possibilities empowering high-rate systems through the real-time estimation of their structural health, here termed high-rate structural health monitoring (HRSHM).

There have been some efforts in enabling HRSHM through the generation of algorithms with convergence times approaching the high-rate time scales. An experimental testbed, the Dynamic Reproduction of Projectiles in Ballistic Environments for Advanced Research (DROPBEAR), was used to generate data to test various HRSHM algorithms and is described in more detail later in this chapter [14]. Some methods based on model reference adaptive systems (MRASs) have shown promise.

For example, an MRAS was applied to the DROPBEAR to identify the position of a moving cart mimicking some high-rate features. The algorithm identified the position of the cart online through the estimation of the system's fundamental frequency using a sliding mode observer [14]. The authors in [25] used MRAS to estimate the stiffness of the same moving cart in real time, also leveraging sliding mode theory. An average computing time of 93 μs was reported. DROPBEAR was used in [8], where the authors proposed to estimate the cart position by matching extracted frequencies to pre-generated finite element models. The authors implemented the real-time algorithm in a field-programmable gate array and evaluated experimentally, with a reported 4.04 ms computing time with minimum error and an improved iteration time minimizing errors and computation time of 0.83 ms to match the pre-generated models.

Other work was conducted on data generated from the droptower tests shown in Fig. 1. The authors in [12] proposed a wavelet neural network to conduct signal estimations. A particularity of the algorithm was its on-the-edge learning capability, without pre-training, provided by a self-organizing input mechanism that permitted adaptation of the neuro inputs to local stationarities. While the algorithm showed promise for online applications, its computing time was not suitable for high-rate applications. Inspired by this work, a fast deep learning algorithm was proposed in [1], but with real-time HRSHM applicability. An average computing time of 25 μs was reported.

In this chapter, opportunities and limitations in using real-time learning for HRSHM are discussed, with an emphasis on physics-informed methods that carefully craft the input of the representation enabling time series feature extraction. The discussion starts by reviewing opportunities in crafting the input space [10] for accelerating convergence speed in estimating non-stationary time series. After, the deep learning strategy developed in [1] for fast real-time applications is presented. Lastly, a path in conducting rapid real-time state estimation is discussed.

2 Opportunities in the Input Space

On a simplistic perspective, one can initially view the problem of HRSHM as one of non-stationary time series estimation and forecast. A critical challenge in building a black-box representation for a non-stationary system is in the time-varying nature of dynamic characteristics, whereas representative features are expected to evolve with time. Some machine learning techniques lend themselves to characterizing time series signals, in particular recurrent neural networks (RNNs) that can learn temporal dependencies, and their infamous long short-term memory (LSTM) architecture capable of preserving long-term knowledge. However, machine learning techniques typically heavily rely on pre-training, and curated training data related to high-rate system are highly limited due to the important costs involved in experimentation and highly uncertain dynamic environment. A successful machine learning algorithm for HRSHM is one that can be trained on these very limited datasets, and with strong

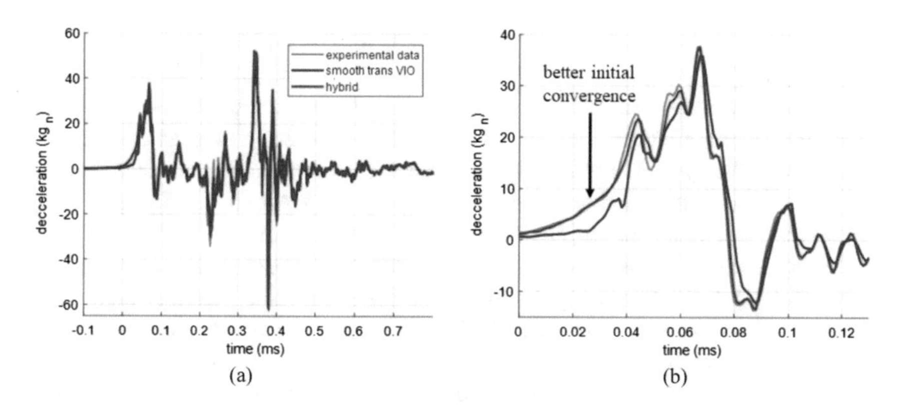

Fig. 2 Pure on-the-edge machine learning algorithm ("smooth trans VIO") versus hybrid algorithm ("hybrid"): **a** overall time series; and **b** zoom on the first section of the time series

on-the-edge learning capabilities. In addition, the algorithm must be lean in order to yield fast computing, yet capable of representing important dynamics to ensure fast convergence.

Physical knowledge can be integrated with the machine learning technique to ameliorate convergence and computing performance. An obvious choice is the utilization of an approximate physical representation that runs in parallel, where the machine learning algorithm focuses on mapping the unmodeled dynamics. Such a technique is popular, because it can yield a significant increase in estimation accuracy [13] and convergence speed. For example, Fig. 2 plots the time series estimation of a dataset from the droptower experiment (Fig. 1), where time series data from accel 1 were mapped to time series data from accel 2. The figure compares results from the pure on-the-edge learning algorithm ("smooth trans VIO") presented in [12] and discussed later, against those obtained from the same algorithm but using a six degrees-of-freedom approximation of the system running in parallel ("hybrid"). It can be observed that the use of physical knowledge substantially increases the quality of the estimation.

While this parallel architecture has merit, it does not necessarily affect the computing speed of the learning algorithm. A solution is to inject physical knowledge directly into the machine learning algorithm in order to accelerate both its convergence and computing time. The authors have studied opportunities in doing so at the input space level, by manipulating what set of features goes into the black-box model. Intuitively, by selecting features that preserve the essential dynamics of the time series of interest, the representation would be leaner and more efficient.

There exists a multitude of techniques that can be used in analyzing a non-stationary time series and extracting appropriate features [15]. A method of interest to the authors is based on the embedding theory, based on the celebrated Takens embedding theorem [22]. The theorem states that the phase space of an autonomous system can be reconstructed topologically using a vector of delayed measurements called delay vector ν, with

$$\nu = [\, y(t)\ y(t-\tau)\ \cdots\ y(t-(d-1)\tau)\,] \tag{1}$$

where $y(t)$ is a measurement taken at time t, τ the time delay, and d the embedding dimension. In other words, there exists a delayed vector that preserves the essential dynamics of the system. While originally developed for autonomous systems, the theorem has been extended to non-autonomous systems with deterministic forcing [20], state-dependent forcing [5], and stochastic forcing [21].

The embedding theorem does not apply to non-stationary systems. It was demonstrated in [10] that, while using ν constructed based on the embedding theorem to estimate a non-stationary system did yield adequate performance, the method showed greatly unstable with respect to the choice of τ and d, where a mere over-embedding by one dimension could result in close to 100% underperformance in accuracy and convergence. Yet, the authors in [17] proposed one such application for neurocontrol. The idea was to use a sliding window through the time series measurement to extract the delayed feature vectors (i.e., τ and d), assuming that each time series window was stationary, and use these features as the input to the representation. Thus, for each new window, new features were extracted and the input space adapted accordingly. This algorithm was termed "self-organizing input space." Results yielded a leaner neuro-representation, computationally faster, and of better convergence compared with the same representation built with the best non-variable input space (i.e., ν constant) found through brute force.

The method was used by the authors in [12] to evaluate its promise for HRSHM applications. This was done by constructing an observer equipped with such self-organizing input space, termed "variable input observer" or VIO. The architecture of the VIO is illustrated in Fig. 3, where the τ and d parameters are sequentially selected by analyzing the dynamics of the local time series and are used to construct ν that serves as the input delay vector to the wavelet network. Here, the wavelet network has self-adaptive capabilities (i.e., nodal weights, bandwidths, and centers), and the adaptation is conducted using the estimation error as the cost function.

Numerical simulations using the VIO were conducted on the droptower dataset (Fig. 1), and results also compared against those obtained from a non-variable input space found through brute force. Results showed that the use of the VIO yielded a neuro-representation that was approximately 50% leaner (i.e., fewer nodes), showed a reduction of approximately 50% in the root mean square error (RMSE), and an approximate 20% improvement in convergence speed. Conclusions of the study pointed toward important opportunities in the input space, similar to conclusions

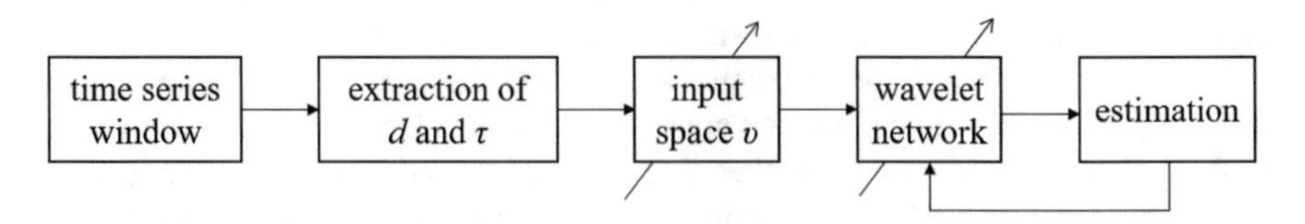

Fig. 3 Architecture of the VIO

drew in [10] that also demonstrated that the selection of the wrong input space for a high-rate system could have strong negative consequence on the quality of the representation. Nevertheless, the VIO had two important limitations impeding its application to HRSHM. Firstly, the architecture of the representation was incapable of convergence given the constantly changing input space. Secondly, the computing time used in finding the optimal input space features was quite significant. Based on these findings, the authors developed a deep learning strategy specifically crafted for HRSHM applications, presented in what follows.

3 Deep Learning Strategy for HRSHM

Opportunities and limitations identified in the research on the VIO led to a deep learning algorithm to build a stable predictor for non-stationary time series [1]. The strategy consists of utilizing n RNNs organized in parallel, termed ensemble of RNNs, each ith RNN associated with an input space ν_i constructed with a unique combination of τ and d (i.e., τ_i and d_i). The architecture of the proposed ensemble of RNNs is illustrated in Fig. 4. The selection of the n delay vectors will be discussed later in this section. The outputs of each RNNs are assembled through an attention layer that assigns weights to each output and passed through an linear neural to yield the prediction. The combination of the attention layer and linear neuron forms the function that maps the feature vectors $\mathbf{h}$ to the prediction. The RNNs are constructed with short-sequence LSTM cells to better cope with non-stationarities and empower faster computing. Here, the pure on-the-edge capabilities constraint assigned to the VIO is relaxed. Yet, it is assumed that only limited training data is available, and that the available training data does not represent most of the dynamics that could be experienced by the system. Thus, transfer learning is utilized to train the LSTM cells in real-time, using the one-step ahead prediction error for the cost function.

Using a discrete time k notation for simplicity, the ith delay vector at time k is written

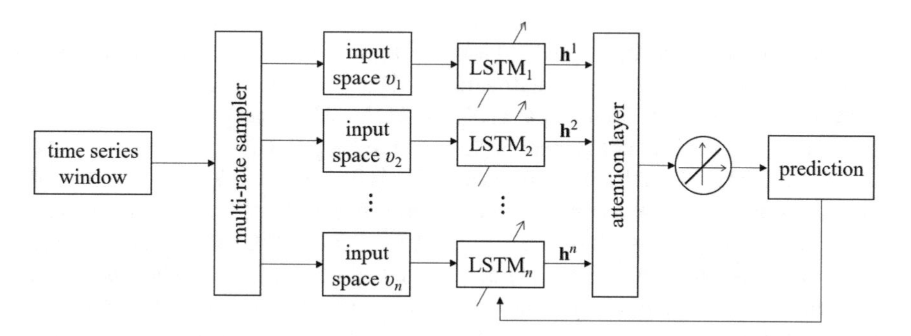

Fig. 4 Architecture of the ensemble of LSTMs

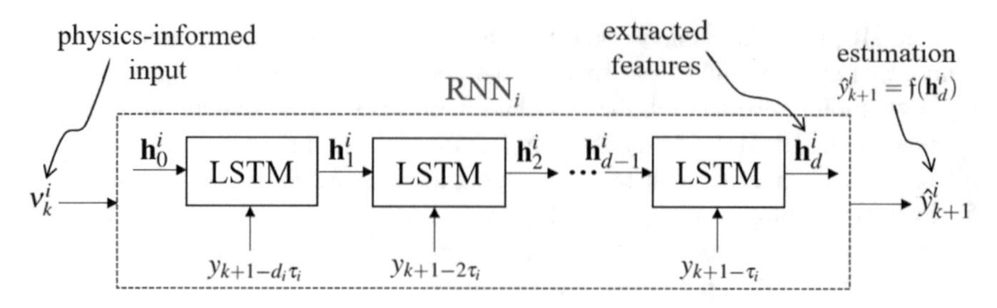

Fig. 5 Unfolder architecture of the ith RNN

$$v^i_k = [y_{k+1-d_i\tau_i} \; y_{k+1-(d_i-1)\tau_i} \; \cdots \; y_{k+1-2\tau_i} \; y_{k+1-\tau_i}] \tag{2}$$

where τ is a positive integer. The role of each LSTM cell is to recursively update a hidden state vector $\mathbf{h}$. Here, these hidden states represent features extracted from the time series, and these extracted features are mapped to the prediction. This process is illustrated in Fig. 5 for the ith RNN that maps the ith delay vector v^i_k to the step ahead prediction $\hat{y}^i + k + 1$ as a function $\mathfrak{f}$ of the last extracted feature vector $\mathbf{h}^i_d$. The role of the attention layer is to combine all the individual predictions to compute the ensemble prediction $\hat{y}_{k+1}$. Remark that, as implicitly shown in Eq. 2, the algorithm predicts the $k+1$ output $\hat{y}_{k+1}$, consisting of the one-step ahead prediction. For multiple steps ahead, the one-step ahead prediction is iterated using predicted values as synthetic measurements. Formal equations characterizing LSTM networks can be found in [9]. The physics-informed input space v^i is the corner stone of the algorithm, enabling more targeted feature extraction and thus improved prediction accuracy and horizon for non-stationary time series. Its construction is discussed in what follows.

3.1 Physics-Informed Input Space

The construction of the input space is based on physical knowledge and is conducted in two steps. The first step is the extraction of data structure from the training time series through principal component analysis (PCA). The strategy is to decompose the signal into principal components (PCs), and to keep the first PCs that represent most of the signal. For a time series dataset, PCA is typically conducted by a singular value decomposition of the signal's autocorrelation matrix. See [2] for a mathematical description of the process. Here, the first n PCs that account for 95% of the signal are taken, and n different input space are constructed, given raise to n RNNs running in parallel. As an example, consider test data generated from the droptower shown in Fig. 1a taken from accelerometer 1 (test 1—TS_1). Figure 6 shows 95% of the signal reconstructed with only the first five PCs that are individually plotted in Fig. 7.

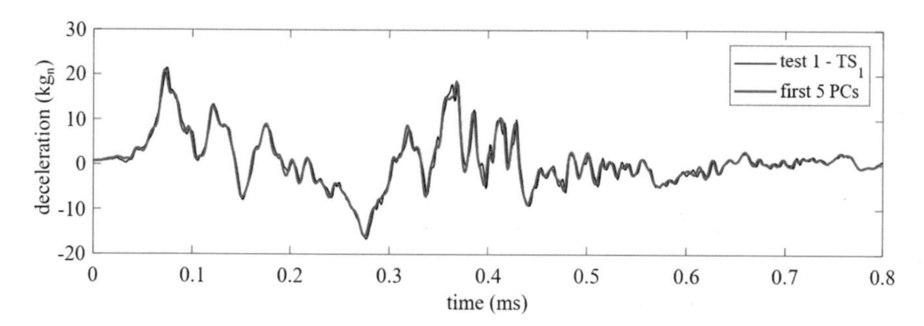

Fig. 6 Reconstructed time series TS_1 using the first five PCs

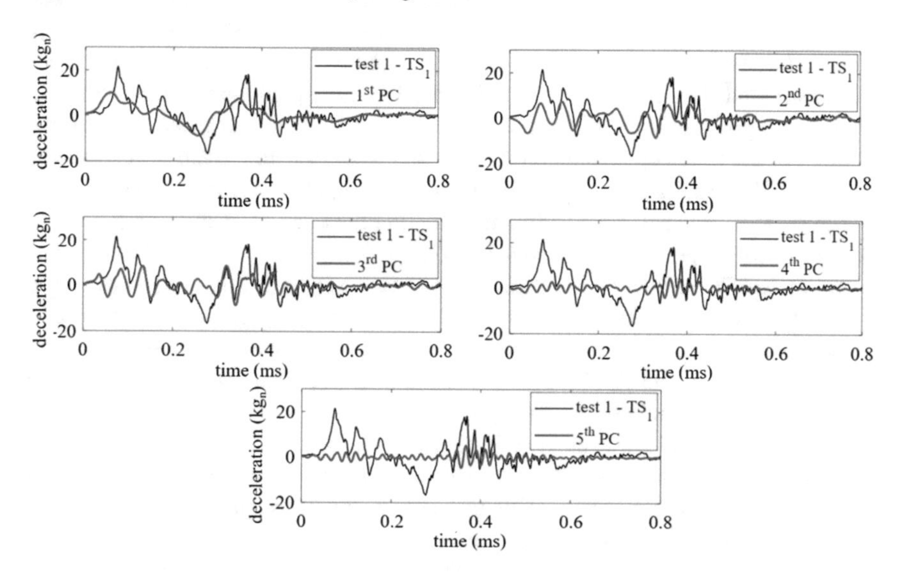

Fig. 7 Individual PCs used in reconstructing the signal

Once the data structure is extracted through PCA, the topology of the ith PC (Fig. 7 for instance) is investigated to select parameters τ_i and d_i that will be used to construct ν^i. This is done through well-established numerical techniques, in particular the mutual information (MI) test [3] to select τ, and the false nearest neighbors (FNN) test [12] to select d. The MI tests is based on information theory and is a measure of the nonlinear dependence of measurements for different sampling periods and selects a delay value that adds most information to the sequence. The FNN test searches for the embedding dimension d that minimizes the number of false neighbors when altering the dimension.

Once constructed, the delay vectors are used to train the RNNs. The RNNs are trained individually, on the full training dataset, using the associated delay vector (Fig. 5). After, the trained LSTM cell is extracted and becomes the feature extractor that will map v_k^i to $\mathbf{h}^i$ (Fig. 4). During the real-time prediction phase, the neuro-paremeters are sequentially updated based on a cost function based on the one-step ahead prediction error.

3.2 Numerical Demonstration

A numerical demonstration is conducted on the droptower dataset. The training data (i.e., source domain) is a single time series measurement taken from accelerometer 1 (test 1—TS_1). The target domain are five different time series measurements taken from accelometer 2 during five consecutive tests (tests 1–5, TS_2). Using the strategy discussed in the last subsection, here termed "PCA inputs," five distinct input spaces are constructed. Performance of the algorithm is compared against that with five other input spaces selected through brute force such that the root mean square error (RMSE) of the one-step ahead prediction is minimized, here termed grid search or "GS inputs." The parameters used to construct each input spaces are listed in Table 1.

The performance is the algorithm at predicting the time series over an horizon of q discrete time steps is investigated using four performance metrics. The first metric is the RMSE. The second metric is the mean absolute error (MAE). The third metric is the naive prediction length, consisting of the total length of naive predictions that are defined as predicted value taken as approximately equal to the last prediction. The fourth metric is the similarity between the extracted features and target domain measured using dynamic time warping (DTW).

Figure 8 plots the RMSE and MAE metrics as a function of the prediction horizon q (in discrete time steps). Results also show the line of indifference (LOI) that corresponds to the performance using the mean value of the time series as the prediction, and the performance of the VIO reported in [12] obtained for a one-step ahead prediction horizon. Results show that the GS strategy outperforms the PCA over short prediction horizons, which is expected given the procedure in selecting the input space. Nevertheless, the PCA strategy quickly starts to outperform GS after approximately five prediction steps, and remains below the LOI over a much longer

Table 1 Input space hyper-parameters selected using the "GS inputs" and "PCA inputs" techniques

LSTM	GS inputs		PCA inputs	
	τ	d	τ	d
1	14	8	25	5
2	11	10	15	6
3	8	12	11	8
4	5	14	7	12
5	4	15	5	15

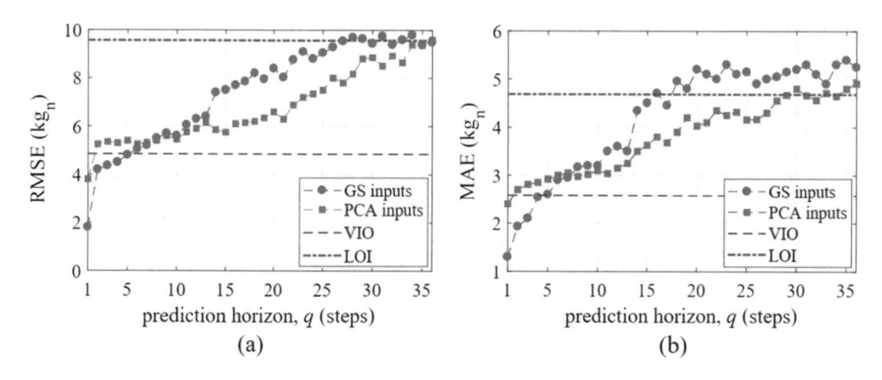

Fig. 8 **a** RMSE; and **b** MAE as a function of prediction horizon q

horizon, where the PCA strategy shows to outperform the LOI up to 34 steps ahead with respect to the RMSE, and 29 steps head with respect to the MAE.

Figure 9 plots the naive prediction length and DTW metrics as a function of the prediction horizon q. A study of the naive prediction length shows that the prediction by the PCA strategy is more stable and richer in information compared with the GS strategy. This can also be observed through the DTW plot where the PCA strategy significantly outperforms the GS strategy over mid- and long-term prediction horizons, demonstrating that the proposed physics-informed technique generates features (**h**) that are well representative of the non-stationary time series, thus explaining the better quality of predictions over long prediction horizons.

Overall, through this numerical demonstration, we demonstrated that a deep learning tool with a physics-informed input space could be used to quickly learn and predict a non-stationary time series. Hand-coded in Python and without leveraging a parallel processing architecture for the RNNs, the average computing time of the algorithm was 25 μs per time step. With the 1 MHz sampling rate used in acquiring data on the droptower test, this is equivalent to 25 steps ($q = 25$). This would govern what should be the minimum data sampling rate, unless the algorithm is coded in a way that step ahead predictions could be processed in parallel. Remark that it is anticipated that hardware implementation of the algorithm along with parallel processing of the RNNs themselves will significantly increase computing speed, thus improving performance with respect to the minimal sampling rate and/or required prediction horizon for HRSHM applications.

4 Path to Rapid State Estimation

The algorithm presented in the previous section showed direct applicability to HRSHM through adequate prediction and computing performance. However, a key limitation is that time series predictions cannot be used directly as actionable data. It

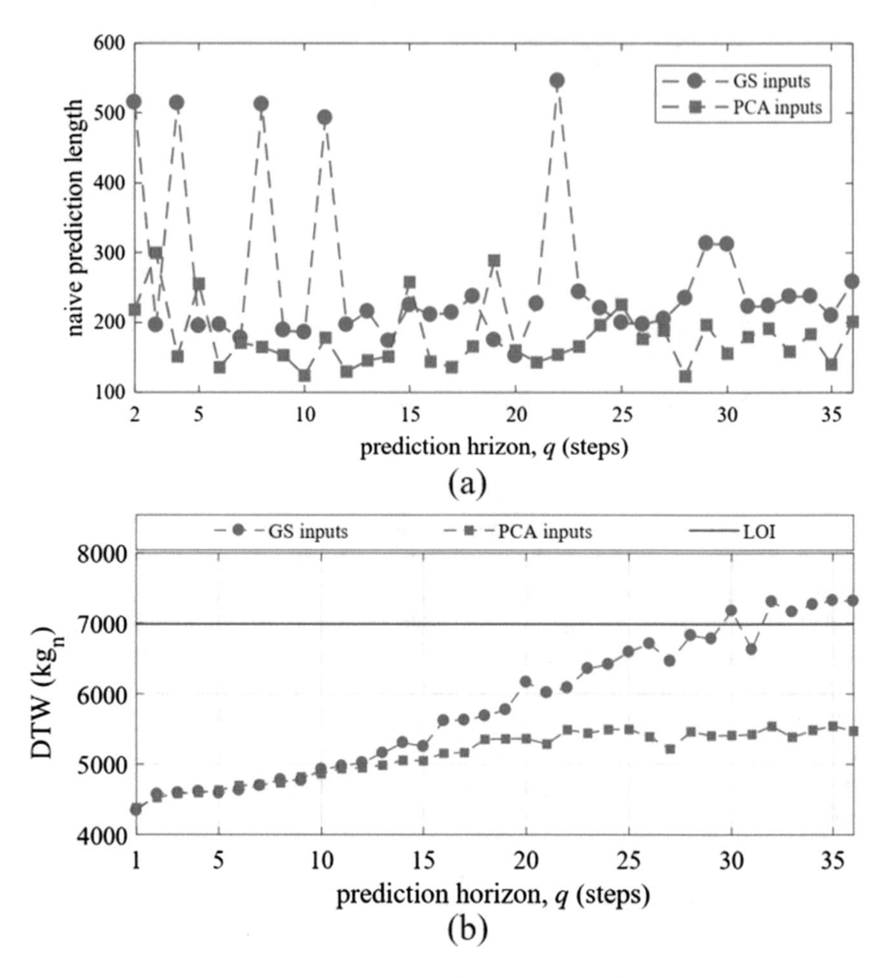

Fig. 9 **a** Naive prediction horizon; and **b** DTW as a function of prediction horizon q

follows that a complete HRSHM algorithm is one that yields information that can be used within a feedback system to make rapid decisions. A simple application of time series prediction to decision making is the detection of prediction errors to detect changes in dynamics, although this method is limited to a binary output "change" versus "no change" that is difficult to integrate in a decision system. Often, state estimation rather than measurement prediction is preferred, whereas physical properties such as stiffness and damping are estimated and can be used as a direct way to detect, localize, and quantify changes in dynamics (e.g., damage). For example, work presented in [6] proposed an algorithm for online estimation of bridge stiffness over a series of discrete elements in order to localize and quantify damage, where the estimated stiffness values can be used in a decision loop to detect the presence

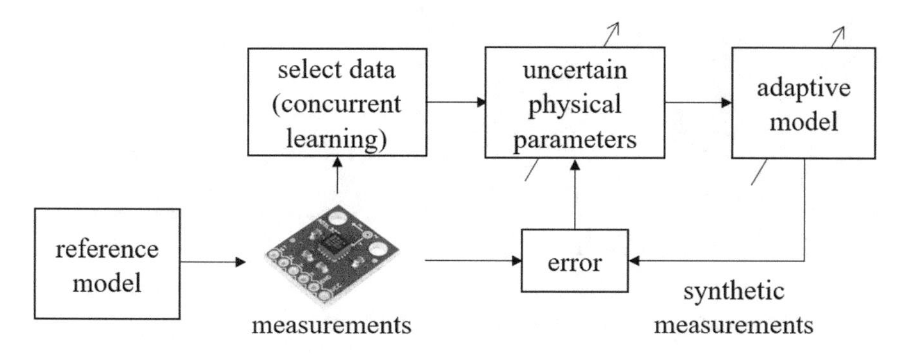

Fig. 10 MRAS-based algorithm for high-rate state estimation

of important damage post-seismic events and decide upon urgent inspections and/or bridge closure.

State estimation inherently relies on physical representations, which complicates the task in the high-rate dynamics realm given the lack of physical knowledge combined with the complexity of the dynamics. As discussed in the introduction, attempts have been made in HRSHM using MRAS approaches, in particular in [25] where the authors showed that the position of a moving cart could be identified with an average computing time of 93 μs. The algorithm is schematized in Fig. 10. It consists of an adaptive model running in parallel with a reference model (i.e., "real" system). The adaptive model is used to generate synthetic measurements, and the errors on measurements are used to update the uncertain parameters of the adaptive model. A critical contribution of the work is that it integrates the notion of concurrent learning to cope with the lack of persistent excitation, as it is often the case for high-rate events consisting of impulse loads. Concurrent learning consists of concurrently using real-time measurements augmented with a temporally limited historical data set to guarantee exponential convergence.

In this section, we summarize the demonstration of the algorithm on experimental data acquired on the DROPBEAR testbed found in [25]. After introducing the experimental methodology that includes the construction the MRAS algorithm, we present and discuss key results that show rapid state estimation capabilities and conclude by discussing how this method could be integrated with our proposed deep learning approach.

4.1 Methodology

The DROPBEAR testbest, shown in Fig. 11, was designed to validate online parameter estimation algorithms for high-rate dynamic systems. It is discussed in detail in [14]. Briefly, it consists of a 505 mm cantilever steel beam equipped with a sliding cart to produce a gradual or sudden change in stiffness, along with a mass attached

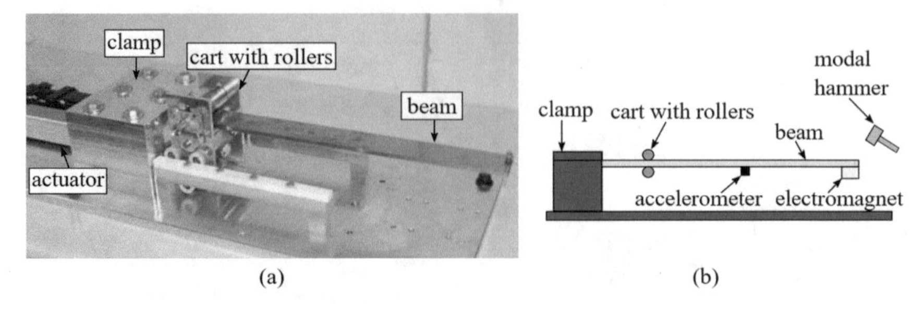

Fig. 11 DROPBEAR testbed: **a** picture; and **b** schematic

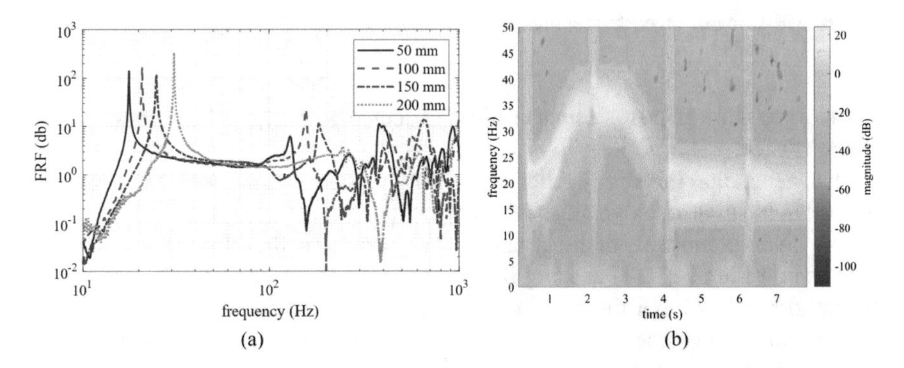

Fig. 12 **a** Frequency response functions (FRFs) for the beam with mass at various fixed cart positions; and **b** wavelet transfer for the beam without mass with the moving cart

at the tip using an electromagnet to produce a sudden change in mass. An impact hammer is used to excite the beam.

The experiment consisted of moving the cart along the beam at different fixed locations: 50 mm, 100 mm, 150 mm, and 200 mm away from the clam, and moving the cart back and forth from 50 mm to 200 mm away from the clamp. Data were acquired using an accelerometer sampled at 25 kHz. Figure 12a plots the frequency response functions (FRFs) obtained under the fixed cart positions of interest (the mass is attached), showing a dominating frequency between 17.7 Hz (at 50 mm) and 31.0 Hz (at 200 mm). Figure 12b shows the wavelet transform of the signal under the moving cart experiment (the mass is not attached), showing the dominating frequency moving approximately between 20 and 35 Hz. The peaks in frequency contents correspond to times when the beam was hit by an impact hammer. Note that the cart location can be approximated through a linear relationship with the measured dominating frequency, whereas the testbed is typically utilized as a starting point to benchmark algorithms in terms of computing speed to conduct high-rate state estimation.

The first step in formulating the MRAS algorithm is in constructing a reduced order representation. Here, this is done through the system equivalent reduction expansion process [6, 19] to find a state-space representation that mimics the modal properties of the system based on the sensor placement strategy. Thus, because only

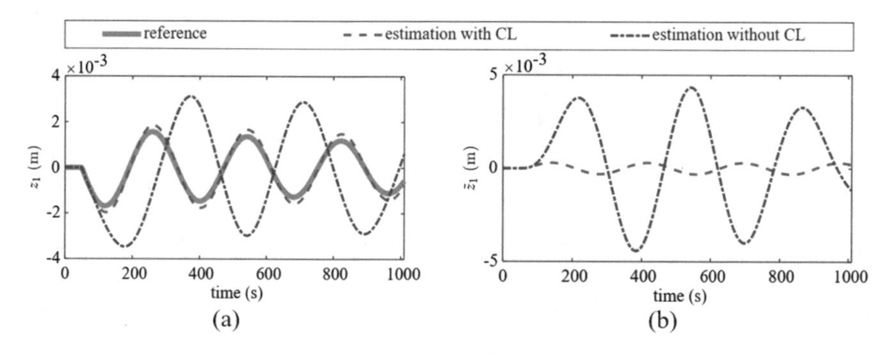

Fig. 13 Estimation versus reference with and without CL: **a** displacement z_1; and **b** displacement error $\tilde{z}_1$

one sensor is utilized in the experimental setup, the reduced order representation only has one degree-of-freedom. After, the adaptive dynamic parameters of the state-space representation, which consist of the equivalent mass, stiffness, and damping, are updated sequentially during the acquisition of measurements. This is done through sliding mode theory. As discussed above, a key issue in using impact loads is the lack of persistent excitation, where the adaptive dynamic parameters are not expected to converge. Concurrent learning is used as a solution, where only a short sequence representative of the input, termed history stack, is used in the adaptive mechanism. At each time step, the algorithm investigates whether the new input adds information to the history stack, and if it does, the new input replaces the less informative value in the sequence. Figure 13 demonstrates the use of concurrent learning on a toy single degree-of-freedom example, plotting the estimation of displacement z_1 and its estimation error $\tilde{z}_1$ over time. Results show that without concurrent learning, the estimation does not converge to the reference value.

4.2 Numerical Demonstration

Results from the fixed cart experiment are first described, where five tests were conducted under each cart location. Under these tests, the mass was taken as known, and both the stiffness and damping properties estimated, with both values taken initially as null. Figure 14a, b plot typical results for the estimation of stiffness and damping, respectively, under each cart location, with the blue vertical line showing the convergence point. Results show that both the stiffness and damping converge with the stiffness converging to its real value and damping to an adequate value (the real value of damping was unknown). Average results are tabulated in Fig. 14c, where the estimated frequency computed from the estimated stiffness shows a great match. It can also be observed that the convergence time decreases significantly with the increasing frequency of the system. Also, while the convergence time appears

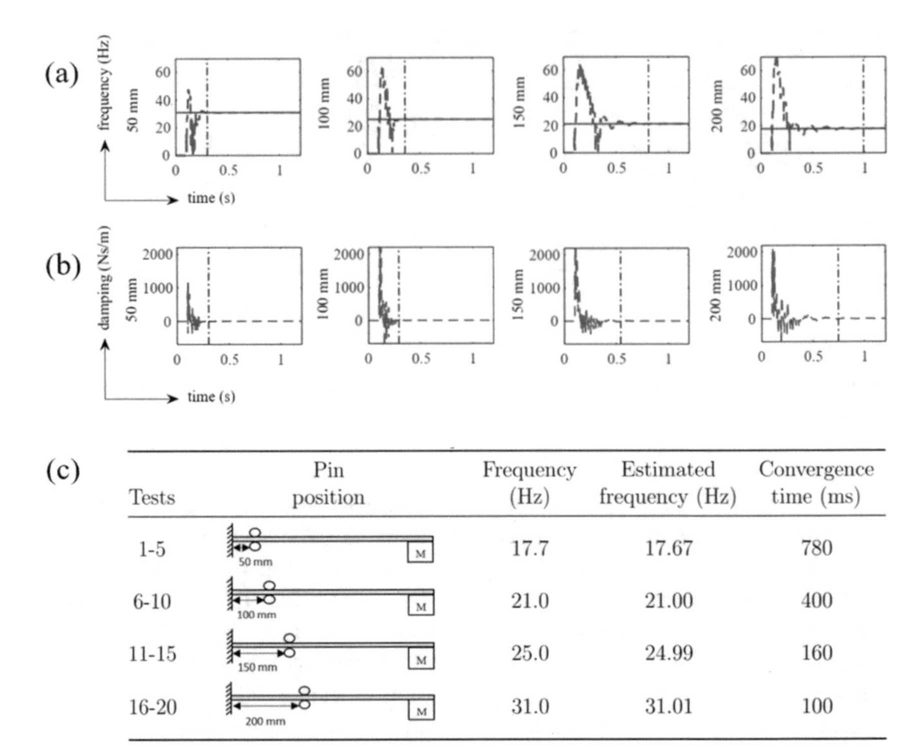

Tests	Pin position	Frequency (Hz)	Estimated frequency (Hz)	Convergence time (ms)
1-5	50 mm	17.7	17.67	780
6-10	100 mm	21.0	21.00	400
11-15	150 mm	25.0	24.99	160
16-20	200 mm	31.0	31.01	100

Fig. 14 Key results from fixed cart experiments: **a** typical results for stiffness estimation; **b** typical results for damping estimation; and **c** average frequency estimate per series of tests and convergence time

important in terms of HRSHM, it is attributable to the initial conditions on the dynamic parameters (i.e., null) that start far from the real values.

After the verification of the algorithm on the fixed cart experiments, we evaluated the capability of the algorithm at tracking the cart position. The beam was hit with four impulses during the experiments, as shown in Fig. 15c. Figure 15a, b plot the performance of the algorithm at estimating the displacement z_1 along with the estimation error $\tilde{z}_1$. Results show that the algorithm converges quickly, with higher errors at the time of impulse and while the cart is moving. Figure 15d plots the cart position tracking through the frequency estimation, assuming a direct mapping to the position. Results show good agreement with the real position (black line), with large errors at the time of impulses. There is also high chattering during movement of the cart, which may be attributed to the lack of excitation input. The algorithm appears to converge faster when the cart is close to 200 mm (higher frequency) than when it is getting closer to 50 mm (lower frequency), consistent with the results from the fixed cart experiment.

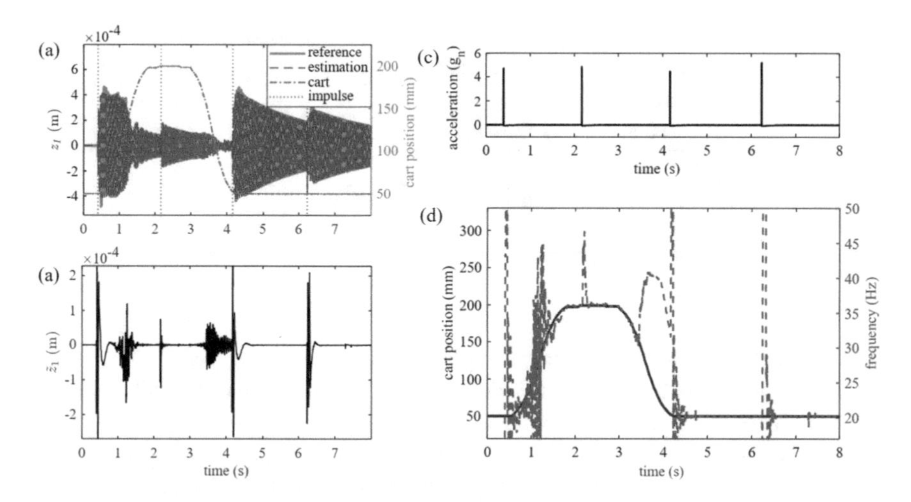

Fig. 15 **a** Displacement z_1 estimation versus reference during cart movement; **b** error on displacement estimation $\tilde{z}_1$; **c** impulse load inputs; and **d** cart position estimation (red-dashed line) versus real (black line)

4.3 Discussion

State estimation (i.e., cart position) of the moving cart was conducted with an average computing time per step of 95 μs using MATLAB. This compares well with results obtained from Downey et al. [8], where the authors reported an average computing time per step of 4.04 ms using a Fourier transform to identify the fundamental frequency and conduct model matching. Analogous to the deep learning algorithm, we anticipate that hardware implementation would significantly decrease computing time and thus increase HRSHM applicability.

While both this work and the one on the ensemble of RNNs discussed in the previous section are yet to be combined, we view the use of the MRAS-based technique as a path to providing rapid state estimation for HRSHM. An obvious method to combine both methods would be to produce time series predictions and train the MRAS algorithm on the predictions, thus enabling predictive capabilities on the cart position. Alternatively, the MRAS algorithm could be integrated in parallel with the ensemble of RNNs to yield a physics-informed machine learning approach, where the deep learning algorithm would learn and leverage a physical representation to improve predictive performance. As compared with purely data-driven machine learning models, the resulting physics-informed machine learning models would likely generalize better to scenarios unseen during model training, especially those that are not well represented by the training data.

Obviously, there exists a vast number of different algorithms and hybrid permutations that could be used to enable rapid state estimation for HRSHM. Regardless of the selected algorithms, computing speed is a critical metric enabling real-time applications, and convergence speed will define the accuracy of the algorithm. As briefly

Table 2 Comparison of convergence time after second, third, and fourth impulse under $\overline{J} = 10$, 30, and 50, along with average computation time per sample

$\overline{J}$	Convergence time (ms)			Computation
	Second impulse	Third impulse	Fourth impulse	time/sample (μs)
10	161	432	327	64
30	144	289	264	95
50	131	249	136	112

discussed in the introduction, computing time and convergence speed will often be conflicting goals. As an example, Table 2 lists the performance of the MRAS algorithm on the moving cart experiments as a function of the length of data kept in the history stack $\bar{J}$. While the convergence time after each impulse decreases with $\bar{J}$ increasing, the average computing time per sample increases, showing the important trade-off between computing time and accuracy. It follows that the minimum computing time in an HRSHM application will be governed by the required accuracy, which itself will depend on the required performance of the feedback system through the quality of closed-loop decisions.

5 Conclusion

In this chapter, we have introduced the problem of high-rate structural health monitoring (HRSHM) and discussed the use of real-time machine learning to empower rapid state estimation, in the sub-millisecond realm. We have started by defining the HRSHM challenge, where the dynamics of interest are importantly characterized by (1) large uncertainties in the external loads; (2) high non-stationarities and heavy disturbance; and (3) unmodeled dynamics generated from changes in system configurations, and argued that a successful HRSHM algorithm is one capable of leading to closed-loop decisions under 1 ms. Given the complexity of the dynamics, we found that adaptive mechanisms are well suited for the problem, as long as they are capable of fast convergence and short computation time. This can be done through the proper design of machine learning algorithms.

Among solutions, we reviewed opportunities in crafting the input space to yield lean and effective representations. In particular, it was discussed that being capable of adapting the inputs to the local dynamics could dramatically improve performance of the algorithm. We presented a machine learning algorithm based on an ensemble of recurrent neural networks (RNNs) that leveraged the idea of adaptive input spaces and demonstrated that a physics-informed selection of inputs yielded significant improvement in predictive performance. Importantly, the algorithm was capable of time series predictions in with an average computation time of 25 μs.

After, in an effort to produce more actionable quantities using real-time learning mechanisms, we presented a model reference adaptive system (MRAS) algorithm capable of conducting state estimation with an average computation time of 95 μs. Physical knowledge was used to generate and adapt a reduced order representation of the system of interest. While the algorithm was applied to a much simpler dynamics compared to the ensemble of RNNs, results showed that it was possible to quickly conduct state estimation for HRSHM. A path to rapid state estimation for more representative high-rate dynamic systems was discussed. It consisted of integrating the MRAS with the ensemble of RNN algorithm to improve the level of physical information present in the machine learning algorithm. We ended the discussion by warning of the conflict between convergence and computation time, where the accuracy of a given HRSHM algorithm will come at the expense of computation time.

References

1. Barzegar V, Laflamme S, Hu C, Dodson J (2021) Ensemble of recurrent neural networks with long short-term memory cells for high-rate structural health monitoring. MSSP (under review)
2. Barzegar V, Laflamme S, Hu C, Dodson J (2021) Multi-time resolution ensemble lstms for enhanced feature extraction in high-rate time series. Sensors (under review)
3. Belghazi MI, Baratin A, Rajeshwar S, Ozair S, Bengio Y, Courville A, Hjelm D (2018) Mutual information neural estimation. In: International conference on machine learning, PMLR, pp 531–540
4. Bodson M, Chiasson J, Novotnak RT (1995) Nonlinear speed observer for high-performance induction motor control. IEEE Trans Ind Electron 42(4):337–343
5. Caballero V (2000) On an embedding theorem. Acta Math Hung 88(4):269–278
6. Cancelli A, Laflamme S, Alipour A, Sritharan S, Ubertini F (2020) Vibration-based damage localization and quantification in a pretensioned concrete girder using stochastic subspace identification and particle swarm model updating. Struct Health Monit 19(2):587–605
7. Dodson J, Downey A, Laflamme S, Todd M, Moura A, Wang Y, Mao Z, Avitabile P, Blasch E (2021) High-rate structural health monitoring and prognostics: an overview. In: IMAC proceedings
8. Downey A, Hong J, Dodson J, Carroll M, Scheppegrell J (2020) Millisecond model updating for structures experiencing unmodeled high-rate dynamic events. Mech Syst Signal Process 138:106551
9. Greff K, Srivastava RK, Koutnik J, Steunebrink BR, Schmidhuber J (2017) LSTM: a search space odyssey. IEEE Trans Neural Netw Learn Syst 28(10):2222–2232. https://doi.org/10.1109/tnnls.2016.2582924
10. Hong J, Laflamme S, Dodson J (2018) Study of input space for state estimation of high-rate dynamics. Struct Control Health Monit 25(6):e2159
11. Hong J, Laflamme S, Dodson J, Joyce B (2018) Introduction to state estimation of high-rate system dynamics. Sensors 18(1):217
12. Hong J, Laflamme S, Cao L, Dodson J, Joyce B (2020) Variable input observer for nonstationary high-rate dynamic systems. Neural Comput Appl 32(9):5015–5026
13. Hu Z, Hu C, Mourelatos ZP, Mahadevan S (2019) Model discrepancy quantification in simulation-based design of dynamical systems. J Mech Des 141(1)
14. Joyce B, Dodson J, Laflamme S, Hong J (2018) An experimental test bed for developing high-rate structural health monitoring methods. Shock Vib

15. Kantz H, Schreiber T (2004) Nonlinear time series analysis. Cambridge University Press
16. Kettle RA, Dick AJ, Dodson JC, Foley JR, Anton SR (2016) Real-time state detection in highly dynamic systems. In: Rotating machinery, hybrid test methods, vibro-acoustics & laser vibrometry, vol 8. Springer, Berlin, pp 27–34
17. Laflamme S, Slotine JE, Connor J (2012) Self-organizing input space for control of structures. Smart Mater Struct 21(11):115015
18. Lee SJ, Jang MS, Kim YG, Park GT (2011) Stereovision-based real-time occupant classification system for advanced airbag systems. Int J Automot Technol 12(3):425–432
19. Qu ZQ (2013) Model order reduction techniques with applications in finite element analysis. Springer Science & Business Media
20. Stark J (1999) Delay embeddings for forced systems. I. Deterministic forcing. J Nonlinear Sci 9(3):255–332
21. Stark J, Broomhead D, Davies M, Huke J (2003) Delay embeddings for forced systems. II. Stochastic forcing. J Nonlinear Sci 13(6):519–577
22. Takens F (1980) Detecting strange attractors in turbulence. In: Dynamical systems and turbulence, Warwick, pp 366–381
23. Wadley HN, Dharmasena KP, He M, McMeeking RM, Evans AG, Bui-Thanh T, Radovitzky R (2010) An active concept for limiting injuries caused by air blasts. Int J Impact Eng 37(3):317–323
24. Xu Z, Rahman F, Xu D (2007) Comparative study of an adaptive sliding observer and an ekf for speed sensor-less dtc ipm synchronous motor drives. In: Power electronics specialists conference, IEEE, pp 2586–2592
25. Yan J, Laflamme S, Hong J, Dodson J (2021) Online parameter estimation under non-persistent excitations for high-rate dynamic systems. MSSP (under review)
26. Zhang Y, Zhao Z, Lu T, Yuan L, Xu W, Zhu J (2009) A comparative study of luenberger observer, sliding mode observer and extended kalman filter for sensorless vector control of induction motor drives. In: Energy conversion congress and exposition, IEEE, pp 2466–2473

Development and Validation of a Data-Based SHM Method for Railway Bridges

Ana Cláudia Neves, Ignacio González, and Raid Karoumi

Abstract Despite several successful applications, structural health monitoring (SHM) of bridges is still in its exploratory phase and, despite the increase in research, many challenges remain in order for it to become a commonplace practice in civil engineering. New SHM approaches have emerged sparked by the massive amount of acquired experimental monitoring data and breakthroughs in technology, computing capability and data storage solutions. To this end, the data-based approaches, mostly by resorting to machine learning techniques, have shown to be promising. This work proposes an unsupervised learning approach based on feedforward artificial neural networks for damage identification and condition monitoring of railway bridges. The inputs and output of the algorithm typically consist of measured accelerations in the bridge deck due to train passages, measurements which can be acquired easily with few installed sensors. Based only on data and statistical analysis, alarms with reference to early damage in the bridge can be triggered by the deployed SHM system. The implementation of the proposed approach is demonstrated and validated with both numerical and experimental case studies, where different aspects with relevance to SHM are as explored.

Keywords Artificial neural network · Data-based method · Unsupervised learning · Damage detection

1 Introduction

The concerns about maintenance and monitoring of structures have become a big challenge for engineers, researchers, and the civil engineering community. There

A. C. Neves (✉) · I. González · R. Karoumi
Department of Civil Engineering, KTH-Royal Institute of Technology, 10044 Stockholm, Sweden
e-mail: acneves@kth.se

I. González
e-mail: ignaciog@kth.se

R. Karoumi
e-mail: raidk@kth.se

A. Cury et al. (eds.), *Structural Health Monitoring Based on Data Science Techniques*, Structural Integrity 21, https://doi.org/10.1007/978-3-030-81716-9_5

has never been a more appropriate time to develop robust and reliable structural damage detection systems as aging civil engineering structures, specifically bridges, continue to be used past their life expectancy and well beyond their original design load capacity. The increasing monetary pressure on bridge authorities to extend the life span of the existing structures as far as possible is mainly driven by the considerable expenses associated with building a completely new bridge. In addition to this are also the costs associated with the physical removal of the old bridge and consequent costs from withdrawing the bridge from service, which affects passengers and freight transportation in several different ways. The reasonable thing to do is to keep the present structures in operation while ensuring public and structural safety at minimum cost. Concurrently, this requires managers and decision makers to prioritize rehabilitation and replacement programs. The problem is that when a significant damage to the structure is discovered, the deterioration has often already progressed far, and required repair is substantial and costly. One way to tackle this issue is by defining clever maintenance strategies that make use of structural monitoring in real time, thus enabling the detection of damage in its earliest stage and providing accurate remaining life predictions.

Structural health monitoring (SHM) aims at providing support for these strategies, through the collection of reliable data on the real condition of a bridge, the observation of its evolution over time, and characterization of the degradation. By permanently installing a number and variety of sensors, which continuously measure structural and environmental parameters, it is possible to obtain a real-time representation of the structure's current state. However, this information is only useful to the decision-making process if it is not misleading. This can be avoided by assuring that reliable SHM systems, methods for data analysis, and statistics tools are put into place.

2 Non-destructive Evaluation, Physics-Based and Data-Based Methods in SHM and Damage Detection in Bridges

Non-destructive evaluation (NDE) methods, commonly integrated with visual inspections, can be an effective tool for the inspection of bridge structures. NDE is particularly advantageous for evaluating bridges in service, since these can remain intact and open to traffic under the evaluation period. However, these methods can only be applied locally and may require the access to specific components of the structure that are not easy to reach. Besides, the results are usually subjective. A means to alleviate these drawbacks has emerged with advanced sensor and materials technologies, allowing sensors to be integrated into structural components for continued long-term use.

The methods used in the assessment of structural health can be said to be split into two main classes, according to their approach. The classical physics-based approach to damage detection in bridges typically presupposes the development of a finite element (FE) model of the target structure. These models are often of gradually increasing complexity during the development stage, so as to make sure that the measured responses can be reproduced as accurately as possible by the FE model. The well-known vibration-based damage identification (VBDI) methods consist of measuring and evaluating the dynamic behavior of the structure often by comparing it to the behavior simulated by numerical models, for instance FE models. Natural frequencies are the most fundamental vibration parameter, and methods directly measuring shifts in natural frequency can be used for identifying damage. Since many algorithms of damage detection are based on the difference between the models before and after occurrence of damage (Axiom II of SHM [1]), problems such as parameter identification and damage detection are closely related to model updating. Over the recent years, a second approach has emerged sparked by the massive amount of acquired experimental SHM data and breakthroughs in technology, computing capability, and storage solutions. The data-based approach is free of geometrical and material information, allowing to circumvent the burden of having to develop a detailed FE model of the target structure. With this approach, time series analysis, often along with signal processing techniques, is employed to extract damage-sensitive features from measured signals. In these regards, choosing an appropriate damage feature is crucial for the success of damage detection as these features are used to establish baseline statistics and to monitor changes in the normal structural behavior. This procedure makes the data-based approach well suited for permanent automated long-term monitoring.

Machine learning (ML) is an application of AI that provides systems the ability to automatically learn and improve from experience without explicitly programming them to do so. The idea behind using the data-based approach for SHM, and more specifically for damage detection, is to use the data sets of signals obtained from a structure over time and to use soft computing methods to warn about damage and its characteristics. Pattern recognition is a particularly useful branch, wherein labels, such as "healthy" or "damaged," are assigned to a given input value with the help of an algorithm. The type of learning performed by the computer usually falls into two major categories—supervised and unsupervised learning—depending on what type of data is used. In the context of damage detection, supervised learning implies that both data from the undamaged (reference) and damaged (novel) conditions of the structure are provided to the algorithm. Since prior knowledge about possible damage scenarios is available, classification and regression analysis can be carried out. This approach thus allows to characterize damage up to the level 3 [2] of damage identification (i.e., quantification of damage). On the other hand, unsupervised learning implies that only data from the undamaged (reference) condition of the structure is provided to the algorithm. This means that the data is not labeled, and therefore, it is only possible to group and interpret data based on input data. It is at most possible to reach level 2 [2] of damage identification (i.e., localization of damage) with this approach, the main goal being to carry out novelty detection and eventually also clustering and dimensionality reduction. In this manner, a reference condition of the structure is first established by the algorithm based on appropriate features extracted from data, which can be either measured or obtained from a numerical model of the

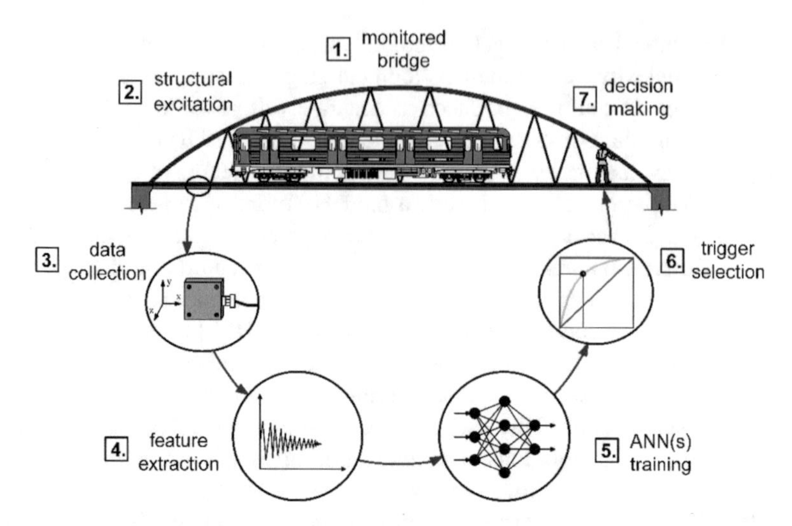

Fig. 1 Seven-step closed loop cycle of the proposed data-based approach to bridge SHM

structure. Afterward and through a process of statistical pattern recognition, features extracted from newly acquired data over time are compared with the ones observed in the reference condition. If there are significant deviations, the algorithm is said to indicate novelty.

Based on the aforementioned advantages, it would be reasonable to choose supervised over unsupervised learning for training the algorithm for damage detection. However, due to the nature of civil engineering structures, it is unlikely that the damaged data is readily available. It is also impractical to introduce damage to a seemingly healthy structure in operation in order to acquire the damaged data. Even if it is possible, the structures being dealt with are fundamentally unique, making it difficult to use the previous experience and knowledge gathered from even seemingly similar structures. For these reasons, ML algorithms used for SHM of civil engineering structures often rely on unsupervised learning approaches. In line with this, a seven-step unsupervised learning approach [3] based on feedforward ANN for damage identification and condition monitoring of railway bridges (Fig. 1) is proposed in the next section, followed by case studies where the respective findings and results are presented.

3 An Unsupervised Data-Based Approach to SHM Based on Feedforward ANN

ANNs are among the most common ML techniques used for novelty and damage detection, as well as to solve classification and regression problems. The ANN has the capability of modeling and predicting [4] the behavior of intricate linear and nonlinear systems exclusively based on data. For this reason, even the most basic feedforward ANN has shown to be a suitable method for predicting the behavior of civil engineering structures, as demonstrated by plenty of research works [3, 5–10].

When creating the ANN model, the total amount of train data is divided into three data subsets (Fig. 2), each to be used at different stages of the training process. The first data subset corresponds to the *training subset*, in which a number of examples are used to fit the parameters (primarily weights of connections between neurons) of the ANN model. The second data subset corresponds to the *validation subset*, which provides an unbiased evaluation of the model fit to the training subset while tuning the hyperparameters of the model. The last data subset corresponds to the *testing subset*, which provides an unbiased evaluation of the final model fit to the training dataset. In the context of damage detection, a portion of the recorded healthy acceleration data sets during train passages are used as train data for the ANN. Since the train data stems from one single state of the structure, learning can be said to be carried out in an unsupervised manner. After the ANN is trained, new acceleration data relating to healthy and damaged states can be used for testing the performance of the network.

A portion of the available healthy acceleration data sets recorded during train passages will be used as the damage-sensitive feature to train an ANN. For simplification, only part of the recorded acceleration time history of a train passage with duration T is plotted in Figs. 3 and 4.

During the training process, inputs (i) and matching targets (t) are assigned within each sample (train passage) of the *training subset*. After the training process is completed, the ANN will be asked to make predictions in the form of outputs (o). The inputs, targets, and outputs correspond to the discrete acceleration values registered by one sensor at any given time instant n. Training can be conducted, for example, by giving four accelerations $\{i_{n-4}, i_{n-3}, i_{n-2}, i_{n-1}\}$ (■) as input accelerations to

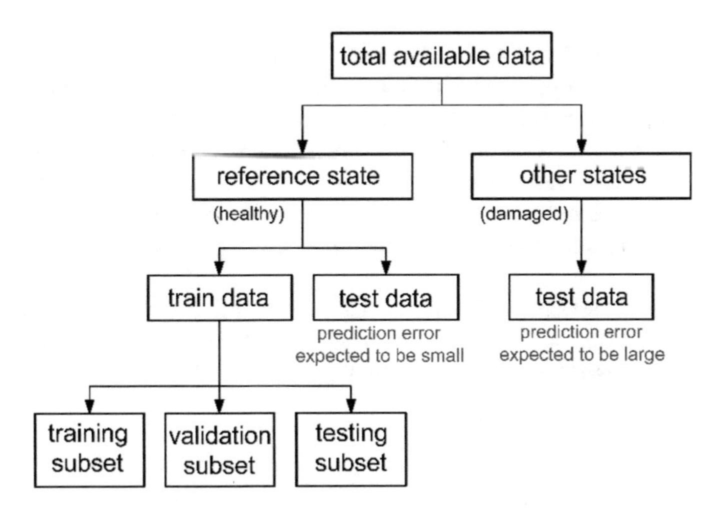

Fig. 2 Flowchart representing the data partition for the ANN model designed for damage detection

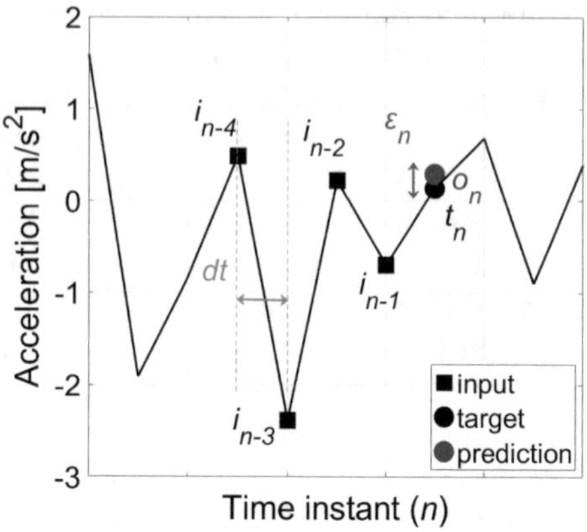

Fig. 3 Generic made-up graphic example of the input, target, and predicted accelerations when acceleration at time instant ***n*** is being predicted

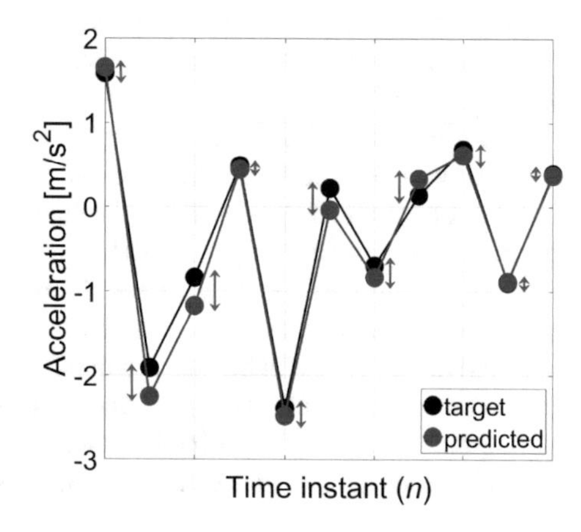

Fig. 4 Generic made-up graphic example showing the target, and predicted accelerations for all time instants within part of the recorded acceleration time history

the target acceleration t_n (●) as depicted in Fig. 3. Once the ANN is trained to learn the relationship between inputs and target, the output acceleration o_n (●) can be predicted based on the inputs of the *test data*. Ideally, target and predicted output would coincide if the condition of the structure does not change over time. However, predictions are not perfect and hence the discrepancy ($\varepsilon_n = o_n - t_n$) observed between ●and ●for all time instants (Fig. 4), even for the reference state. These discrepancies are expected to become much larger once damage appears in the structure.

For each train passage p with duration T and sampling period dt, giving a total amount of samples $N = T/\mathrm{d}t$, the total error—root mean square error (RMSE)—is given in Eq. (1) based on the difference between target and predicted output at every time instant n according to

$$\mathrm{RMSE}_{\mathrm{prediction}} = \sqrt{\frac{\sum_{n=1}^{N-N_w} (o_n - t_n)^2}{N - N_w}}, \tag{1}$$

where N_w are the very first four accelerations (in this example) for which the predictions cannot be calculated, and hence, the average is made for $N - N_w$. According to a detection threshold that is defined for the system, the damage detection process consists in verifying Eq. (2),

$$\begin{aligned} &\text{RMSE}_{\text{prediction}} < \text{threshold} = \text{healthy} \rightarrow \text{no alarm}, \\ &\text{RMSE}_{\text{prediction}} \geq \text{threshold} = \text{damaged} \rightarrow \text{alarm}. \end{aligned} \tag{2}$$

The information coming from each sensor individually may be more or less reliable depending on the sensor characteristics and their location on the structure, but when the information coming from all sensors is combined, the reliability of the detection method is expected to increase. A damage index (DI) provides a tool to combine the information obtained from multiple sensors into one single number, which works as a detection threshold set for the system as a whole. One possible way to carry out this combination is through the Mahalanobis distance, a covariance-weighted distance metric shown to successfully allow for condition discrimination in several damage detection studies [11]. The distance between a point P and a distribution D is measured in standard deviations. Each newly obtained DI point is compared to the cloud of DI points that is considered as the reference. For each train passage p, the damage index DI_p is formulated according to Eq. (3) as

$$\text{DI}_p = \sqrt{(\text{RMSE}_{\text{prediction}} - \mu)^T . C^{-1} . (\text{RMSE}_{\text{prediction}} - \mu)}, \tag{3}$$

where $\text{RMSE}_{\text{prediction}}$ (obtained according to Eq. (1)) is the prediction error in any state, μ is the mean prediction error of the distribution within the reference state, and C is the covariance matrix of the distribution within the reference state.

Naturally, since reality is much more complex, this binary discrimination reliant on a hard threshold can lead to instances of false diagnostics (Table 1). In a way, damage detection can be said to be based on hypothesis testing where the null hypothesis states that the structure is healthy. Incorrect diagnosis comprises False Positives (FPs) and False Negatives (FNs), where a healthy structure is perceived to be damaged and where existing damage in a structure goes undetected, respectively. On the other

Table 1 Damage detection hypothesis test inference matrix

Damage detection hypothesis test inference matrix		Reality	
		H_0—null hypothesis Bridge is not damaged	H_1—alternative hypothesis Bridge is damaged
Inference	Bridge is not damaged	TN No error	FN Type II error
	Bridge is damaged	FP Type I error	TP No error

hand, correct diagnosis comprises True Positives (TPs) and True Negatives (TNs), where a damaged structure and a healthy structure are properly assessed, respectively. FPs and FNs are also commonly recognized as Type I and Type II errors and are to be avoided: FP could be linked to extra inspections or unnecessary repairs. FN could be linked to particularly undesired consequences due to the continuous accumulation of undetected damage and ensuing failure of the structure, with life-threatening safety implications and casualties to society. Figure 5 shows the information contained in Table 1 in a graphical form. Given a chosen detection threshold (– – –) for the system, the obtained data sample errors from the ANN in light of healthy (O) and damaged (✱) data according to Eq. (1) can be classified into FP, TP, TN, and FN.

The receiver operating characteristic (ROC) curve is an easily interpretable two-dimensional graphic that illustrates the performance of a binary classifier system as a function of an adjustable threshold. The area under the curve (AUC), which measures the entire two-dimensional area underneath the entire ROC curve, is routinely used to compare the classifiers' accuracy and hence its usefulness. For instance, in Fig. 6, system is the ideal system with AUC = 1.0, whereas system —is the worst with

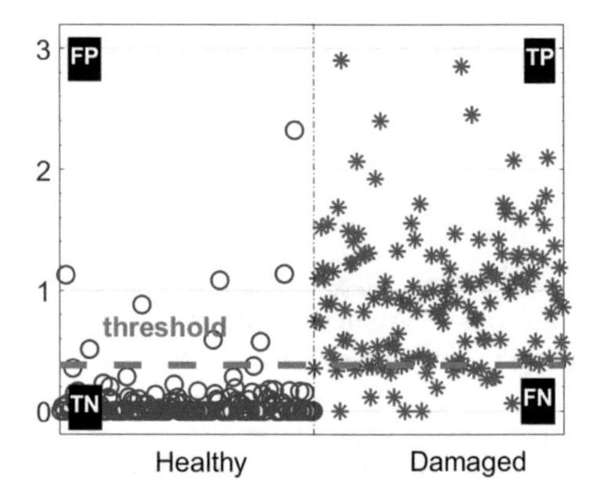

Fig. 5 Classification of data sample errors ($\boldsymbol{RMSE}_{prediction}$) into FP, TP, TN, and FN in relation to a threshold: —healthy data, —damaged data

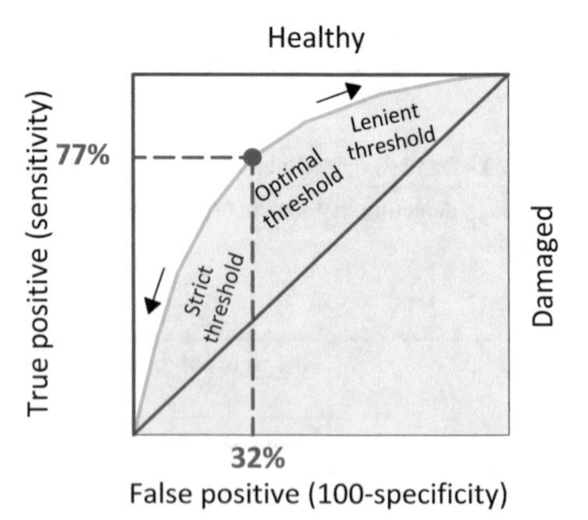

Fig. 6 Generic ROC curves: —excellent, —typical, —worthless classifier. The depicted chosen threshold (— — —) for a typical classifier yields TP = 77% and FP = 32%

AUC = 0.5. The green hatched area (O) represents the AUC for an average ROC curve that is more likely to occur for real systems ✱. In a ROC curve, the TP rate is plotted against the FP rate for all possible values of detection threshold, the points along which the curve is drawn. It is understandable that a very high threshold (strict threshold) will result in 0% of FPs and TPs. On the contrary, a very low threshold (lenient threshold) will result in 100% of FPs and TPs. It is important to define the detection threshold that finds a balance between the two types of error, as it is impossible to minimize them simultaneously. It is important to stress that the ROC curves cannot generally be obtained in practice. Firstly because it requires the data from damaged condition to also be available. Second because the true condition of the structure at any moment has to be known, which is precisely the point of employing the monitoring system. Nonetheless, ROC curves can become very useful in the developmental stage of methods for damage detection as their performances can be compared on the same grounds.

4 A Numerical Case Study

A numerical 3D finite element (FE) model of a single-track railway bridge was developed using the FEM software ABAQUS. The bridge consists of longitudinal steel beams that support the concrete deck and steel cross bracings that connect the girders. The deck and the girder beams were modeled using shell elements, and the cross bracings were modeled using truss elements. All the elements of the bridge are assumed to be rigidly connected to each other. The relevant material and geometric properties of the structural parts are given in Tables 2 and 3 in [8].

Table 2 Possible types of input features and values for the different ANN configurations

Possible parameters	Possible values
Input features	Acceleration, strain, temperature, train speed
Decimation factor	None, 1.5, 2, 3
Times steps	20, 40, 60, 80
Hidden neurons	10, 20, 30, 40

Table 3 Time periods for each state of the structure

2018	2019					
Nov	… Mar	Apr	May	Jun	… Sep	Oct
	before-train	before-test		during-test		after-test

Damage in the bridge is simulated considering two damage scenarios. In damage case 1 (*DC1*), a section of the bottom flange of one steel beam is removed by some length l (Fig. 3 in [8]), in an attempt to represent a damage situation where a fatigue crack exists. In damage case 2 (*DC2*), one cross bracing is malfunctioning (Fig. 4 in [8]), which equivalently corresponds to reducing to approximately zero its elastic modulus in the model. The measurement setup is made of accelerometers that are installed on the top of the bridge deck: three aligned with the railway track and three aligned with the steel beam in which damage in *DC1* takes place.

The proposed method for structural assessment is intended to detect existing damage from the measured vibration (vertical accelerations) of the bridge. The most common dynamic loads come from traffic, which is expected to be continuous while the bridge is in service. Therefore, traffic-induced vibration was generated by simulating the passage trains with a certain constant configuration, crossing the bridge at speeds ranging between [70–100] km/h, in increments of 0.1 km/h. Data was collected at a sampling frequency of about 130 Hz during 4 s and for a total of 300 different train passages. The moving axle loads were modeled as a series of impulse forces with short time increments conforming to vehicle motion. The properties of the train can be found in [8]. The train model was based on the high-speed load model (*HSLM*) train with two assumptions: the train has the dimensional properties (coach length, D, and bogie axle spacing, d) of the *HSLM-A4* train but is composed of only two intermediate coaches (N) instead of the fifteen accordingly to [12]; maintaining the train configuration just described, three different axle loads [170 180 190] kN were assumed.

The ANNs were developed and trained [8] by means of the neural network toolbox available in MATLAB. The total amount of available data was, for this study, generated and collected from simulations of the FE model above described for 300 train passages. In an initial trial, the data from 150 train passages was used for training the network, and the 150 remaining ones were used for testing and validation of the trained network. The measured signal is in reality distorted by various sources of error and, to consider that fact, Gaussian white noise with a constant standard deviation of 0.0005 m/s^2 was added to the uncorrupted acceleration time histories obtained from the FE model, before these were used to train the ANN. Despite appearing little, considering the acceleration magnitude under the moving load, this level of noise is actually substantial. This level of noise corresponds to approximately 2% of the standard deviation of the measured signals, a value that is commonly considered for such analysis. Another measurement uncertainty considered was the axle load magnitude and, in that sense, 5% random oscillations of the axle load were considered. A limited number of ANNs groups were trained with controlled variation of input variables. To better evaluate the neural network's performance and attempt its optimization, the authors recommend the use of validation techniques such as K-fold cross-validation [13]. In each group, six different networks were trained, ANN$_n$ with $n \in \{1, \ldots, 6\}$, each predicting for one of the six sensors. The chosen training algorithm was the Levenberg–Marquardt backpropagation algorithm. The following configuration of the neural network was explored and proved to work out well:

- 49 input neurons: The number of neurons equals the number of features, which are the 30 accelerations a_{t-i}^n registered by the 6 sensors $n \in \{1, \ldots, 6\}$ in the last 5 samplings $i \in \{1, \ldots, 5\}$, the 18 axle loads and 1 axle position relative to a reference point;
- 30 neurons for the hidden layer based on empirical rules-of-thumb [14];
- 1 output neuron: the current acceleration a_t^n at time t predicted by sensor n.

All the input parameters were assembled into one single-input matrix. The only input variable that remained present in all the trials of ANN configurations was the acceleration. Any other deemed relevant parameter can also be given as input. What the network does is to predict a new acceleration at a certain instant in time based on previous accelerations (■) as shown in Figs. 3 and 4. The number of delays has to be chosen before the training phase. Perceptibly, it cannot be a small number as that will give little information for the training of the network, and it cannot be a large number since that leads to the increase in computation time. In this case study, the five previous accelerations $\{i_{n-5}, i_{n-4}, i_{n-3}, i_{n-2}, i_{n-1}\}$ preceding the acceleration to be predicted o_n were chosen as the time window for the input matrix.

One way of evaluating the performance of the trained network is by determining the deviation in the predictions. The use of the root mean squared error (RMSE) is very common and makes an outstanding general purpose error metric for numerical predictions. For each sensor and for each train passage (or each speed), in a similar way for both healthy and damaged scenarios, one can estimate the $RMSE_{prediction}$ as expressed in Eq. (1). Figure 7 illustrates the RMSE of the predicted accelerations by the six sensors, in the presence of an undamaged structure (○) and for a damaged

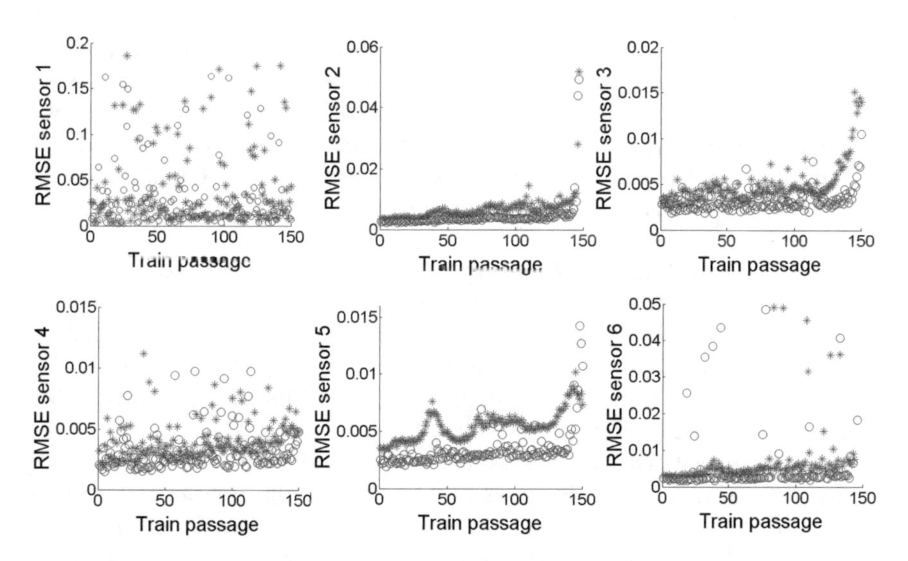

Fig. 7 RMSE against increasing speed of the train. DC1: damage extension of $0.9 \times 0.4m^2$. ○Data from healthy structural condition; ✱Data from damaged structural condition

structure (✱) in DC 1 with a $0.9 \times 0.4\text{m}^2$ section reduction. An axle load of 180 kN was considered, and the train passages in the x-axis are ordered by increasing speed.

There seems to be a tendency for the error to increase with increasing train speed. Moreover, for the highest speeds in the considered range, it seems like even the response of the bridge in the reference structural condition is poorly predicted (large RMSE). This could be justified by the fact that the maximum considered speeds excite the structure to frequencies close to its natural frequencies. Also noticeable is that the sensors situated closer to the geometric middle of the bridge (sensors 2 and 5) seem to be associated with networks that yield a better separation between structural states, while sensors placed nearby the end supports (sensors 1 and 4) are not as efficient. This may be due to the fact that the response of the structure is more prominent in the middle of the span than in its extremities, favoring a clearer distinction between different states. It should also be noted that the nonlinear input–output relating function that the network uses can be quite complex and for that reason training an ANN that covers all the train load cases and speeds is unmanageable. Thankfully, many bridges are routinely crossed by trains of the same configuration, with similar axle loads and within a restricted range of speeds. Therefore, the ANN is trained for those circumstances.

Even if the bridge is found to be in its original condition, the recorded dynamic responses will be different for each train passage, depending on the speed of the train for instance. Hence, the distribution of errors within the reference condition should be characterized stochastically, and it is the errors that significantly depart from this distribution that will work as an indication of damage. The prediction errors from 150 randomly selected train passages in the healthy condition of the bridge are used to fit a statistical distribution (Fig. 8) that will work as a baseline for each sensor. The Gaussian process (GP) [15] consists in assigning a normally distributed random variable to every point in some continuous domain. For each train speed, the associated predicted errors are normally distributed, and the mean (—) and two standard deviations (▮) of the error can be different for each speed. The idea is to compute discordancy measures for data and then compare the discordancy with a threshold, from which one is able to discriminate between healthy and damaged structural condition (Eq. (2)).

Once the DIs for different train passages are determined according to Eq. (3), the ROC curves corresponding to different damage extents within DC1 can be obtained as shown in Fig. 9. In the manifestation of such a damage scenario, for example, a fixed FPr of 8% has an associated TPr of 86%, 90.7%, 92%, 96%, and 99.3% with damage severities of 20 cm, 40 cm, 70 cm 100 cm, and 160 cm, respectively. The ROC curve associated with DC2 can be found in [8].

The enhancement in detection capability with increasing damage, as expected, is mostly but not always verified. For instance, one can see in Fig. 9 that the performance of the classifier for very low values of FPr is worse in the presence of the largest damage (−160 cm), since it is related to a lower TPr, when compared with any other smaller damage. The fact that some ROC curves regarding different extents of damage intersect each other at certain points makes it more difficult to choose the optimal threshold. Furthermore, the process encompasses statistical reasoning, thus

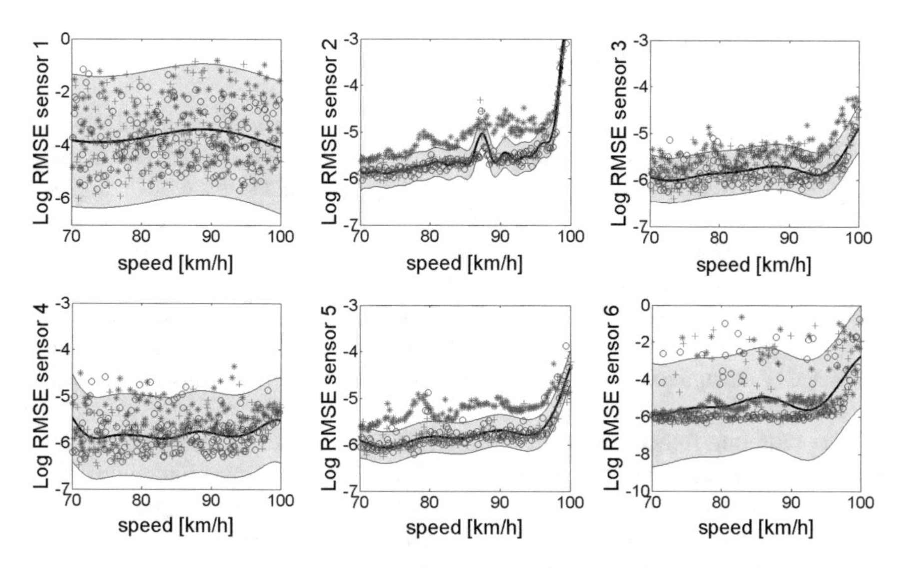

Fig. 8 Gaussian Process fitted by prediction errors against increasing train speed. Here, a log-normal distribution of the error is considered. +Mean; ○Standard deviation; ✳Data to fit the GP; —Data from healthy condition; ▒ Data from damaged condition, considering DC1 with a $0.9 \times 0.4 \boldsymbol{m}^2$ section

Fig. 9 ROC curves for different damage extensions l of DC1: damage resulting from cutting off a section of extension l from the bottom flange of one girder beam: —20 cm; —40 cm; —70 cm; —100 cm; —160 cm

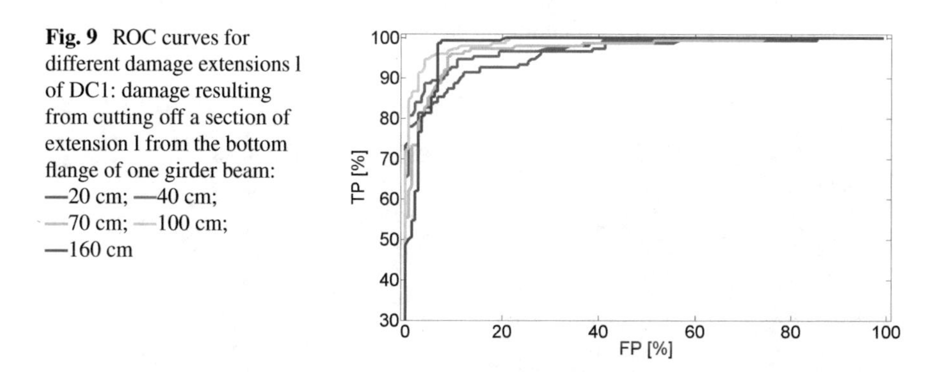

yielding slightly different results every time a ROC is regenerated. In any case, the desired virtues of a good model such as reliability and robustness are proven by the above-presented ROC curves. The model is considered reliable if damage is early detected with a high probability of detection, i.e., if small damage can be identified. The model is considered robust if two conditions are satisfied: the probability of detection increases with increasing damage severity, and changing an input parameter by a small amount does not lead to failure or unacceptable variation of the outcome but rather to proportional small changes.

5 A Field Case Study

The railway bridge KW51 (Fig. 10) is a steel bowstring bridge on railway line L36N located in Leuven, Belgium. The 115 m long bridge enables the crossing of the Leuven Mechelen canal and is only used by passenger trains with an imposed maximum speed of 160 km/h. The railway line consists of two tracks, hereafter referred to as track A at the north-side and track B at the south-side. At the crossing of the bridge, both tracks are curved. The monitoring campaign started on October 2, 2018, and under the period May 15 to September 14, 2019, the bridge was retrofitted (Fig. 11) to solve a construction error noticed during inspection. The retrofitting consisted of strengthening the connections between the diagonals and the arches and the bridge deck.

The sampling frequency is 1651 Hz. Regarding the data processing, the strain signals are low-pass filtered with an eighth-order Chebyshev Type I filter with 0.1 dB of passband ripple and a cut-off frequency of 16 Hz. The acceleration signals are high-pass filtered with a fourth-order Butterworth filter with cut-off frequency of 0.5 Hz. The measurements include acceleration and strain measurements on the bridge, strain measurements on the rails, and temperature and relative humidity measurements. Detailed information about the measurement setup can be found in [16]. Although not all sensors are used for this study, they are here mentioned for completeness.

The optimal selection of the ANN training and architectural parameters can be an exhaustive process. Judging which input (damage sensitive) features are relevant is

Fig. 10 Railway bridge KW51

Fig. 11 Bridge with scaffolding during retrofitting

an open problem as well, and finding the optimal ANN model turns out to be a trial–error process. Nonetheless, some empirical rules exist, and these can be followed as a starting point for suggesting reasonable models and ad-hoc parameters. Table 2 [17] lists relevant ANN key parameters to be defined, such as input features, the number of neurons in the hidden layer, the training algorithm, *et cetera.*

Several ANN configurations were tested, and each was obtained from different combinations of the parameters as presented in Table 2. The ideal hyperparameter search procedure would consist in searching all possible combinations of influencing features. Realistically, this would be impractical due to major computational resources and the curse of dimensionality [18]. This being said that the objective of this work is to provide some sort of check list for the implementation of the proposed method for SHM of railway bridges. As such, a limited number of different manually selected configurations are studied where, for each, the effects of making changes within one hyperparameter/feature while keeping the remaining constant are analyzed. Two aspects are kept invariable throughout all configurations: only one hidden layer is used and the training function is the Levenberg–Marquardt backpropagation algorithm.

The idea is to start simple by defining an ANN that has, for example, 10 neurons in the hidden layer and that uses the 20 previous raw accelerations from sensor A as input (e.g., a_{1A} to a_{20A} in m/s^2) to predict the current acceleration (e.g., a_{21A} in m/s^2). In other words, the dataset is broken into windows of size 20. Progressively, more complex ANNs (Fig. 12) are then defined and tested. For example, increasing the amount of training data so that the ANN uses the 30 previous (e.g., a_{1A} to a_{30A}

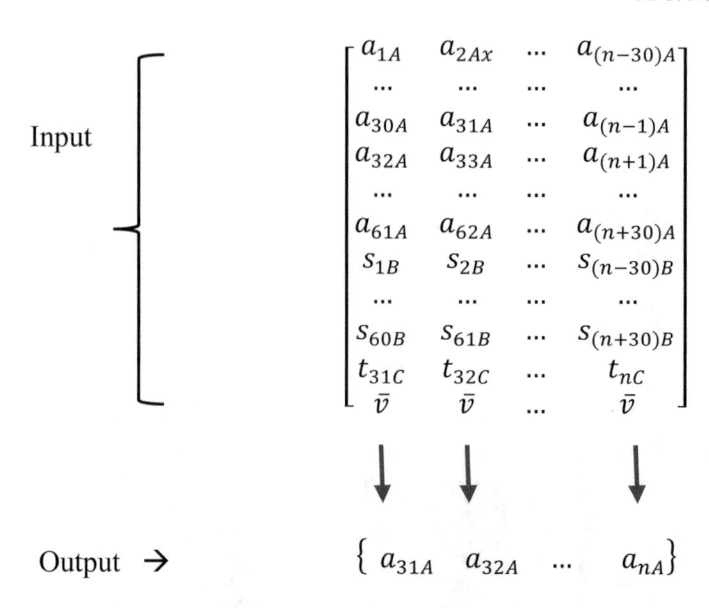

Fig. 12 A more complex input matrix and output vector for ANN

in m/s^2) but also the 30 posterior (e.g., a_{32A} to a_{61A} in m/s^2) raw accelerations from sensor A as input to predict the current acceleration (a_{31A} in m/s^2). Additionally, the ANN can also consider strains from sensor B (e.g., s_{1B} to s_{60B}), temperature from sensor C (t_{31C} in °C), and average train speed ($\overline{v}$ in m/s) as inputs. It is to be noted that the matrix and vector presented in Fig. 12 are related with one input data corresponding to one train passage, but training of the ANN is carried out by looking at many train passages.

In the following figures in this and in next section, the represented prediction errors for each train passage p are normalized relative to the RMS of the target response signal. The chronological dates of the train passages are presented in Table 3. The labels "before-," "during-," and "after-" that appear in this and following sections concern the structural states in reference to retrofitting. The data sets collected under the different states are labeled according to Table 4. If optimization is the goal, the

Table 4 Number of samples and labels for data points collected under different states and uses

State	Datasets	Label	Number samples (j)
Before retrofitting	train	+	84
	test	o	28
During retrofitting	test	* (grey)	93
After retrofitting	test	*	40

authors recommend the use of validation techniques (e.g., K-fold) to evaluate which data sets from the "before-retrofitting" period are better used as training and test data sets; however, this process was left out of the scope of this case study.

In this case study, two criteria are used to evaluate the performance of the algorithm: accuracy and sensitivity. The first concerns comparing the magnitude of the prediction errors between train and test data in the reference state (before-retrofitting); the second concerns comparing the magnitude of the prediction errors between test data in the reference state (before-retrofitting) and test data in other states (during- and after-retrofitting). The accuracy criterion is a reflection of how good of an approximator the ANN. But in the end, it is all the more important that a novel state of the structure (e.g., damage) is distinguishable from the reference one, which is expressed by the sensitivity criterion. Even if the prediction errors in the reference state are high, it is sufficient that the prediction errors in the novel state result even higher.

The following results focus on the prediction errors related to accelerometer 3. Furthermore, the prediction errors represented in Figs. 13, 14, 15, and 16 concern the average normalized prediction errors [19] for all the considered train passages within each of the structural states (Table 3). The errors in the just mentioned figures are represented in relation to the train samples (before-retrofitting) errors: for example, a small ε means that the taken test sample yields prediction errors with the same order of magnitude of the ones obtained for the train samples. The general thinking adopted when visualizing these figures is that the smaller the error for before-test

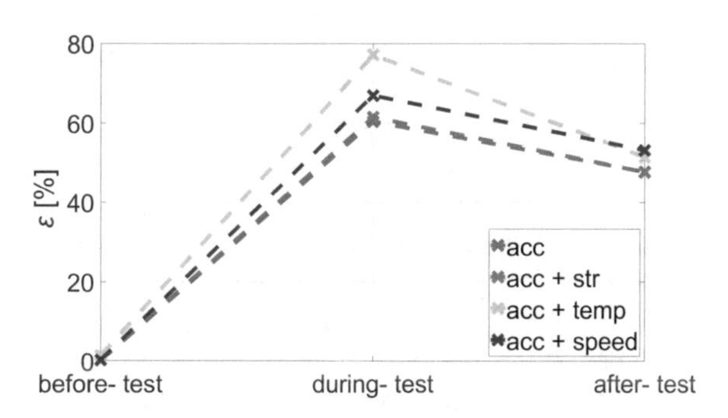

Fig. 13 Different natures of input data: prediction errors in linear scale

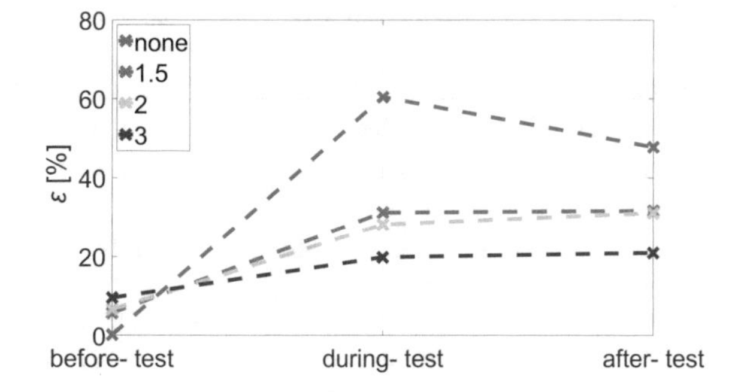

Fig. 14 Different amounts of down sampling of input data

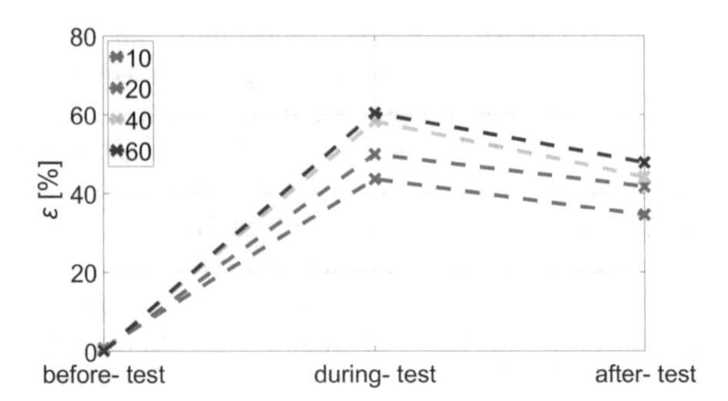

Fig. 15 Different time steps of input data

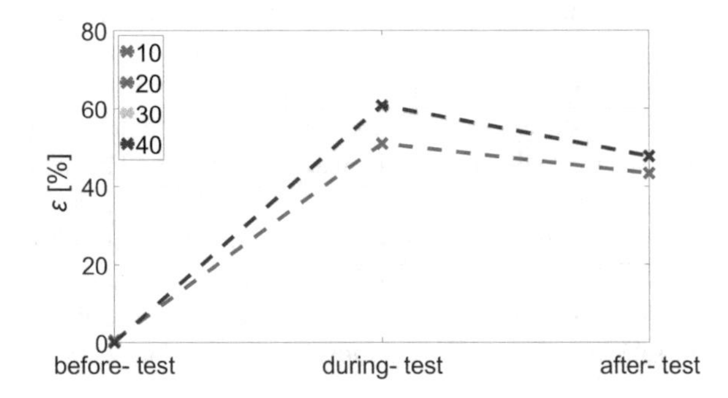

Fig. 16 Different number of hidden neurons

and the larger the error for during-test and after-test the better the performance of the system.

Even though the monitoring spans over periods of ambient, free, and forced vibrations, only the forced vibration part of the signal in which the train is on the bridge was used. Not only because the bridge vibration activated by the crossing vehicle can produce higher response amplitudes that can ease the interpretation of results but also because the use of higher amplitude loads may be required in order to detect some types of damage (e.g., cracks in concrete). Moreover, the ANN is trained and tested for a specific loading that is frequent for this bridge, a 52-axle train. The information regarding the train configuration (e.g., number of axles) was derived from the strain measurements on the rails. The axle loads for this type of train are accepted to vary within a relatively limited range of values. Therefore, including them as an input for the feature is thought to yield little to no improvement of the ANN performance, as shown in [8]. Furthermore, it was assumed that there is only one train on the bridge at a time. Giving data collected with deck accelerometers 1–6 as input to train the ANN and then asking it to predict for accelerometer 3 seems to lead to similar results if only the input of accelerometer 3 is given to train the ANN. This suggests that only the data collected by accelerometer 3 seems to be necessary as input to train the ANN to predict to said sensor. The same reasoning is followed for the remaining sensors, and accordingly, six different ANNs, one for each sensor, were trained.

Besides acceleration data, which alone was used as the original input for the ANN, strain, temperature, and average speed of the train were also added to see if these improved the ANN. Nearly no improvements are perceived (Fig. 13) if strain data is added. Strain gauges are not normally used in high frequency applications, and it may be necessary to further down sample the original signal (1651 Hz) since the strains only exceed the noise floor below 10 Hz. Adding temperature or train speed data, however, seems to increase the sensitivity of the ANN. However, it may not be worth the effort required to have the extra sensors in the bridge, especially given that the ANN is performing well without that extra information. Therefore, the input matrix used for training the ANNs can comprise solely accelerations.

Measuring, processing, and storing of massive amounts of data can provide valuable information in order to manage structures [20]. At the same time, the sampling frequency with which the input data is given to the ANN (i.e., the amount of data) affects the ANN's performance. As such, different versions of ANNs were trained by having the input data sampled with the original sampling frequency and also by down sampling it by a factor of 1.5, 2, and 3. Resampling is performed by using a polyphaser anti-aliasing filter. The advantage with down sampling is the reduced amount of resultant input data and thus faster training process of the ANN. On the other hand, by doing so, less information is fed into the ANN during the training process. That seems to reflect in higher prediction errors within test data collected under the reference state but also in the reduced ability of the ANN to distinguish between the different states (Fig. 14). Therefore, input data collected at the original frequency was preferred.

Different configurations of ANNs were tried where the minimum number of given previous acceleration data points to predict the current one starts at 10. Increasing the number of total time steps from 10 to 20 and from 20 to 40 seems to significantly improve the results (Fig. 15). Increasing the number of time steps from 40 to 60 marginally improves the results and over that the improvement is negligible. Therefore, to save in computational effort spent on training, 60 time steps are deemed as the most suitable for this case study. The 60 chosen time steps correspond to giving 30 previous and 30 posterior acceleration data points to calculate the current acceleration. Providing not only earlier accelerations as input but also the following accelerations to the time step to be predicted seems to yield both a more accurate and discriminative ANN.

A compromise must be reached for the number of neurons in the hidden layers. Two main problems can arise from having too few and too many neurons in the hidden layer(s): underfitting and overfitting, respectively. The former occurs when there are too few neurons in the hidden layers to adequately detect the signals in a complex data set. The latter occurs when the ANN has so much information processing capacity that the limited amount of information enclosed in the training set is not enough to train all of the hidden neurons. In light of the present case study, increasing the number of hidden neurons from 10 to 20 seems to significantly improve the results, whereas increasing it from 20 to 30 or even 30 to 40 produces no changes in the results (Fig. 16). Therefore, 20 hidden neurons are reasoned as adequate for this case study.

Figure 17 depicts the distribution of normalized prediction errors ε_p (Eq. 4) associated with accelerometer 3 against each train passage p ordered chronologically

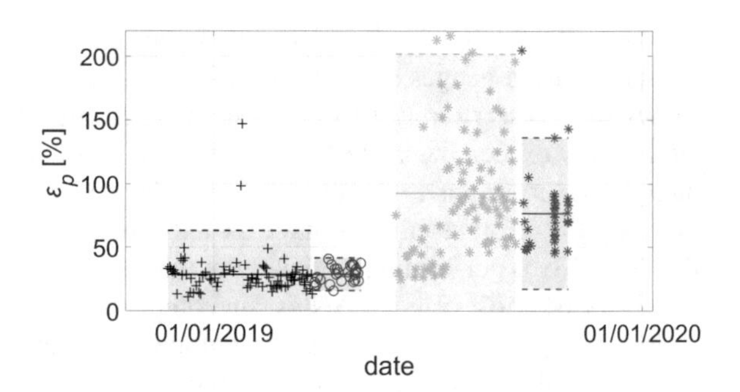

Fig. 17 Prediction errors for sensor 3 VS date: before-retrofitting train (+), before-retrofitting test (○), during-retrofitting test (✻), and after-retrofitting test (✻). All available number of samples j within each state are plotted

(Table 3). The data sets collected under the different states of the structure are labeled according to Table 4. It is possible to see with the naked eye that most of the lowest prediction errors correspond to the before-retrofitting period. That conclusion is more obvious by looking at the mean of the prediction errors (μ) for each period (—, —, —, —) and how they compare to each other. The dashed lines (- - -, - - -, - - -, - - -) represent the boundaries obtained by summing the mean with the two standard deviations of the prediction errors for each data set, i.e., $[\mu - 2\sigma, \mu + 2\sigma]$, which delimit the colored areas ■, ■, ■ and ■. Additionally, it is possible to perceive the progressive retrofitting taking place with the gradual increase of the prediction errors within the duration of the retrofitting period (✻). That also explains the larger variance of errors within the period. In the after-retrofitting period, any interventions come to an end and the scaffolding and necessary equipment are removed from the bridge. The new condition of the bridge is noticeably different from the reference one as the ANN predicts consistently worse.

The ROC curves associated with a system made up of accelerometer 3 alone (- - -, - - -) are depicted in Fig. 18. It is obvious that such a simple damage detection system robustly identifies the new states of the structure. The same outstanding performance may not persist for other sensors, for example if accelerometer 6 alone constitutes the damage detection system (····, ····). The ROC curves (—, —) that are

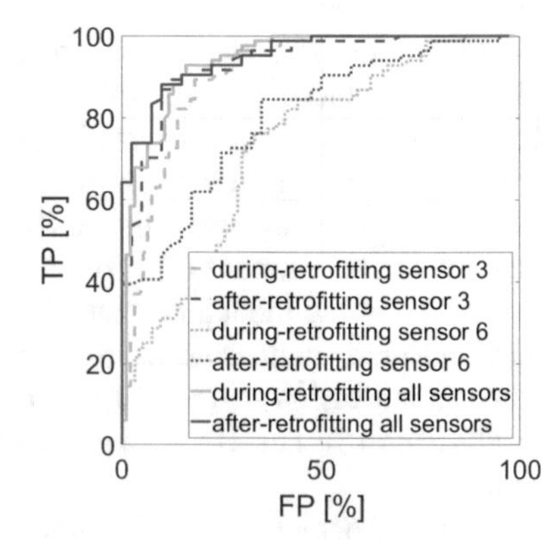

Fig. 18 ROC curves associated with different systems: one that considers only sensor 3, one that considers only sensor 6 and one that considers all sensors

obtained when the information from all the 6 accelerometers placed under the bridge deck is combined, according to Eq. (3).

6 Conclusions

The work here presented aims at contributing to an enhanced understanding of the possibilities, implications, and limitations of using a data-based approach for SHM in bridges. The following conclusions can be made based on the results obtained from two case studies:

- If the train is the main source of bridge excitation and the data is recorded during train passages, the speed of the train seems to have a significant influence on the predictions by the ANN. Ideally, the interval of observed speeds should be narrow and the speeds that can cause resonance of the bridge should be excluded from analysis. Similarly it may be important to consider variations in the environmental conditions, namely temperature, during the training phase of the algorithm.
- Most SHM systems are centered on measuring accelerations and strains, which makes it reasonable to use them as the principal sources of data. However, in the second case study, adding the features directly drawn from strain time series in the training phase seem to not improve the performance of an ANN. The authors want to stress that this conclusion is nevertheless case specific and that strains can help in the process of damage detection depending on aspects such as the localization of the strain gauges, the sampling frequency, *et cetera.*
- If damage detection is the goal and it is performed in a novelty detection fashion, good performance of the ANN in terms of the sensitivity criterion (how much larger the prediction errors are in damaged condition compared to healthy condition) is in principle preferred over the accuracy criterion (the magnitude of the prediction errors in healthy condition).
- Even though the monitoring is realized by several sensors, results show that it is sufficient to train one ANN for each sensor, only with input data provided by itself. Only in the final stage, after establishing a predictive model for each ANN and describing their performance with a ROC curve, the information from all sensors should be combined through a damage index.

Acknowledgements The authors are grateful for the data provided by Dr. Kristof Maes from KU Leuven.

References

1. Farrar C, Worden K (2013) Structural health monitoring—a machine learning perspective. Wiley, New York

2. Rytter A (1993) Vibration based inspection of civil engineering structures
3. Malekjafarian A, Golpayegani F, Moloney C, Clarke S (2019) A machine learning approach to bridge-damage detection using responses measured on a passing vehicle. Sensors 19(18)
4. Lapedes A, Farber R (1987) How neural nets work. In: Neural information processing systems, Denver, Colorado, USA
5. González I, Karoumi R (2015) BWIM aided damage detection in bridges using machine learning. J Civ Struct Health Monit 5(5):715–725
6. Hakim SJ, Razak HA (2013) Structural damage detection of steel bridge girder using artificial neural networks and finite element models. In: Structural damage detection of steel bridge girder using artificial neural networks and finite element models, vol 14, no 4
7. Lee Y, Cho S (2016) SHM-based probabilistic fatigue life prediction for bridges based on FE model updating. Sensors 16(3):317
8. Neves A, Gonzalez I, Leander J, Karoumi R (2017) Structural health monitoring of bridges: a model-free ANN-based approach to damage detection. J Civ Struct Health Monit 7(3)
9. Pandey P, Barai S (1995) Multilayer perceptron in damage detection of bridge structures. Comput Struct 54(4):597–608
10. Weinstein J, Sanayei M, Brenner B (2018) Bridge damage identification using artificial neural networks. J Bridge Eng 23(11)
11. Moughty J, Casas J (2017) Performance assessment of vibration parameters as damage indicators for bridge structures under ambient excitation. Procedia Eng 199:1970–1975
12. 1991-2, CEN-EN (1991) Eurocode 1: actions on structures—Part 2: traffic loads on bridges. European Committee for Standardization
13. Rodriguez J, Perez A, Lozano J (2010) Sensitivity analysis of k-fold cross validation in prediction error estimation. IEEE Trans Pattern Anal Mach Intell 32(3):569–575
14. Swingler K (1996) Applying neural networks: a practical guide. Academic Press, London
15. Rasmussen C, Williams C (2006) Gaussian processes for machine learning. The MIT Press
16. Maes K, Lombaert G (2020) Monitoring railway bridge KW51 before, during, and after retrofitting. ASCE J Bridge Eng
17. Neves A, Maes K, Karoumi R, Gonzalez I (2020) Application of a model free artificial neural network based novelty detection method: the case study of the KW51 railway bridge in Leuven
18. Donoho D (2000) High-dimensional data analysis: the curses and blessings of dimensionality. In: Lecture delivered at the “mathematical challenges of the 21st century” conference of the American Math. Society, Los Angeles
19. Neves AC (2020) Structural health monitoring of bridges: data-based damage detection method using machine learning (PhD dissertation, KTH Royal Institute of Technology)
20. Cremona C, Santos J (2018) Structural health monitoring as a big-data problem. Struct Eng Int 28(11):1–11

Real-Time Unsupervised Detection of Early Damage in Railway Bridges Using Traffic-Induced Responses

Andreia Meixedo, Diogo Ribeiro, João Santos, Rui Calçada, and Michael D. Todd

Abstract This chapter addresses unsupervised damage detection in railway bridges by presenting a novel AI-based SHM strategy using traffic-induced dynamic responses. To achieve this goal a hybrid combination of wavelets, PCA, and cluster analysis is implemented. Damage-sensitive features from train-induced dynamic responses are extracted and allow taking advantage not only of the repeatability of the loading, but also, of its large magnitude, thus enhancing sensitivity to small-magnitude structural changes. The effectiveness of the proposed methodology is validated in a long-span bowstring-arch railway bridge with a permanent structural monitoring system installed. A digital twin of the bridge was used, along with experimental values of temperature, noise, trains loadings, and speeds, to realistically simulate baseline and damage scenarios. The methodology proved highly sensitive in detecting early damage, even in case of small stiffness reductions that do not impair structural safety, as well as highly robust to false detections. The ability to identify early damage, imperceptible in the original signals, while avoiding observable changes induced by environmental and operational variations, is achieved by carefully defining the modelling and fusion sequence of the information. A damage detection strategy capable of characterizing multi-sensor data while being sensitive

A. Meixedo (✉) · R. Calçada
CONSTRUCT-LESE, Faculty of Engineering, University of Porto, Porto, Portugal
e-mail: ameixedo@fe.up.pt

R. Calçada
e-mail: ruiabc@fe.up.pt

D. Ribeiro
CONSTRUCT-LESE, School of Engineering, Polytechnic of Porto, Porto, Portugal
e-mail: drr@isep.ipp.pt

J. Santos
LNEC, Laboratório Nacional de Engenharia Civil, Lisbon, Portugal
e-mail: josantos@lnec.pt

M. D. Todd
Department of Structural Engineering, University California San Diego, San Diego, CA, USA
e-mail: mdtodd@ucsd.edu

A. Cury et al. (eds.), *Structural Health Monitoring Based on Data Science Techniques*, Structural Integrity 21, https://doi.org/10.1007/978-3-030-81716-9_6

to identify local changes is proposed as a tool for real-time structural assessment of bridges without interfering with the normal service condition.

Keywords Railway bridges · Structural health monitoring · Traffic-induced dynamic responses · Damage detection · Unsupervised learning · Data-driven · Artificial intelligence

1 Introduction

Modern societies are critically dependent upon transport infrastructure such as roadway or railway bridges and tunnels, which has motivated active research to reduce the costs of inspection and maintenance. A large number of bridges are nearing the end of their life cycle, and since these infrastructures cannot be economically replaced, techniques for damage detection are being developed and implemented so that their safe operation may be extended beyond the design basis for service life. Structural health monitoring (SHM) based on artificial intelligence (AI) represents a promising strategy in this ongoing challenge of achieving sustainable infrastructural systems since it has the potential to identify structural damage before it becomes critical, enabling early preventive actions to be taken to minimize costs. The main goal of SHM should not be to replace the traditional inspection techniques but to complement them with quantitative information. Proactive conservation strategies based on long-term monitoring are increasingly recommended for special structures such as long-span bridges. In fact, disruption or even the collapse of a bridge can lead to important and irreversible negative consequences for society and the economy. In short, SHM offers economical, efficient and intelligent technologies to manage the operation and maintenance of infrastructure, thereby improving safety, increasing longevity, and reducing maintenance (Fig. 1).

SHM techniques can follow model-updating or data-driven approaches for damage detection. Model updating consists of fitting a numerical model to experimental data to infer damage-related information that cannot be directly measured on site. Despite their reported accuracy, these techniques have an inherent computational complexity, and the need for user judgement makes them less suitable for real-time SHM [1, 2]. On the other hand, data-driven approaches rely on data mining techniques to extract meaningful information from time-series acquired on site. The computational simplicity of these approaches renders them more attractive and cost-effective to implement online damage detection in large-scale structures.

Damage detection strategies have been widely classified by the literature within a five-level hierarchy [3] (i) damage detection, (ii) localization, (iii) type, (iv) severity, and (v) lifetime prediction. The present chapter addresses the first level of the aforementioned hierarchy through data-driven methods based on train-induced dynamic responses. To fulfil this goal, four main operations need to be employed after the acquisition of data: (i) feature extraction, (ii) feature modelling, (iii) data fusion, and (iv) feature classification.

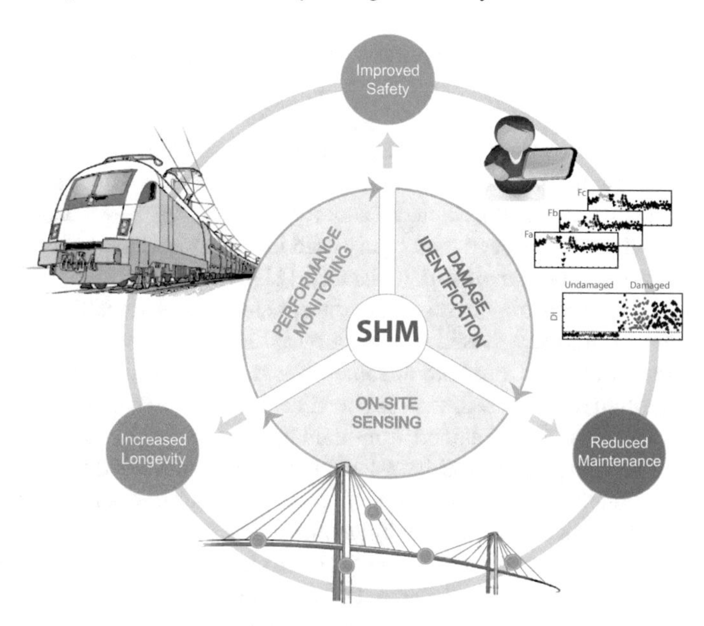

Fig. 1 Structural health monitoring cycle and its advantages

Feature extraction refers to the process of transforming the time-series acquired on site into an alternative information, where the correlation with the damage is more readily observed. Modal or modal-based features are the most common in the literature [4] due to the advantage of being directly associated with the mass and, more importantly, with structural stiffness, which is expected to change in the presence of damage. Nevertheless, operational modal analysis (OMA)-based information can also be considered not sensitive to early damage due to the need of identifying high order modes shapes, which proves very challenging for real structure monitoring. Symbolic data [5], Continuous Wavelet Transform (CWT) [6], and autoregressive models (AR) [7] are examples of techniques successfully applied as extractors of damage-sensitive features for both static and dynamic monitoring.

Effective SHM techniques for damage detection face the challenge of distinguishing the measured effects caused by environmental and operational variations (EOVs) from those triggered by damage [8]. Hence, SHM methods that can overcome this issue must necessarily resort to feature modelling. This operation is crucial for false alarm prevention since environmental (such as temperature) and operational effects (like trains crossing at different speeds) may impose greater variations than those due to damage. Two approaches are generally found in the literature and in the practice of feature modelling: (i) input–output, based on regression methods such as multiple linear regression (MLR) [9] or (ii) output-only, based on latent variable methods such as principal component analysis (PCA) [10]. The first removes the effects of the EOVs, establishing relationships between measured actions (e.g. temperature, traffic, wind) and measured structural responses. When monitoring systems do not include the measurement of EOVs, latent variable methods can be employed. These methods can suppress independent actions using only structural measurements.

Data fusion focuses on reducing the volume of data while preserving its most relevant information. The fusion process may combine features from a single sensor, and features from spatially distributed sensors or even heterogeneous data types. The Mahalanobis distance has been thoroughly used in this context due to its capacity to describe the variability in multivariate datasets [11].

Feature classification aims at discriminating the features into healthy or damaged. It can be divided into supervised or unsupervised learning algorithms [12]. When training data is available from both undamaged and damaged structures, supervised learning algorithms can be used, such as statistical process control [6] or MLP neural networks [13]. Since data obtained from damaged structures is rare or inexistent, unsupervised learning algorithms have been increasingly observed in the literature. Novelty detection methods are the primary class of algorithms used in this situation. This type of algorithm is a two-class problem that indicates if the acquired data comes from normal operating conditions or not. Due to its simplicity and effectiveness, outlier analysis is a broadly implemented damage detection technique [14]. In spite of the SHM feature classification resorting to clustering methods has been reported mainly following the supervised strategy of pre-defining cluster partitions to describe one or more known structural behaviours, and subsequently compare them with new ones, and this type of techniques have an unsupervised nature [15]. The major advantage of cluster-based strategies, over those previously described, consists of the greater sensitivity exhibited by these algorithms, which is related to their capacity to analyses data compactness and separation instead of defining boundaries between or around data objects. The works describing cluster-based classification for damage detection refer its high sensitivity to structural changes and associate it with the ability of these methods to analyses compactness and separation within feature sets.

While most of SHM works rely on responses derived from ambient vibrations or static responses, recent works have also been using the structural responses generated by traffic on bridges to take advantage of the repeatability of these actions, their known behaviour, and their large magnitude, which imposes a greater excitation of the bridge in a short time, when compared with ambient or static loads [16, 17]. However, robust and effective implementations of SHM in bridges based on traffic-induced dynamic responses are still scarce. In most damage detection methodologies that have been proposed, the EOVs in the structural response is often disregarded, the type of damages is limited, or the loading scenarios are very specific, which limits their usability in real and complex bridges.

In this context, the present research work aims at implementing and validating a real-time unsupervised data-driven SHM strategy for early damage detection in railway bridges using traffic-induced dynamic responses.

2 SHM Procedure for Early Damage Detection

The process of implementing a damage detection strategy involves the observation of a structure over a period of time using periodically spaced measurements, the extraction of features from these measurements, and the analysis of these features to determine the current state of health of the system [18].

A schematic representation of the proposed SHM strategy for early damage detection followed in this chapter is depicted in Fig. 2. The first step to develop a SHM strategy is performing an operational evaluation and defining a data acquisition system for the selected structure to set limitations on what will be monitored and how the monitoring will be accomplished. In this research work, a long-span bowstring-arch railway bridge was selected as case study. Section 3 details the bridge, the monitoring system installed, and the undamaged and damaged scenarios considered.

In order to accomplish, a fully autonomous and real-time SHM system the following four main steps regarding damage detection are implemented in Sect. 4: (i) Feature extraction, (ii) Feature modelling, (iii) Data fusion, and (iv) Feature classification.

The feature extraction is accomplished implementing a hybrid combination of CWT and PCA to the vibration-based measurements acquired by the monitoring system installed in the railway bridge. During this step, data compression is achieved by transforming the thousands of points from each dynamic response of the structure into a few hundreds of features. Subsequently, feature modelling is performed to reduce the influence of operational and environmental conditions. A latent variable method (PCA) is implemented to remove EOVs influence without measuring the actions, that is, based on structural measurements alone. To enhance sensitivity, a pattern-level data fusion is performed afterwards by implementing a Mahalanobis distance to merge the features without losing damage-related information. Finally,

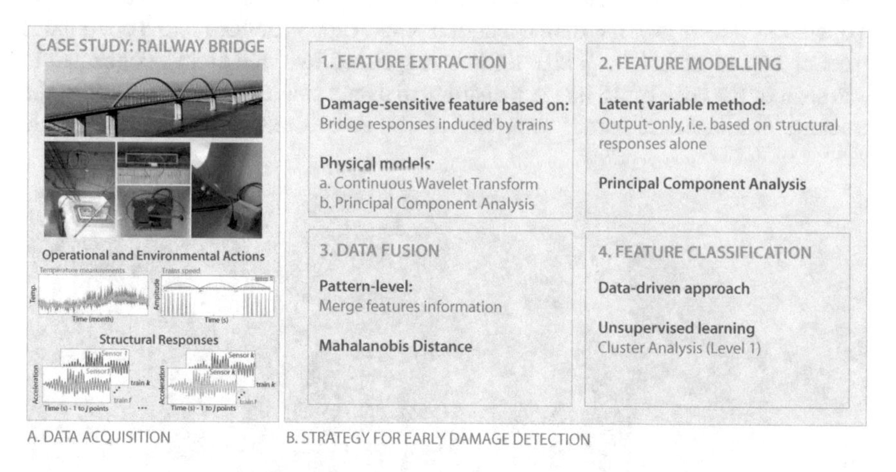

Fig. 2 Schematic representation of the SHM strategy for early damage detection

feature classification is performed as a data-driven approach and implementing unsupervised machine learning algorithms, namely cluster analyses.

As data-driven approaches are usually less computationally complex, they are better suited for early damage detection. Moreover, in civil engineering structures the most important question to answer is if there is or not a damage. The questions about location and severity are usually less important in this type of structures since the simple existence of damage will trigger other management procedures. For those reasons, data-driven approaches are followed in the present research with the aim of detecting damage (level 1). However, finite element models can be used to simulate damage scenarios that are not possible to obtain in any other way. These data can be then used to test the validity and robustness of the methodologies proposed for damage detection. This approach was followed, and a progressive numerical model validation of the railway bridge over the Sado River was performed, in order to, afterwards, simulate damage scenarios and demonstrate the efficiency of the developed strategy.

3 Data Acquisition from a Bowstring-Arch Railway Bridge

3.1 The Railway Bridge Over the Sado River

A bowstring-arch railway bridge over the Sado River was selected as the case study used throughout this research work. It is located on the southern line of the Portuguese railway network that establishes the connection between Lisbon and Algarve (Fig. 3). The bridge is prepared for conventional and tilting passenger trains with speeds up to 250 km/h, as well as for freight trains with a maximum axle load of 25 t. Even though the bridge accommodates two rail tracks, only the upstream track is currently in operation. The bridge has a total length of 480 m and is divided into 3 continuous spans of 160 m each. The bridge deck is suspended by three arches connected to each span of the deck by 18 hangers distributed over a single plane on the axis of the structure. The superstructure is composed of a steel–concrete composite deck, while

Fig. 3 Overview of the bridge over the Sado River

the substructure, which includes the piers, the abutments, and the pile foundations, is built with reinforced concrete. The deck is fixed on pier P1, whereas on piers P2, P3, and P4 only the transverse movements of the deck are restrained, while the longitudinal movements are constrained by seismic dampers.

3.2 SHM Monitoring System

The structural health condition of the railway bridge over the Sado River has been controlled with a comprehensive autonomous online monitoring system, as detailed in Fig. 4a, since the beginning of its life cycle. This monitoring system was defined based on an operational evaluation and allowed the acquisition of data necessary to

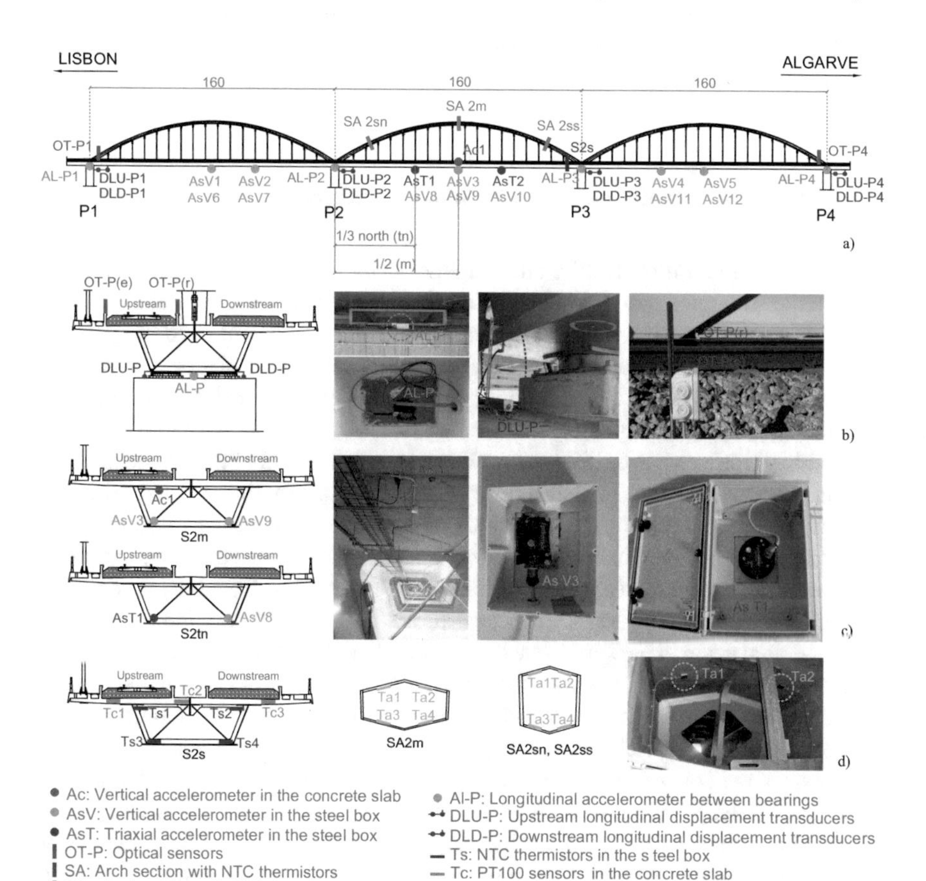

Fig. 4 SHM system installed in the railway bridge over the Sado River: **a** overview, **b** longitudinal accelerometer, displacement transducer, and optical sensor, **c** vertical and triaxial accelerometers, and **d** NTC thermistors and PT100 sensors

implement the strategy for early damage detection (Fig. 2). To identify each train that crosses the bridge and compute its speed, two pairs of optical sensors are installed at both ends of the bridge (Fig. 4b). The structural temperature action is measured using PT100 thermometers and NTC thermistors. Three sections of the arch were instrumented with twelve NTC thermistors. Additionally, four NTC thermistors are fixed to the steel box girder and three PT100 thermometers are embedded in the concrete slab (Fig. 4d). To control the behaviour of the bearing devices, the responses from longitudinal displacement transducers are obtained from eight sensors, each adjacent to a bearing device (Fig. 4b). The set of sensors also includes one vertical piezoelectric accelerometer fixed at the mid-span of the concrete slab, two triaxial force balance accelerometers at the thirds of the mid-span steel box girder, and twelve vertical force balance accelerometers fixed along each span of the steel box girder (Fig. 4c). Four longitudinal MEMS DC accelerometers are also installed at the top of each pier (Fig. 4b). Data acquisition is carried out continuously, at a sampling rate of 2000 Hz, by a locally deployed industrial computer to save the time history during the passage of the trains.

3.3 Baseline and Damage Scenarios Simulation

A realistic simulation of healthy and damage scenarios was conducted to test and validate the strategies proposed herein since it was not possible to simulate damage scenarios experimentally. After a successful validation of the methodology, it can be directly applied to experimental data from different types of bridges, where a baseline scenario is defined, and further experimental data can be tested to detect the occurrence of eventual structural changes.

For this purpose, a 3D finite element (FE) numerical model of the bridge is developed in ANSYS software [19] and fully validated with experimental data (Fig. 5). Among the modelled structural elements, those defined as beam finite elements consist of piers, sleepers, ballast-containing beams, rails, arches, hangers, transverse stiffeners, diaphragms, and diagonals. Shell elements were used to model the concrete slab and the steel box girder, while the pads, the ballast layer, and the foundations were modelled using linear spring-dashpot assemblies. The mass of the non-structural elements and the ballast layer was distributed along the concrete slab. Concentrated mass elements were used to reproduce the mass of the arches' diaphragms and the mass of the sleepers, which were simply positioned at their extremities. The connection between the concrete slab and the upper flanges of the steel box girder, as well as the connection between the deck and the track, were performed using rigid links. Special attention was paid to the bearings supports, as they can strongly influence the performance of the bridge. Hence, in order to simulate the sliding behaviour of the bearings, nonlinear contact elements were applied. Moreover, constraint elements located between the bearings were used to restrict the transversal movement in each pier, and the longitudinal and transversal movements in the case of the first pier.

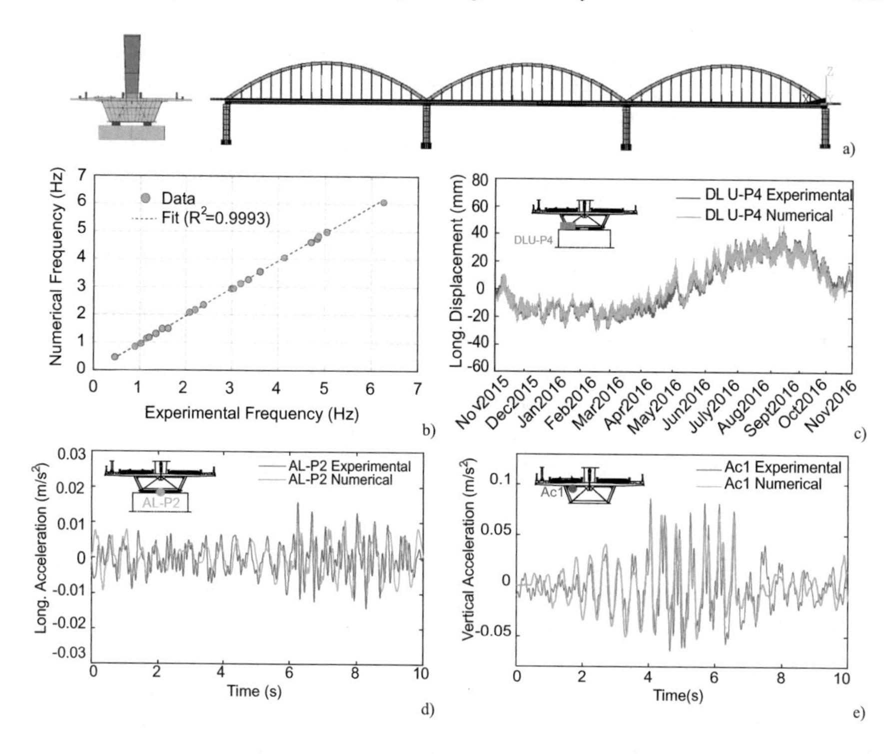

Fig. 5 Numerical modelling and validation: **a** 3D FE numerical model of the bridge over the Sado River, **b** agreement between numerical and experimental modal frequencies, **c** static validation of the displacements measured on pier P4, **d** dynamic validation of longitudinal accelerations at pier P2 with the AP at 216 km/h, and **e** dynamic validation of vertical accelerations at the concrete slab (Ac1) with the AP at 216 km/h

To ensure that the numerical model accurately simulates the structural behaviour of the bridge, the responses obtained from modal, static, and dynamic analyses were compared with those measured by the SHM system [20]. The numerical natural frequencies were compared with those obtained experimentally during an ambient vibration test [21]. Figure 5b shows a very high coefficient of determination ($R^2 = 0.9993$) between the numerical and the experimental results. To validate the static behaviour of the numerical model, the response of the structure to the action of temperature was studied. The structural static behaviour of the bridge was simulated in the FE model by running a time-history analysis using experimental data as input. The simulation procedure consisted of using the temperatures acquired every hour on site over the course of one year. Figure 5c presents a very good agreement between the numerical and experimental displacements of pier P4 for the temperature measured on site between November 2015 and November 2016. Regarding the dynamic behaviour, numerical simulations were conducted considering the Portuguese Alfa Pendular (AP) train as a set of moving loads crossing the bridge over the Sado River at a speed of 216 km/h. Figure 5d, e shows a very good agreement between the experimental and numerical responses, in terms of the longitudinal accelerations measured on pier P2, and the vertical accelerations acquired on the concrete slab at the second mid-span (Ac1), respectively. Before the comparison, the time-series were filtered based on a low-pass digital filter with a cut-off frequency equal to 15 Hz. A detailed description of the numerical model and its validation com be found in Meixedo et al. [20].

The dynamic numerical simulations implemented in the present research work aimed at replicating the structural quantities measured in the exact locations of the 23 accelerometers installed on site (Fig. 4) during the passage of a train in the bridge. To correctly reproduce these structural responses, the temperature action measured precisely during the passage of each train was introduced as input in the numerical model. The measurements of the optical sensors' setup were used to obtain the train speed and axle configuration, as well as the type of train [22]. The dynamic analyses mentioned hereafter were carried out for two of the passenger trains that typically cross the bridge over the Sado River, namely the AP train and the Intercity (IC) train. Their frequent speeds on the bridge are 220 km/h for the AP train and 190 km/h for the IC train. The nonlinear problem was solved based on the full Newton–Raphson method, and the dynamic analyses were performed by the Newmark direct integration method, using a methodology of moving loads [9]. The integration time step (Δt) used in the analyses was 0.005 s.

Figure 6a summarizes the 100 simulations of the baseline (undamaged) condition that aim at reproducing the responses of the bridge taking into account the variability of temperature, speed, loading schemes (LS), and type of train. These baseline scenarios compose the training dictionary and do not include any damage on any location. During each simulation, real temperatures measured by the SHM system were introduced in the elements of the bridge. The average values for each season were 21 °C for spring, 30 °C for summer, 16 °C for autumn, and 10 °C for winter, but the dispersion across the structure was considered by measuring and using temperature values in all elements of the bridge. The simulations included the AP and IC trains crossing the bridge with ten different loading schemes, according to the experimental observations previously made by Pimentel et al. [23]. Three train speeds were considered for each type of train, as observed in Fig. 6, thus resulting in 100 time-history simulations for the baseline condition, each taking approximately 10 h on a 4.2 GHz Quad-Core desktop with 32.0 GB of RAM.

On the other hand, the damage scenarios were chosen based on possible vulnerabilities identified for the type of structural system, taken into account its materials, behaviour, loadings, and connections. As shown in Fig. 6b, damage scenarios were simulated according to different groups: (i) damage in the bearing devices (type D1), (ii) damage in the concrete slab (type D2), (iii) damage in the diaphragms (type D3), and (iv) damage in the arches (type D4). Each scenario was simulated considering only one damage location. Nevertheless, if, by any chance, two or more damage scenarios in different locations are observed at the same time, the effects from multiple damage locations are expected to superimpose, and the influence on the features extracted from the data will be greater. Therefore, the multiple damage scenario will be more observable than the scenarios tested here. Regarding the group

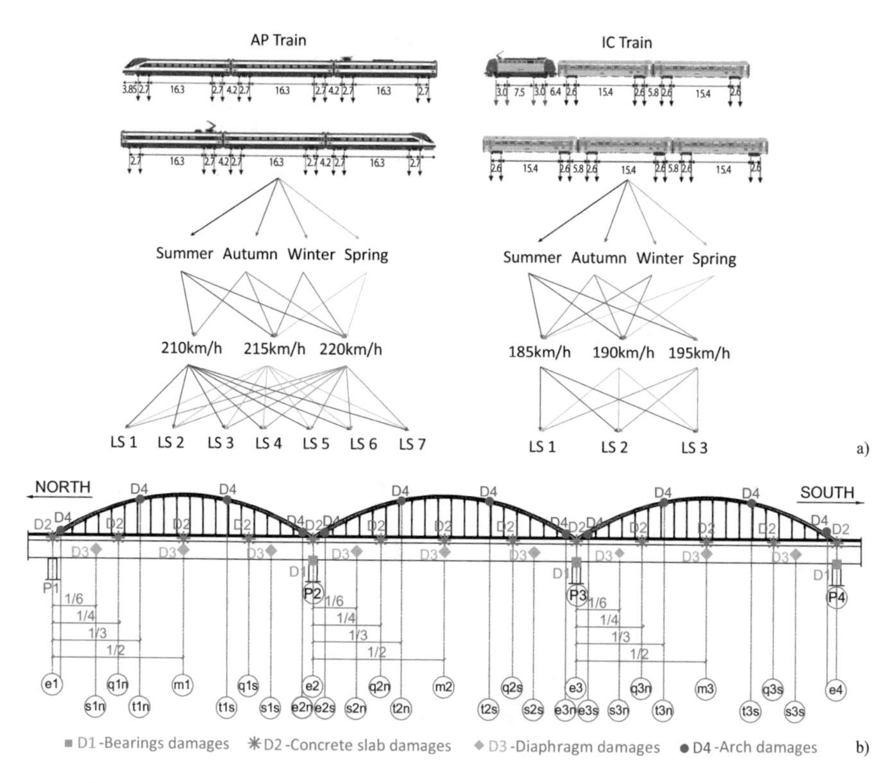

Fig. 6 Baseline and damage scenarios: **a** combination of 100 simulations for the baseline condition, **b** types of damages and their location on the bridge over the Sado River

of type D1, four severities of damage were included, namely increases of the friction coefficient from a reference value of 1.5% to 1.8%, 2.4%, 3.0%, as well as to a full restrain of the movements between the pier and the deck. The remaining damage scenarios consisted of 5%, 10%, and 20% stiffness reductions in the chosen sections of the bridge (Fig. 6b) on the concrete slab (D2), the diaphragms (D3), and arches (D4). These structural changes were simulated by reducing the modulus of elasticity of concrete (type D2) and of steel (types D3 and D4). A total of 114 damage scenarios were simulated for AP train crossings at 220 km/h and adding as input the temperatures measured on site during a summer day. Additional damage scenarios could have been simulated for different combinations of EOVs. However, as observed in Sect. 4.2, the proposed methodology is effective in removing these effects and keeping only those generated by structural changes.

The time-series illustrated in Fig. 7 are examples of simulated responses for baseline and damage conditions, acquired from the accelerometer Ac1.

To obtain the most reliable reproduction of the real SHM data, the noise measured on site by each accelerometer was added to the corresponding numerical output. These noise distributions were acquired while no trains were travelling over the bridge and under different ambient conditions. Each simulation was corrupted with different

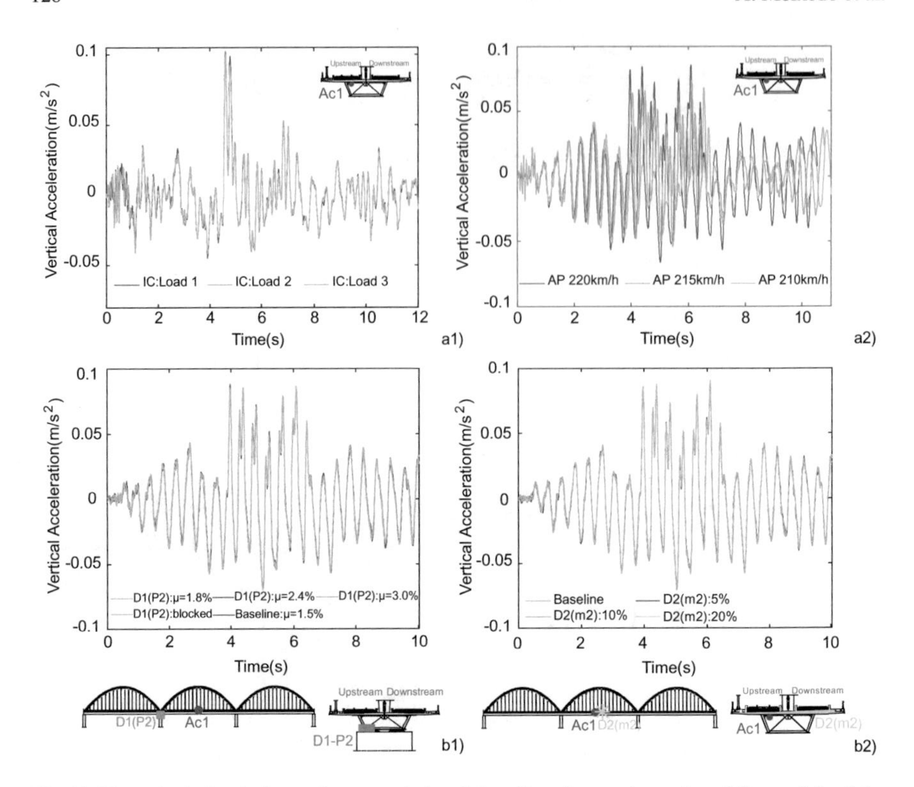

Fig. 7 Numerical simulations of sensor Ac1: **a1** baseline time-series using different LS of the IC train at 190 km/h, **a2** baseline time-series using different speeds of the AP train with LS5, **b1** damage time-series considering friction increase D1 (P2), and **b2** damage time-series considering stiffness reduction D2 (m2)

noise signals acquired at different days, thus ensuring the most representative validation for the techniques developed herein.

The variations associated with different train types, loading schemes and train speeds are shown in Fig. 7a1, a2. A clear distinction between the bridge responses for the IC (Fig. 7a1) train and the AP train (Fig. 7a2) passages can be observed, thus displaying the necessity of considering different train types for implementing damage detection strategies. Contrariwise, Fig. 7a1 allows observing that different LS generate smaller changes in the dynamic responses. The train speed also has an important influence in the structural response induced by trains crossing the bridge, as shown in Fig. 7a2.

The influence of damage scenarios in the signal obtained for the train crossings appears to be much smaller than that observed for EOVs, even when regarding sensors adjacent to the damages and for the biggest magnitudes considered (20% stiffness reductions). This conclusion can be easily observed in Fig. 7b1, b2, where the bridge responses considering friction increments in the bearing devices of pier P2 and stiffness reductions in the concrete slab are, respectively, presented.

4 Strategy for Early Damage Detection Using Train-Induced Responses

4.1 Features Extraction Based on CWT and PCA

The first step of the strategy is to extract damage-sensitive features based on the hybrid combination of Continuous Wavelet Transform (CWT) and principal component analysis (PCA). After extracting the wavelets coefficients by applying a CWT to the acceleration measurements, a PCA is performed to significantly compress information.

Wavelet functions are composed of a family of basic functions that can describe a signal in localized time (or space) and frequency (or scale) domain. The main advantage achieved by using wavelets is the ability to perform local analysis of a signal, i.e. zooming on any interval of time or space. Signal-based damage detection techniques that involve wavelet analysis take advantage of this to be capable of revealing some hidden aspects of measured signals [24].

The CWT is a well-established method of implementing multiscale signal analysis. This technique will only be introduced here, and the reader is referred to Cohen and Ryan [25], amongst others, for a more detailed mathematical explanation. The CWT decomposes the analysed signal into a set of coefficients in two dimensions, shift and scale, where scale is approximately inversely proportional to frequency. A basis function is translated (shift) and stretched (scale), and compared against the signal. High coefficients represent a good match between signal and wavelet at a particular instant in time and related frequency [26]. Hereupon, the CWT provides variable resolution and delivers a map of the energy content of the signal in time and frequency.

Let $f(t)$ be the acceleration response of the system, where t denotes time. The wavelet coefficients are described as the inner product of the function f and the wavelet $\psi_{a,b}$ corresponding to parameters a (scale) and b (shift) [26]:

$$W_{\psi}f(a,b) = \langle f, \psi_{a,b} \rangle = \frac{1}{\sqrt{a}} \int_{-\infty}^{+\infty} f(t)\psi^{*}\left(\frac{t-b}{a}\right)\mathrm{d}t \text{ for } a > 0 \quad (1)$$

In the previous equation, $\psi(t)$ is the mother wavelet, in which the superscript asterisk indicates complex conjugation. The wavelet functions $\psi_{a,b}(t)$ are constructed by a translated and dilated version of the mother wavelet, using the two parameters a and b (Eq. 2). The parameter b localizes the basis function at $t = b$ and its neighbourhood by windowing over a certain temporal stretch depending on the parameter, a.

$$\psi_{a,b}(t) = \frac{1}{\sqrt{a}}\psi\left(\frac{t-b}{a}\right) \text{ for } a, ba, b \in R^{+} \quad (2)$$

The frequency content can be controlled by varying the parameter a, as shown by the Fourier transform of the wavelet function (Eq. 3). Therefore, the wavelet transform coefficient for any particular a and b characterizes the contribution to the function $f(t)$ in the neighbourhood of $t = b$ and in the frequency band corresponding to the dilation factor of a.

$$\hat{\psi}_{a,b}(\omega) = \sqrt{a}\hat{\psi}(a\omega)\mathrm{e}^{i\omega b} \tag{3}$$

Usually, the wavelet coefficients are shown in terms of scale a. However, it is common in engineering practice to work in the frequency domain. Since there is not a direct relationship between scale and frequency, the results illustrated herein are given in terms of pseudo-frequency. A pseudo-frequency f that corresponds to the scale a can be defined by Eq. (4) where Δt is the sampling period of the analysed signal and f_c is the centre frequency given by the maximizer of $\left|\hat{\psi}\right|$ [26, 27].

$$f = \frac{f_c}{a\Delta t} \tag{4}$$

Considering a vector of 2112 acceleration measurement points, the present analysis used the Morlet mother wavelet [28] to extract matrices of *2112-by-82* features (wavelets coefficients) for each of the 23 sensors and for each of the 214 structural conditions. To illustrate the feature extraction procedure, Fig. 8 shows these matrices, plotted as images with scaled colours, for two different sensors: (1) AL-P2—longitudinal accelerometer located in pier P2, (2) Ac1—vertical accelerometer in the mid-span section of the concrete slab; and for four structural conditions: (a) undamaged scenario AP|220 km/h|AUT—the AP train crossing the bridge at 220 km/h during an autumn day, (b) undamaged scenario IC|190 km/h|SPR—the IC train crossing the bridge at 190 km/h in a spring day, (c) damaged scenario D1 (P2: restrained)—the full restraint of the bearing devices in pier P2, and (d) damaged scenario D2 (e2: 20%)—the 20% stiffness reduction in a section of the concrete slab aligned with pier P2.

In Fig. 8 is possible to clearly observe different energy concentrations depending on the sensor. Sensor AL-P2 seems to be more sensitive to the different structural conditions analysed since the CWT coefficients from this sensor (Fig. 8a1, b1, c1, d1) provide a much clearer image of the evolution of the energy content from the undamaged scenarios to the damaged scenarios. On the contrary, the energy concentration on the images from sensor Ac1 seems to be very similar between the undamaged scenario AP|220 km/h|AUT (Fig. 8a2) and both damage scenarios D1 and D2 (Fig. 8c2, d2). The main variation in frequency values is observed for the undamaged scenario IC|190 km/h|SPR (Fig. 8b2), which indicates that the influence of varying environmental and operational conditions (such as different temperature, train, and speed) may be greater than the damage occurrence. These results also show that different sensors can store different information about the bridge structural condition, thus combining this information may enhance the features sensitiveness.

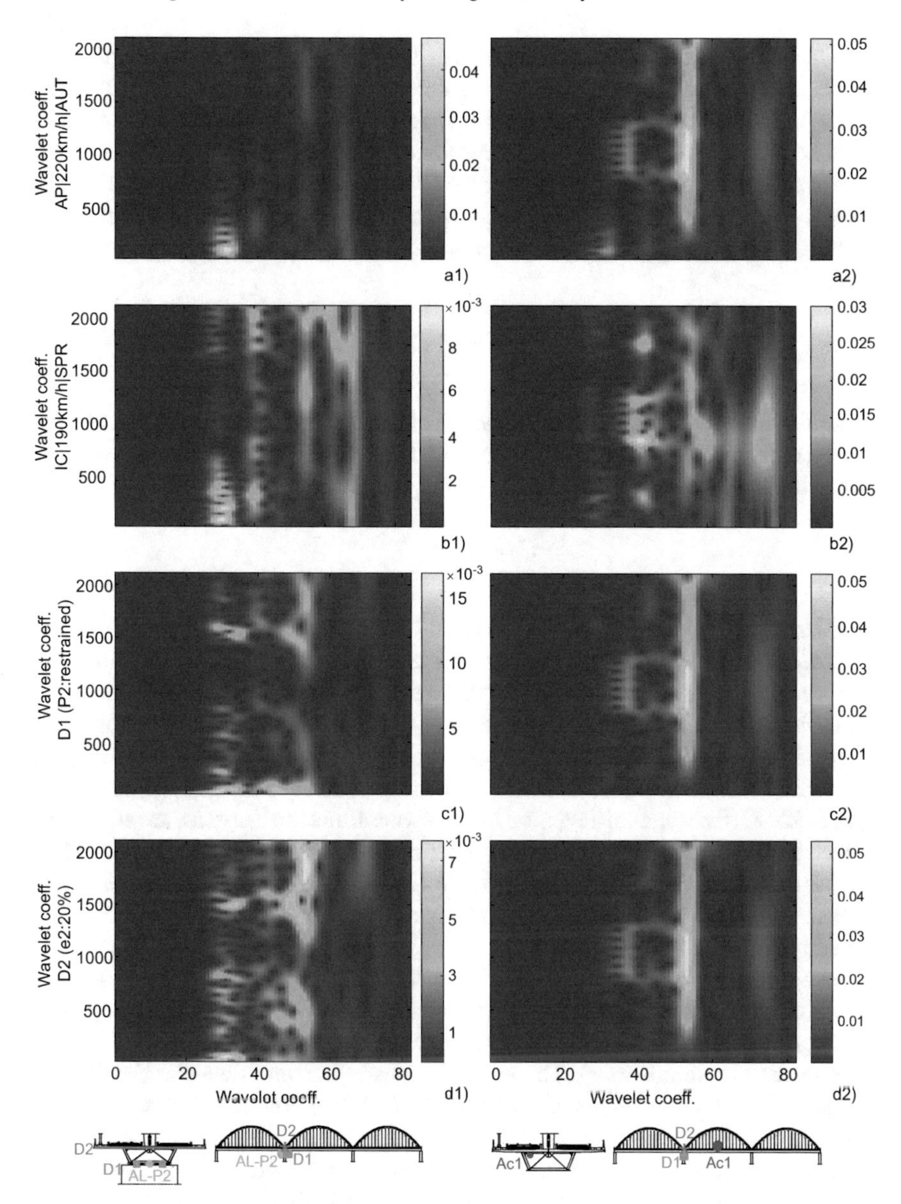

Fig. 8 Wavelet coefficients extracted from the acceleration responses according to the structural condition of the bridge and the sensor location

PCA is a multivariate statistical method that produces a set of linearly uncorrelated vectors called principal components, from a multivariate set of vector data [29]. This operation intends to extract damage-sensitive features from the CWT coefficients while performing important data compression. Considering an *n-by-m* matrix X with the wavelets coefficients obtained after the CWT implementation, being in this case study $n = 2112$ and $m = 82$, a transformation to another set of m coefficients, Y, designated principal components or scores, can be achieved by the following equation:

$$Y = X \cdot T \quad (5)$$

where T is an *m-by-m* orthonormal linear transformation matrix that applies a rotation to the original coordinate system. The covariance matrix of the coefficients, C, is related to the covariance matrix of the scores, Λ, as follows:

$$C = T \cdot \Lambda \cdot T^T \quad (6)$$

in which T and Λ are matrixes obtained by the singular value decomposition of the covariance matrix C. The columns of T are the eigenvectors, and the diagonal matrix Λ comprises the eigenvalues of the matrix C in descending order. Hence, the eigenvalues stored in Λ are the variances of the components of Y and express the relative importance of each principal component in the entire dataset variation [30].

To allow data compression, four statistical parameters, namely the root mean square (RMS), the standard deviation, the Skewness, and the Kurtosis, are afterwards extracted from the scores, Y. Thereby, the information presented in a matrix of *2112-by-82* is transformed into a matrix of *4-by-82*. A total of 328 features are thus extracted from each of the 23 accelerometers. This operation is implemented for each of the 214 structural conditions.

Four of the 328 features (12, 47, 116, and 323) obtained for two of the 23 sensors, AL-P2 and Ac1, are represented in Fig. 9. These features are divided according to the structural condition in two main groups: baseline (first 100 simulations) and damage (subsequent 114 simulations). The main changes in the amplitudes of the features are induced by the type and speed of the trains. In addition, for each speed value, the changes observed in the amplitude of the statistical parameters are generated by changes in the structural temperature values (chosen for autumn, spring, summer, or winter). The different LS (the seven symbols in a row in the case of the AP and three symbols in a row in the case of the IC) considered for each train type and speed, and each temperature, are the operational factors with the smallest influence on the feature variability regarding the baseline simulations. The analysis of the features shown in Fig. 9 that allows drawing some conclusions about the difficulty in distinguishing undamaged and damage scenarios since the variations caused by environmental and operational effects result in similar or greater changes in the parameters.

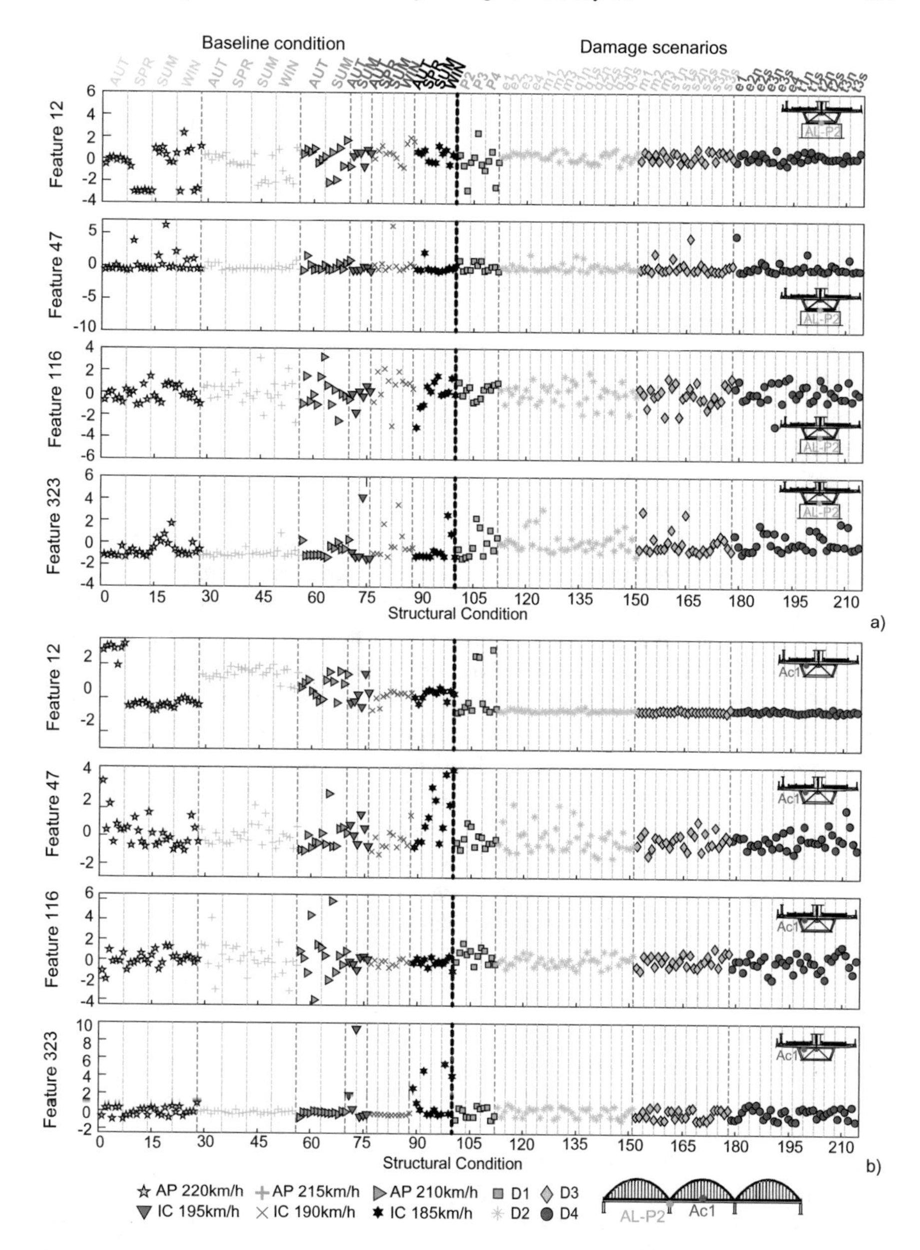

Fig. 9 Amplitude of four of the 328 statistical features extracted based on CWT and PCA, for all 214 structural conditions and from two of the 23 sensors: **a** longitudinal accelerometer located on pier P2 (AL-P2), **b** vertical accelerometer located in the mid-span section of the concrete slab (Ac1)

4.2 Features Modelling Based on PCA

The analysis of the features is presented in Fig. 9 that revealed the necessity to adequately model these statistical parameters to remove the changes generated by EOVs and highlight those generated by damage. Assuming that environmental conditions has a linear effect on the identified features, the implementation of a latent variable method as PCA to the extracted features may efficiently remove environmental and operational effects, without need to measure these actions [11, 30].

Considering now an *n-by-m* matrix X with the features extracted from the dynamic responses, where n is the number of structural conditions and m is the number of features from all the sensors (i.e. 328), a transformation to another set of m parameters, Y, can be achieved by applying Eq. (5).

As demonstrated by Santos et al. [10], the PCA is able to cluster meaningful information related to EOVs in the first components, while variations related to other small-magnitude effects, such as early damage, may be retained in latter components. Since the purpose of the present research work is to detect damage, which has generally a local character, the feature modelling operation consists of eliminating the most important principal components (PCs) from the features and retaining the rest for subsequent statistical analysis. Bearing this in mind, the matrix Λ from Eq. (6) can be divided into a matrix with the first e eigenvalues and a matrix with the remaining m-e eigenvalues. Defining the number of e components remains an open question with regard to the representation of the multivariate data; although several approaches have been proposed, there is still no definitive answer. In this work, the value of e (or the number of PCs to discard) is determined based on a rule of thumb in which the cumulative percentage of the variance reaches 80% [31]. After choosing e, the m-e components of the matrix Y can be calculated using Eq. (5) and a transformation matrix $\hat{T}$ built with the remaining m-e columns of T. Those m-e components can be remapped to the original space using the following:

$$F_{\text{PCA}} = X \cdot \hat{T} \cdot \hat{T}^T \tag{7}$$

where F_{PCA} is the *n-by-m* matrix of CWT-double PCA-based features, expected to be less sensitive to environmental and operational actions and to be more sensitive to the damage scenarios. This procedure is repeated for each sensor.

Since the cumulative percentage of the variance of the sum of the first nine principal components was higher than 80% for different structural conditions, these nine PCs were discarded during the modelling process (i.e. $e = 9$). Figure 10 shows the series of four features (12, 47, 116, and 323) across the 214 scenarios obtained for AL-P1 and Ac1 accelerometers, after the application of the double PCA. The direct comparison of these action-free damage-sensitive features with those shown before the feature modelling (Fig. 9) allows observing that, in fact, the feature modelling enabled removing the variations generated by the temperature, as well as by the type and speed of the train, but not those generated by damage. Moreover, the features sensitivity to the damage scenarios was increased.

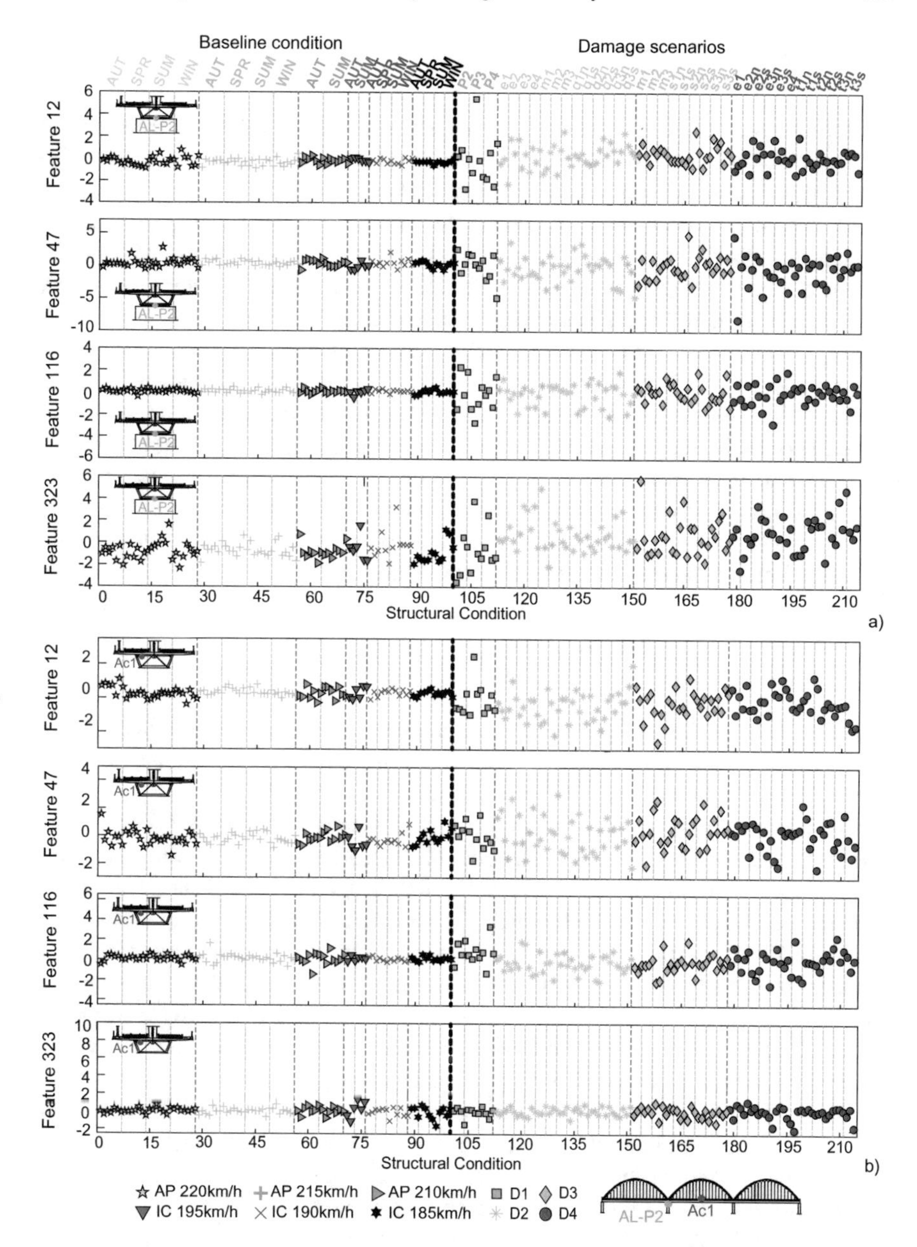

Fig. 10 Amplitude of four of the 328 features modelled based on PCA, for all 214 structural conditions and from two of the 23 sensors: **a** longitudinal accelerometer located on pier P2 (AL-P2), **b** vertical accelerometer located in the mid-span section of the concrete slab (Ac1)

4.3 Features Fusion

To improve the features' discrimination sensitivity, data fusion was performed. A Mahalanobis distance was implemented to the modelled features in order to describe their variability and allowing their effective fusion.

The Mahalanobis distance measures the distance between the baseline features and the damage-sensitive features to express the similarities between them, with shorter distances representing greater similarities. The Mahalanobis distance is generic enough to be used to detect any damage scenario, while providing a weighting that is entirely unsupervised, and therefore independent of human intervention, the type of structure, and the actions imposed on it. It consists of a weighted damage indicator in which the weights are determined by the covariance structure. In addition, and more importantly, the weighting proportional to the covariance structure provides an additional layer of feature modelling which, when defined for regular actions, allows outlining with high sensitivity those that were not used for the definition of the covariance structure. The analytical expression of the Mahalanobis distance for each simulation i, denoted as MD_i, is the following:

$$\mathrm{MD}_i = \sqrt{(x_i - \overline{x}) \cdot S_x^{-1} \cdot (x_i - \overline{x})^T} \tag{8}$$

where x_i is a vector of m features representing the potential damage/outlier, $\overline{x}$ is the matrix of the means of the features estimated in the baseline simulations, and S_x is the covariance matrix of the baseline simulations.

The Mahalanobis distance is computed for each simulation and each sensor resulting in a matrix with n Mahalanobis distances for k sensors, where n is the total number of structural conditions. When data from a structural state that differs from the baseline is tested, the MD value is expected to increase substantially.

Hence, the Mahalanobis distance allowed transforming, for each sensor and train crossing, the 328 features into one single feature (a distance in the feature space), which exhibits higher values for different structural conditions and null (or near-null) values for identical structural scenarios. The outcome of this procedure is a vector of *214-by-1*, of distances, one for each of the 23 sensors.

Figure 11 shows the results achieved for two of the 23 sensors (AL-P2 and Ac1). The two plots in Fig. 11 clearly show the difference in sensitivity for different sensors in each structural condition. The longitudinal accelerometer on pier P2 (Fig. 11a) is more sensitive to damages on the bearing devices of piers P3 and P4. Conversely, the accelerometer located at the second mid-span of the concrete slab (Fig. 11b) exhibits an important global sensitivity to damage since there is a distinction between the baseline simulations and the damage scenarios, but it is not efficient in distinguish the different types of simulated damages.

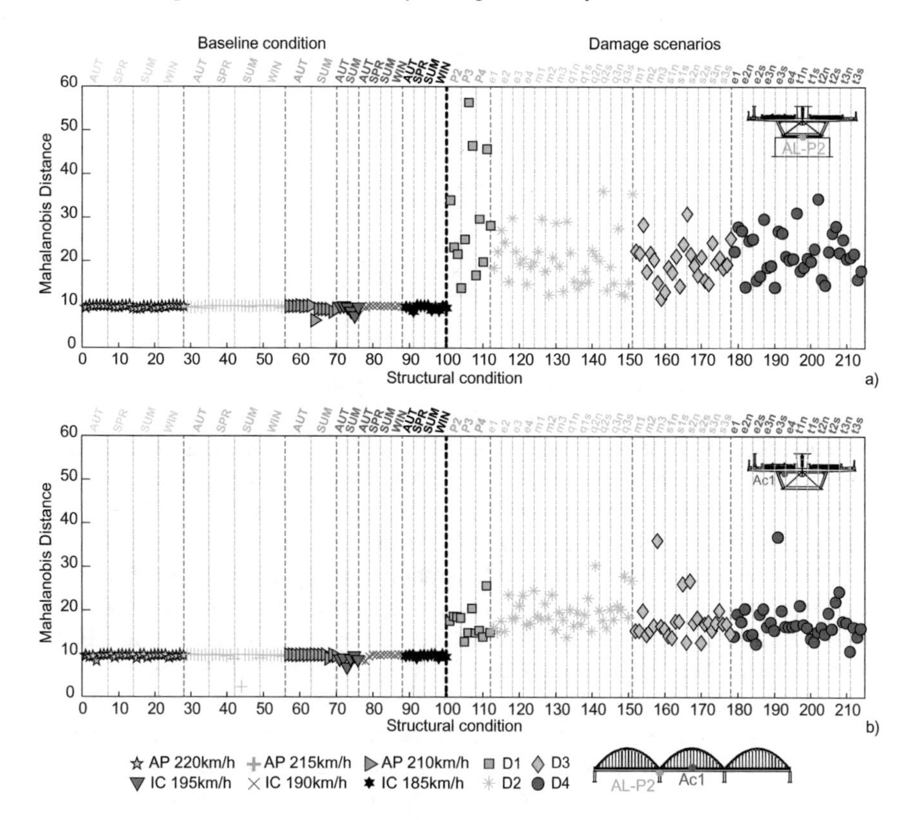

Fig. 11 Features fusion based on the Mahalanobis distance for all 214 structural conditions, considering the responses from accelerometers: **a** AL-P2, **b** Ac1

4.4 Clustering-Based Classification

Time-series analysis and distance measures can help perform data analysis and suggest the existence of different structural behaviours within a dataset, as shown in the previous sections. However, the development of real-time SHM strategies should resort to machine learning algorithms that can autonomously decide whether one or more distinct structural behaviours are being observed from patterns in the features. Hence, feature discrimination is addressed herein using unsupervised classification algorithms.

Cluster analysis was the data mining technique chosen to address feature classification. The aim of the clustering process is to divide a dataset into groups, which must be as compact and separate as possible. This can be mathematically posed as an attempt to minimize the dissimilarity between features assigned to the same cluster (within-cluster distance), which, consequently, maximizes the dissimilarity between the features assigned to different clusters (between-cluster distance) [15]. Considering a given partition containing K clusters, $P_k = \{C_1, \ldots, C_k\}$, the overall within-cluster dissimilarity $W(P_k)$ and the overall dissimilarity OD can be defined as:

$$W(P_k) = \frac{1}{2} \sum_{k=1}^{K} \sum_{c(i)=k} \sum_{c(j)=k} d_{ij} \tag{9}$$

$$\mathrm{OD} = \frac{1}{2} \sum_{i=1}^{N} \sum_{j=1}^{N} d_{ij} \tag{10}$$

in which the between cluster dissimilarity is given by the subtraction $B(P_k) = \mathrm{OD} - W(P_k)$. Here, N is the total number of features and $c(i)$ is a many-to-one allocation rule that assigns feature i to cluster k, based on a dissimilarity measure d_{ij} defined between each pair of features i and j. The best-known clustering algorithm is iterative and called k-means [32]. The k-means requires that the number of $K < N$ clusters be initially defined along with a randomly defined set of K clusters' prototypes. This task is called initialization. Afterwards, each iteration starts by allocating the features to the clusters according to an allocation rule, $c(i)$, that assigns each feature to the least dissimilar (closest) cluster prototype. The second step of each k-means' iteration is called representation and consists of defining the centroids of the K clusters as their prototypes and assuming that each feature belongs to the cluster whose prototype is closest. These two steps, allocation and representation, are subsequently repeated until an objective function, which depends on the compactness and separation of the cluster, reaches its global minimum value. The k-means considers the squared within-cluster dissimilarity measured across the K clusters as an objective function [32]. Clusters' dissimilarities are generally defined as distance metrics. Among these, the Euclidean (square root of the sum-of-squares) is used here.

As previously mentioned, the k-means clustering method requires that the number of clusters is defined in advance and provided as input (in the initialization phase). For damage detection, there is no way of knowing this number in advance, which requires that multiple partitions, comprising different numbers of clusters, be tested and their outcomes analysed using cluster validity indices [32]. Numerous validity indices have been proposed and tested, not only in specific literature but also in SHM applications. Herein, the global silhouette index (SIL) is used since it revealed a superior performance in previous studies [33], in which its formulation is carefully described.

The application of the k-means along with the SIL index is exemplified here using the features extracted from the sample time-series. For the present work, it is important to note that, among the K tested, the partition that generates the highest SIL value is the one that is expected to best describe the analysed feature set and should, therefore, be considered for SHM purposes. Using the CWT-double PCA-based features after fusion from all sensors installed on site, the SIL indices extracted from five cluster partitions, shown with 'o' marks in Fig. 12a, exhibit a maximum for $k = 2$ clusters. The corresponding features' allocations were automatically generated

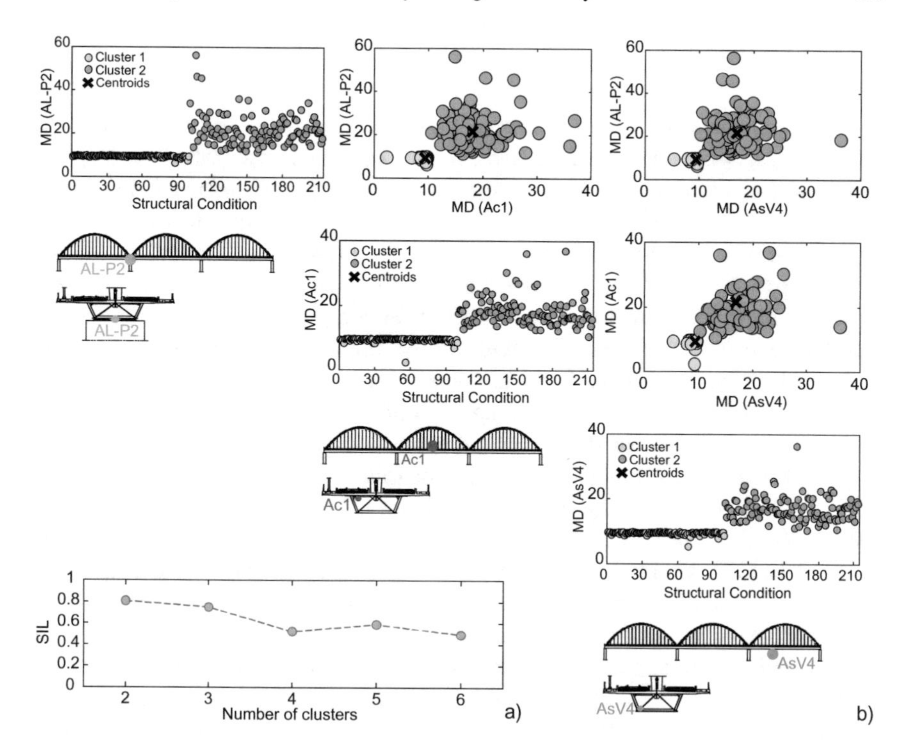

Fig. 12 Allocation of damage-sensitive features into clusters: **a** silhouette index (SIL), **b** clusters defined for all structural conditions and their centroids for three of the 23 sensors

by the *k*-means method and are shown in Fig. 12b for three of the 23 sensors: (i) AL-P2, (ii) Ac1, and (iii) AsV4. These plots demonstrate that the clustering method can divide the features without any human interaction or input. In Fig. 12b can be observed the dissimilarity between the two centroids of three different combinations of sensors, while the plots in the diagonal of this figure show that the two clusters found for each couple of sensors are compact over time and separated when the simulated damages start. This result undeniably shows that the *k*-means method is capable of analysing the feature set and, in a fully automated manner, separating it according to the structural conditions observed on site. Also, it is demonstrated that the clusters have de advantage of allowing a multidimensional representation of the features, which, in this case study, led to a classification without false detections.

After the definition of the baseline, which can be promptly achieved after one day of trains crossings, each new train crossing can be used to test the bridge structural condition based on the proposed strategy.

5 Conclusions

This research presents a data-driven AI-based SHM strategy for conducting real-time unsupervised early damage detection in railway bridge vibration response from traffic-induced excitation, applying time-series analysis and machine learning techniques. The strategy consists of fusing sets of acceleration measurements to improve sensitivity and combines: (i) CWT and PCA for feature extraction, (ii) PCA for feature modelling, (iii) Mahalanobis distance for feature fusion, and (iv) clustering algorithms for feature classification. A comprehensive dataset of baseline and damaged scenarios was simulated using a highly reliable digital twin of the Sado Bridge tuned with experimentally obtained actions as input, namely temperature, train loadings, and speeds. Damage severities of 5%, 10%, and 20% stiffness reductions in the concrete slab, diaphragm, and arches were simulated, as well as friction increases in the movements of the bearing. The damage-sensitive features were extracted from the bridge accelerations induced by train crossings by combining CWT with PCA. The wavelets coefficients were first extracted from the time-series. Afterwards, PCA was implemented to the wavelets coefficients and statistical parameters were extracted from the PCs to allow data compression. The study of the wavelet-PCA-based features extracted from different structural conditions allowed drawing conclusions about the supremacy of the EOVs when compared with damage, proving the importance of feature modelling. Moreover, the information obtained from each feature was different depending on the sensor location and the statistical parameter. PCA was once again implemented to modelling the features. This latent variable method proved its importance and effectiveness in removing observable changes induced by variations in train speed or temperature without the need to measure them and without losing sensitivity to damage. To describe the variability present in the modelled features, a Mahalanobis distance was implemented to the 328 features extracted from each sensor signal. This implementation allowed corroborated that different sensors have greater or lesser sensitivity, depending on the location of the damage. Moreover, this step proved to be crucial to achieve the highest possible level of information fusion and to obtain a clear distinction between undamaged and damaged conditions. In order to automatically detect the presence of damage, a clustering-based classification was performed. The robustness and effectiveness of the proposed strategy were demonstrated by automatically detecting the damage scenarios as different from those belonging to undamaged structural conditions. Using features modelled based only on structural responses, no false detections occurred. An additional important conclusion obtained from this work is that, even with an SHM system not capable of measuring EOVs, it is possible to successfully detect different types of damage using the bridge's responses to train crossings. This achievement renders the strategy the ability to be less dependent on spatial actions very difficult to characterize, thus contributing for the normalization of SHM procedures. This strategy also has the advantages of minimizing the number of sensors that need to be installed and, consequently, the cost of the SHM system, as well as allowing for a more automatic and straightforward implementation.

Acknowledgements This work was financially supported by the Portuguese Foundation for Science and Technology (FCT) through the PhD scholarship SFRH/BD/93201/2013. The authors would like to acknowledge the support of the R&D project RISEN through the H2020|ES|MSC—H2020|Excellence Science|Marie Curie programme, the Portuguese Road and Railway Infrastructure Manager (I.P), the Portuguese National Laboratory for Civil Engineering (LNEC), and the Base Funding—UIDB/04708/2020 of the CONSTRUCT—Instituto de I&D em Estruturas e Construções—financed by national funds through the FCT/MCTES (PIDDAC).

References

1. Melo LRT, Ribeiro D, Calçada R, Bittencourt TN (2020) Validation of a vertical train–track–bridge dynamic interaction model based on limited experimental data. Struct Infrastruct Eng 16(1):181–201. https://doi.org/10.1080/15732479.2019.1605394
2. Meixedo A, Ribeiro D, Calçada R, Delgado R (2014) Global and local dynamic effects on a railway viaduct with precast deck. In: Proceedings of the second international conference on railway technology: research, development and maintenance. Civil-Comp Press, Stirlingshire. https://doi.org/10.4203/ccp.104.77
3. Rytter A (1993) Vibrational based inspection of civil engineering structures. Dept. of Building Technology and Structural Engineering, Aalborg University, Aalborg
4. Meixedo A, Alves V, Ribeiro D, Cury A, Calçada R (2016) Damage identification of a railway bridge based on genetic algorithms. In: Maintenance, monitoring, safety, risk and resilience of bridges and bridge networks—proceedings of the 8th international conference on bridge maintenance, safety and management, IABMAS 2016, Foz Do Iguaçu; Brazil
5. Cury A, Cremona C (2012) Assignment of structural behaviours in long-term monitoring: application to a strengthened railway bridge. Struct Health Monit 11(4):422–441. https://doi.org/10.1177/1475921711434858
6. Posenato D, Kripakaran P, Smith IFC (2010) Methodologies for model-free data interpretation of civil engineering structures. Comput Struct 88(7–8):467–482. https://doi.org/10.1016/j.compstruc.2010.01.001
7. Meixedo A, Santos J, Ribeiro D, Calçada R, Todd M (2021) Damage detection in railway bridges using traffic-induced dynamic responses. Eng Struct 238(112189). https://doi.org/10.1016/j.engstruct.2021.112189
8. Mujica LE, Gharibnezhad F, Rodellar J, Todd M (2020) Considering temperature effect on robust principal component analysis orthogonal distance as a damage detector. Struct Health Monit 19(3):781–795. https://doi.org/10.1177/1475921719861908
9. Cavadas F, Smith IFC, Figueiras J (2013) Damage detection using data driven methods applied to moving-load responses. Mech Syst Signal Process 39(1–2):409–425. https://doi.org/10.1016/j.ymssp.2013.02.019
10. Santos JP, Crémona C, Orcesi AD, Silveira P (2013) Multivariate statistical analysis for early damage detection. Eng Struct 56:273–285. https://doi.org/10.1016/j.engstruct.2013.05.022
11. Hu WH, Moutinho C, Caetano E, Magalhães F, Cunha Á (2012) Continuous dynamic monitoring of a lively footbridge for serviceability assessment and damage detection. Mech Syst Signal Process 33(November):38–55. https://doi.org/10.1016/j.ymssp.2012.05.012
12. Farrar CR, Worden K (2013) Structural health monitoring: a machine learning perspective. Wiley, New York, pp 1–45
13. De LOR, Omenzetter P (2010) Damage classification and estimation in experimental structures using time series analysis and pattern recognition. Mech Syst Signal Process 24(5):1556–1569. https://doi.org/10.1016/j.ymssp.2009.12.008
14. Gonzalez I, Karoumi R (2015) BWIM aided damage detection in bridges using machine learning. J Civ Struct Heal Monit 5(5):715–725. https://doi.org/10.1007/s13349-015-0137-4

15. Cardoso R, Cury A, Barbosa F (2019) Automated real-time damage detection strategy using raw dynamic measurements. Eng Struct 196(109364). https://doi.org/10.1016/j.engstruct.2019.109364
16. Azim R, Gül M (2019) Damage detection of steel girder railway bridges utilizing operational vibration response. Struct Control Health Monit 26(e2447):1–15. https://doi.org/10.1002/stc.2447
17. Nie Z, Lin J, Li J, Hao H, Ma H (2019) Bridge condition monitoring under moving loads using two sensor measurements. Struct Health Monit 19(3):917–937. https://doi.org/10.1177/1475921719868930
18. Farrar CR, Doebling SW, Nix DA (2001) Vibration–based structural damage identification. Philos Trans R Soc London A: Math Phys Eng Sci 359(1778):131–149. https://doi.org/10.1098/rsta.2000.0717
19. ANSYS. Academic Research. Release 17.1 2016
20. Meixedo A, Ribeiro D, Santos J, Calçada R, Todd M (2021) Progressive numerical model validation of a bowstring-arch railway bridge based on a structural health monitoring system. J Civ Struct Heal Monit 11(2):421–449. https://doi.org/10.1007/s13349-020-00461-w
21. Min X, Santos L (2011) Ensaios dinâmicos da ponte ferroviária sobre o rio sado na variante de alcácer. Lisboa [Portuguese]
22. Meixedo A, Gonçalves A, Calçada R, Gabriel J, Fonseca H, Martins R (2016) On-line monitoring system for tracks. In: exp.at 2015—3rd experiment international conference, Sao Miguel Island, Azores. https://doi.org/10.1109/EXPAT.2015.7463240
23. Pimentel R, Ribeiro D, Matos L, Mosleh A, Calçada R (2020) Bridge weigh-in-motion system for the identification of train loads using fiber-optic technology. Structures 2021(30):1056–1070. https://doi.org/10.1016/j.istruc.2021.01.070
24. Ren WX, Sun ZS (2008) Structural damage identification by using wavelet entropy. Eng Struct 30:2840–2849. https://doi.org/10.1016/j.engstruct.2008.03.013
25. Cohen A, Ryan RD (1995) Wavelets and multiscale signal processing. Chapman & Hall, Boundary Row, London
26. Cantero D, Ülker-kaustell M, Karoumi R (2016) Time–frequency analysis of railway bridge response in forced vibration. Mech Syst Signal Process 76–77:518–530
27. Ülker-kaustell M, Karoumi R (2012) Influence of non-linear stiffness and damping on the train-bridge resonance of a simply supported railway bridge. Eng Struct 41:350–355. https://doi.org/10.1016/j.engstruct.2012.03.060
28. Teolis A (1998) Computational signal processing with wavelets. Birkhauser
29. Ribeiro D, Leite J, Meixedo A, Pinto N, Calçada R, Todd M (2021) Statistical methodologies for removing the operational effects from the dynamic responses of a high-rise telecommunications tower. Struct Control Health Monit 28(4):e2700. https://doi.org/10.1002/stc.2700
30. Yan A, Kerschen G, De BP, Golinval J (2005) Structural damage diagnosis under varying environmental conditions—Part I: a linear analysis. Mech Syst Signal Process 19(4):847–864. https://doi.org/10.1016/j.ymssp.2004.12.002
31. Jolliffe IT (2002) Principal component analysis, 2nd edn. Springer, New York, pp 112–147
32. Hastie T, Tibshirani R, Friedman J (2011) The elements of statistical learning, data mining inference, and prediction, 2nd edn. Springer, Stanford, pp 460–462
33. Santos J, Crémona C, Calado L (2016) Real-time damage detection based on pattern recognition. Struct Concrete 17(3):338–354. https://doi.org/10.1002/suco.201500092

Fault Diagnosis in Structural Health Monitoring Systems Using Signal Processing and Machine Learning Techniques

Henrieke Fritz, José Joaquín Peralta Abadía, Dmitrii Legatiuk, Maria Steiner, Kosmas Dragos, and Kay Smarsly

Abstract Smart structures leverage intelligent structural health monitoring (SHM) systems, which comprise sensors and processing units deployed to transform sensor data into decisions. Faulty sensors may compromise the reliability of SHM systems, causing data corruption, data loss, and erroneous judgment of structural conditions. Fault diagnosis (FD) of SHM systems encompasses the detection, isolation, identification, and accommodation of sensor faults, aiming to ensure the reliability of SHM systems. Typically, FD is based on "analytical redundancy," utilizing correlated sensor data inherent to the SHM system. However, most analytical redundancy FD approaches neglect the fault identification step and are tailored to specific types of sensor data. In this chapter, an analytical redundancy FD approach for SHM systems is presented, coupling methods for processing any type of sensor data and two machine learning (ML) techniques, (i) an ML regression algorithm used for fault detection, fault isolation, and fault accommodation, and (ii) an ML classification algorithm used for fault identification. The FD approach is validated using an artificial neural network as ML regression algorithm and a convolutional neural network as ML classification algorithm. Validation is performed through a real-world

H. Fritz (✉) · M. Steiner
Computing in Civil Engineering, Bauhaus University Weimar, Weimar, Germany
e-mail: henrieke.fritz@uni-weimar.de

M. Steiner
e-mail: maria.steiner@uni-weimar.de

J. J. Peralta Abadía · K. Dragos · K. Smarsly
Institute of Digital and Autonomous Construction, Hamburg University of Technology, Hamburg, Germany
e-mail: joaquin.peralta@tuhh.de

K. Dragos
e-mail: kosmas.dragos@tuhh.de

K. Smarsly
e-mail: kay.smarsly@tuhh.de

D. Legatiuk
Applied Mathematics, Bauhaus University Weimar, Weimar, Germany
e-mail: dmitrii.legatiuk@uni-weimar.de

A. Cury et al. (eds.), *Structural Health Monitoring Based on Data Science Techniques*, Structural Integrity 21, https://doi.org/10.1007/978-3-030-81716-9_7

SHM system in operation at a railway bridge. The results demonstrate the suitability of the FD approach for ensuring reliable SHM systems.

Keywords Structural health monitoring (SHM) · Fault diagnosis (FD) · Machine learning (ML) · Artificial neural network (ANN) · Convolutional neural network (CNN) · Signal processing · Wavelet transform

1 Introduction

Structural health monitoring (SHM) is a nondestructive evaluation strategy that uses data obtained by sensors to assess the condition of structures over time. In recent advancements of SHM, smart SHM systems that automatically make decisions and take actions based on the structural response have become popular [1]. However, the reliability and accuracy of sensors in SHM systems may be compromised by sensor faults, caused by hardware malfunctions, battery exhaustion, or environmental impacts [2].

A sensor fault can be defined as a defect of a sensor, leading to an error [3] that may result in failure of an SHM system. Failures are detectable in sensor data and vary according to the type of sensor fault. In general, fault diagnosis (FD) includes the following tasks [4]:

1. *Fault detection*: Recognizing the adverse operation of the system,
2. *Fault isolation*: Specifying the exact location of the fault,
3. *Fault identification*: Determining the type (or nature) of the fault, and
4. *Fault accommodation*: Compensating for the effects of the fault.

Fault diagnosis in SHM systems has been a topic of ongoing research for more than 40 years [5]. FD approaches usually build upon either physical redundancy or analytical redundancy. Physical redundancy approaches base fault detection on the comparison between sensor data collected by SHM systems and readings of "redundant" sensors collocated with the sensors of SHM systems. In analytical redundancy approaches, virtual sensor data is calculated based on inherent correlations of structural response data obtained from different sensors on the structure [6]. For effectively mapping these correlations, the full length of sets of sensor data is utilized, only part of which is normally used for SHM-related objectives, the rest being characterized as "redundant." As a result, FD relies on residuals between virtual and actual sensor data, which is typically structural response data [7].

Theoretically, if the structural properties are known, physics-based models can be used for calculating virtual data. However, creating physics-based models for real structures is largely based on assumptions on structural properties, which are likely to introduce epistemic uncertainty to virtual data and compromise the FD approach. Therefore, data-driven models, based on machine learning (ML) techniques, have become popular for FD in recent years. Because of the ability to recognize and classify patterns in large data sets, ML techniques are of increasing interest in SHM-related research. Machine learning helps SHM systems to adapt to new circumstances (e.g., changes in environmental conditions) by processing and analyzing data, extrapolating patterns, and making predictions [8].

Several data-driven ML techniques have been applied to FD problems [9]. Artificial neural networks as ML regression algorithms, for example, have been widely used for fault detection, fault isolation, and fault accommodation in different engineering fields, e.g., in wind turbine structures [10], aircraft engines [11], fossil-fuel power plants [12], flight control systems [13], unmanned airborne vehicles [14], and robotic systems [15]. Feed forward neural networks also have been employed for fault accommodation [16]. Furthermore, a decentralized approach toward fault detection and fault isolation in wireless SHM systems using artificial neural networks for predicting virtual data in the time domain has been presented [7]. The aforementioned approach has been extended to the frequency domain, i.e., using correlations between the Fourier amplitude of peaks at resonant frequencies [3]. Various topologies of artificial neural networks for detecting process failures, including sensor faults, have been tested in [17]. Furthermore, a distributed recurrent artificial neural network (ANN) has been employed for sensor fault detection [18], while some researchers have applied (bidirectional) recurrent neural networks accounting for both the spatial and the temporal correlation among sensor data [19]. Apart from artificial neural network, other ML techniques have been applied for FD in SHM. For example, support vector regression has been used for fault detection and isolation [20], and support vector machines based on chaos particle swarm optimization have been proposed for FD in SHM systems [21]. Finally, fault identification also has been studied by applying support vector machines [22].

Most FD approaches focus on detecting, isolating, and accommodating sensor faults in SHM systems. However, identifying (i.e., classifying) fault types is essential in FD, as fault types, in general, are related to the origin of faults in SHM systems. For fast maintenance, fault types should be automatically identified by FD approaches, indicating possible reasons for malfunctions in SHM systems. Furthermore, most FD approaches proposed so far have been developed only for one type of sensor data, mostly accelerations, which significantly limits the applicability of FD approaches.

To provide a general FD approach, including sensor fault identification, this chapter presents a generalization and extension of the work proposed in [3], representing a novel approach toward autonomous FD based on analytical redundancy. For fault detection and fault isolation, a ML regression algorithm is used for predicting virtual sensor data of each sensor, based on correlations with data from neighboring sensors, which is used as input data to the ML regression algorithm. Then, the virtual data is compared to the actual sensor data, and if the deviation between the virtual and the actual sensor data is greater than a pre-defined threshold, a sensor alert is triggered, and fault identification and fault accommodation are initiated. Fault accommodation is performed using the virtual data of the faulty sensors calculated by the ML regression algorithm. For fault identification, patterns in sensor data representing sensor fault types are used, which in this approach are exposed by a wavelet transform. For autonomous fault identification, patterns of wavelet-transformed sensor data are used as input to a ML classification algorithm, classifying the sensor fault types. For validation, an ANN as ML regression algorithm and a convolutional neural network

(CNN) as an ML classification algorithm are applied. Validation of the approach is performed by showcasing the ability of the FD approach to work with real-world data recorded by an SHM system in operation on a railway bridge.

The remainder of this chapter is structured as follows: First, the background of fault identification through signal processing and ML techniques is illuminated. Subsequently, the methodology and implementation of the FD approach proposed are described, and validation tests are performed. Finally, the results of the FD approach are discussed, and an outlook on potential future work is provided.

2 Fault Identification Through Signal Processing and Machine Learning

In SHM systems, generally, seven basic sensor fault types occur, which are depicted in Fig. 1 together with formal descriptions as functions over time. Bias is a deviation by a constant value between sensor data and actual structural response. Drift is the incrementing deviation between sensor data and actual structural response over time. In the case of a gain, sensor data is scaled by a constant value; while in precision degradation, sensor data is contaminated with white noise. For a complete failure, sensor data consists of a constant value or noise regardless of changes in the actual structural response. Outliers are isolated, non-continuous changes ("dropouts") of the signal at individual points in time, with signals returning to their real values after each single dropout.

Table 1 presents an overview of studies available in literature analyzing reasons for the occurrence of individual sensor fault types in SHM systems focusing only on failures in the sensing mechanism and not on failures coming from synchronization faults. In addition to the causes in Table 1, it should be noted that the mechanism and recurrence frequency of individual sensor fault types are highly dependent on the physical design of the sensors as well as the environmental conditions.

Fault identification of sensors deployed in SHM systems is a crucial step toward FD, as fault types may allow inferring causes of sensor faults in SHM systems. For autonomous fault identification, ML provides promising classification algorithms. ML generally represents learning processes of computer systems, often described as the conversion of experience into expertise or knowledge [37]. ML algorithms may help analyze large amounts of data, recognize data patterns, and adapt to the patterns autonomously. Compared to traditional algorithms (such as expert systems), ML algorithms offer two advantages, (i) ML algorithms work with previously unknown (i.e., new) data for which the system has not been trained, and (ii) ML algorithms can adapt to environmental conditions and resulting changes in the data.

Theoretically, automatic identification of sensors faults can be performed by using ML classification algorithms with raw sensor data as input. However, because raw sensor data typically contains noise and random components that may introduce aleatory uncertainties, patterns in the data indicative of sensor fault types ("fault

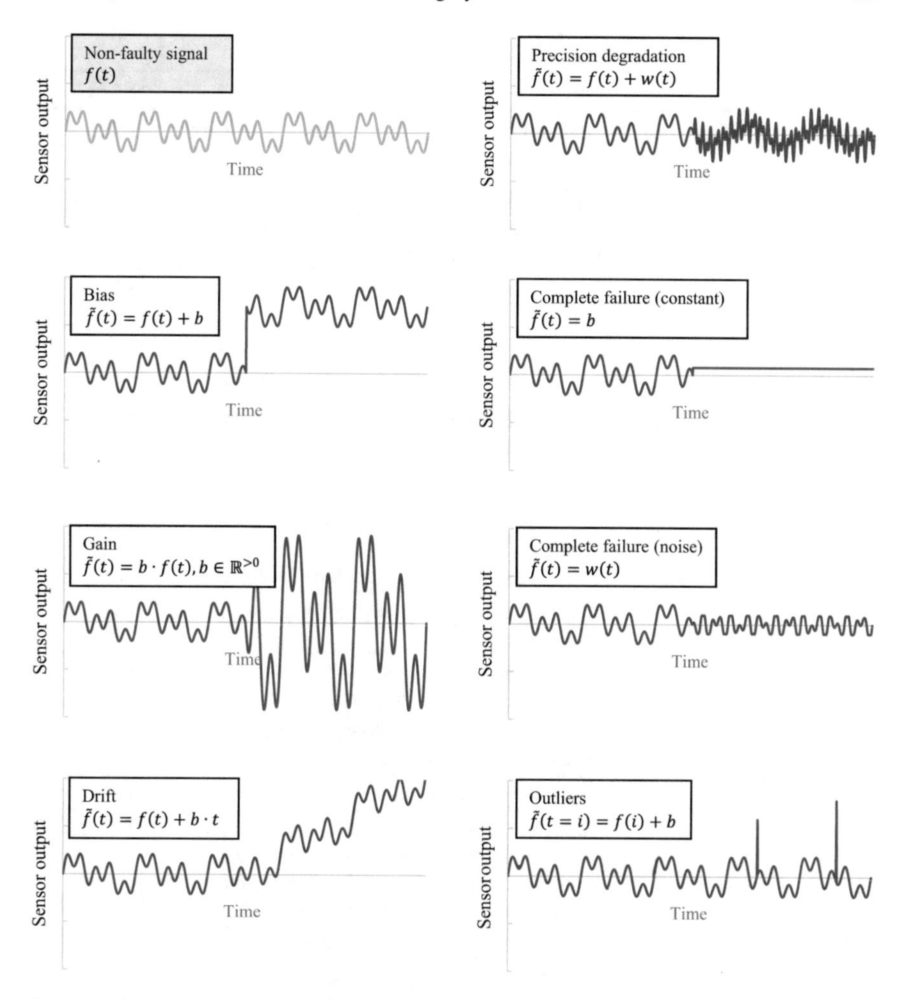

Fig. 1 Basic sensor fault types in SHM systems

patterns") may remain undetected when applying ML classification algorithms to raw sensor data. Therefore, most ML classification algorithms use pre-processing techniques on raw sensor data to enhance fault patterns or, if not visible before, to unveil fault patterns existing in the sensor data. This pre-processing of raw sensor data is typically done using techniques of signal processing.

The techniques employed for signal processing for analyzing one-dimensional time-dependent signals (representing sets of sensor data) can be categorized into two groups, (i) time-domain techniques and (ii) frequency-domain techniques. The choice of technique depends on the signals and varies from problem to problem. For example, a common approach for analyzing signals with a dynamic frequency spectrum is the wavelet transform. In contrast to classical Fourier transform, wavelet transform has a high resolution in both the time domain and the frequency domain

Table 1 Sensor fault types and causes

Fault type	Cause	Source
Bias	• Degradation, corrosion, or breakage of junction	[23]
	• Incorrect calibration	[23]
	• Physical changes in the sensor system (e.g., temperature variations that lead to changes in the mechanical properties of sensors)	[23, 24]
	• Partial loss of connectivity of the sensor to the structure	[25, 26]
	• A short-circuit (or degradation) in the lead wires	[27]
	• Changes in resistance due to surface stresses	[23]
Complete failure (constant)	• Loss of contact with the lead wires over time (e.g., due to fatigue or shock)	[23]
	• Battery failure	[28]
	• Sensitive core insulation resistance of sensor, drops, or damage	[26]
Complete failure (noise)	• Stray magnetic fields (particularly strong fields)	[23]
	• Weather conditions, external attacks, and unstable wireless connection	[28]
Drift	• Inhomogeneous changes in composition of the material (e.g., due to long exposures to high temperatures or fatigue of the material)	[27, 29, 30]
	• Aging of the sensor (node)	[31]
	• A short (or degradation) in the lead wires	[27]
	• Strain on resistive wire based on shock and vibration, chemical reactions of sensing materials, transmission issues (e.g., deformation), and weather changes (temperature, humidity)	[26, 28, 37]
Gain	• Debonding on the interface between sensor and surface	[23, 32]
	• Unstable voltage supply or nonlinearity of the sensor	[26]
	• Electrical and mechanical fatigue	[24]
	• Short (or degradation) in the lead wires	[27]
	• Cracking of the material of sensors due to fatigue or shock causes	[33]
	• Changes in the orientation of induced magnetic fields in sensors	[23]
Precision degradation	• Stray magnetic fields (particularly strong fields)	[23]
	• Electrical noise through the power supply system	
	• Seismic radiates from an outside source (e.g., airplane)	

(continued)

Table 1 (continued)

Fault type	Cause	Source
Outliers	• Mechanisms such as heavy-tailed distributions or data that are coming from different kinds of distributions	[34]
	• Environmental variations (e.g., cold temperature influence)	[35]
	• Low battery supply, loose electrical contact, and sensor saturations	[36]

[38]. The wavelet transform uses a series of functions, the so-called wavelets, each with a different scale. Since wavelets are localized in time, the wavelet transform can extract coupled time–frequency information from a signal.

Definition 1 Let $u, \psi \in L^2(\mathbb{R}, \mathbb{R})$. For $b \in \mathbb{R}$ and $a > 0$, the wavelet transform of u with ψ is defined by

$$L_\psi u(a, b) := \int_{\mathbb{R}} u(t) \frac{1}{\sqrt{a}} \psi\left(\frac{t-b}{a}\right) \mathrm{d}t, \tag{1}$$

where a is a scale parameter, b is a spatial parameter, and ψ is the mother wavelet. Depending on the choice of ψ, wavelet transforms with different properties can be constructed. Figure 2 shows wavelet transforms of an artificial signal of the form

$$f(t) = \sin \pi t + \sin 2\pi t + \sin 5\pi t, \tag{2}$$

representing a multicomponent periodic signal. The output of the wavelet transform of function (2) is exemplarily shown, forming a matrix of wavelet coefficients on different scales.

When transforming the fault types of Fig. 1 through wavelets, all sensor fault types, except outliers, are recognizable because of unique patterns the fault types follow. The reason outliers are not identified is that they correspond to point singularities in a continuous signal, which are, in general, not detectable by standard integral transforms. However, outliers can be easily detected and removed from the original signal by thresholding, and therefore, outlier identification is not critical to the complete FD approach. Thus, outliers will not be considered in this work.

After pre-processing faulty sensor data, ML algorithms may be used to allow automatic identification of sensor faults. To integrate fault identification into the overall FD process, built upon the wavelet transform, a generalized FD approach is proposed in the following section.

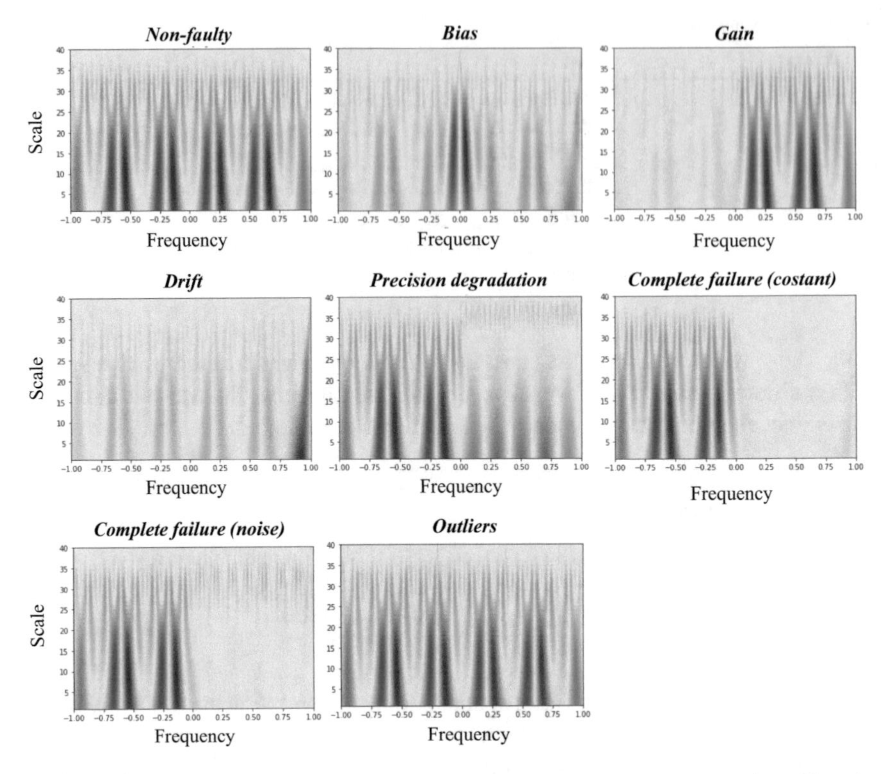

Fig. 2 Plots of wavelet transform of artificial signal (2) with different fault types injected into the signal

3 Fault Diagnosis Based on Signal Processing and Machine Learning Techniques

This section presents the methodology for the autonomous FD using wavelet transform and machine learning. The methodology for autonomous FD presented in this chapter is realized via two ML algorithms, (i) a ML regression algorithm for fault detection, fault isolation, and fault accommodation, and (ii) a ML classification algorithm for fault identification. The ML approach for fault detection and isolation has already been introduced in previous work [3] and is generalized herein. For training the ML regression algorithm, correlations among sensor data at non-faulty sensor operation are exploited. Correlations between sensor data may occur as a result of sensor placement on the structure and may be readily visible in the raw data or exposed through data-pre-processing (e.g., for acceleration data through using the Fourier amplitudes of correlated sensor data in the frequency domain). Fault isolation is accomplished by designing a separate instance of the ML regression algorithm for each sensor. For the ML classification algorithm of the FD approach, patterns indicative of faulty sensor data, as described in the previous section, is used to automate the fault identification process. The algorithmic representation of the proposed approach combining the two ML algorithms can be defined as follows:

1. Use the sensor data $f_{Sj}(t)$ of correlated sensors $j = 1, \ldots, k$ at non-faulty sensor operation as input data for the ML regression algorithm of sensor i.
2. For sensor i, approximate sensor data $f_{Si}(t)$ by the output of the ML regression algorithm in the form of virtual data $\hat{f}_{Si}(t)$.
3. Perform fault detection and fault isolation by calculating the deviation between $\hat{f}_{Si}(t)$ and $f_{Sj}(t)$

   (a) If the deviation between $\hat{f}_{Si}(t)$ and $f_{Si}(t)$ is smaller than a pre-defined threshold a, trigger a sensor alert and initiate fault identification and fault accommodation.
      (i) Perform fault accommodation by replacing faulty sensor data $f_{Si}(t)$ with virtual data $\hat{f}_{Si}(t)$.
      (ii) Perform fault identification by calculating the wavelet transforms $L_\psi f_{Si}(t)$ and classify the results using the ML classification algorithm.
   (b) If the deviation between $\hat{f}_{Si}(t)$ and $f_{Si}(t)$ is greater or equal to the threshold a, repeat steps 1–3.

A conceptual implementation framework is presented in Fig. 3.

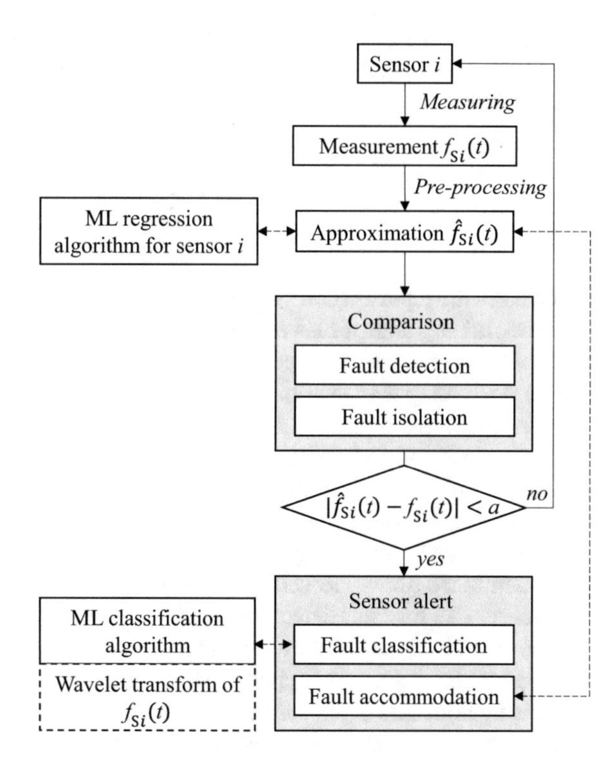

Fig. 3 Implementation framework for the FD approach

4 Validation Tests

In this section, validation tests are performed to showcase the ability of the FD approach to diagnose sensor faults reliably and accurately on real-world data recorded by an SHM system in operation on a railway bridge. First, the ML algorithms used in the validation tests are briefly described. Next, the railway bridge is presented. Then, the data pre-processing and the determination of ML algorithms parameters are explained. Finally, the validation test results of the FD approach are shown and discussed.

4.1 Description of the Machine Learning Algorithms

The validation tests are performed using an ANN as a ML regression algorithm and a CNN as a ML classification algorithm. In this subsection, the ANN and the CNN are briefly described, with emphasis on how each algorithm is applied to the respective FD tasks.

Fault detection, fault isolation, and fault accommodation using artificial neural networks. The ANN employed for the validation tests of this study performs fault detection, fault isolation, and fault accommodation. The ANN concept is based on the function of biological neurons, which "fire" (become activated) upon receiving stimuli. Mathematically, the activation (i.e., output) of neuron M is described as follows [39]

$$N_M := f\left(\sum_{i=1}^{K} w_i x^i + b\right), \tag{3}$$

which takes the input x^i (components of some input vector $\boldsymbol{x}$), makes the summation with weights w_i, adds a bias b, and passes it with a transfer function f. Typically, "sigmoid" functions (e.g., hyperbolic tangent) are used as activation functions. The overall ANN output is then obtained as a weighted combination of activations of individual neurons organized in layers, constituting an input layer, one (or more) hidden layer(s) and an output layer, see [39] for details. Further information on the concept of artificial neural networks for fault detection and fault isolation used in this study may be found in [3].

Figure 4 shows the topology of an ANN for fault detection and isolation in sensor i. As mentioned above, to achieve fault isolation, one ANN model is created for each sensor. As for fault detection, the ANN model for sensor i goes through three phases, the training phase, the testing phase, and the application phase. In the *training* phase, a training data set is formed in a SHM system state designated as "non-faulty," comprising both data of sensor i (target output) and corresponding (i.e., at the same time points) sensor data correlated with the data of sensor i (input data), collected

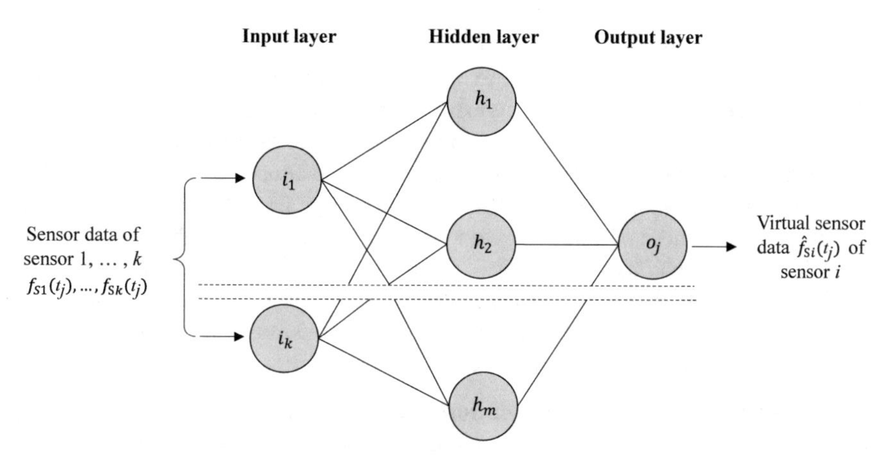

Fig. 4 ANN for FD of sensor *i*

from $1, \ldots, k$—usually neighboring—sensors of the SHM system. Since one ANN instance is created for each sensor, the output layer of the ANN contains one neuron that produces virtual data $\hat{f}_{Si}(t)$, for sensor i. The ANN weights are defined based on the training data set, with training being considered complete once the deviation between the target output and the virtual data is at a minimum. In the *testing* phase, a testing data set is formed and used in a similar way as in the training data set, albeit with new sensor data independent from the training data set. The purpose of the testing phase is to verify that the ANN has been properly trained in terms of weights defined. Based on the quality reached in the testing phase, the threshold for the detection phase is defined. In contrast to Dragos and Smarsly [3], no signal characteristics applicable to specific types of sensor data are used as input data to the ANN; rather, sensor data is used directly with only rudimentary pre-processing (e.g., scaling), thus making the approach applicable for different types of sensor data (e.g., strains and temperatures). The nonlinear relationship between the inputs and the output, drawing from the ANN theory, is accounted for through use of at least one hidden layer. In the *application* phase, sensor data from sensors $1, \ldots, k$ corresponding to an unknown SHM system state (i.e., faulty, or non faulty) are fed to the ANN instance to produce virtual outputs $\hat{f}_{Si}(t)$, and the existence of sensor faults is judged upon comparing virtual outputs with the actual sensor data of sensor i.

The measurand for comparing the deviation between the actual sensor data $f_{S_i}(t_j)$ and the virtual data $\hat{f}_{S_i}(t_j)$ for all time steps t_j in an observation period T is the coefficient of determination R^2 [40]. The coefficient of determination describes the part of the variation of a function f that can be mapped by an approximation function $\hat{f}$ and is given by:

$$R^2 = 1 - \frac{\sum_j (f(t_j) - \hat{f}(t_j))^2}{\sum_j (f(t_j) - \overline{f})^2} \tag{4}$$

where $\overline{f}$ denotes the mean value of $f(t_j)$ for $t_j \in T$. Typically, values of R^2 lie between 0 and 1, with values close to 1 indicating a good approximation quality. Values less than zero are possible when the approximation function fits the data worse than the mean value of the data. A sensor fault detection alert is issued if the R^2 value differs more than a tolerance ε from the threshold already defined at the testing phase. For differentiating between sensor faults and structural damage, the virtual data of ANN instances of several sensors need to be analyzed collaboratively. Fault detection alerts issued by the majority of sensors may indicate structural damage.

If a sensor fault alert is issued, fault identification and fault accommodation are initiated. In case of a faulty sensor, for fault accommodation, the faulty sensor data $f_{Si}(t)$ is replaced by the virtual data $\hat{f}_{Si}(t)$.

Fault identification using convolutional neural networks. In what follows, autonomous fault identification is implemented through a CNN, representing the supervised ML classification algorithm in this study, for the sensor fault types presented in Sect. 2. Convolutional neural networks represent a class of ML algorithms for classification problems, commonly applied to image and pattern recognition [41]. In contrast to artificial neural networks, convolutional neural networks combine three characteristics, making CNN algorithms suitable for classification:

- *Local receptive fields*, allowing the first layer to extract features from segments of the input data by means of filters. The subsequent layers combine the features extracted, detecting higher-order abstractions.
- *Shared weights*, which derive from the idea that filters can be used across the entire input data to extract features irrespective of the section of the input data. The exact location of a feature is not important, as only its relative position to other features is relevant. Thus, by applying the weights of a filter to different sections of the input data, a feature map is generated. Input data of the same type may by roughly different; but, if the relative position of the features is approximately the same, a CNN is capable of classifying the input data as belonging to the same type.
- *Subsampling*, reducing the spatial resolution of a filter and the sensitivity of the output to variations, shifts, and noise.

The concept of using a CNN for fault identification is presented in Fig. 5. A CNN takes as input a signal $x(u)$, which is here an image obtained after a wavelet transform of the faulty sensor data. Neuron values for x_j for layer j are computed from values of

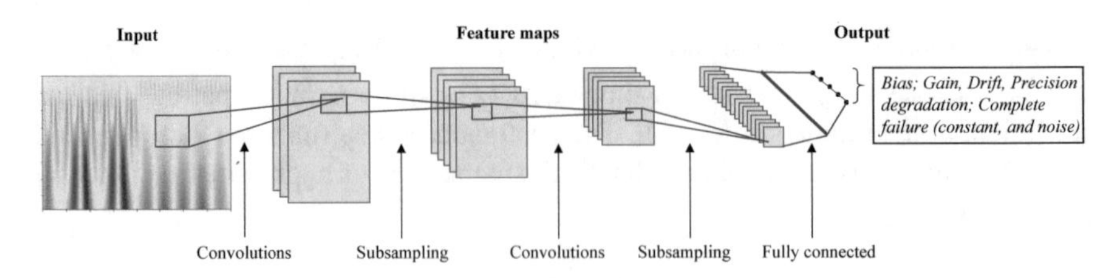

Fig. 5 Fault identification using a CNN

the previous layer x_{j-1} by applying a linear operator W_j and a pointwise nonlinearity rectifier ρ, see [42] for details:

$$x_j = \rho W_j x_{j-1}. \tag{5}$$

Details on the parameters of the CNN used in the validation tests are given below. By help of (5), a higher abstraction of the wavelet transforms is obtained ("abstracted wavelet transforms"), generalizing the main features of the different fault types, as indicated in Fig. 5. The complete CNN is then obtained by constructing a cascade of several convolutions of the form (5). Additionally, the abstracted wavelet transforms are flattened, generating a one-dimensional vector that is interpretable by the final output layer of the CNN. The output layer returns a percentage for every possible fault type, representing the probability of the abstracted wavelet transform to belong to each fault type. Thus, the CNN is able to classify the abstracted wavelet transforms into each fault type, based on the highest percentage obtained from the activation function.

The six fault types that, as previously mentioned, are considered in this study constitute six classes for the ML classification. It is important to note that a seventh class for non-faulty data is not required, because non-faulty data is sieved out by the ANN on the fault detection step of the FD approach proposed in this paper. If the accuracy of the ANN in the FD process is not sufficient for application, the seventh class of "non-faulty" data may be added to the CNN, for avoiding error accumulation.

4.2 Description of the Structure Used for Validation

The validation is performed using sensor data recorded by an SHM system installed on a reinforced concrete railway bridge that includes monitoring data of one year, collected as part of a previous research project [43]. The bridge is a double-track railway overpass. The bridge deck consists of ten spans each resting on four piers with circular cross-sections, which are monolithically connected to the deck. The piers are founded on bored piles, whose heads are connected with rigid beams (pile head beams), and the sensors of the SHM system are installed in the pile head beam. The data set contains sensor data from one data acquisition unit of one pile head beam of the bridge comprising 11 sensors, whose types and labels are listed in Table 2. The top view of the pile head beam and the sensor positions of sensor S1–S11 are illustrated in Fig. 6. Further figures and information of the validation structure and the monitoring data may be found in [43].

The data set contains data from eight strain sensors and three temperature sensors from the year 2017, each data point being collected every 10 min, i.e., at a sampling frequency of 1.7 MHz, corresponding to $n = 52{,}560$ data points per sensor. The validation of the FD approach proposed is shown for faults artificially injected into the sensor data of strain gauge S1.

Table 2 Sensor types, measuring units, and sensor labels of the selected sensors

Sensor type	Measuring unit	Label
Strain gauges on steel reinforcement	mV/V	S1, S2, S3, S4
Strain gauges on concrete	mV/V	S5, S6, S7, S8
Temperature sensor	°C	S9, S10, S11

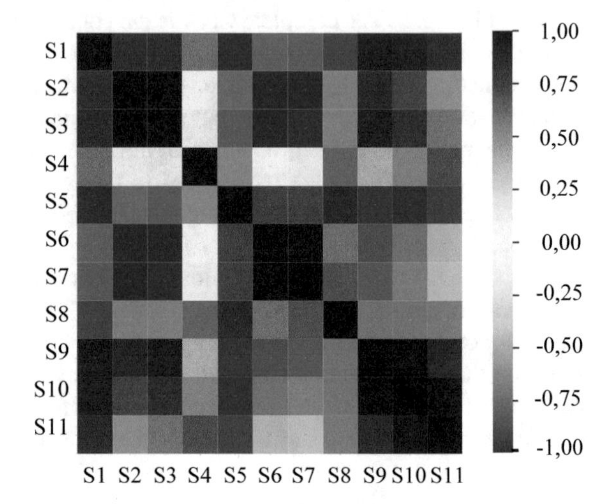

Fig. 6 Top view of the pile head beam showing the positions of the selected measuring unit and sensors

4.3 Data Preparation and Determination of the Machine Learning Algorithm Parameters

Prior to training the ANN, the Pearson correlation coefficients, as shown in Fig. 7, are calculated to unveil correlations between sensor data, in particular between the temperature sensors and the strain gauges on the steel reinforcement. Not all sensor data in the data acquisition unit is correlated. If sufficient training data is available, a lack of correlation in the training data between the input layer and the output layer does not degrade the precision of the ANN. However, when having limited training data, only correlated sensor data should be used.

Before training, the sensor data serving as input to the ANN (i.e., sensor data from sensors S2–S11) is normalized because the magnitude of the values of the sensor data is different from sensor to sensor due to the different types (e.g., temperature values are much higher than strain values). The ANN topology as well as the neuron activation functions are defined through trial-and-error, as listed in Table 3. The input layer of the ANN consists of ten neurons, representing data points at time step t_j for sensors S2–S11. The output layer of the ANN has one neuron, which yields virtual data at time step t_j for sensor S1. The activation function selected for the input layer and the hidden layers is the rectified linear unit (ReLU). For the output layer, the

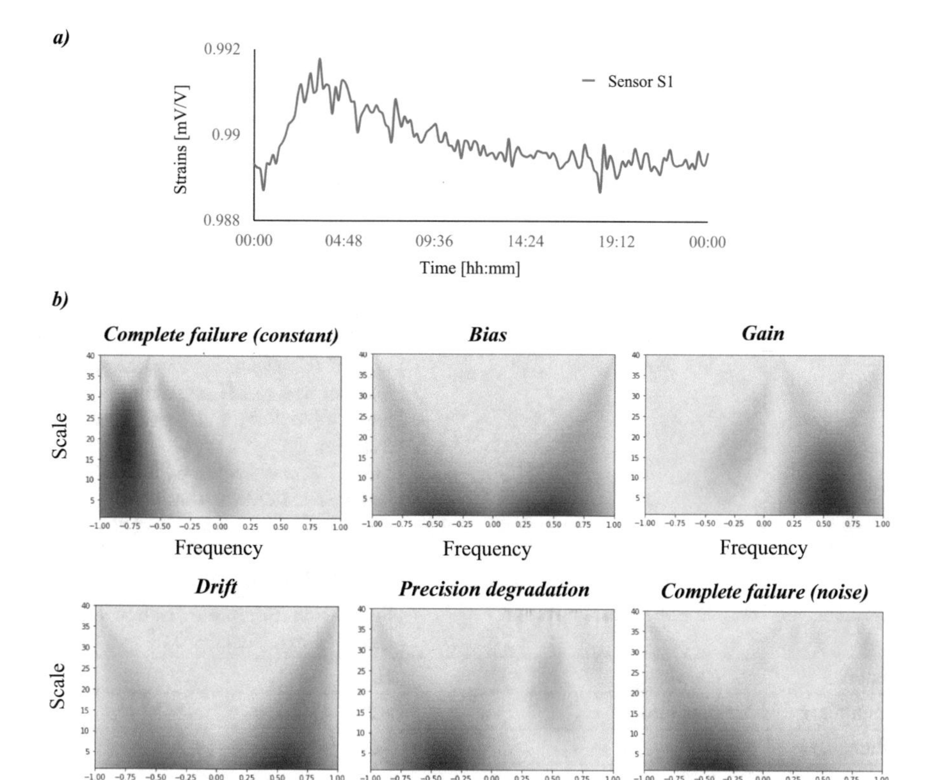

Fig. 7 Pearson correlation coefficients among sensors S1–S11

linear activation function is used. Each ANN is trained in 15 "epochs" (complete training cycles using the entire training data set), with a "batch size" (number of data points propagated through the ANN before updating the weights) of 32 and the "Adam" gradient-descent optimization algorithm for updating the weights [44].

For training the CNN for fault identification, sensor data from one year is split into one-day-sensor data sets (each containing 144 data points). Instances of each fault type, following the equations of the fault types from Sect. 2, are simulated and introduced (with variable magnitude) in the middle of the signal and transformed through wavelet transform as presented in Sect. 2. In this work, the classical Mexican hat wavelet, which is the second derivative of a Gaussian function, has been used:

$$\psi(t) = \frac{2}{\pi^{\frac{1}{4}}\sqrt{3\sigma}}\left(\frac{t^2}{\sigma^2} - 1\right)e^{-\frac{t^2}{2\sigma^2}} \tag{6}$$

Table 3 ANN and CNN topology for validation

	ANN	CNN
Input layer	10 neurons, ReLU activation	One image (39 × 144), ReLU activation
Hidden layer	Layer 1: 32 neurons, ReLU activation Layer 2: 64 neurons, ReLU activation Layer 3: 256 neurons, ReLU activation	Layer 1 (*Convolution*): 16 filters, 5 × 5 kernel size, ReLU activation Layer 2 (*Pooling*): 2 × 2 pool size, 1 × 1 strides Layer 3 (*Convolution*): 32 filters, 3 × 3 kernel size, ReLU activation Layer 4 (*Pooling*): 2 × 2 pool size, 2 × 2 strides Layer 5 (*Convolution*): 256 filters, 5 × 5 kernel size, ReLU activation Layer 6 (*Pooling*): 2 × 2 pool size, 2 × 2 strides Layer 7 (*Convolution*): 64 filters, 3 × 3 kernel size, ReLU activation Layer 8 (*Pooling*): 2 × 2 pool size, 2 × 2 strides, dropout = 0.2% Layer 9 (*Flattening*)
Output layer	One neuron, linear activation	Six neurons, Softmax activation, L2 quadratic regularizer = 0.01
Epochs	15	20
Optimizer	Adam optimizer	Adam optimizer
Batch size	32	32

setting parameter $\sigma = 40$. Exemplarily, a one-day sensor data set used for training the CNN is presented in Fig. 8a. The strain curve of the one-day-sensor data set exhibits fluctuations due to the changes in ambient temperature.

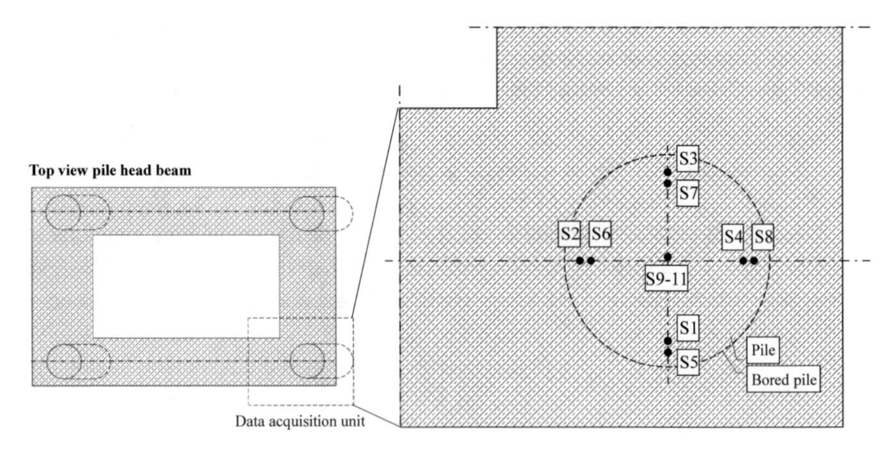

Fig. 8 **a** Example sensor data of sensor S1 and **b** corresponding wavelet transforms

Each input for the CNN in the training process is one 2D image of wavelet-transformed one-day sensor data of faulty signals, as exemplarily shown in Fig. 8b. In total, a number of $n = 5000$ training images per fault type are fed to the CNN. The hyperparameters and properties of the CNN, selected through trial-and-error, are listed in Table 3, namely the network topology (the number of neurons per layer) and the neuron behavior (the activation functions). The ReLU activation function is used for the input and intermediate layers; while for the output layer, the "softmax" activation function is selected. "Pool size" refers to the factor by which pooling layers reduce the feature map, and "strides" represent the frame size with which the filters of convolution layers "convolve" around their input. Finally, "dropout" is a probability value below which neurons of a layer are ignored ("dropped").

It should be noted that since the ANN is able to adapt to the condition of the monitored structure, the ANN properties are case-specifically defined for the validation tests presented herein, i.e., the ANN properties depend on the characteristics of the monitored structure and on the SHM system setup. The CNN, however, due to identifying artificial faults within a wide range of fault magnitudes is measurement-independent.

4.4 Validation Tests and Results

For the validation tests, the fault types previously mentioned (bias, drift, gain, precision degradation, and complete failure) are simulated and injected into the sensor data of sensor S1. Faulty data is characterized by function $\tilde{f}$ defined by the faults compromising non-faulty data f. Tests are performed by varying the magnitude of the faults and observing the change in the coefficient of determination. The results of the ANN performance with increasing fault magnitudes are shown in Table 4. The ANN for sensor S1 is trained with $n = 20{,}000$ data points as training data. In the testing phase, using a testing set of $n = 2000$ data points, the high precision of the training is showcased through the coefficient of determination equal to $R^2 = 0.992$. The magnitude of sensor faults b is based on the mean value $\bar{x}$ of the data set to be tested. Since the training precision of the ANN is high, the threshold for indicating a fault is accordingly set high to $a = 0.98$.

Furthermore, tests have been performed to identify above which percentage of a faulty signal the ANN is capable of detecting faults. Here, three percentage levels of faulty data are considered, 1%, 3%, and 5%. For all fault types, except drift, it is possible to detect the fault within under 1% of faulty data. Since drift requires time to fully manifest, detecting drift with 5% of faulty data is not possible. With 30% of faulty data in the sensor data, a drift is detectable (marked with * in Table 5).

For validating the CNN, a "confusion matrix" is calculated, as presented in Table 6. The confusion matrix presents the classification outputs of the CNN for the test data. The CNN classifies correctly most of the fault types, achieving an accuracy of 99.39%. Furthermore, Table 7 presents the evaluation of the CNN during testing. For each fault type, precision, recall, and F1-score are presented. Precision represents the

Table 4 Comparison results of the validation examples of the sensor faults

Fault type	Fault parameter	Comparison result R^2
Non-faulty $f(t)$		0.992
Bias	$b = 0.01\overline{x}$	0.856
$\tilde{f} = f(t) + b$	$b = 0.02\overline{x}$	0.618
	$b = 0.05\overline{x}$	0.085
	$b = 0.10\overline{x}$	−0.323
Gain	$b = 1.01\overline{x}$	0.851
$\tilde{f} = b \cdot f(t)$	$b = 1.05\overline{x}$	0.072
	$b = 1.10\overline{x}$	−0.333
Drift	$b = 0.01\overline{x}$	0.938
$\tilde{f} = f(t) + b \cdot t$	$b = 0.02\overline{x}$	0.805
	$b = 0.05\overline{x}$	0.404
	$b = 0.10\overline{x}$	0.030
Precision degradation	$\sigma^2 = 0.02$	0.473
(PD)	$\sigma^2 = 0.05$	0.125
$\tilde{f} = f(t) + w(t)$	$\sigma^2 = 0.1$	0.037
Complete failure		
(a) Noise $\tilde{f} = w(t)$	$\sigma^2 = 0.1$	−1.066
(b) Constant $\tilde{f} = b$	$b = 0\overline{x}$	−1.087

Table 5 Prediction results of the ANN depending on percentage of faulty data

	1% faulty data (*10%)	3% faulty data (*30%)	5% faulty data (*50%)
Bias ($b = 0.01$)	$R^2 = 0.625$	$R^2 = 0.442$	$R^2 = 0.383$
Drift ($b = 0.1$)*	$R^{2*} = 0.973$	$R^{2*} = 0.818$	$R^{2*} = 0.608$
Gain ($b = 1.1$)	$R^2 = 0.617$	$R^2 = 0.434$	$R^2 = 0.375$
PD ($\sigma^2 = 0.1$)	$R^2 = 0.557$	$R^2 = 0.321$	$R^2 = 0.232$
Complete failure			
(a) Noise ($\sigma^2 = 0.1$)	$R^2 = -0.036$	$R^2 = -0.063$	$R^2 = -0.088$
(b) Constant ($b = 0$)	$R^2 = -0.031$	$R^2 = -0.064$	$R^2 = -0.089$

ratio between correctly predicted fault types and the total predictions of the fault type. Recall represents the ratio between correctly predicted fault types and the actual total observations of the fault type. Finally, the F1-score represents the weighted average of precision and recall. For example, the bias fault has 394 correctly predicted fault types, eight incorrectly predicted fault types as drift, and one observation incorrectly

Table 6 Confusion matrix of the CNN

	Bias	Gain	Drift	PD	CF a	CF b
Bias	394	0	8	0	0	0
Gain	0	383	0	0	0	0
Drift	0	0	371	0	0	4
PD	0	0	0	369	1	0
CF a	0	0	0	0	403	0
CF b	1	0	0	0	0	376

Table 7 Evaluation of the CNN during testing

Fault type	Precision (%)	Recall (%)	F1-score (%)
Bias	99.74	98.00	98.86
Drift	97.89	98.93	98.41
Gain	100.00	100.00	100.00
PD	100.00	99.73	99.86
CF a	99.75	100.00	99.87
CF b	98.94	99.73	99.33

classified as bias. Therefore, bias has a precision of 99.74%, a recall of 98.00%, and a F1-score of 98.86%. It can be observed that the evaluation of CNN gives results close to 100% in all scores, thus denoting reliable results.

5 Summary and conclusions

Reliable operation of SHM systems is of increasing importance, as sensor faults may compromise the quality of monitoring. To ensure reliable operation of SHM systems, this chapter has proposed a novel approach toward autonomous FD using signal processing and ML techniques.

An ML regression algorithm has been proposed for fault detection and isolation that uses structural response data from correlated sensors of SHM systems as input and predicts virtual sensor data. In the case of discrepancies between virtual and actual sensor data, a sensor alert is issued, and fault identification and fault accommodation are initiated. Fault accommodation has been performed using the virtual sensor data of the ML regression algorithm, providing reliable results for the sensor failure duration. Through wavelet transform, as a signal processing technique, patterns indicative of sensor faults has been analyzed. To identify the sensor fault types autonomously, an ML classification algorithm has been proposed that uses faulty sensor data, analyzed with wavelet transform, as input and classifies the corresponding sensor fault type. Validation tests of the FD approach have been conducted using an ANN as ML regression algorithm and a CNN as ML classification algorithm. The validation tests have been performed using real-world data from a SHM system in operation on a railway bridge. The results have demonstrated the ability of the ANN to predict virtual sensor data and the ability of the CNN to identify (classify) sensor faults correctly.

In summary, the FD approach proposed in this chapter enables autonomous sensor FD in SHM systems by performing fault detection, fault isolation, fault identification, and fault accommodation, ensuring reliable operation of sensors in SHM systems. Further research will focus on embedding the FD approach into sensor nodes of wireless SHM systems for decentralized FD.

Acknowledgements The authors gratefully acknowledge the support offered by the German Research Foundation (DFG) under grants SM 281/9-1, SM 281/12-1, SM 281/14-1, SM 281/15-1, and LE 3955/4-1. This work is also partially sponsored by the German Federal Ministry of Transport and Digital Infrastructure (BMVI) under grant VB18F1022A. Any opinions, findings, conclusions, or recommendations expressed in this chapter are those of the authors and do not necessarily reflect those of DFG or BMVI.

References

1. Glisic B, Inaudi D, Casanova N (2010) SHM process as perceived through 350 projects. In: SPIE 7648, smart sensor phenomena, technology, networks, and systems. San Diego, CA, USA, 04/08/2010, pp 76480P
2. Zhang Z, Mehmood A, Shu L, Huo Z, Zhang Y, Mukherjee M (2018) A survey on fault diagnosis in wireless sensor networks. IEEE Access 6:11349–11364
3. Dragos K, Smarsly K (2016) Distributed adaptive diagnosis of sensor faults using structural response data. Smart Mater Struct 25(10):105019
4. Patton RJ (1991) Fault detection and diagnosis in aerospace systems using analytical redundancy. Comput Control Eng J 2(3):127–136
5. Willsky AS (1976) A survey of design methods for failure detection systems. Automatica 12:601–611
6. Kraemer P, Fritzen C-P (2007) Sensor fault identification using autoregressive models and the mutual information concept. Key Eng Mater 347:387–392
7. Smarsly K, Law KH (2014) Decentralized fault detection and isolation in wireless structural health monitoring systems using analytical redundancy. Adv Eng Softw 73(2014):1–10
8. Luckey D, Fritz H, Legatiuk D, Dragos K, Smarsly K (2020) Artificial intelligence techniques for smart city applications. In: Proceedings of the international ICCCBE and CIB W78 joint conference on computing in civil and building engineering 2020. Sao Paolo, Brazil, 08/18/2020
9. Dai X, Gao Z (2013) From model, signal to knowledge: a data-driven perspective of fault detection and diagnosis. IEEE Trans Industr Inf 9(4):2226–2238
10. Zaher A, McArthur SDJ, Infield DG, Patel Y (2009) Online wind turbine fault detection through automated SCADA data analysis. Wind Energy 12(6):574–593
11. Tolani DK, Yasar M, Ray A, Yang V (2006) Anomaly detection in aircraft gas turbine engines. J Aerosp Comput Inf Commun 3(2):44–51
12. Xu X, Hines JW, Uhrig RE (1999) Sensor validation and fault detection using neural networks. In: Proceedings of the maintenance and reliability conference, Gatlinburg, TN, USA, 05/10/1999

13. Campa G, Fravolini ML, Seanor B, Napolitano MR, Del Gobbo D, Yu G, Gururajan S (2002) Online learning neural networks for sensor validation for the flight control system of a B777 research scale model. Int J Robust Nonlinear Control 12(11):987–1007
14. Cork L R, Walker R, Dunn S (2005) Fault Detection, Identification and Accommodation Techniques for Unmanned Airborne Vehicle. In: Proceedings of the Australian International Aerospace Congress. Melbourne, Australia, 03/15/2005.
15. Vemuri AT, Polycarpou MM (1997) Neural network-based robust fault diagnosis in robotic systems. IEEE Trans Neural Networks 8(6):1410–1420
16. Eski I, Erkaya S, Savas S, Yildirim S (2011) Fault detection on robot manipulators using artificial neural networks. Robot Comput-Integr Manuf 27(1):115–123
17. Venkatasubramanian V, Vaidyanathan R, Yamamoto Y (1990) Process fault detection and diagnosis using neural networks—1. Steady-state processes. Comput Chem Eng 14(7):699–712
18. Obst O (2009) Distributed fault detection using a recurrent neural network. In: 2009 International conference on information processing in sensor networks. San Francisco, CA, USA, 04/13/2009
19. Jeong S, Ferguson M, Hou R, Lynch JP, Sohn H, Law KH (2019) Sensor data reconstruction using bidirectional recurrent neural network with application to bridge monitoring. Adv Eng Inform 42:100991
20. Steiner M, Legatiuk D, Smarsly K (2019) A support vector regression-based approach towards decentralized fault diagnosis in wireless structural health monitoring systems. In: The 12th international workshop on structural health monitoring (IWSHM). Stanford, CA, USA, 10/09/2019
21. Zhao C, Sun X, Sun S, Jiang T (2011) Fault diagnosis of sensor by chaos particle swarm optimization algorithm and support vector machine. Expert Syst Appl 38(8):9908–9912
22. Yu CB, Hu JJ, Li R, Deng SH, Yang RM (2014) Node fault diagnosis in WSN based on RS and SVM. In: Proceedings 2014 international conference on wireless communication and sensor network (WCSN). Wuhan, China, 12/13/2014
23. Balaban E, Saxena A, Bansal P, Goebel KF, Curran S (2009) Modeling, detection, and disambiguation of sensor faults for aerospace applications. IEEE Sens J 9(12):1907–1917
24. Huang H-B, Yi T-H, Li H-N (2017) Bayesian combination of weighted principal-component analysis for diagnosing sensor faults in structural monitoring systems. J Eng Mech 143(9):04017088
25. Dragos K, Jahr K, Smarsly K (2016) Nonlinear sensor fault diagnosis in wireless sensor networks using structural response data. In: Proceedings of the 23rd international workshop of the European group for intelligent computing in engineering (EG-ICE). Krakow, Poland, 06/29/2016
26. Li L, Liu G, Zhang L, Li Q (2019) Sensor fault detection with generalized likelihood ratio and correlation coefficient for bridge SHM. J Sound Vib 442:445–458
27. Webster JG (1999) The measurement, instrumentation, and sensors handbook. CRC/IEEE Press, Boca Raton
28. Teh HY, Kempa-Liehr AW, Wang KI-K (2020) Sensor data quality: a systematic review. J Big Data 7(11):1–49
29. Ternan JG (1983) Thermoelectric drift of thermocouples due to inhomogeneous changes in composition. J Appl Phys 55:199–209
30. Hamada T, Suyama Y (2004) EMF drift and inhomogeneity of type K thermocouples: In: SICE 2004 annual conference. Sapporo, Japan, 08/06/2004
31. Wang D, Fotinich Y, Carman GP (1998) Influence of temperature on the electromechanical and fatigue behavior of piezoelectric ceramics. J Appl Phys 83(10):5342–5350
32. Tong L, Sun D, Atluri SN (2001) Sensing and actuating behaviours of piezoelectric layers with debonding in smart beams. Smart Mater Struct 10:713–723
33. Li C, Weng GJ (2002) Antiplane crack problem in functionally graded piezoelectric materials. J Appl Mech 69(4):481–488
34. Hawkins DM (1980) Identification of outliers. Springer, Dordrecht

35. Dervilis N, Antoniadou I, Barthorpe RJ, Cross EJ, Worden E (2015) Robust methods for outlier detection and regression for SHM applications. Int J Sustain Mater Struct Syst 2(1/2):78354
36. Fu Y, Peng C, Gomez F, Narazaki Y, Spencer BF (2019) Sensor fault management techniques for wireless smart sensor networks in structural health monitoring. Struct Control Health Monit 26(7):e2362
37. Russel SJ, Norvig P (2014) Artificial intelligence: a modern approach, 3rd edn. Pearson Education, New Jersey
38. Najmi AH, Sadowsky J (1997) The continuous wavelet transform and variable resolution time-frequency analysis. J Hopkins APL Tech Dig 18(1):134–140
39. Kanevski M, Pozdnoukhov A, Timonin V (2009) Machine learning for spatial environmental data. Theory, applications and software. EPFL Press, Lausanne
40. Netter J, Kutner MH, Nachtsheim CJ, Wasserman W (1996) Applied linear statistical models. Irwin Professional Publishing, Burr Ridge
41. LeCun Y, Bottou L, Bengio Y, Haffner P (1998) Gradient-based learning applied to document recognition. Proc IEEE 86(11):2278–2324
42. Mallat S (2016) Understanding deep convolutional networks. Phil Trans R Soc A 374:20150203
43. Bauhaus-Universität Weimar (2020) Messdaten eines Monitoringsystems einer Eisenbahnbrücke (online). Available at: https://www.mcloud.de/zh/web/guest/suche/-/results/detail/F9BFABA3-29A9-48A2-973B-E1C29E8555B0
44. Kingma DP, Ba J (2014) Adam: a method for stochastic optimization. In: 3rd International conference for learning representations. San Diego, CA, USA, 05/07/2015

A Self-adaptive Hybrid Model/data-Driven Approach to SHM Based on Model Order Reduction and Deep Learning

Luca Rosafalco, Matteo Torzoni, Andrea Manzoni, Stefano Mariani, and Alberto Corigliano

Abstract Aging of structures and infrastructures urges new approaches to ensure higher safety levels without service interruptions. Structural health monitoring (SHM) aims to cope with this need by processing the data continuously acquired by pervasive sensor networks, handled as vibration recordings. Damage diagnosis of a structure consists of detecting, localizing, and quantifying any relevant state of damage. Deep learning (DL) can provide an effective framework for data processing, regression, and classification tasks used for the aforementioned damage diagnosis purposes. Within this framework, we propose an approach that exploits a deep convolutional neural network (NN) architecture. The training of the NN is carried out by exploiting a dataset, numerically built through a physics-based model of the structure to be monitored. Parametric model order reduction (MOR) techniques are then exploited to reduce the computational burden related to the dataset construction. Within the proposed approach, whenever a damage state is detected, the physical model of the structure is adaptively updated, and the dataset is enriched to retrain the NN, allowing for the previously detected damage state as the new baseline.

L. Rosafalco · M. Torzoni · S. Mariani (✉) · A. Corigliano
Dipartimento di Ingegneria Civile e Ambientale, Politecnico di Milano,
Piazza L. da Vinci 32, 20133 Milano, Italy
e-mail: stefano.mariani@polimi.it

L. Rosafalco
e-mail: luca.rosafalco@polimi.it

M. Torzoni
e-mail: matteo.torzoni@polimi.it

A. Corigliano
e-mail: alberto.corigliano@polimi.it

A. Manzoni
MOX, Dipartimento di Matematica, Politecnico di Milano,
Piazza L. da Vinci 32, 20133 Milano, Italy
e-mail: andrea1.manzoni@polimi.it

A. Cury et al. (eds.), *Structural Health Monitoring Based on Data Science Techniques*,
Structural Integrity 21, https://doi.org/10.1007/978-3-030-81716-9_8

Keywords Structural health monitoring · Physics-based models · Reduced order modeling · Digital twins · Deep learning · Transfer learning · Damage identification

1 Introduction

Civil infrastructures are crucial in our digital and smart society. However, they are exposed to aging and to a progressive accumulation of damage, also caused by exceptional loading conditions. As a prompt detection of structural damage can prevent catastrophic events, SHM is becoming more and more an active field of research [8]. Global damage detection methods are looked for, to be robust against operational (e.g., in terms of the load amplitude) and environmental (e.g., due to varying thermal and hygrometric conditions) variability.

Customary vibration-based monitoring techniques rely on the fact that modal parameters, and therefore, the overall dynamic response of a structure depends on its mass and stiffness properties. Any change in their values due to the inception of a structural damage accordingly affects the dynamic response of the structure [15]. Within such a perspective, through the SHM system the observations in time (e.g., acceleration recordings shaped as multivariate time series) recorded by a sensor network are analyzed to extract some damage-sensitive features able to discriminate virgin and damaged states. Signal processing is to be carried out via effective procedures, able to handle the raw vibration signals, and retrieve the aforementioned damage-sensitive features (from now on simply referred to as *features*), to allow the SHM procedure to detect damage.

Local damage detection methods, based on visual inspections or non-destructive testings, are widespread in civil engineering; however, they require some a-priori knowledge on the position of damage, so a proper expertise [12]. Thanks to the recent advancements in sensor technology, global monitoring techniques based on continuous vibration measurements are now more often used. Among the vibration-based techniques, two main approaches can be distinguished: the *model-based* approach and the *data-driven* approach. Within a model-based approach, the discrepancy between data and model output is minimized via a model update procedure, e.g., by means of Kalman filters [9, 10]. Exploiting the measurements from real structures, damage localization, and quantification can be achieved. However, a model-based approach may be affected in its accuracy and efficacy when a large amount of noisy data has to be processed. Moreover, the associated inverse problem to be solved for damage identification is usually ill-posed. On the other hand, a data-driven approach [11] does not typically rely upon any physics-informed model. Because of its ability to also handle big data acquired on-the-fly through deployed sensors, this approach is increasingly attracting interest. Within a standard implementation, damage-sensitive features are extracted from raw vibration signals by exploiting their statistical characteristics. According to the statistical pattern recognition paradigm [12], the SHM task can be arranged into four sequential steps: operational evaluation, data acquisition, features extraction, and statistical modeling for features discrimination. Since civil structures are always subject to varying operational and environmental conditions, potentially hampering the monitoring task, the SHM strategy should account for such variability within the dataset or exploit data normalization techniques [22] to distinguish the effects of the aforementioned variability from those due to damage.

The pattern recognition paradigm is well suited for implementation through machine learning algorithms. Indeed, machine learning allows to statistically handle data by *learning* the functional link between the damage-sensitive features and the structural states. Recently, DL algorithms have been allowed to further empower the approach; thanks to DL, it is possible to accomplish features extraction and discrimination in one single shot. DL algorithms can deal with high dimensional data to automatically catch temporal and spatial correlations, within and across time recordings. This allows to extract and exploit damage-sensitive features in an end-to-end efficient way.

According to Rytter [21], damage identification consists of four levels of increasing complexity: damage detection, localization, quantification, and prognosis. Data-driven approaches can work within a *supervised* [25, 27] or *unsupervised* [14, 17] setting. Unsupervised methods exploit unlabeled data relevant to a reference condition, often the damage-free baseline, to test the current state and basically accomplish early damage detection. Vice versa, supervised methods exploit labeled data related to both undamaged and damage scenarios, to be possibly undergone by the structure, and are therefore well suited to accomplish also damage localization and quantification.

When dealing with civil structures, data related to damage conditions are often unavailable. To solve this drawback, a commonly adopted paradigm is the simulation-based classification [1, 20, 24], which aims to fuse the model-based and the data-driven approaches into a hybrid one. With this novel paradigm, real data are replaced by the output of numerical experiments where the effect of damage on the structural response is simulated through a physics-based model of the structure. Accordingly, the inception and subsequent growth of damage can be framed in a way similar to model update, wherein the tuning/training of the model itself is not carried out to allow for epistemic uncertainties, but to track the time-varying health of the structure. In this work, we propose a NN-based, mixed offline-online SHM procedure aiming to first detect and localize damage, and then quantify it in a dynamic environment, so as to promptly react in case of detection of any variation of the structural state.

The remainder of this chapter is organized as follows. In Sect. 2, the proposed procedure is detailed to highlight the features of the mixed offline–online strategy, together with the potentiality offered by transfer learning (TL). In Sect. 3, dataset assembling is addressed together with the adopted MOR strategy. In Sect. 4, two NN architectures are discussed for, respectively, simultaneous damage detection and localization, and damage quantification. The capability of the proposed method to track damage evolution is assessed in Sect. 5. In Sect. 6, conclusions on the present work and future developments are finally discussed.

2 Monitoring Procedure

We propose a supervised data-driven SHM approach, tailored to provide outcomes regarding damage detection, localization, and also quantification. The procedure is routed by a damage localization tool followed by a damage quantification tool and is able to self-adapt under time-evolving damage states.

The damage detection/localization algorithm relies on a deep NN-based classification model [18], through which damage, if present, is localized within a classification scheme. Structural damage patterns are modeled as local stiffness reductions; hence, the initial baseline is allowed for an additional (damage-free) structural configuration. Varying operational and environmental conditions are accounted for through a suitable parametrization of the structural model [23]. Vibration signals mimicking the recordings of a sensor network are obtained from the numerical simulations as time histories of nodal displacements and/or accelerations. These data are exploited to train the NN offline, to link any response with the corresponding damage condition (considered as a categorical label). A second deep NN is used to perform a regression on the damage level, quantified in terms of stiffness reduction, to the associated damage condition. The same data are used to detect and localize the damage and therefore to perform the training of this further NN.

The entire procedure consists of offline and online stages, as reported in Fig. 1. In the preliminary *offline* phase, the pseudo-experimental dataset is generated by collecting synthetic vibration recordings under different combinations of operational and damage conditions. Each damage condition is characterized by a magnitude assigned within a pre-defined range, held constant within the observation time interval; accordingly, damage growth does not need to be numerically modeled. The two NNs are then trained to handle both the damage classification and quantification tasks. In the *online* phase, every time new measurements are acquired, they are first processed by the classification model, which provides the localization of damage, if any, and gives a label to the current structural state. If a (new) damage is identified, the monitoring system generates a first warning and the measurements are passed onto the regression model, which aims at estimating on-the-fly the associated stiffness reduction. The procedure accordingly self-adapts on the basis of the identified damage condition, keeping in mind that only low damage evolution rates are of interest. The baseline model is finally updated based on the identified damage state, and the procedure is restarted to process the new incoming measurements.

The computational costs associated with the proposed procedure are mainly related to the simulations to obtain the pseudo-experimental dataset, and to the training of the NNs. The simulations required for the dataset must ensure a fine exploration of the parametric space of the model input, and a good trade-off between the amount of training data and NNs parameters to be tuned, to avoid overfitting of the same data. By relying upon a finite element (FE) model of the system, the reduced basis method [16] is adopted as a parametric MOR strategy. The high-fidelity full-order model (FOM) is thus replaced by a less resource-demanding, yet accurate reduced-order model (ROM) to speedup the dataset construction.

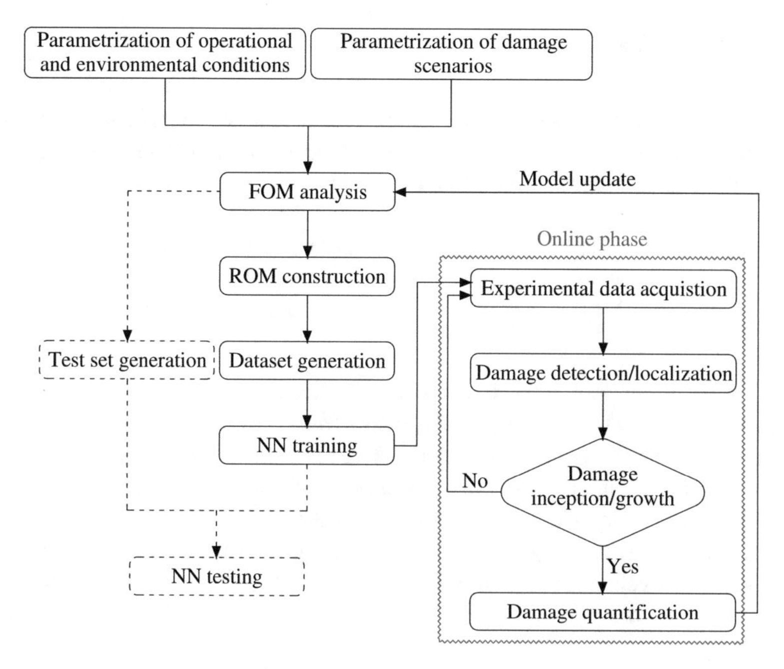

Fig. 1 Methodology flowchart

The two NNs share the same fully convolutional network (FCN) architecture, already successfully adopted, e.g., in [18, 20, 25]. This class of NNs has been reported to be powerful in detecting hierarchical patterns in the data, being also computationally efficient thanks to their shared-weights architecture. By adopting the same architecture for the classification and the regression tasks, TL [4, 7, 13] can be exploited too. TL takes advantage of the knowledge gained during the training of a NN, to train another NN aimed at solving a different task. The present scenario is known as *inductive transfer learning*, which assumes that labeled data for related but different tasks are available. This framework allows reducing both the training time and the amount of training data, without compromising the performance of the procedure.

3 Numerical Modeling

In the following, the procedure to build a FOM of civil structures or buildings is first described; then, we specify the features of the dataset used to train the NNs; finally, we discuss how to speedup the dataset construction, exploiting a MOR technique.

3.1 Full-Order Model

Resting on a linearized kinematics assumption, the monitored structure is modeled as a continuum and discretized in space via a FE triangulation. The semi-discretized form of the elasto-dynamic problem then reads:

$$\begin{cases} \mathbf{M}\ddot{\mathbf{v}}(t) + \mathbf{K}(g,\delta)\mathbf{v}(t) = \mathbf{f}(\boldsymbol{\eta}), & t \in (0, T_f) \\ \mathbf{v}(0) = \mathbf{v}_0 \\ \dot{\mathbf{v}}(0) = \dot{\mathbf{v}}_0 \end{cases} \tag{1}$$

where $t \in (0, T_f)$ denotes time; $\mathbf{v} = \mathbf{v}(t) \in \mathbb{R}^M$ is the nodal displacement vector, with M denoting the number of degrees of freedom (dofs); $\ddot{\mathbf{v}}$ is the vector of nodal accelerations; $\mathbf{M} \in \mathbb{R}^{M \times M}$ is the mass matrix; $\mathbf{K}(g, \delta) \in \mathbb{R}^{M \times M}$ is the stiffness matrix, with g and δ the parameters providing its dependence on damage as specified below; $\mathbf{f}(\boldsymbol{\eta}) \in \mathbb{R}^M$ is the vector of external loads; $\boldsymbol{\eta}$ is the vector of parameters ruling the operational conditions; $\mathbf{v}_0$ and $\dot{\mathbf{v}}_0$ are the initial conditions, in terms of nodal displacements and velocities at $t = 0$. Due to the small relevance of damping in the identification of continuously excited systems [6], dissipation effects have been disregarded.

Damage is modeled as a local stiffness reduction, assumed frozen within the time interval $(0, T_f)$ of interest. This assumption involves a timescale separation between damage growth and health assessment; the considered scenarios are thus characterized by a low damage evolution rate and allow keeping the linear structural response valid. According with the classification framework, only a finite set of damage states are considered: those have to be preliminary determined in order to allow covering relevant structural failure modes. For the problem at hand, G damage conditions $g \in \{1, \ldots, G\}$ have been then accounted for, in addition to the undamaged one $g = 0$. Each condition is characterized by a damage in the corresponding subdomain Ω_g, wherein the stiffness reduction has a time-invariant magnitude taking value in a specified range for δ.

3.2 Dataset Assembly

Synthetic vibration recordings can be built starting from the dynamic response of the model through nodal displacements mimicking the SHM sensing network output. Regarding the optimal location of sensors, even in the present pseudo-experimental setting, to understand how to maximize the sensitivity of measurements to the damage scenarios to be detected, readers are referred to [5]. The monitoring system is assumed to consist of N_u sensors recording the structural displacements $\mathbf{u}_n(t)$, $n = 1, \ldots, N_u$. Each recording $\mathbf{u}_n(t)$ consists on its own of L_M data points, which are gathered in the matrix $\mathbf{U}(\boldsymbol{\eta}, g, \delta) = [\mathbf{u}_1, \ldots, \mathbf{u}_{N_u}] \in \mathbb{R}^{L_M \times N_u}$. Next, data are corrupted with a

Gaussian noise featuring a specific signal-to-noise ratio, in order to mimic the sensor self-noise that inevitably affects the experimental data.

The dataset, denoted by $\mathbf{D}$, is built by assembling I instances according to:

$$\mathbf{D} = \begin{Bmatrix} \mathbf{U}_1(\boldsymbol{\eta}_1, g_1, \delta_1) \ , & & \mathbf{U}_I(\boldsymbol{\eta}_I, g_I, \delta_I) \\ g_1 \ , & \dots & g_I \\ \delta_1 \ , & & \delta_I \end{Bmatrix}. \tag{2}$$

Each instance in $\mathbf{D}$ is composed of recordings $\mathbf{U}_i(\boldsymbol{\eta}_i, g_i, \delta_i)$, $i = 1, \dots, I$, alongside the associated labels g_i and δ_i, which are the corresponding damage condition and damage level. During the training of the NNs, I_{tr} instances are effectively employed for the learning process, while $I_{val} = I - I_{tr}$ instances are used to validate it.

The parametric space, namely the input space for the model, is defined by combining $\{\boldsymbol{\eta}, g, \delta\}$. Each entry is characterized by a uniform probability distribution function, being continuous for $\boldsymbol{\eta}$ and δ, and discrete for g. The Latin hypercube sampling rule is adopted to efficiently explore the parametric space by means of the I instances. In the present work, $I = 10,000$ instances, with a ratio $80:20$ between I_{tr} and I_{val}, have been collected in order to train and validate the two NNs. To simplify the notation, the index i will be dropped in the following.

3.3 Parametric Model Order Reduction for Dataset Generation

To speedup the generation of $\mathbf{D}$, the number of dofs in Eq. (1) can be reduced through MOR techniques for parametrized problems [1, 20]. The reduced basis method is here adopted to also control the approximation error.

The FOM solution $\mathbf{v} = \mathbf{v}(t, \boldsymbol{\eta}, g, \delta)$ is approximated as $\mathbf{v} \approx \mathbf{W}\mathbf{v}_R$, by linearly combining $W \ll M$ basis functions $\mathbf{w}_w \in \mathbb{R}^M$, $w = 1, \dots, W$ collected into $\mathbf{W} = [\mathbf{w}_1, \dots, \mathbf{w}_W] \in \mathbb{R}^{M \times W}$, where W has to be set to attain the target accuracy. $\mathbf{v}_R = \mathbf{v}_R(t, \boldsymbol{\eta}, g, \delta)$ thus becomes the vector gathering the ROM dofs.

To build $\mathbf{W}$, the proper orthogonal decomposition (POD), see, e.g., [16], is adopted. POD provides the projection bases via a singular value decomposition of the snapshot matrix $\mathbf{S} = [\mathbf{v}_1, \dots, \mathbf{v}_S] \in \mathbb{R}^{M \times S}$, collecting S snapshots of the FOM, according to:

$$\mathbf{S} = \mathbf{P}\boldsymbol{\Sigma}\mathbf{Z}^\top, \tag{3}$$

where $\mathbf{P} = [\mathbf{p}_1, \dots, \mathbf{p}_M] \in \mathbb{R}^{M \times M}$ is an orthogonal matrix, whose columns are the left singular vectors of $\mathbf{S}$; $\boldsymbol{\Sigma} \in \mathbb{R}^{M \times S}$ is a pseudo-diagonal matrix collecting the singular values of $\mathbf{S}$, arranged so that $\sigma_1 \geq \sigma_2 \geq \dots \geq \sigma_r \geq 0$, $r = \min(S, M)$ being the rank of $\mathbf{S}$; $\mathbf{Z} = [\mathbf{z}_1, \dots, \mathbf{z}_S] \in \mathbb{R}^{S \times S}$ is an orthogonal matrix, whose columns are the right singular vectors of $\mathbf{S}$.

By retaining the first $W \leq r$ left singular vectors, the POD bases $\mathbf{W} = [\mathbf{p}_1, \ldots, \mathbf{p}_W]$ are obtained. The dimension W can be set by prescribing the reconstruction error in reducing the order of the problem from M to W, to be smaller than a tolerance ϵ according to:

$$\frac{\sum_{l=1}^{W}(\sigma_l)^2}{\sum_{l=1}^{r}(\sigma_l)^2} \geq \epsilon. \tag{4}$$

Through a Galerkin projection of the FOM onto the space spanned by $\mathbf{W}$, we obtain:

$$\begin{cases} \mathbf{M}_R\ddot{\mathbf{v}}_R(t) + \mathbf{K}_R(g,\delta)\mathbf{v}_R(t) = \mathbf{f}_R(\boldsymbol{\eta}) \ , \quad t \in (0, T_f) \\ \mathbf{v}_R(0) = \mathbf{W}^\top \mathbf{v}_0 \\ \dot{\mathbf{v}}_R(0) = \mathbf{W}^\top \dot{\mathbf{v}}_0 \end{cases} \tag{5}$$

where

$$\mathbf{M}_R \equiv \mathbf{W}^\top\mathbf{M}\mathbf{W} \ , \qquad \mathbf{K}_R(g,\delta) \equiv \mathbf{W}^\top\mathbf{K}(g,\delta)\mathbf{W}, \qquad \mathbf{f}_R(\boldsymbol{\eta}) \equiv \mathbf{W}^\top\mathbf{f}(\boldsymbol{\eta}). \tag{6}$$

Once integrated in time, the reduced order solution $\mathbf{V}^R = [\mathbf{v}_1^R, \ldots, \mathbf{v}_{L_M}^R] \in \mathbb{R}^{W \times L_M}$, can be back-projected to recover the FOM solution at each time instant.

4 Deep Learning

In this section, we aim at detailing the two NNs used to perform the damage detection and localization first, and the damage quantification next. We describe how TL can be exploited to speedup and also improve the training of the NN adopted in the second stage of the procedure.

The two NNs aim at approximating a target function, in our case linked to the structural response to the external loading, respectively, mapping $\mathbf{U}$ onto the discrete set g of damage patterns for the joint detection and localization task (working as a classification model), and onto the damage level δ for the quantification task (working as a regression model).

4.1 Damage Detection and Localization

The supervised training of a NN concretely consists of a tuning of the network parameters, usually performed via gradient-descent algorithms in an iterative manner, to minimize the discrepancy between the network output and the dataset $\mathbf{D}$, processed a certain number of times or epochs. At each iteration, B instances, called mini-batches, are simultaneously processed; in the present analysis, we have assumed $B = 32$.

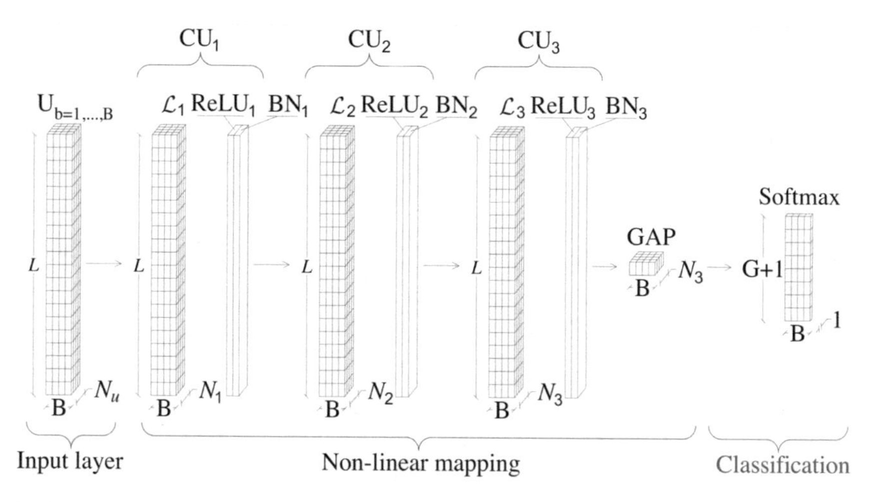

Fig. 2 FCN classification model architecture

As anticipated, training aims at minimizing a so-called loss function, to tune the weights that rule the (nonlinear) mappings performed by the NN. For the classification model, the loss function C_c that quantifies the classification error is given by the cross-entropy:

$$C_c(\mathbf{\Phi}, \mathbf{\Psi}) = -\frac{1}{B}\sum_{b=1}^{B}\sum_{g=0}^{G}\varphi_{bg}\log\left(\psi_{bg}\right) \tag{7}$$

between $\varphi_{bg} \in \{0, 1\}$, stating if the g-th damage class has to be associated with the b-th instance $\mathbf{U}_b$ of the processed mini-batch, and $\psi_{bg} \in [0, 1]$, providing the confidence by which the b-th instance $\mathbf{U}_b$ is associated with g. The values of φ_{bg} and ψ_g are collected in the matrices $\mathbf{\Phi} \in \{0, 1\}^{B\times(G+1)}$ and $\mathbf{\Psi} \in [0, 1]^{B\times(G+1)}$, respectively.

In Fig. 2, a schematic representation of the classification model is reported. To compute $\mathbf{\Psi}$, B input instances are processed by three convolutional units (CUs), featuring, respectively, $N_1 = 32$, $N_2 = 64$ and $N_3 = 32$ kernel filters, followed by a global average pooling (GAP) layer and by a linear projection whose output is the score matrix $\mathbf{\Theta} = (\boldsymbol{\vartheta}_0, \ldots, \boldsymbol{\vartheta}_G) \in \mathbb{R}^{B\times(G+1)}$. $\mathbf{\Theta}$ is next handled as the argument of a softmax function in the following form:

$$\psi_{bg} = \frac{e^{\vartheta_{bg}}}{\sum_{g=0}^{G} e^{\vartheta_{bg}}} \tag{8}$$

to finally obtain the matrix $\mathbf{\Psi}$ associated with the input mini-batch. Each CU consists of a convolutional layer $\mathcal{L}$, a rectified linear unit (ReLU) activation layer, and batch normalization (BN) layer. Readers are referred to, e.g., [18, 20, 25], for further details on how CUs and GAP perform signal processing and feature extraction.

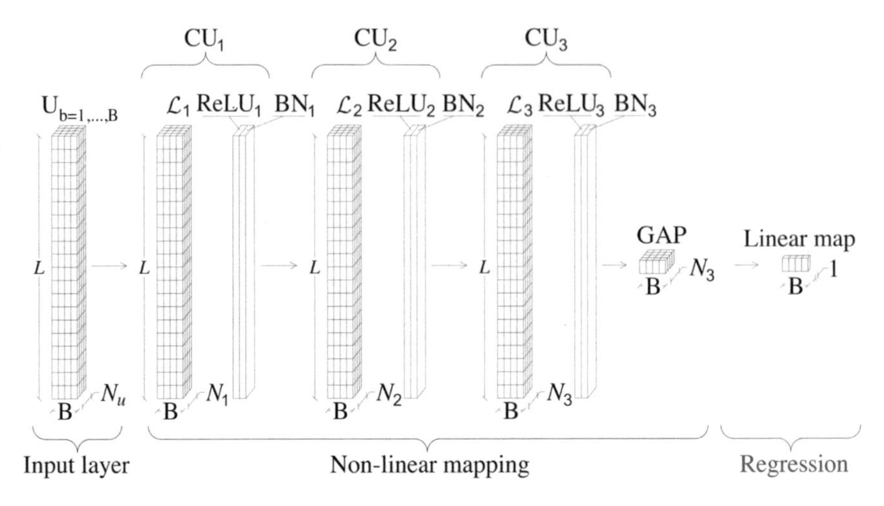

Fig. 3 FCN regressor architecture

4.2 Damage Quantification

To perform damage quantification, the FCN architecture shown in Fig. 3 is employed. Its part carrying out features extraction is identical to the one employed for classification; the most important difference is represented by the linear projection operated after the GAP, whose outcome is the predicted damage level $\boldsymbol{\delta}_r = [\delta_{r1}, \ldots, \delta_{rB}] \in \mathbb{R}^B$ for the processed mini-batch. The adopted loss function C_r, according to the discussion in [19] regarding regression, is the mean square error (mse), so that:

$$C_r\left(\boldsymbol{\delta}_r, \boldsymbol{\delta}\right) = \frac{1}{B} \sum_{b=1}^{B} \left(\delta_{rb} - \delta_b\right)^2, \tag{9}$$

where $\boldsymbol{\delta} = [\delta_1, \ldots, \delta_B] \in \mathbb{R}^B$ are the target damage levels associated with the processed mini-batch instances.

It is to note that the information regarding damage localization is not allowed for by the regression model. Running several tests, we have figured out that this choice has no impact on the procedure performance in the present case; anyhow, we cannot exclude that, for more complex applications, allowing for the outcome of damage localization as a further input to the damage quantification stage may be beneficial.

4.3 Transfer Learning

The weights of the regression model can be randomly initialized or, alternatively, their initialization can account for the weights of the trained classification model. In this way, the regression model can exploit the signal processing and feature extraction capacity of the classification model. Indeed, features are synthetic descriptions of data obtained for a specific task, but they may be also adopted for other tasks if there is redundancy across the learnt and new tasks, see [26].

Two benefits are expected by the use of TL: speeding up of training; reduction of the data required for training. All these outcomes will be investigated next with reference to a specific case test.

5 Results

To assess the capability and the performance of the proposed SHM procedure, we apply here our strategy to the case study as shown in Fig. 4 and referred to a two-dimensional portal frame.

The adopted FE mesh shown in Fig. 4a contains 1884 dofs. The frame is perfectly clamped at the bases and has a thickness of 0.1 m. It is assumed to be made of concrete, whose mechanical properties are: Young's modulus $E = 30$ GPa, Poisson's ratio $\nu = 0.2$, and density $\rho = 2500$ kg/m^3. The structure is excited by seven different loading conditions $\mathcal{C} \in \{1, \ldots, 7\}$, obtained by combining the three distributed loads q_k, $k = 1, 2, 3$ as shown in Fig. 4b. Each load q_k varies in time according to $q_k(t) = Q_k \sin(2\pi f_k t)$, where Q_k and f_k are the relevant amplitude and frequency that vary in the ranges [10–50 kPa] and [50–95 Hz], respectively. These parameters, which embody the variability of the operational conditions, are collected in the vector $\boldsymbol{\eta} = \{\mathcal{C}, Q_1, Q_2, Q_3, f_1, f_2, f_3\}^\top$.

The $G = 4$ damageable regions Ω_g are shaded in Fig. 4b. The considered damage scenarios are built by assuming that stiffness reduction can vary in each domain within the range $\delta \in$ [10–25%]. In the same Fig. 4b, also the $N_u = 7$ sensed displacements $\mathbf{u}_n(t)$ are shown. The sensors are scheduled to gather the measurements with a sampling frequency of 200 Hz; this allows avoiding aliasing regarding the first 7 vibration modes of the structure. The signal-to-noise ratio of such measurements is assumed to be equal to 80.

$Y = 400$ samples of $\{\boldsymbol{\eta}, g, \delta\}$ have been adopted to assemble $\mathbf{S}$; for each sample, the solutions at $X = 121$ time instants have been collected. Therefore, a total of $S = X \times Y = 48{,}400$ snapshots are handled. By prescribing an error tolerance $\epsilon = 10^{-3}$, $W = 70$ POD bases get selected, to replace the original 1884 dofs of the FOM: the first ten of such bases are sketched in Fig. 5.

As far as damage detection and localization are concerned, the evolution of the loss function $\mathcal{L}_c$ and the prediction accuracy during the training of the NN are reported in Fig. 6 against the number of epochs, both for training and validation. The training is stopped either when a maximum number of epochs has been attained, or when the loss C_c does not decrease during 15 epochs in a row. The loss function plot shows that the minimization of C_c is successfully carried out; this assures the increasing accuracy of classification, which represents the percentage of instances for which the damage is correctly detected and localized. The training has been early stopped and overfitting is shown to be avoided, since the outcomes regarding the training and the validation datasets are very similar. Once trained, the classification model has been tested by employing another dataset collecting pseudo-experimental instances simulated through the FOM. The relevant classification outcome is reported in the confusion matrix of Fig. 7, which shows an overall accuracy of around 99%, with just a few misclassification errors. The mentioned misclassification errors affect the damage localization task, while damage detection, to distinguish the undamaged state $g = 0$ from all the other scenarios, is perfectly accomplished.

As far as damage quantification is instead concerned, Fig. 8 provides the evolution of the loss C_r against the training epochs. The training has been early stopped after 5 epochs without a decrease of the loss on the validation set. The plot highlights a consistent reduction of such loss, especially in the first stage of training. By comparing the evolutions relevant to the training and validation sets, it can be stated that overfitting is avoided again.

The trained regression model has been next adopted to perform damage quantification on a test set provided as before by the FOM. The relevant outcomes are shown in Fig. 9, in terms of a parity plot to compare the NN predictions (reported along the vertical axis) and the target damage levels (reported instead along the horizontal axis), and in terms of the corresponding prediction errors. The reported values of the damage level are distributed in a rather narrow range around the line bisecting the quadrant, to testify the NN generalization capacity for unseen instances generated by the FOM. In the bar chart, the counts are proportional to the number of instances

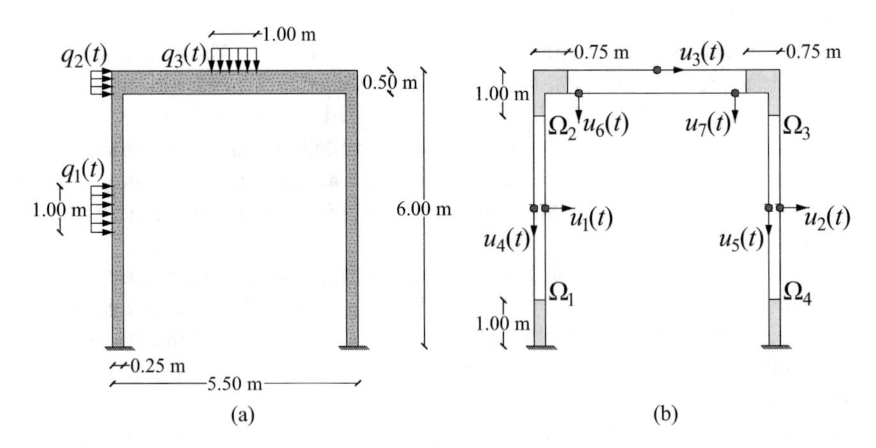

Fig. 4 Portal frame. **a** Geometry, FE discretization and applied loads; **b** subdomains $\{\Omega_1, \ldots, \Omega_4\}$ where damage may occur, and sensed displacement components.

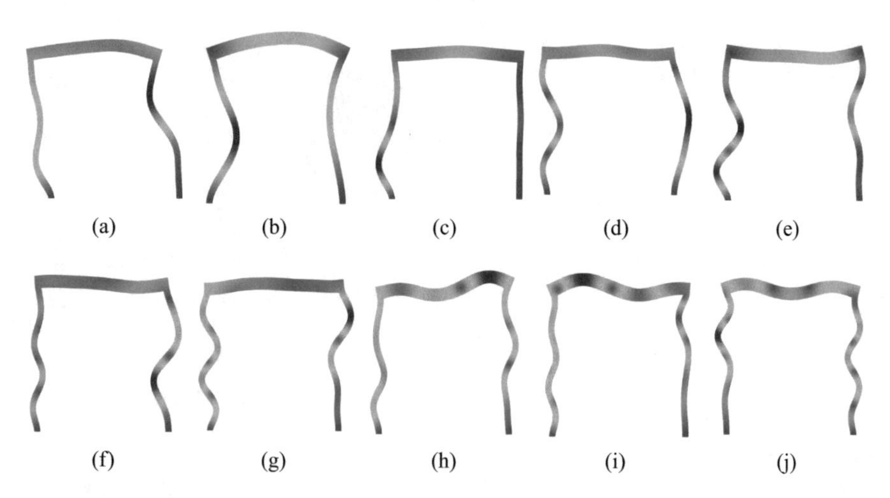

Fig. 5 Portal frame—first ten POD bases

Fig. 6 Evolution across epochs **a** of the loss C_c and **b** of the global accuracy of the classification model

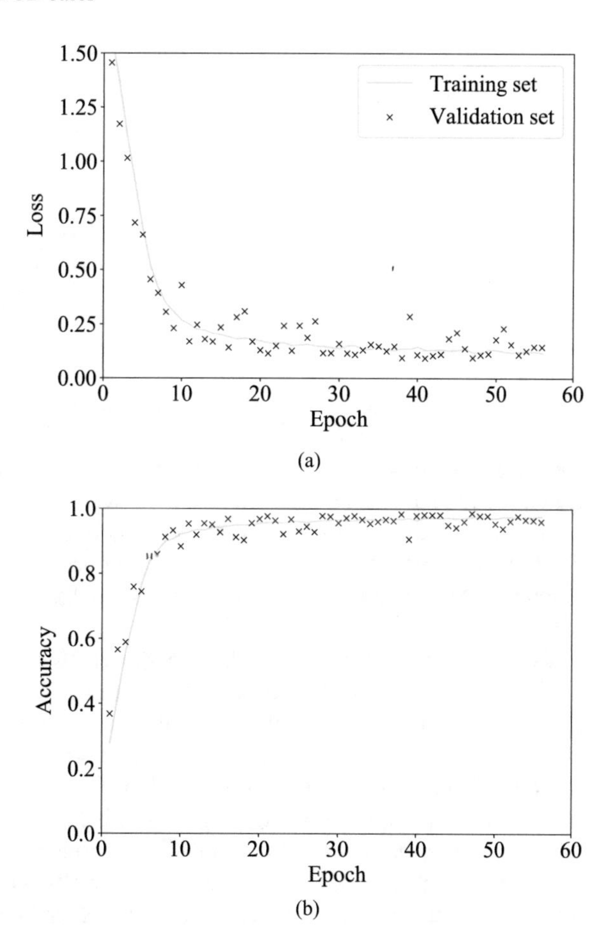

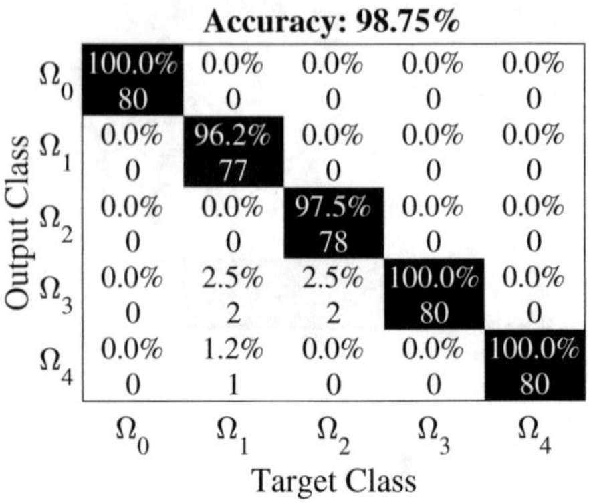

Fig. 7 Confusion matrix relevant to the classification model testing

Accuracy: 98.75%

Output Class \ Target Class	Ω_0	Ω_1	Ω_2	Ω_3	Ω_4
Ω_0	100.0% 80	0.0% 0	0.0% 0	0.0% 0	0.0% 0
Ω_1	0.0% 0	96.2% 77	0.0% 0	0.0% 0	0.0% 0
Ω_2	0.0% 0	0.0% 0	97.5% 78	0.0% 0	0.0% 0
Ω_3	0.0% 0	2.5% 2	2.5% 2	100.0% 80	0.0% 0
Ω_4	0.0% 0	1.2% 1	0.0% 0	0.0% 0	100.0% 80

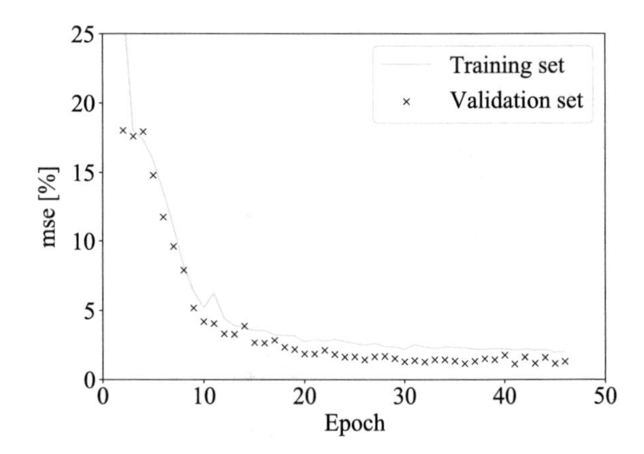

Fig. 8 Evolution across epochs of the loss C_r of the regression model

for which a certain prediction error has been observed a-posteriori; a Gaussian-like distribution of the mentioned prediction error can be seen. This outcome is actually connected to the definition of C_r: when the mse is employed as the loss function, the regression model tries to predict the expected value $\mathbb{E}\left[\delta_b|\mathbf{U}_b\right]$ of the sequence δ_b, with $b = 1, \ldots, B$, conditioned on $\mathbf{U}_b$, see [3]; due to the central limit theorem, the distribution asymptotically tends to a Gaussian one if the number of samples grows to infinity. Accordingly, the prediction errors tend to inherit a Gaussian distribution too.

Regarding the exploitation of TL, from the results reported in Fig. 10 it is possible to appreciate the induced speedup of the learning process: the number of training epochs turns out to be more than halved than that necessary for the case without TL, as shown in Fig. 8. The reduced number of epochs is obtained together with a reduction of the training time, as reported in Fig. 11. In the same plot, the training time required when the training dataset is halved is also reported; by halving the dimension of $\mathbf{D}$, we aimed at assessing if TL may enable the use of a smaller dataset, assuring the same or similar performances of the NN. TL is expected to enhance the NN performance because it eases the training procedure, via a better initialization

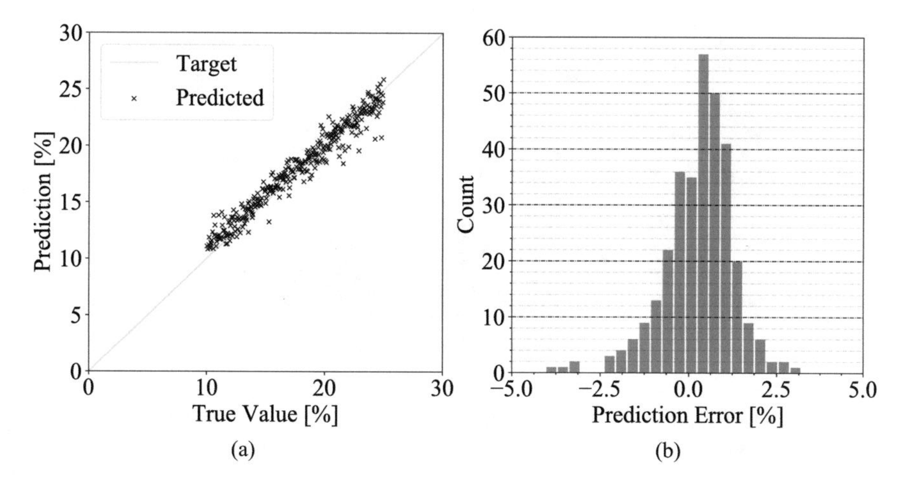

Fig. 9 Regression outcomes in terms of damage level: **a** Parity plot comparing the predicted values to the ground truth ones; **b** histogram of the relevant prediction errors

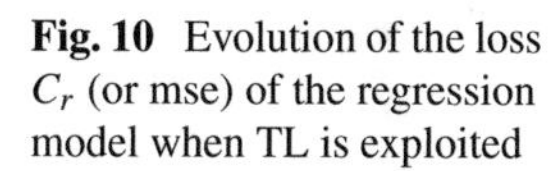

Fig. 10 Evolution of the loss C_r (or mse) of the regression model when TL is exploited

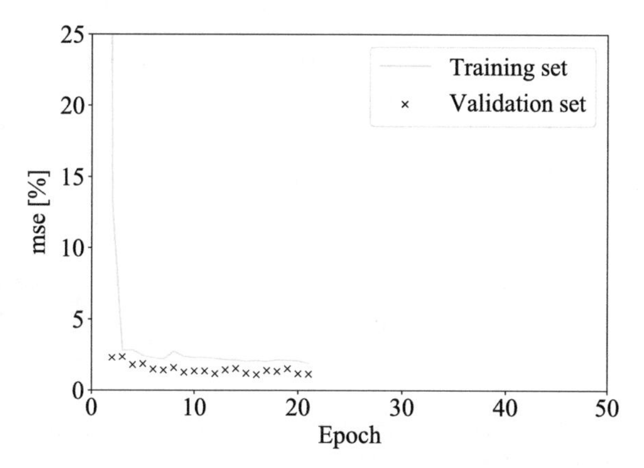

of the weights, and therefore allows obtaining optimal results even if **D** is narrowed. The performance improvement can be qualitatively assessed through Fig. 12, and specifically comparing the prediction error distribution with that reported in Fig. 9: the error results to be smaller if TL is employed.

A quantitative assessment of the effect of TL on the regression performance is reported in Table1, by adopting the mean absolute error (mae) on the test set as the evaluation metric. TL leads to an improvement of the regression performance when the entire dataset **D** is considered. The same improvement has not been obtained when the dimension of **D** has been halved, as an almost identical mae value is obtained for the analyses with and without TL. On the other hand, in all the cases TL provides a considerable speedup of training, with a reduction of the number of epochs from 46 to 21 for the entire dataset, and from 79 to 25 for the halved one. In this table, the mean μ and standard deviation ς of the prediction error, as evaluated on the

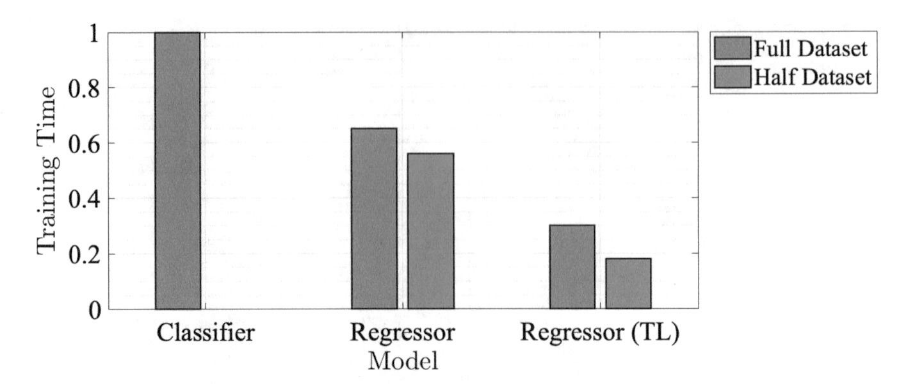

Fig. 11 Comparison among the training times of the three adopted models, at varying number of instances in the processed dataset

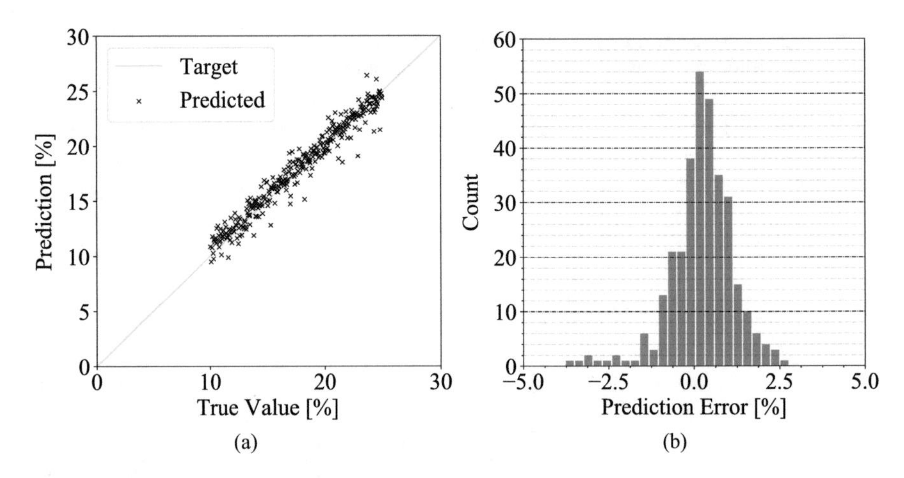

Fig. 12 Regression outcomes when TL is exploited, in terms of damage level: **a** Parity plot comparing the predicted values to the ground truth ones; **b** histogram of the relevant prediction errors

test set, are also reported. Having adopted the mse as loss function, these statistic features are employed to assess the performance of the regression model, as they fully characterize a Gaussian distribution. The error mean μ shows a small bias in the prediction of δ, probably because the solution has got entrapped into a local minimum during the optimization procedure. The greater accuracy of the regression model trained by exploiting TL is highlighted by the smaller values of ς in both the analyses.

The proposed classification and regression models are finally adopted to track the evolution of the damage level δ, under the loading conditions detailed in Table2. In the same Table, also the time-evolving value of δ to be identified on-the-fly is reported. During the analysis, each time damage is detected and localized by the classification model, and the regression model is plugged in to predict its magnitude.

Table 1 Regression model: comparison of the performances if TL is exploited or not, and if dataset is reduced in size, in terms of training time, generalization capability, and mean μ and standard deviation ς of the prediction error

Model	Dataset (%)	Training epochs	Mae (test set)	μ	ς
Regression	100	46	0.80	0.26	0.98
Regression	50	79	0.79	0.14	1.06
Regression (TL)	100	21	0.70	0.22	0.92
Regression (TL)	50	25	0.81	−0.27	1.02

Table 2 Data regarding the damage level and damage class, and output of the evolutionary classification model. The considered test case features: $g = 1$, $Q_1 = Q_3 = 0$, $Q_2 = 20$ kPa, $f_2 = 60$ Hz

Damage level (%)	Output class	Target class
5	0	1
6	0	1
7	0	1
8	0	1
9	1	1
10	1	1
⋮	⋮	⋮
25	1	1

At this point, the dataset **D** is updated by dropping the instances featuring a value of δ in the domains where damage has been localized, smaller than the identified one, and by adding new instances featuring a value of δ larger than or equal to the previously mentioned one. Within the present approach, those instances are provided by the FOM to keep the accuracy of the online identification task the highest possible and avoid a bias in the estimations.

Regarding the time necessary to update the baseline, generate new data and re-train the NNs every time damage grows, some results are provided next as obtained with a laptop featuring an Intel Core i5 CPU @ 2.6 GHz and 8 GB RAM. The CPU time required by each ROM analysis is 0.2 s, and by each FOM analysis is 1.3 s, respectively. The duration of the subsequent training of the classifier amounts to around 10 min, while that of the regressor amounts to 4 min if TL is exploited. Overall, the update stage requires less than 1 h of outage for the SHM procedure. If the FOM analyses become too time-taking for baseline update to be handled close to online, techniques for the dynamic update of the ROM bases may be exploited. This duration can be obviously (by far) reduced if a more powerful computer is used to run the procedure.

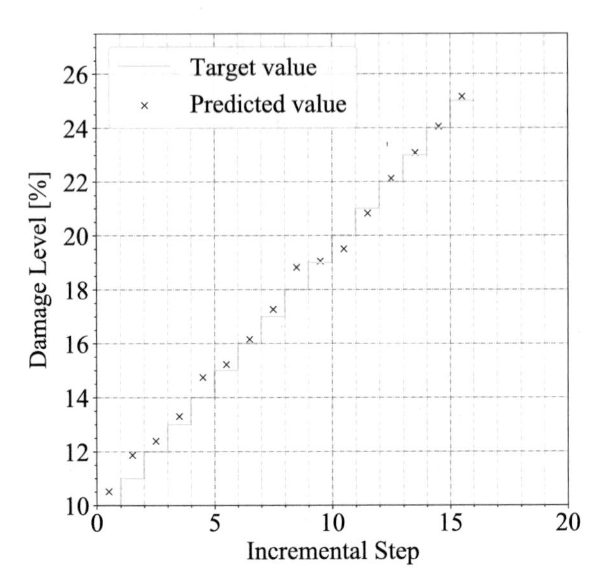

Fig. 13 Regression outcome in terms of step-wise damage growth: comparison between predicted evolution and groud-truth data

As shown in Fig. 13, the proposed procedure is able to self-adapt and to track damage evolution. For the sake of brevity, results are shown for this damage case only, but the present framework can be successfully applied also to other cases characterized by different damage patterns.

6 Conclusions

In this work, we have proposed a self-adaptive, evolutionary-like procedure to detect, localize and quantify a growing damage in structural systems. According to a supervised learning strategy, a full-order numerical model is assumed to play the role of a digital-twin of the structure, allowing to parametrize the loading conditions and the damage state.

A dataset of pseudo-experimental vibration recordings, assumed to be sensed by a monitoring system, is simulated via a reduced-order model built upon the full order one through the reduced basis method. The dataset is used to train offline first a NN-based classification model for damage detection and localization and then a NN-based regression model for damage quantification. Transfer learning has been exploited to improve and speedup the training of the regression model, by allowing for the tuned parameters of the NN obtained for the classification model. Once trained, the two models have been adopted to track the evolution of damage in the monitored structure, showing an extremely good capability to allow an online update of the estimates while damage is growing.

In future activities, we aim to address the optimal placement of sensors in the network, as studied, e.g., in [2, 5], since the relevant outcomes can positively affect the accuracy of the NN-based approach and allow reducing the amount of data required to attain the sought accuracy. The environmental variability is going to be also allowed for. The procedure will be then applied to real experimental settings.

Acknowledgements M. T. acknowledges the financial support by Politecnico di Milano through the interdisciplinary Ph.D. Grant "Physics-informed deep learning for structural health monitoring."

References

1. Bigoni C, Hesthaven JS (2020) Simulation-based anomaly detection and damage localization: an application to structural health monitoring. Comput Methods Appl Mech Eng 363:112896. https://doi.org/10.1016/j.cma.2020.112896
2. Bigoni C, Zhang Z, Hesthaven JS (2020) Systematic sensor placement for structural anomaly detection in the absence of damaged states. Comput Methods Appl Mech Eng 371:113315. https://doi.org/10.1016/j.cma.2020.113315
3. Bishop CM (1994) Mixture density networks. Aston University
4. Bull L, Worden K, Dervilis N (2020) Towards semi-supervised and probabilistic classification in structural health monitoring. Mech Syst Signal Process 140:106653. https://doi.org/10.1016/j.ymssp.2020.106653
5. Capellari G, Chatzi E, Mariani S (2018) Structural health monitoring sensor network optimization through Bayesian experimental design. ASCE-ASME J Risk Uncertain Eng Syst, Part A: Civ Eng 4(2):04018016. https://doi.org/10.1061/AJRUA6.0000966
6. Corigliano A, Mariani S (2004) Parameter identification in explicit structural dynamics: performance of the extended Kalman filter. Comput Methods Appl Mech Eng 193(36–38):3807–3835. https://doi.org/10.1016/j.cma.2004.02.003
7. De S, Britton J, Reynolds M, Skinner R, Jansen K, Doostan A (2020) On transfer learning of neural networks using bi-fidelity data for uncertainty propagation. Int J Uncertain Quantif 10(6):543–573
8. Doebling SW, Farrar CR, Prime MB, Shevitz DW (1996) Damage identification and health monitoring of structural and mechanical systems from changes in their vibration characteristics: a literature review. Tech Rep LA-13070-MS, Los Alamos National Lab., NM, USA. https://doi.org/10.2172/249299. https://www.osti.gov/biblio/249299
9. Eftekhar Azam S, Chatzi E, Papadimitriou C (2015) A dual Kalman filter approach for state estimation via output-only acceleration measurements. Mech Syst Signal Process 60–61:866–886. https://doi.org/10.1016/j.ymssp.2015.02.001
10. Eftekhar Azam S, Mariani S (2018) Online damage detection in structural systems via dynamic inverse analysis: a recursive Bayesian approach. Eng Struct 159:28–45. https://doi.org/10.1016/j.engstruct.2017.12.031
11. Entezami A, Sarmadi H, Behkamal B, Mariani S (2020) Big data analytics and structural health monitoring: a statistical pattern recognition-based approach. Sensors 20(8):2328. https://doi.org/10.3390/s20082328
12. Farrar C, Worden K (2013) Structural health monitoring: a machine learning perspective. Wiley, Hoboken. https://doi.org/10.1002/9781118443118
13. Goodfellow I, Bengio Y, Courville A (2016) Deep learning. MIT Press. http://www.deeplearningbook.org
14. Munir M, Siddiqui SA, Dengel A, Ahmed S (2019) DeepAnT: a deep learning approach for unsupervised anomaly detection in time series. IEEE Access 7:1991–2005. https://doi.org/10.1109/ACCESS.2018.2886457
15. Pandey A, Biswas M (1994) Damage detection in structures using changes in flexibility. J Sound Vib 169(1):3–17. https://doi.org/10.1006/jsvi.1994.1002

16. Quarteroni A, Manzoni A, Negri F (2015) Reduced basis methods for partial differential equations: an introduction. Unitext, vol 92. Springer, Berlin
17. Rafiei MH, Adeli H (2018) A novel unsupervised deep learning model for global and local health condition assessment of structures. Eng Struct 156:598–607. https://doi.org/10.1016/j.engstruct.2017.10.070
18. Rosafalco L, Manzoni A, Mariani S, Corigliano A (2020) Fully convolutional networks for structural health monitoring through multivariate time series classification. Adv Model Simul Eng Scie 7:38. https://doi.org/10.1186/s40323-020-00174-1
19. Rosafalco L, Manzoni A, Mariani S, Corigliano A (2021) An autoencoder-based deep learning approach for load identification in structural dynamics. Submitted
20. Rosafalco L, Torzoni M, Manzoni A, Mariani S, Corigliano A (2021) Online structural health monitoring by model order reduction and deep learning algorithms. Submitted
21. Rytter A (1993) Vibrational based inspection of civil engineering structures. Ph.D. thesis, University of Aalborg, Denmark
22. Sohn H, Worden K, Farrar CR (2002) Statistical damage classification under changing environmental and operational conditions. J Intell Mater Syst Struct 13(9):561–574. https://doi.org/10.1106/104538902030904
23. Sudret B, Defaux G, Pendola M (2007) Stochastic evaluation of the damage length in rc beams submitted to corrosion of reinforcing steel. Civ Eng Environ Syst 24(2):165–178. https://doi.org/10.1080/10286600601159305
24. Taddei T, Penn J, Yano M, Patera A (2018) Simulation-based classification; a model-order-reduction approach for structural health monitoring. Arch Comput Methods Eng 25(1):23–45
25. Torzoni M, Rosafalco L, Manzoni A (2020) A combined model-order reduction and deep learning approach for structural health monitoring under varying operational and environmental conditions. Eng Proc 2(1). https://doi.org/10.3390/ecsa-7-08258
26. Zamir AR, Sax A, Shen W, Guibas LJ, Malik J, Savarese S (2018) Taskonomy: disentangling task transfer learning. In: 2018 IEEE/CVF conference on computer vision and pattern recognition, 18 June–23 June, pp 3712–3722. Salt Lake City, UT. https://doi.org/10.1109/CVPR.2018.00391
27. Zhang T, Biswal S, Wang Y (2020) Shmnet: condition assessment of bolted connection with beyond human-level performance. Struct Health Monit 19(4):1188–1201. https://doi.org/10.1177/1475921719881237

Predictive Monitoring of Large-Scale Engineering Assets Using Machine Learning Techniques and Reduced-Order Modeling

Caterina Bigoni, Mengwu Guo, and Jan S. Hesthaven

Abstract Structural health monitoring techniques aim at providing an automated solution to the threat of unsurveilled aging of structures that can have tremendous consequences in terms of fatalities, environmental pollution, and economic loss. To assess the state of damage of a complex structure, this paper proposes to fully characterize its behavior under multiple environmental and operational scenarios and compare new sensor measurements with the baseline behavior. However, the repeated simulations of a nonlinear, time-dependent structural model with high-dimensional input parameters represent a severe computational bottleneck for large-scale engineering assets. This chapter presents how to use efficient reduced-order modeling techniques to mitigate the computational effort of many-query simulations without jeopardizing the accuracy. To compare new sensor measurements with the natural behavior of synthetic solutions, the proposed methodology uses hierarchical semi-supervised learning algorithms on a small amount of extracted damage-sensitive features, thus allowing one to assess the state of damage in real time. Using the inexpensive simulations, one can also optimally place sensors to maximize the observability of discriminant features. The all-round methodology is validated on a numerical example.

Keywords Simulation-based anomaly detection · Reduced order models · One-class classification · Sensor placement · Structural health monitoring · Digital twins

C. Bigoni (✉) · J. S. Hesthaven
Computational Mathematics and Simulation Science (MCSS), Ecole Polytechnique Fédérale de Lausanne (EPFL), Lausanne, Switzerland
e-mail: caterina.bigoni@alumni.epfl.ch

J. S. Hesthaven
e-mail: Jan.Hesthaven@epfl.ch

M. Guo
Department of Applied Mathematics, University of Twente, Enschede, Netherlands
e-mail: m.guo@utwente.nl

A. Cury et al. (eds.), *Structural Health Monitoring Based on Data Science Techniques*, Structural Integrity 21, https://doi.org/10.1007/978-3-030-81716-9_9

1 Introduction

Many existing private and public assets, such as civil engineering infrastructures, buildings, or aircraft, require reliable damage detection techniques to be safely used, especially during their inevitable aging. When monitoring a structure over its lifecycle, its deterioration and damages represent a great concern and the early detection of critical decay might prevent failures that can cause sudden shutdowns or even catastrophes with severe life-safety and economic repercussions [1]. To prevent these critical failures from happening, techniques of structural health monitoring (SHM) have been developed in recent studies with applications to civil and aerospace engineering, as well as to the conservation of cultural heritage structures. SHM refers to automated monitoring procedures that seek to provide reliable information on the performance and integrity of a structure in real time. In the context of SHM, the combination of sensor measurements, numerical models simulating the underlying behavior of a structure of interest under different environmental and operational conditions, and machine learning techniques has led to the design of structural *digital twins* [2].

The focus of this work is on *wave propagation* approaches to data-driven, predictive SHM, which aim to detect damages by examining the distortions in propagating elastic waves as a result of reflections and amplitude attenuations when intersecting the damage boundary. Featuring a data-driven nature, these approaches, sometimes called *simulation-based* SHM [3–7], are decomposed into an offline phase and an online phase. In the former, a database of synthetic signals is built to represent the structural behavior under different conditions, while in the latter real experimental time signals, collected from sensors placed on a structure, are compared with those simulated offline using a classifier that discriminates between damaged and undamaged states.

Toward an efficient and robust scheme of data-driven predictive monitoring, reduced-order modeling techniques are integrated into a wave propagation approach with cutting-edge machine learning tools. The data-driven SHM setting corresponds to a multiquery problem, where one has to solve high-dimensional, time-dependent, parametric equations, which results in great demands for computational resources. To overcome such a computational burden, model order reduction is employed to project the original full-order system onto a reduced space with a significantly lower dimensionality. In this way, a robust dataset of approximated sensor measurements is generated. In SHM, the task of damage detection is typically reduced to a supervised learning process relying on a fully labeled dataset, obtained from both healthy and damaged structures (either generated with computer-aided procedures or collected experimentally). However, gathering an exhaustive collection of configuration classes anticipating all types of damages is typically unrealistic and the number of different classification labels may grow rapidly. Instead, this work relies on semi-supervised learning techniques, also called *one-class classification* methods, which learn the common features among labeled data belonging to the *normal* class in the training phase. Unlabeled data from both classes are then used in the test phase to identify abnormal data which deviate from the normal mode. One-class algorithms allow to locate the damage and estimate its severity by training a different model for each sensor location. Finally, guided by an appropriate indicator of the damage detection performance, a modified sparse Gaussian process method is applied to the synthetic dataset of healthy configurations to systematically place a fixed number of sensors on a structure of interest.

Following the introduction, a model problem for sensor measurements governed by the acoustic-elastic equation is briefly reviewed in Sect. 2. Basic techniques of reduced-order modeling, which can be used for the multiquery simulations in SHM, are introduced in Sect. 3. A local semi-supervised method is used for automatic anomaly detection in Sect. 4, and the variational sparse Gaussian process model is utilized for optimal sensor placement in Sect. 5. In Sect. 6, the all-round methodology is demonstrated by a numerical example. Finally, conclusions are drawn in Sect. 7.

2 A Model for Sensor Measurements

This section introduces the model problem for sensor measurements, including its governing equations and parametric discrete formulations.

2.1 Governing Equation

Let $\Omega \subset \mathbb{R}^{d_\Omega}$ be a polygonal physical domain with piece-wise smooth boundary $\partial\Omega$, where $d_\Omega = 2, 3$ is the spatial dimension, and let $[0, T]$, with $T \in \mathbb{R}_+$, be a suitable time interval. Here, Ω represents a structure of interest and $[0, T]$ a suitable time window to observe the response of a structure undergoing a predefined excitation (i.e., the effect of an active or passive source) through sensor measurements. Moreover, let $\mathcal{P} \subset \mathbb{R}^{d_\mu}$ be a suitable parameter domain, where d_μ indicates the number of input parameters required to describe the healthy variations that a structure may undergo during its life time, and let $\boldsymbol{\mu} = (\mu_1, \ldots, \mu_{d_\mu})$ be a parameter vector representing one possible healthy variation of the environmental and operational conditions, i.e., μ_i may relate to the material properties, the boundary conditions, the initial conditions, or the source function for $1 \leq i \leq d_\mu$.

Let $\boldsymbol{u} = \boldsymbol{u}(\boldsymbol{x}, t; \boldsymbol{\mu}) : \Omega \times [0, T] \times \mathcal{P} \to \mathbb{R}^{d_\Omega}$ be a vector-valued displacement field, solution of the acoustic-elastic equation equipped with suitable boundary and initial conditions:

$$\begin{cases} \rho(\ddot{\boldsymbol{u}} + \eta\dot{\boldsymbol{u}}) - \nabla \cdot \boldsymbol{\sigma} = h(t; \boldsymbol{\mu})\boldsymbol{b}(\boldsymbol{x}; \boldsymbol{\mu}) & \text{in } \Omega \\ \boldsymbol{u} \cdot \boldsymbol{n} = \boldsymbol{0}, \quad (\boldsymbol{\sigma} \cdot \boldsymbol{n}) \cdot \boldsymbol{\tau} = \boldsymbol{t}_N & \text{on } \partial\Omega \, , \\ \boldsymbol{u}|_{t=0} = \boldsymbol{u}_0, \quad \dot{\boldsymbol{u}}|_{t=0} = \boldsymbol{v}_0 & \text{in } \Omega \end{cases} \tag{1}$$

where $\ddot{\boldsymbol{u}} = \partial^2 \boldsymbol{u}/\partial t^2$ and $\dot{\boldsymbol{u}} = \partial \boldsymbol{u}/\partial t$ are the acceleration and velocity fields. Here, ρ is the density coefficient, η is a dimensionless damping coefficient, $h : [0, T] \times \mathcal{P} \to \mathbb{R}$ and $\boldsymbol{b} : \Omega \times \mathcal{P} \to \mathbb{R}^{d_\Omega}$ are two source functions depending on time and space, respectively, $\boldsymbol{\sigma} = \boldsymbol{\sigma}(\boldsymbol{u}; \boldsymbol{\mu})$ is the stress tensor, $\boldsymbol{n}$ and $\boldsymbol{\tau}$ are the outward normal and tangential (unit) vectors to $\partial\Omega$, respectively, $\boldsymbol{t}_N = \boldsymbol{t}_N(\boldsymbol{x}, t; \boldsymbol{\mu})$ is the traction vector used in the definition of the free-slip boundary conditions, $\boldsymbol{u}_0 = \boldsymbol{u}_0(\boldsymbol{x}; \boldsymbol{\mu})$ and $\boldsymbol{v}_0 = \boldsymbol{v}_0(\boldsymbol{x}; \boldsymbol{\mu})$ describe the initial displacement and velocity in space, respectively.

The ultimate goal is to emulate the real sensor response at m given sensor locations $\boldsymbol{x}_i \in \Omega$ for $1 \leq i \leq m$. To do so, let $\ell : \mathbb{R}^{d_\Omega} \times \mathcal{P} \to \mathbb{R}^{d_g}$ be an input–output function and $g_i : [0, T] \times \mathcal{P} \to \mathbb{R}^{d_g}$ a (parametric) output of interest, i.e., an approximation of the sensor response at time t and location $\boldsymbol{x}_i$:

$$g_i(t; \boldsymbol{\mu}) = \ell(\boldsymbol{u}(\boldsymbol{x}_i, t; \boldsymbol{\mu}); \boldsymbol{\mu}), \quad 1 \leq i \leq m. \tag{2}$$

Before proceeding with classic discretization techniques such as the finite element method, consider the vector space $\mathbb{V} = \{\boldsymbol{w} \in [H^1(\Omega)]^{d_\Omega} : \boldsymbol{w} \cdot \boldsymbol{n} = \boldsymbol{0} \text{ on } \partial\Omega\}$, equipped with a suitable inner product $\langle \cdot, \cdot \rangle_{\mathbb{V}}$ and the corresponding induced norm $\|\cdot\|_{\mathbb{V}}$. Moreover, consider a parametrized linear form $f : \mathbb{V} \times \mathcal{P} \to \mathbb{R}$ where the linearity is with respect to the first argument, and the parametrized bilinear forms $m : \mathbb{V} \times \mathbb{V} \to \mathbb{R}$ and $a : \mathbb{V} \times \mathbb{V} \times \mathcal{P} \to \mathbb{R}$, where the bilinearity is with respect to the first two arguments. Then, the acoustic-elastic problem in abstract form reads: given $t \in [0, T]$ and $\boldsymbol{\mu} \in \mathcal{P}$, find $\boldsymbol{u} = \boldsymbol{u}(t; \boldsymbol{\mu}) \in \mathbb{V}$ such that

$$\begin{cases} \rho \left(m(\ddot{\boldsymbol{u}}, \boldsymbol{\psi}) + \eta\, m(\dot{\boldsymbol{u}}, \boldsymbol{\psi}) \right) + a(\boldsymbol{u}, \boldsymbol{\psi}; \boldsymbol{\mu}) = h(t; \boldsymbol{\mu}) f(\boldsymbol{\psi}; \boldsymbol{\mu}) \quad \forall \boldsymbol{\psi} \in \mathbb{V}, \\ \boldsymbol{u}(t = 0; \boldsymbol{\mu}) = \boldsymbol{0}, \quad \dot{\boldsymbol{u}}(t = 0; \boldsymbol{\mu}) = \boldsymbol{0}, \end{cases} \tag{3}$$

with

$$\begin{aligned} m(\boldsymbol{u}, \boldsymbol{\psi}) &= \int_\Omega \boldsymbol{u} \cdot \boldsymbol{\psi}\, \boldsymbol{\Omega}, \quad f(\boldsymbol{\psi}; \boldsymbol{\mu}) = \int_\Omega \boldsymbol{b}(\boldsymbol{\mu}) \cdot \boldsymbol{\psi}\, \boldsymbol{\Omega}, \\ a(\boldsymbol{u}, \boldsymbol{\psi}; \boldsymbol{\mu}) &= \int_\Omega [\boldsymbol{H}(\boldsymbol{\mu}) : \boldsymbol{\varepsilon}(\boldsymbol{u})] : \boldsymbol{\varepsilon}(\boldsymbol{\psi})\, \boldsymbol{\Omega}, \end{aligned} \tag{4}$$

in which $\boldsymbol{H}$ is the Hooke's stiffness tensor and $\boldsymbol{\varepsilon}(\cdot) = \frac{1}{2}[\nabla \cdot + (\nabla \cdot)^{\mathrm{T}}]$ is the operator of Cauchy strain. Note that, for the sake of simplicity, homogeneous boundary and initial conditions in (3) are considered, i.e., $\boldsymbol{u}_0 = \boldsymbol{0}$, $\boldsymbol{v}_0 = \boldsymbol{0}$, and $\boldsymbol{t}_N = \boldsymbol{0}$.

2.2 Parametric Discrete Problem

This section introduces a discrete approximation space $\mathbb{V}_h \subset \mathbb{V}$ as well as a discretization of the time interval $[0, T]$ in which the approximate solution is sought. The approximation space $\mathbb{V}_h$ is here constructed by a standard finite element method based on piece-wise linear basis functions and a triangulation of Ω, i.e., non-overlapping triangles ($d_\Omega = 2$) or tetrahedra ($d_\Omega = 3$) whose union perfectly coincides with Ω. Alternative discretization strategies include spectral methods or higher-order finite elements. Let $\mathbb{V}_h$ be equipped with a basis $\{\boldsymbol{\varphi}_j(\boldsymbol{x}) \in \mathbb{R}^{d_\Omega}\}_{j=1}^{N_h}$, where $N_h = \dim(\mathbb{V}_h)$ is the number of degrees of freedom (DOFs). Moreover, divide the time interval $[0, T]$ into N_t subintervals of equal length $\Delta t = \frac{T}{N_t}$ and define $t^n = n\Delta t$, $0 \leq n \leq N_t$.

The discrete problem with finite element discretization seeks to find $\boldsymbol{u}_h(t; \boldsymbol{\mu}) \in \mathbb{V}_h$, which can be expressed as $\boldsymbol{u}_h(t; \boldsymbol{\mu}) = \sum_{j=1}^{N_h} (\boldsymbol{u}_h(t; \boldsymbol{\mu}))_j \boldsymbol{\varphi}_j(\boldsymbol{x})$ where $(\boldsymbol{u}_h)_j$ denotes the jth entry of the solution vector $\boldsymbol{u}_h \in \mathbb{R}^{N_h}$. With an additional discretization over time, one can retrieve the solution vector at the nth time step, denoted by $\boldsymbol{u}_h^n(\boldsymbol{\mu}) = \boldsymbol{u}_h(t^n; \boldsymbol{\mu})$, $n = 1, \ldots, N_t$. Moreover, let $\boldsymbol{v}_h^n(\boldsymbol{\mu}) \in \mathbb{R}^{N_h}$ and $\boldsymbol{a}_h^n(\boldsymbol{\mu}) \in \mathbb{R}^{N_h}$ be the velocity and the acceleration vectors, respectively, such that their elements are the multiplicative coefficients of the following expressions: $\dot{\boldsymbol{u}}_h^n(\boldsymbol{\mu}) = \sum_{j=1}^{N_h} (\boldsymbol{v}_h^n(\boldsymbol{\mu}))_j \boldsymbol{\varphi}_j(\boldsymbol{x})$ and $\ddot{\boldsymbol{u}}_h^n(\boldsymbol{\mu}) = \sum_{j=1}^{N_h} (\boldsymbol{a}_h^n(\boldsymbol{\mu}))_j \boldsymbol{\varphi}_j(\boldsymbol{x})$, respectively.

Once the acoustic-elastic equation is spatially discretized by finite elements, the corresponding algebraic formulation is written as follows for a given $\boldsymbol{\mu} \in \mathcal{P}$ and $t \in [0, T]$:

$$\rho \boldsymbol{M}_h \left[\ddot{\boldsymbol{u}}_h(\boldsymbol{\mu}) + \eta \dot{\boldsymbol{u}}_h(\boldsymbol{\mu})\right] + \boldsymbol{A}_h(\boldsymbol{\mu})\boldsymbol{u}_h(\boldsymbol{\mu}) = h(t; \boldsymbol{\mu})\boldsymbol{f}_h(\boldsymbol{\mu}), \tag{5}$$

where $\boldsymbol{M}_h \in \mathbb{R}^{N_h \times N_h}$ is the mass matrix, $\boldsymbol{A}_h(\boldsymbol{\mu}) \in \mathbb{R}^{N_h \times N_h}$ the parametrized stiffness matrix, and $\boldsymbol{f}_h(\boldsymbol{\mu}) \in \mathbb{R}^{N_h}$ the parametrized right-hand side vector with entries

$$\begin{aligned} (\boldsymbol{M}_h)_{ij} &= m(\boldsymbol{\varphi}_j, \boldsymbol{\varphi}_i), \quad (\boldsymbol{A}_h(\boldsymbol{\mu}))_{ij} = a(\boldsymbol{\varphi}_j, \boldsymbol{\varphi}_i; \boldsymbol{\mu}), \\ \text{and} \quad (\boldsymbol{f}_h(\boldsymbol{\mu}))_i &= f(\boldsymbol{\varphi}_i; \boldsymbol{\mu}), \quad 1 \leq i, j \leq N_h. \end{aligned} \tag{6}$$

A classic Newmark method is then applied to the temporal discretization and the governing equation becomes: given $\boldsymbol{\mu} \in \mathcal{P}$, find the acceleration vector $\boldsymbol{a}_h^n(\boldsymbol{\mu}) \in \mathbb{R}^{N_h}$ for $n = 1, \ldots, N_t$ such that

$$\left[\rho(1 + \eta\zeta\Delta t)\boldsymbol{M}_h + \beta(\Delta t)^2 \boldsymbol{A}_h(\boldsymbol{\mu})\right] \boldsymbol{a}_h^n(\boldsymbol{\mu}) = h(t^n; \boldsymbol{\mu})\boldsymbol{f}_h(\boldsymbol{\mu}) - \boldsymbol{q}_h^{n-1}(\boldsymbol{\mu}), \tag{7}$$

in which β and ζ are two constant parameters, here chosen as $\zeta = 2\beta = 2$, which corresponds to a popular second-order method [8, 9], while $\boldsymbol{q}_h^{n-1}(\boldsymbol{\mu}) \in \mathbb{R}^{N_h}$ is given as

$$\begin{aligned} \boldsymbol{q}_h^{n-1}(\boldsymbol{\mu}) = &\boldsymbol{A}_h(\boldsymbol{\mu})\boldsymbol{u}_h^{n-1}(\boldsymbol{\mu}) + (\rho\eta \boldsymbol{M}_h + \Delta t \boldsymbol{A}_h(\boldsymbol{\mu}))\boldsymbol{v}_h^{n-1}(\boldsymbol{\mu}) \\ &+ \left(\rho\eta(1 - \zeta)\Delta t \boldsymbol{M}_h + \tfrac{1-2\beta}{2}(\Delta t)^2 \boldsymbol{A}_h(\boldsymbol{\mu})\right) \boldsymbol{a}_h^{n-1}(\boldsymbol{\mu}). \end{aligned} \tag{8}$$

Finally, the displacement solution vector $\boldsymbol{u}_h^n(\boldsymbol{\mu})$ is obtained using the updating rule of the implicit Newmark method, introduced in [10] and defined as:

$$\boldsymbol{u}_h^n(\boldsymbol{\mu}) = \boldsymbol{u}_h^{n-1}(\boldsymbol{\mu}) + \Delta t \boldsymbol{v}_h^{n-1}(\boldsymbol{\mu}) \left(\beta \boldsymbol{a}_h^n(\boldsymbol{\mu}) + \tfrac{1-2\beta}{2} \boldsymbol{a}_h^{n-1}(\boldsymbol{\mu})\right) \tag{9}$$

$$\boldsymbol{v}_h^n(\boldsymbol{\mu}) = \boldsymbol{v}_h^{n-1}(\boldsymbol{\mu}) + \Delta t \left(\zeta \boldsymbol{a}_h^n(\boldsymbol{\mu}) + (1-\zeta)\boldsymbol{a}_h^{n-1}(\boldsymbol{\mu})\right), \tag{10}$$

Problem (7) is denoted as the *truth problem* and $\boldsymbol{u}_h^n(\boldsymbol{\mu})$ as the *truth solution* at n-th time step, which, in principle, can be achieved with as high accuracy as desired. However, many degrees of freedom may be involved in the problem, thus leading to a computationally expensive method due to inversion of the N_h-dimensional matrix in the left-hand side of (7). In addition, to fully represent the healthy variations of the structure, one needs to estimate an approximation to the output of interest (2) for many input parameter values $\{\boldsymbol{\mu}_1, \boldsymbol{\mu}_2, \dots\}$ over the whole discrete time window $0 = t^0, t^1, \dots, t^{N_t} = T$, i.e.,

$$\boldsymbol{g}_i^k = \left[g_i(t^0; \boldsymbol{\mu}_k), g_i(t^1; \boldsymbol{\mu}_k), \dots, g_i(t^{N_t}; \boldsymbol{\mu}_k)\right], \quad k = 1, 2, \dots, \tag{11}$$

evaluated at all the sensor locations $\boldsymbol{x}_i$, $1 \le i \le m$. For each input parameter value $\boldsymbol{\mu}_k$, the total computational cost involves the resolution of N_t linear systems of dimension N_h.

3 Techniques of Reduced-Order Modeling

When the dimensionality of a full-order system, defined as N_h in Sect. 2, is large, the repeated solution of such a time-dependent problem with varying input parameters can result in great demands on both CPU time and memory, which is often computationally prohibitive. To reduce the computational cost without significantly compromising the overall accuracy, reduced-order models (ROMs) have been developed. In general, reduced-order modeling seeks to find a low-dimensional representation of the full-order solution manifold and hence reduce the dimensionality by projecting the original governing equations onto a low-dimensional space.

The reduced basis (RB) method [11, 12] is a typical projection-based approach to ROMs and features an offline–online framework. With a significantly smaller dimension than the full-order model, a reduced space is spanned by a set of RB modes that are extracted offline from a collection of full-order snapshots at several time-parameter locations. Once the RB space is constructed, the approximate solution for an unseen parameter value is recovered online in the reduced space. Conventionally, a Galerkin projection is adopted to determine the combination coefficients associated with the RB, yielding the reduced-order solutions during the online stage.

For time-dependent problems, due to the traveling-wave behavior of the solution, classic projection-based ROM strategies [13] may pose several challenges, e.g., the manifold of all possible solutions can often not be compressed to a small reduced basis. Furthermore, the sampling strategy is more complicated since it has to combine the solution at different time instants and for different values of the parameter. For example, the readers can refer to the POD-greedy sampling strategy [14] and the randomized SVD algorithms [15]. Recent efforts have been made in the direction of space–time approaches, where projection in space and time is performed simultaneously, see, e.g., [16]. A different effective strategy is to replace the time domain formulation with a frequency domain formulation and to apply a ROM method to replace the full-order problem in frequency domain. In this way, the number of time instances N_t where one expects to solve a linear system equivalent to (7) is reduced to a number of principal frequencies N_z, with $N_z \ll N_t$. In addition, recurring to a ROM strategy reduces the number of degrees of freedom of each linear system to the size of reduced basis, i.e., from N_h to r, with $r \ll N_h$. Without going in too much detail, the reader are referred to [5, 17], where the authors, motivated by the interest of studying the transient response of damaged structures under the effect active sources, construct a reduced-order model of the acoustic-elastic equation in the Laplace domain.

The goal here is to provide a brief introduction to several basic elements of the RB method, which lay the foundation for more advanced techniques. In particular, after introducing a general formulation of the proper orthogonal decomposition (POD) in Sect. 3.1, the construction of RB from full-order snapshots using the POD is also described. The technique to retrieve the RB solution, an approximation of the high-fidelity solution, is ultimately presented in Sect. 3.2.

3.1 Proper Orthogonal Decomposition

General formulation of the POD

In a vector space $\mathbb{X}$, equipped with an inner product $\langle \cdot, \cdot \rangle_{\mathbb{X}}$, consider a collection of snapshot vectors, denoted by $\{p_1, \ldots, p_{N_s}\} \subset \mathbb{X}$. A correlation matrix $\boldsymbol{C} \in \mathbb{R}^{N_s \times N_s}$ of the snapshots is formed as

$$\boldsymbol{C}_{ij} = \langle p_i, p_j \rangle_{\mathbb{X}}, \quad 1 \le i, j \le N_s. \tag{12}$$

The eigenvalue problem of such a correlation matrix $\boldsymbol{C}$ is then written as

$$\boldsymbol{C}\boldsymbol{z}^{(i)} = \lambda^{(i)}\boldsymbol{z}^{(i)}, \quad 1 \le i \le N_s, \tag{13}$$

in which $\lambda^{(1)} \ge \cdots \ge \lambda^{(N_s)} \ge 0$. By taking

$$\phi_i = \sum_{j=1}^{N_s} p_j \left(\boldsymbol{z}^{(i)}\right)_j / \sqrt{\lambda^{(i)}}, \quad 1 \le i \le r \text{with}\, N \le N_s, \tag{14}$$

an orthonormal basis is formed, i.e., $\langle \phi_i, \phi_j \rangle = \delta_{ij}$, $1 \le i, j \le r$, and an r-dimensional subspace is then constructed as $\mathbb{X}_r = \text{span}\{\phi_1, \ldots, \phi_N\} \subset \mathbb{X}$. The projection onto this subspace, denoted by $P_N[\cdot] : \mathbb{X} \to \mathbb{X}_r$, is thus defined as

$$P_r[f] = \arg\min_{\xi \in \mathbb{X}_r} \|f - \xi\|_{\mathbb{X}}^2 = \sum_{i=1}^{r} \langle f, \phi_i \rangle_{\mathbb{X}} \phi_i, \quad f \in \mathbb{X}. \tag{15}$$

It can be shown that the projection error of the snapshots only depends on the truncated eigenvalues, written as

$$\sum_{i=1}^{N_s} \|p_i - P_r[p_i]\|_{\mathbb{X}}^2 = \sum_{i=r+1}^{N_s} \lambda^{(i)}. \tag{16}$$

In addition, $\mathbb{X}_r$ is the optimal subspace of $\mathbb{S} = \text{span}\{p_1, \ldots, p_{N_s}\}$ that minimizes the projection error, i.e.,

$$\sum_{i=1}^{N_s} \|p_i - P_r[p_i]\|_{\mathbb{X}}^2 = \min_{\mathbb{U} \text{ being a subspace of } \mathbb{S}} \left\{ \sum_{i=1}^{N_s} \min_{\xi \in \mathbb{U}} \|p_i - \xi\|_{\mathbb{X}}^2 \right\}. \tag{17}$$

Construction of RB using the POD

At the algebraic level, the solution space for the full-order discrete system (5) is $\mathbb{R}^{N_h}$, i.e., $\mathbb{X} = \mathbb{R}^{N_h}$, and it is correspondingly equipped with the Euclidean inner product. To construct an RB space, one has to collect the solution snapshots of N_t time instances $\{t^0, t^1, \ldots, t^{N_t}\}$ at N_μ parameter locations $\{\boldsymbol{\mu}_1, \boldsymbol{\mu}_2, \ldots, \boldsymbol{\mu}_{N_\mu}\}$, i.e., $\{p_i\}_{i=1}^{N_s} = \{\boldsymbol{u}_h^n(\boldsymbol{\mu}_k) : 1 \le n \le N_t, 1 \le k \le N_\mu\}$ and $N_s = N_t N_\mu$. Let $\boldsymbol{S} \in \mathbb{R}^{N_h \times N_s}$ denote the snapshot matrix collecting all the N_s snapshot vectors as columns.

Using the POD, r basis vectors are obtained and collected in a matrix $\boldsymbol{V}_r \in \mathbb{R}^{N_h \times r}$, whose i-th column, i.e., the i-th basis vector, corresponds to the i-th eigenvalue $\lambda^{(i)}$ of the correlation matrix $\boldsymbol{C} = \boldsymbol{S}^{\mathrm{T}}\boldsymbol{S}$, $1 \le i \le r$. In fact, given the SVD of the snapshot matrix $\boldsymbol{S}$, written as

$$\boldsymbol{S} = \boldsymbol{U}\boldsymbol{\Sigma}\boldsymbol{Z}^{\mathrm{T}}, \tag{18}$$

the basis vectors in $\boldsymbol{V}_r$ are the first r columns of $\boldsymbol{U}$, i.e., $\boldsymbol{V}_r = \boldsymbol{U}[:, 0 : r-1]$ in a Python notation, and $\boldsymbol{\Sigma}$ is a diagonal matrix of singular values, i.e., $\boldsymbol{\Sigma} = \text{diag}(\sqrt{\lambda^{(1)}}, \sqrt{\lambda^{(2)}}, \ldots, \sqrt{\lambda^{(N_s)}})$. Especially when the singular values decay fast, a small number of basis vectors can achieve a small projection error according to (16).

In this way, a reduced basis $\boldsymbol{V}_r$ is obtained, reducing the N_h-dimensional, full-order solution space $\mathbb{R}^{N_h}$ to an r-dimensional, reduced space $\text{Col}(\boldsymbol{V}_r)$, Col representing the column space. With a rapid decay of singular values, the dimensionality reduction is significant ($r \ll N_h$) but the accuracy is still under control.

3.2 Reduced-Order Solutions

The discrete solution is approximated as a linear combination of RB vectors $\boldsymbol{V}_r$, written as $\boldsymbol{u}_h \approx \boldsymbol{V}_r \boldsymbol{q}_r$ with $\boldsymbol{q}_r \in \mathbb{R}^r$ denoting the RB coefficients, and project the full-order system (5) onto the reduced space $\mathrm{Col}(\boldsymbol{V}_r)$. A reduced-order system is thus obtained as

$$\rho \boldsymbol{M}_r \left[\ddot{\boldsymbol{q}}_r(\boldsymbol{\mu}) + \eta \dot{\boldsymbol{q}}_r(\boldsymbol{\mu})\right] + \boldsymbol{A}_r(\boldsymbol{\mu})\boldsymbol{q}_r(\boldsymbol{\mu}) = h(t; \boldsymbol{\mu})\boldsymbol{f}_r(\boldsymbol{\mu}), \tag{19}$$

in which the reduced-size matrices $\boldsymbol{M}_r \in \mathbb{R}^{r \times r}$, $\boldsymbol{A}_r \in \mathbb{R}^{r \times r}$ and $\boldsymbol{f}_r \in \mathbb{R}^r$ are defined as $\boldsymbol{M}_r = \boldsymbol{V}_r^{\mathrm{T}} \boldsymbol{M}_h \boldsymbol{V}_r$, $\boldsymbol{A}_r = \boldsymbol{V}_r^{\mathrm{T}} \boldsymbol{A}_h \boldsymbol{V}_r$ and $\boldsymbol{f}_r = \boldsymbol{V}_r^{\mathrm{T}} \boldsymbol{f}_h$, respectively. Such an r-dimensional reduced system is solved in the online stage for any new parameter value $\boldsymbol{\mu}$.

If the full-size, parameter-dependent stiffness matrix $\boldsymbol{A}_h(\boldsymbol{\mu})$ and source term vector $\boldsymbol{f}_h(\boldsymbol{\mu})$ can be expressed as a linear combination of parameter-independent matrices/vectors with scalar-valued, parameter-dependent coefficients, often referred to as an *affine* form, i.e., $\boldsymbol{A}_h(\boldsymbol{\mu}) = \sum_j \omega_j^a(\boldsymbol{\mu}) \boldsymbol{A}_j$ and $\boldsymbol{f}_h(\boldsymbol{\mu}) = \sum_j \omega_j^f(\boldsymbol{\mu}) \boldsymbol{f}_j$, one can evaluate their reduced-size counterparts offline as $\boldsymbol{A}_{r,j} = \boldsymbol{V}_r^{\mathrm{T}} \boldsymbol{A}_j \boldsymbol{V}_r$ and $\boldsymbol{f}_{r,j} = \boldsymbol{V}_r^{\mathrm{T}} \boldsymbol{f}_j$, $j = 1, 2, \ldots$, and the online assembly only requires linear combinations $\boldsymbol{A}_r(\boldsymbol{\mu}) = \sum_j \omega_j^a(\boldsymbol{\mu}) \boldsymbol{A}_{r,j}$ and $\boldsymbol{f}_r(\boldsymbol{\mu}) = \sum_j \omega_j^f(\boldsymbol{\mu}) \boldsymbol{f}_{r,j}$, respectively. In this case, the online assembly is conducted in the reduced dimensionality and guarantees a good online efficiency. However, if an affine form of the full-size matrix/vector is not available, one has to recall the them during the reduced-size assembly, which can often compromise the online efficiency. To overcome the difficulties stemming from the non-affinity, hyper-reduction techniques have been developed to recover an affine approximation of the non-affine operators, see [18, 19] for example.

An alternative approach to recover reduced-order solutions is through non-intrusive surrogate modeling. In addition to the construction of an RB space using the POD, one has to train a regression model to approximate $\boldsymbol{q}_r :]0, T] \times \mathcal{P} \to \mathbb{R}^r$, $(t, \boldsymbol{\mu}) \mapsto \boldsymbol{V}_r^{\mathrm{T}} \boldsymbol{u}_h(t; \boldsymbol{\mu})$, mapping the time-parameter inputs to the projection coefficients onto the RB. The training data of input–output pairs are derived from a set of collected full-order snapshots. Gaussian process regression has been used for the non-intrusive, reduced-order surrogate modeling in [20–23].

4 Automatic Anomaly Detection with Unbalanced Datasets

This section presents a data-driven technique to detect, localize, and estimate the severity of structural anomalies by observing healthy configurations only. One-class classification learning methods offer the possibility of training a set of samples all belonging to the same *class* and test if a new sample is abnormal, i.e., it belongs to a different class with respect to the training data. Typical one-class classification methods, sometimes called semi-supervised methods, include one-class support vector machines (oc-SVMs), isolation forests, and local outlier factors. During the offline phase, these methods learn a description of the salient features that the training data have in common to ultimately detect if a previously unseen object reflects this description by means of an online anomaly (or novelty) score. If the new unseen sample is associated with an anomaly score close to the ones observed in the training phase, the new object is classified as healthy, otherwise it is classified as damaged. The crucial part is to define what *close to* means from a mathematical standpoint. Let x be an unseen object, then the outcome of all one-class classification methods can be summarized as follows:

$$\begin{cases} \text{score}(x) \geq \theta & \text{damaged/outlier} \\ \text{score}(x) < \theta & \text{healthy/inlier} \end{cases}, \tag{20}$$

where $\text{score}(x)$ is the anomaly score associated with x and θ is an ad hoc threshold to be estimated by observing the anomaly score value of healthy data only. From a practical perspective, in the semi-supervised context, θ is heuristically chosen by observing the highest anomaly score value in the training data, i.e., θ should be equal to the anomaly score of the most outlier sample among all the inliers. Consider $\mathcal{D}$ the dataset of healthy measurements and $\varepsilon \in \mathbb{R}$, then the threshold value is fixed as

$$\theta = \max_{x \in \mathcal{D}} \text{score}(x) + \varepsilon, \tag{21}$$

where a positive value of ε indicates the user accepts a higher false alarm rate, while a negative value implies a higher miss detection rate. The trade-off between false positive and false negative errors should guide the choice of ε and ultimately of θ. It becomes clear that to choose an effective threshold value, the training set has to be the most comprehensive as possible, covering several healthy environmental and operational scenarios.

An alternative approach to detect anomalies is to include sensor measurements belonging to damaged structures in the training set, which leads to using traditional two-class supervised learning methods to distinguish healthy scenarios from damaged ones. In this approach, the choice of the threshold value benefits form the availability of two (or more) classes in the training phase. However, an increasing trend toward the assumption that it would be unreasonable to describe all types of damages is observed in the literature; as a consequence, representing only some damaged configurations would lead to a bias toward certain types and therefore to possible misdetections with high probability (see, e.g., [5, 24]). For this reason, in this chapter, one-class classification methods are used instead.

The general one-class approach is introduced in Sect. 4.1 and explain the need for feature selection in subsection 4.2. For a detailed description of the classic one-class models, the reader is referred to [25, 26] for the oc-SVMs, [27] for the isolation forest and [28] for the local outlier factor. A Python implementation of these methods can be found for example in scikit-learn library [29].

4.1 Local Semi-supervised Method

Considering that the time signals, i.e., the output of interests (2), are collected at multiple sensor locations, one has to decide on how to best aggregate these data. There exists two typical approaches in the literature to combine sensor data: decision-level fusion or feature-level fusion. The latter combines data after feature extraction and considers one global classifier (sensor independent), thus exploiting the correlations across sensors. On the other hand, in decision-level fusion, the signals are classified for each sensor location by a local classifier (sensor dependent) and the results are then combined into a decision output. The two strategies are summarized in Fig. 1. While the superiority of one method over the other one depends strongly on the problem at hand, to exploit the local aspect of the data the authors propose to use the decision-level fusion approach which facilitates the use of a hierarchical classification approach where increasing levels of damage identification can be defined to ultimately gain information on the existence, localization, and severity of the damage.

In a decision-level fusion approach, one has to train as many one-class algorithms as the number of sensor locations. Thus, the global classification model (20) is replaced with m local detection models, where m is the number of sensors:

$$\begin{cases} \text{score}_i(x) \geq \theta_i & \text{damage in the proximity of the } i^{th} \text{ sensor,} \\ \text{score}_i(x) < \theta_i & \text{health in the proximity of the } i^{th} \text{ sensor,} \end{cases} \tag{22}$$

for $1 \leq i \leq m$. From a computational cost point of view, note that the process can be run in parallel since the local models are independent. Moreover, in the feature-based fusion approach, aggregating the local features leads to high-dimensional input data, while the dimensionality of the input data for the classifiers in the decision-based

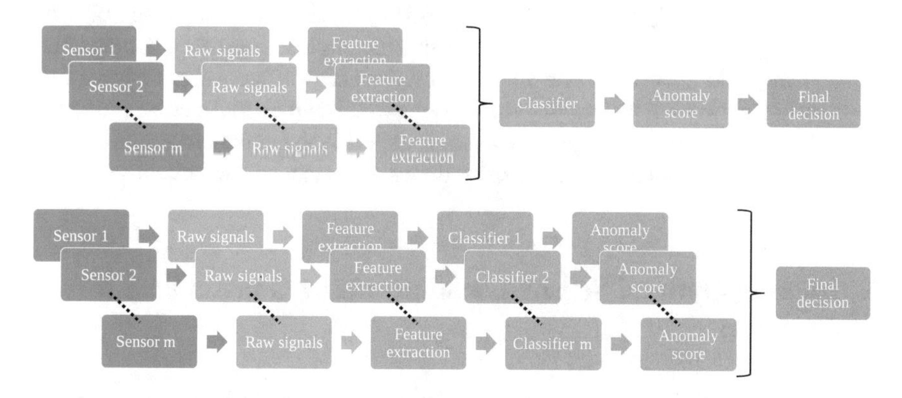

Fig. 1 Flowchart to compare the feature-level (*top*) and the decision-level (*bottom*) fusion approaches for the semi-supervised damage detection strategy with multidimensional training data captured by m sensors

strategy remains small. As further explained in Sect. 4.2, high-dimensional data may lead to overfitting, a well-known problem in machine learning.

4.2 Damage-Sensitive Features

Overfitting refers to the phenomenon observed when the model performs well on training data, but fails to generalize well to new observation. An overfitted model is described by more parameters than can be justified by the data and is typically associated with high-dimensional input data. While sometimes adding more training samples may be a remedy to overfitting, a more effective solution to prevent this behavior is to reduce the dimensionality of the data, i.e., to perform what is called *feature selection*.

Recall that the input data correspond to N_t-dimensional sensor signals, with N_t the number of time steps, which may be very large. Therefore, one needs to express these high-dimensional time signals by means of few variables, extracted from the signals themselves. Ideal features should be sensitive to damage and, at the same time, robust toward noise and healthy variations. Common choices for the engineering-based, damage-sensitive features can be found for example in [24, 30]. When studying the acoustic-elastic equation, i.e., in the context of guided-waves, relevant features are the crest factor, which indicates how extreme the peaks are in a waveform, the maximum and minimum values of the time response, the corresponding arrival times, i.e., the onset, and the number of peaks and troughs in the signals. Without further detail, the reader is referred to [5] and references therein for a thorough description of damage-sensitive features for a guided-wave monitoring approach.

Finally, note that autoencoders, a particular type of neural networks, trained to attempt to copy their inputs to their outputs, have gained particular interest in the framework of anomaly detection, see, e.g., [31–33]. The main advantage of using autoencoders for anomaly detection is that specific engineering-based damage indicator features do not need to be specified by the user, different from classic one-class methods. Instead, by learning the features which suffice to describe and reconstruct the input, autoencoders provide a purely data-driven feature extraction method. Hence, raw measurements such as sensor time signals can be used directly.

5 Finding Optimal Sensor Locations Using Gaussian Processes

Until this point, this work has covered the problem of how to best detect damages given a fixed network of sensors. The specular research question is how to choose the location (and the number) of sensors in order to best detect defaults. Sensor placement strategies are extremely important to optimally equip structures, whose monitoring performance depends critically on the quality of the information collected by sensors. Hence, it is no surprise that this problem has been extensively addressed in the SHM literature, see, e.g., the thorough review [34] and references therein. For most of the sensor placement strategies, the objective is to optimize a suitable cost function with respect to some operational parameters, e.g., the candidate sensor locations and the available number of sensors. However, classic cost functions are usually formulated in terms of damage detectability, which poses a problem when one wishes to make no assumption on the potential damages. Thus, a procedure to place a fixed budget of sensors in the context of anomaly detection is proposed, i.e., when only healthy scenarios are included in the training phase. The proposed strategy relies on sparse Gaussian process to identify the spacial positions that minimize the reconstruction error of an output of interest at all "unsensed" locations. The quantity of interest that defines the cost function for the sensor placement optimization algorithm is the same quantity used to train the anomaly detection classifier, i.e., the damage-sensitive features extracted from the synthetic sensor measurements (11), as described in Sect. 4. As such, the proposed placement strategy is based on an appropriate indicator of the damage detection performance of a given network. Note that, while this approach requires the number and type of sensors to be fixed, it can be easily extended to help the user to identify the minimum number of sensors to achieve a preset coverage.

This section presents the sensor placement strategy introduced in [35], to which the reader is referred for a more in-depth discussion. In particular, after a brief introduction to Gaussian process (GP) regression and sparse GP in Sect. 5.1, the description of how to leverage this technique to systematically place sensors on a structure of interest is provided in Sect. 5.2.

5.1 (Sparse) Gaussian Process Regression

A GP is a collection of random variables, any finite number of which obeys a joint Gaussian distribution. In Gaussian process regression (GPR), the prior of regression function is assumed to be a GP corrupted by an independent Gaussian noise term, i.e., for $(\boldsymbol{x}, \boldsymbol{x}') \in \Omega \times \Omega$ with $\Omega \subset \mathbb{R}^{d_\Omega}$ denoting the domain of regression,[1]

$$f(\boldsymbol{x}) \sim \mathrm{GP}(\mathbf{0}, \kappa(\boldsymbol{x}, \boldsymbol{x}')), \quad y = f(\boldsymbol{x}) + \epsilon, \quad \epsilon \sim \mathcal{N}(0, \chi^2). \tag{23}$$

There are many different options for the covariance/kernel function $\kappa : \Omega \times \Omega \to \mathbb{R}$, a typical form of which is written as

$$\kappa(\boldsymbol{x}, \boldsymbol{x}') = \sigma^2 \phi(r), \tag{24}$$

[1] Here f denotes a generic regression function and should not be confused with the linear functional $f(\cdot; \boldsymbol{\mu})$ in Sect. 2.

where $\phi(\cdot)$ is a radial basis function and r [2] can be defined as

$$r = \|\boldsymbol{x} - \boldsymbol{x}'\|/\ell \quad \text{or} \quad r = \sqrt{\sum_{k=1}^{d_\Omega} \frac{(x_k - x_k')^2}{\ell_k^2}},$$

the former for a stationary kernel with isotropic lengthscale ℓ, while the latter for an automatic relevance determination (ARD) kernel that considers an individual correlated lengthscale ℓ_k for each input dimension and allows for differentiated relevances of input features to the regression.

Given a finite number of training input locations in the domain Ω, a prior joint Gaussian is defined for the corresponding outputs:

$$\boldsymbol{y}|\boldsymbol{X} \sim \mathcal{N}(\boldsymbol{0}, \boldsymbol{K}_y), \quad \boldsymbol{K}_y = \text{cov}[\boldsymbol{y}|\boldsymbol{X}] = \kappa(\boldsymbol{X}, \boldsymbol{X}) + \chi^2 \boldsymbol{I}_M, \tag{25}$$

where $\boldsymbol{y} = \{y_1, y_2, \ldots, y_M\}^{\mathrm{T}}$, $\boldsymbol{X} = [\boldsymbol{x}_1|\boldsymbol{x}_2|\cdots|\boldsymbol{x}_M]^{\mathrm{T}}$, $\boldsymbol{I}_M$ is the M-dimensional unit matrix, and M is the number of training samples.

The goal of a regression model is to predict the noise-free output $f^*(\boldsymbol{s})$ at any new, unseen input location $\boldsymbol{s} \in \Omega$. By the standard Bayesian rule $p(f^*(\boldsymbol{s})|\boldsymbol{X}, \boldsymbol{y}) = p(f^*, \boldsymbol{y}|\boldsymbol{s}, \boldsymbol{X})/p(\boldsymbol{y}|\boldsymbol{X})$, the posterior distribution conditioned on the training data $(\boldsymbol{X}, \boldsymbol{y})$ can be obtained as a new GP:

$$\begin{aligned} f^*(\boldsymbol{s})|\boldsymbol{X}, \boldsymbol{y} &\sim \text{GP}(m^*(\boldsymbol{s}), c^*(\boldsymbol{s}, \boldsymbol{s}')), \\ m^*(\boldsymbol{s}) &= \kappa(\boldsymbol{s}, \boldsymbol{X})\boldsymbol{K}_y^{-1}\boldsymbol{y}, \quad c^*(\boldsymbol{s}, \boldsymbol{s}') = \kappa(\boldsymbol{s}, \boldsymbol{s}') - \kappa(\boldsymbol{s}, \boldsymbol{X})\boldsymbol{K}_y^{-1}\kappa(\boldsymbol{X}, \boldsymbol{s}'), \end{aligned} \tag{26}$$

The values of hyperparameters $\boldsymbol{\theta} = \{\ell \text{or}(\ell_1, \ldots, \ell_d), \sigma^2, \chi^2\}$ make significant difference on the predictive performance. In this chapter, an empirical Bayesian approach of maximizing marginal likelihood is adopted to determine a set of optimal values of the parameters. Using a standard gradient-based optimizer, the optimal hyperparameters $\boldsymbol{\theta}^*$ can be estimated via the maximization problem as follows:

$$\begin{aligned} \boldsymbol{\theta}^* &= \arg\max_{\boldsymbol{\theta}} \log p(\boldsymbol{y}|\boldsymbol{X}, \boldsymbol{\theta}) = \arg\max_{\boldsymbol{\theta}} \log \left[\mathcal{N}(\boldsymbol{y}|\boldsymbol{0}, \boldsymbol{K}_y(\boldsymbol{\theta}))\right] \\ &= \arg\max_{\boldsymbol{\theta}} \left\{-\frac{1}{2}\boldsymbol{y}^{\mathrm{T}}\boldsymbol{K}_y^{-1}(\boldsymbol{\theta})\boldsymbol{y} - \frac{1}{2}\log\left|\boldsymbol{K}_y(\boldsymbol{\theta})\right| - \frac{M}{2}\log(2\pi)\right\}, \end{aligned} \tag{27}$$

where $p(\boldsymbol{y}|\boldsymbol{X}, \boldsymbol{\theta})$ is the density function of $\boldsymbol{y}$ given $\boldsymbol{X}$ under hyperparameters $\boldsymbol{\theta}$, considered as the marginal likelihood $p(\boldsymbol{y}|\boldsymbol{X}, \boldsymbol{\theta}) = \int p(\boldsymbol{y}|\boldsymbol{f}, \boldsymbol{X}, \boldsymbol{\theta})p(\boldsymbol{f}|\boldsymbol{X}, \boldsymbol{\theta})\boldsymbol{f}$.

It is important to remark that the computational complexity of generating a GP model is $\mathcal{O}(M^3)$ and the associated storage requirement $\mathcal{O}(M^2)$, which becomes intractable for large datasets. To overcome the computational limitation, the corresponding sparse methods rely on a small set of $m \ll M$ points, called *inducing*

[2] Here r denotes a (scaled) radius and should not be confused with the reduced dimensionality in Sect. 3.

points, to facilitate the information gain of the whole dataset, thus allowing for a complexity reduction, i.e., $\mathcal{O}(Mm^2)$. An overview of well-known sparse GP regression methods can be found for example in [36], where each sparse method is described as an exact inference with a specific approximated prior, different from the true GP prior (23). A different approach is presented in [37], where both the m inducing points, indicated as $\boldsymbol{D} = [\boldsymbol{d}_1|\boldsymbol{d}_1|\cdots|\boldsymbol{d}_m]^\mathrm{T}$, and the hyperparameters $\boldsymbol{\theta}$ are considered as variational parameters to be estimated by minimizing the Kullback–Leibler (KL) divergence between the true posterior (26) and a variational posterior. This is equivalent to maximize the following variational lower bound:

$$\begin{aligned}(\boldsymbol{D}^*, \boldsymbol{\theta}^*) &= \arg\max_{\boldsymbol{D},\boldsymbol{\theta}} \mathcal{L}(\boldsymbol{D}, \boldsymbol{\theta}) \\ &= \arg\max_{\boldsymbol{D},\boldsymbol{\theta}} \left\{ \log\left[\mathcal{N}(\boldsymbol{y}|\boldsymbol{0}, \boldsymbol{Q}(\boldsymbol{D},\boldsymbol{\theta}) + \chi^2 \boldsymbol{I}_M)\right] - \frac{1}{2\chi^2}\mathrm{Tr}(\kappa(\boldsymbol{X},\boldsymbol{X}) - \boldsymbol{Q}(\boldsymbol{D},\boldsymbol{\theta}))\right\},\end{aligned} \tag{28}$$

where $\boldsymbol{Q} = \kappa(\boldsymbol{X}, \boldsymbol{D})(\kappa(\boldsymbol{D}, \boldsymbol{D}))^{-1}\kappa(\boldsymbol{D}, \boldsymbol{X})$ is the Nystrom approximation of the true prior covariance. Note that the trace term in (28) acts as a regularization term of the marginal log likelihood, which can be viewed as an accuracy indicator of how well the inducing points summarize the overall statistics.

5.2 *Variational Approximation for Systematic Sensor Placement*

The aforementioned variational sparse GP model together with the numerical approach defined in the previous sections can be used to systematically place a network of sensors on a structure of interest. Following the description in Sect. 4.2, let $\boldsymbol{y} = \{y_1, \ldots, y_{n_\mathrm{dof}}\}^\mathrm{T}$ be the damage-sensitive features extracted from the synthetic time signals (11), collected at the n_dof points of a coarse mesh of the input domain Ω, which is denoted as $\boldsymbol{X} = [\boldsymbol{x}_1|\boldsymbol{x}_2|\cdots|\boldsymbol{x}_{n_\mathrm{dof}}]^\mathrm{T}$, where $n_\mathrm{dof} \ll N_h$. Given m the number of sensors that the user wishes to place on the structure, one can apply the variational sparse GP strategy to this collection of data, with m being the number of desired inducing points. Ultimately, one can identify the sought sensor locations with the inducing points obtained by variational inference.

Although the procedure is quite simple, some remarks ought to be made. First of all, observe that the hyperparameters and the inducing inputs are estimated by maximizing the variational lower bound (28), which is in general an unconstrained, non-convex optimization problem. This may be problematic because one needs to impose some locality constraints on the inducing points to prevent them from being outside the input domain, especially when this is non-convex. Therefore, the standard variational approximation should be replaced with a constrained optimization:

$$(\boldsymbol{D}^*, \boldsymbol{\theta}^*) = \arg\max_{\boldsymbol{D}\in\Omega_s,\boldsymbol{\theta}} \mathcal{L}(\boldsymbol{D}, \boldsymbol{\theta}), \tag{29}$$

where $\Omega_s \subset \Omega$ indicates the admissible domain for sensor locations and, with a slight abuse of notation, $\boldsymbol{D} \in \Omega_s$ means that each inducing point $\boldsymbol{d}_i$, $1 \le i \le m$ is constrained to belong to Ω_s. For real-world problems, the complexity of the domain may be such that the boundaries of Ω_s cannot be easily specified analytically and, in such cases, it may be worth to replace Ω_s with a discrete counterpart. If that is the case, instead of gradient-based optimization techniques, one could opt for discrete optimization methods such as the genetic algorithm [38].

A second point to notice is that the output of interest $\boldsymbol{y}$ is in general parameter dependent, i.e., $\boldsymbol{y} = \boldsymbol{y}(\boldsymbol{\mu})$. Hence, choosing the sensor locations as the optimal inducing points obtained for one specific input configuration may not be optimal for another context, described by a different parameter. To overcome this, this work proposes to apply the variational sparse GP approach to N_μ outputs of interest $\boldsymbol{y}(\boldsymbol{\mu}_i)$, with $\boldsymbol{\mu}_i \in \mathcal{P}$ for $1 \le i \le N_\mu$. To summarize the information from the so-obtained $N_\mu m$ inducing points, the K-medoids algorithm, a well-known unsupervised clustering technique, is employed to find m clusters and their corresponding centers, called centroids. As a last step, the desired sensor locations will be chosen as the clusters' centers.

To quantify the quality of the placements, the simplest choice is to compare the relative reconstruction error of the high-fidelity quantity of interest at unsensed locations with respect to the mean of the posterior distribution of the sparse model based on the estimated variational parameters. Alternatively, the pointwise relative variance reduction, defined as

$$V_i = \frac{\kappa(\boldsymbol{x}_i, \boldsymbol{D}^*)(\kappa(\boldsymbol{D}^*, \boldsymbol{D}^*))^{-1}\kappa(\boldsymbol{D}^*, \boldsymbol{x}_i)}{\kappa(\boldsymbol{x}_i, \boldsymbol{x}_i)}, \quad \text{for } 1 \le i \le n_{\text{dof}}, \tag{30}$$

provides an indicator on how much variance reduction can be achieved by including $\boldsymbol{x}_i$ to the set of selected sensor locations. When the relative variance reduction is close to one, it means that the inducing variables alone can well reproduce the full GP prediction.

6 Numerical Example

In this section, a numerical problem in 2D is used to illustrate the results in terms of damage detection and sensor placement. Similar results for more complex 3D problems can be found in [5, 35].

Damage detection

The first step consists in generating a synthetic database of sensor observations. Apply (1) with homogeneous free slip boundary conditions and homogeneous initial conditions, i.e., $\boldsymbol{u}_0 = \boldsymbol{v}_0 = \boldsymbol{t}_N = 0$, to the healthy geometry illustrated in Fig. 2a and equipped with $m = 15$ sensors. The high-fidelity numerical solutions are computed using the FE approximation with $\mathbb{P}_1$ elements over a discretized domain with

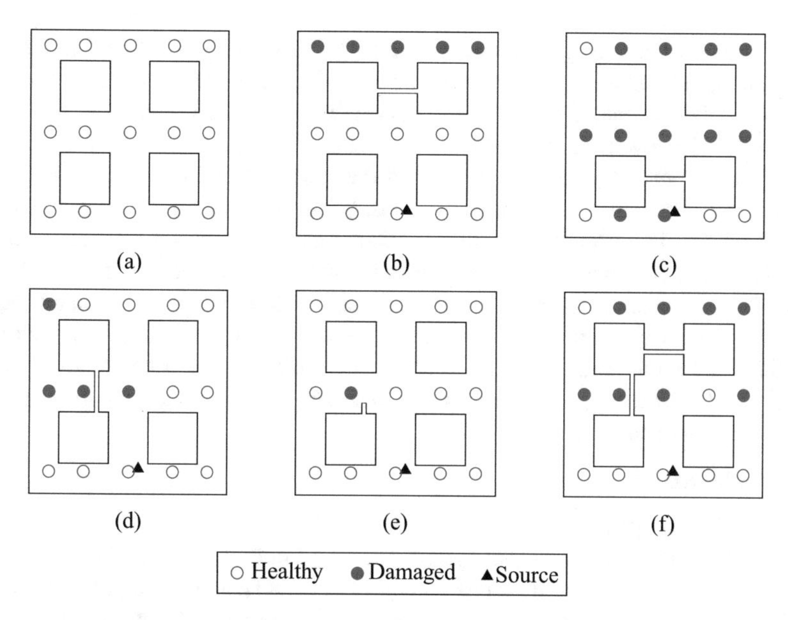

Fig. 2 Summary of the one-class SVMs classification results for one healthy geometry (**a**) and 5 damaged ones (**b**–**f**)

a total of $N_h = 30'912$ degrees of freedom, while for the RB solver the model relies on 267 basis. The natural variations are described by $d_\mu = 3$ parameters, i.e., $\boldsymbol{\mu} = (E, \nu, k) \in \mathcal{P} = [0.999, 1.001] \times [0.329, 0.331] \times [1.9, 2.1]$, where E is the Young's modulus and ν the Poisson's ratio, defining the stress tensor $\boldsymbol{\sigma}$, while k is a parameter representing the number of cycles before attenuation of the source impulse. The position of the active source as well as the density and damping coefficients are fixed, i.e., $\rho = 1, \eta = 0.1$, respectively.

For each sensor location, one should follow the pipeline introduced in Sect. 4 and generate a training dataset of $N_{\text{train}} = 1000$ samples, obtained by extracting a few damage-sensitive features from the discrete time signal (11), in their turn obtained with a suitable reduced-order modeling approach, as described in Sect. 3. Then, one has to train m separate one-class SVM models to learn the common traits of the local healthy features and test the results on sensor measurements belonging to both healthy and damaged geometries. In this example, the test signals are generated synthetically using the high-fidelity model. Different from the training set, to approximate experimental measurements, a Gaussian noise term $\varepsilon_i \sim \mathcal{N}(0, \gamma_i^2)$ is added to the test time signal (11), with γ being the 0.01% of the maximum amplitude of 30 randomly chosen training healthy signals in the training set. Damages are obtained by modifying the geometry of the structure to include discontinuities, as shown in Fig. 2b–f. The crack-modeling approach is a common choice in the literature, see, e.g., [39] where artificial damages on the blade of a wind turbine are implemented via a trailing edge opening. Anomaly detection results are shown in Fig. 2, where for each geometry the average outcome of the oc-SVM for 10 simulations (i.e., for 10 different input parameters) is presented. For each damaged scenario, at least one sensor is classified as damaged, while for the healthy scenario all sensors are classified as healthy. Moreover, for most of the damaged scenarios, one can observe a certain level of proximity between the cracks and the sensors classified as damaged, thus guiding the localization of damages. An exception corresponds to the geometry in Fig. 2c, where almost all sensors are classified as damaged, thus preventing localization. This issue is attributable to the relative position of the source and the crack, i.e., to localize damage (c) the source should be placed differently. A reasonable solution is to consider different locations for the active source, for example by defining an additional input parameter in the model. This approach is already employed in SHM with Lamb wave propagation where piezoelectric transducers are used as both sensors and actuators (see e.g., [40]).

Sensor placement

To identify the optimal sensor locations, one has to create a new database of $N_\mu = 100$ synthetic observations sampled from a Sobol's sequence [41]. Keep the same parameters as in the previous paragraph, but, instead of computing the synthetic time signals at few predefined points in space (the sensor locations), collect the measurements at all the nodes of a new coarse mesh of Ω, for a total of $n_{\text{dof}} = 360$ degrees of freedom. Following the description given in Sect. 5.2, for each one of the N_μ time signals, one has to perform the constrained variational approximation (29) with an ARD-Exponential kernel over $\Omega_s = \Omega \setminus \partial\Omega$ to obtain a set of m inducing points $\boldsymbol{D}^*(\boldsymbol{\mu}_i)$ for $1 \le i \le N_\mu$. Figure 3 sketches the clustering results obtained for different values of N_μ. Observe that the centroids tend to stabilize already after considering a cluster of 10 samples, especially when the sensor budget is high, i.e., for large m values. This can be explained by noticing that when m is small, the algorithm is trying to reconstruct a non-trivial quantity of interest over a complex domain with only a few points, which may lead the sparse model to get stuck in a local minimum without reaching convergence. Finally, the relative variance reduction (30) is used to evaluate the quality of the estimated locations whose results are shown in Fig. 4, where one can observe a variance reduction almost equal to 1 near the sensor locations and a general reduction above 0.7 for all the unsensed locations even for $m = 4$, thus indicating a good sensor placement.

7 Conclusion

This chapter presents how a model-based numerical approach can be integrated with different data-driven techniques in the context of predictive maintenance. A peculiarity of this work is that the authors make no assumption on the type of damages that a structure may undergo during its lifetime, while modeling many environmental and operational healthy scenarios. From a technical point of view, this work describes how reduced-order modeling techniques can be leveraged to generate large and robust

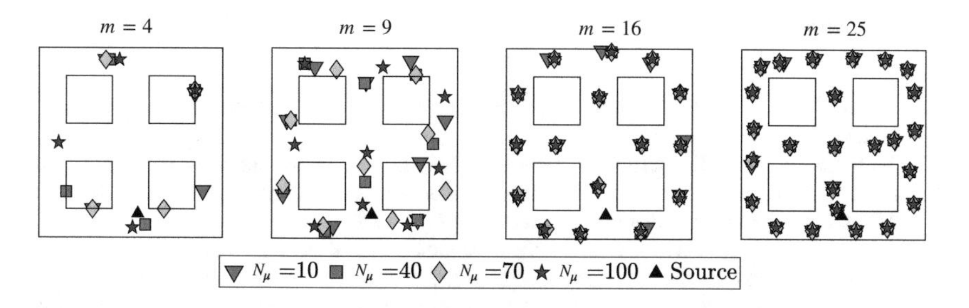

Fig. 3 Comparison of the centroids obtained with the K-medoids clustering algorithm for different sizes of clusters N_{μ}. Each plot shows a fixed number m of inducing points, which increases from left to right

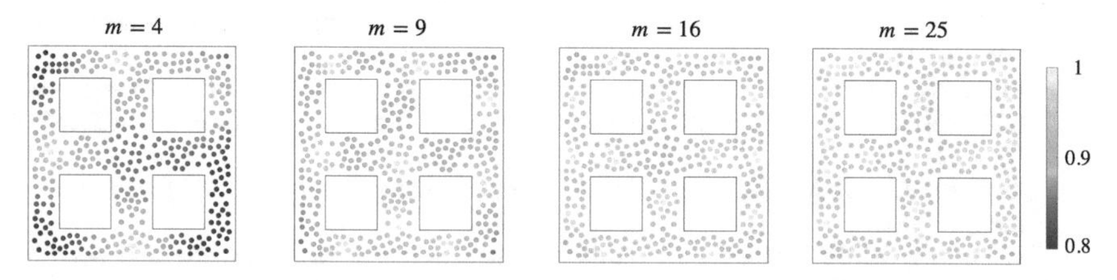

Fig. 4 Relative variance reduction obtained using m centroids and averaged over $N_{\mu} = 100$ samples. Each plot shows a fixed number m of inducing points, which increases from left to right. Values close to 1 indicate a good placement quality

datasets of synthetic sensor measurements and it explains how such datasets can be used to learn the salient features that healthy scenarios have in common to ultimately detect damages. The same damage-sensitive features are also used to guide an automated data-driven sensor placement strategy to increase the detection accuracy for a given budget of sensors.

Although a simple 2D example is used to validate the proposed method, the generalization to more complex, possibly nonlinear problems is possible. Note however that for real-world engineering problems the parameter space describing healthy variations is expected to be high dimensional, thus requiring a high computational effort to elaborate the synthetic time signals. To this end, sensitivity analysis techniques such as the variance-based global sensitivity indices [42] or the derivative-based global sensitivity measures (DGSM) [43] are two popular choices to identify a few parameters that influence the output of interest the most. Finally, note that while the proposed method can be used for real-time predictions thanks to the offline and online decomposition of tasks, a filtering technique to integrate the evolution of the structure and update the model would be a valuable addition.

References

1. Farrar CR, Worden K (2012) Structural health monitoring: a machine learning perspective. Wiley, New York
2. Wagg DJ, Worden K, Barthorpe RJ et al (2020) Digital twins: state-of-the-art and future directions for modeling and simulation in engineering dynamics applications. ASCE-ASME J Risk Uncertainty Eng Syst Part B: Mech Eng 6(3):030901
3. Lecer M, Allaire D, Willcox K (2015) Methodology for dynamic data-driven online flight capability estimation. AIAA J 53(10):3073–3087
4. Taddei T, Penn JD, Yano M, Patera AT (2018) Simulation-based classification; a model-order-reduction approach for structural health monitoring. Arch Comput Methods Eng 25(1):23–45
5. Bigoni C, Hesthaven JS (2020) Simulation-based anomaly detection and damage localization: an application to structural health monitoring. Comput Methods Appl Mech Eng 363:112896
6. Kapteyn MG, Knezevic DJ, Willcox K (2020) Toward predictive digital twins via component-based reduced-order models and interpretable machine learning. In: AIAA Scitech 2020 Forum, p 0418
7. Rosafalco L, Manzoni A, Mariani S et al (2020) Fully convolutional networks for structural health monitoring through multivariate time series classification. Adv Model Simul Eng Sci 7(1):1–31
8. Quarteroni A, Saleri F, Gervasio P (2006) Scientific computing with MATLAB and Octave. Springer, Berlin
9. Zienkiewicz OC, Taylor RL (2005) The finite element method for solid and structural mechanics. Elsevier, Amsterdam
10. Newmark NM (1959) A method of computation for structural dynamics. J Eng Mech Div 85(3):67–94
11. Hesthaven JS, Rozza G, Stamm B (2016) Certified reduced basis methods for parametrized partial differential equations. Springer, Berlin
12. Quarteroni A, Manzoni A, Negri F (2015) Reduced basis methods for partial differential equations: an introduction. Springer, Berlin
13. Benner P, Gugercin S, Willcox K (2015) A survey of projection-based model reduction methods for parametric dynamical systems. SIAM Rev 57(4):483–531
14. Haasdonk B, Ohlberger M (2008) Reduced basis method for finite volume approximations of parametrized linear evolution equations. ESAIM Math Model Numer Anal 42(2):277–302
15. Wang Q, Ripamonti N, Hesthaven JS (2020) Recurrent neural network closure of parametric POD-Galerkin reduced-order models based on the Mori-Zwanzig formalism. J Comput Phys 410:109402
16. Choi Y, Carlberg K (2019) Space-time least-squares Petrov-Galerkin projection for nonlinear model reduction. SIAM J Sci Comput 41(7):A26–A58
17. Bigoni C (2020) Numerical methods for structural anomaly detection using model order reduction and data-driven techniques. Ph.D. thesis No. 7734, EPFL
18. Barrault M, Maday Y, Nguyen NC, Patera AT (2004) An 'empirical interpolation' method: application to efficient reduced-basis discretization of partial differential equations. Comptes Rendus Mathematique 339(9):667–672
19. Chaturantabut S, Sorensen DC (2010) Nonlinear model reduction via discrete empirical interpolation. SIAM J Sci Comput 32(5):2737–2764
20. Kast M, Guo M, Hesthaven JS (2020) A non-intrusive multifidelity method for the reduced order modeling of nonlinear problems. Comput Methods Appl Mech Eng 364:112947
21. Zhang Z, Guo M, Hesthaven JS (2019) Model order reduction for large-scale structures with local nonlinearities. Comput Methods Appl Mech Eng 353:491–515
22. Guo M, Hesthaven JS (2019) Data-driven reduced order modeling for time-dependent problems. Comput Methods Appl Mech Eng 345:75–99
23. Guo M, Hesthaven JS (2018) Reduced order modeling for nonlinear structural analysis using Gaussian process regression. Comput Methods Appl Mech Eng 341:807–826

24. Long J, Buyukozturk O (2014) Automated structural damage detection using one-class machine learning. In: Catbas FN (ed) Dynamics of civil structures, vol 4. Springer, Berlin, pp 117–128
25. Schölkopf B, Williamson RC, Smola AJ et al (2020) Support vector method for novelty detection. Adv Neural Inf Process Syst 12:582–588
26. Cristianini N, Schölkopf B (2002) Support vector machines and kernel methods: the new generation of learning machines. AI Mag 23(3):31
27. Liu FT, Ting KM, Zhou Z-H (2008) Isolation forest. In: 2008 Eighth IEEE international conference on data mining. IEEE, pp 413–422
28. Breunig MM, Kriegel H-P, Ng RT et al. (2000) LOF: identifying density-based local outliers. In: ACM sigmod record, pp 93–104
29. Pedregosa et al (2011) Scikit-learn: machine learning in Python. J Mach Learn Res 12:2825–2830
30. Liu SW, Huang JH, Sung JC (2002) Detection of cracks using neural networks and computational mechanics. Comput Methods Appl Mech Eng 191(25–26):2831–2845
31. Japkowicz N, Myers C, Gluck M (1995) A novelty detection approach to classification. In: International conference on artificial intelligence, vol 1, pp 518–523
32. Marchi E, Vesperini F, Eyben F et al (2015) A novel approach for automatic acoustic novelty detection using a denoising autoencoder with bidirectional LSTM neural networks. In: 2015 IEEE international conference on acoustics, speech and signal processing (ICASSP). IEEE, pp 1996–2000
33. Pathirage CSN, Li J, Li L et al (2018) Application of deep autoencoder model for structural condition monitoring. J Syst Eng Electron 29(4):873–880
34. Ostachowicz W, Soman R, Malinowski P (2019) Optimization of sensor placement for structural health monitoring: a review. Struct Health Monit 18(3):963–988
35. Bigoni C, Zhang Z, Hesthaven JS (2020) Systematic sensor placement for structural anomaly detection in the absence of damaged states. Comput Methods Appl Mech Eng 371:113315
36. Quiñonero-Candela J, Rasmussen CE (2005) A unifying view of sparse approximate Gaussian process regression. J Mach Learn Res 6:1939–1959
37. Titsias M (2009) Variational learning of inducing variables in sparse Gaussian processes. In: Proceedings of the twelfth international conference on artificial intelligence and statistics, PMLR, vol 5, pp 567–574
38. Davis L (1991) Handbook of genetic algorithms. CumInCAD
39. Avendaño-Valencia LD, Chatzi EN, Tcherniak D (2020) Gaussian process models for mitigation of operational variability in the structural health monitoring of wind turbines. Mech Syst Sig Process 142:106686
40. Swartz RA, Flynn E, Backman D et al (2006) Active piezoelectric sensing for damage identification in honeycomb aluminum panels. In: Proceedings of 24th international modal analysis conference
41. Joe S, Kuo FY (2008) Constructing Sobol sequences with better two-dimensional projections. SIAM J Sci Comput 30(5):2635–2654
42. Sobol IM (2001) Global sensitivity indices for nonlinear mathematical models and their Monte Carlo estimates. Math Comput Simul 55(1–3):271–280
43. Kucherenko S, Song S (2016) Derivative-based global sensitivity measures and their link with Sobol'sensitivity indices. In: Owen AB, Glynn PW (eds) Monte Carlo and Quasi-Monte Carlo Methods. Springer, Cham, pp 455–469

Unsupervised Data-Driven Methods for Damage Identification in Discontinuous Media

Rebecca Napolitano, Wesley Reinhart, and Branko Glisic

Abstract Before investing in long-term monitoring or reinforcement of structures, it is essential to understand underlying damage mechanisms and consequences for structural stability. Approaches combining nondestructive evaluation and finite element modeling have been successful in producing qualitative diagnoses for damage to existing structures. However, the real-world impact of such methods will hinge upon a reduced computational burden and improved accuracy of comparison between models and physical infrastructure. This chapter describes a new approach based on unsupervised learning to perform quantitative damage state inversion from sparse datasets. Discrete element modeling was used to simulate the response of masonry walls and other structures under settlement loading. Point cloud representations of the structures, consistent with modern computer vision pipelines used for documentation, were used to generate a low-dimensional manifold based on the Wasserstein metric. This manifold is used to train a Gaussian process model which can then be interrogated to infer loading conditions from the damage state. This method is shown to quantitatively reproduce the loading conditions for masonry structures and was validated against laboratory-scale, experimental masonry walls. Although the approach is demonstrated here for settlement-induced cracking, it has important implications for the broader field of data-driven diagnostics for discontinuous media.

Keywords Damage state inversion · Gaussian process · Discontinuous media · Model reduction · Kriging · Masonry structures

R. Napolitano (✉) · W. Reinhart
The Pennsylvania State University, University Park, PA 16802, USA
e-mail: nap@psu.edu

W. Reinhart
e-mail: reinhart@psu.edu

B. Glisic
Princeton University, Princeton, NJ 08544, USA
e-mail: bglisic@princeton.edu

A. Cury et al. (eds.), *Structural Health Monitoring Based on Data Science Techniques*, Structural Integrity 21, https://doi.org/10.1007/978-3-030-81716-9_10

1 Introduction

1.1 Motivation and Background

Continued advances in computational, networking, and sensor technologies have begun to bring urban-scale structural health monitoring and prognostics into the realm of possibility [1]. The field is starting to address how events can not only impact single buildings, but also how a city can operate as a network of sensors and how urban-scale digital twins can be leveraged to develop smart, sustainable cities. Urban-scale digital twin will be able to provide unprecedented understanding of the state of our communities and predict how they will behave in the future. In that city-scale digital twin however, aging infrastructure introduces a myriad of unknowns such as material condition [2, 3], load paths [4, 5], and so forth [6].

Consider someone developing a digital twin for already existing physical infrastructure with sensors on it; since the structure has a history that might not be entirely known, they can make a digital replica which is only a "best guess." But that does not guarantee that the digital replica and the physical infrastructure will act the same [7–9]. The role of the digital twin is to make the digital replica more than just a model, it should help the model to converge on reality by having it "synchronized" with physical infrastructure. It does this by testing hypotheses about how the structure came to be in its current state. This is paired with physics-based modeling which can then be used to update the model so that the response of the digital replica matches with that of the physical structure. This can then be used to make intelligent interventions on the structure to help predictions and interventions be more accurate [10]. Additionally, this can be used for forecasting analysis once the digital replica has been calibrated. But the crux of the smart interventions and forecasting is on the digital replica. A schematic representation of a digital twin, which combines physics-based and data-driven modeling for diagnostics, prognostics, and long-term management of infrastructure, is shown in Fig. 1.

An open issue of how can different events that could have caused damages can be distinguished. To address this, the present chapter discusses state-of-the-art methodologies for implementing the method of multiple working hypotheses (MMWH) [11] for diagnostics of masonry structures. Masonry structures were selected due to their material complexity, but a similar framework can be applied for structures built of other construction materials. In current approaches, the causative load case is generally assumed; however, this can lead to bias. To reduce bias, MMWH can be applied by laying out the plausible hypotheses, examining the evidence for each, and evaluating which hypotheses can be disregarded and which can remain. This chapter focuses on cracks as they are a symptom of a myriad of problems: moisture, thermal stress, earthquakes, gravity loads, settlement, etc. [10, 12]. For cracks, it is important that the correct cause of damage is determined as that can lead to different monitoring and repair strategies. If repairs are carried out without proper diagnostics, then only the symptoms could be addressed and leave the underlying problem an open issue.

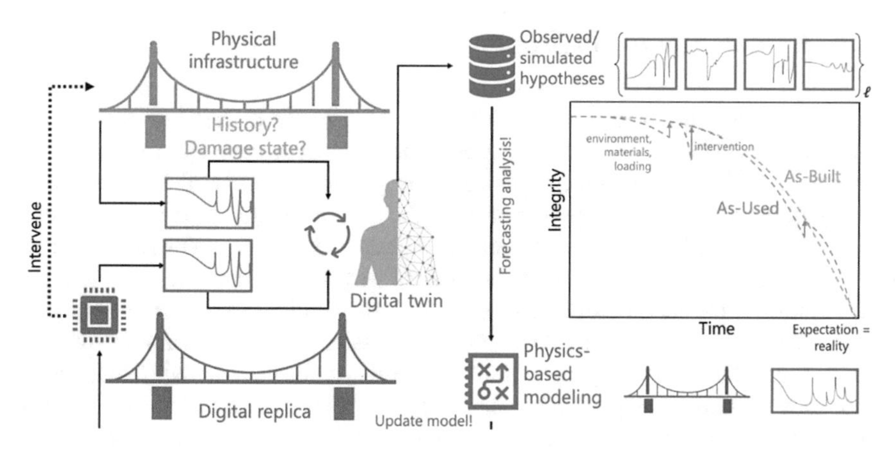

Fig. 1 Schematic of the digital twin constructed with hybrid physics-based and data-driven modeling

1.2 Prior Work

Current crack diagnostic techniques are limited as they are based on manual synthesis of data, structural intuition, and manual crack mapping. Combining methods for acoustic emissions, flat jack tests, and finite element modeling, Anzani et al. [13] assessed the existing damage on a history masonry tower. In that study, quantitative comparisons were made between the results of acoustic monitoring and the time of seismic events; only qualitative comparison was used to compare the finite element model with existing damage; furthermore, only one potential cause of the damage was hypothesized and therefore simulated [11]. Continuing this trend, Alessandri et al. [14] combined structural health monitoring and numerical methods, but again only qualitatively compared cracks on the existing structure to the results of a single hypothesized simulation. Similarly, when Milani et al. [15] used numerical methods to evaluate the origins of crack patterns on the narthex of the Church of the Nativity in Bethlehem, the existing conditions were only qualitatively compared to a single load case simulation.

The concept of using qualitative methods to compare existing crack patterns to simulations based on singular load cases is ubiquitous in the literature [7, 16–20] even though assuming only one specific load is culpable increases bias in the results [11]. Maps of cracks on an existing structure do not record underlying problems, rather they capture symptoms of these problems. This can pose issues for diagnostics and rehabilitation as damage may have already been stabilized and not be related to current risks [21]. Therefore, understanding how cracks on a structure could have evolved is vital to understanding the structures potential for future stability [21].

1.3 Research Aim and Scope

The aim of this work is to create workflow for inferring the cause of deformed (i.e., damaged) regions of masonry structures using automated methods. As a proof of concept, consider only settlement-induced damage from a known initial state and seek to determine a parametric representation of the settlement condition quantitatively. Thus, the proposed workflow may require additional expansions to be successful on a wide range of damage or loading conditions which differ qualitatively. Here the earth mover's distance (EMD) is used to quantify similarity between a library of known damage states and Gaussian process regression (GPR) to accurately interpolate between them to identify the most likely cause of deformation. Manifold learning strategies are also considered to handle extending the method to higher-dimensional spaces, a challenge which must be overcome in order to address real-world structural health monitoring scenarios where damage states cannot be parameterized in a simplified manner.

2 Methods

2.1 Supervised Gaussian Process Regression

This first step in creating the workflow is to develop a metric for how different scenarios will be compared. This step is schematically represented in Fig. 2. Since the proposed workflow involves numerous steps, a conceptual summary of the method is provided first. First, a point cloud representing an existing structure is acquired. Based on historical record or expert knowledge about the construction and history of the structure, a hypothetical undamaged or native state is produced (i.e., manually using CAD) and supplied to the algorithm in the form of a triangulated mesh (state P in Fig. 2). Then, a library of damaged conditions (states Q_l in Fig. 2) for the structure

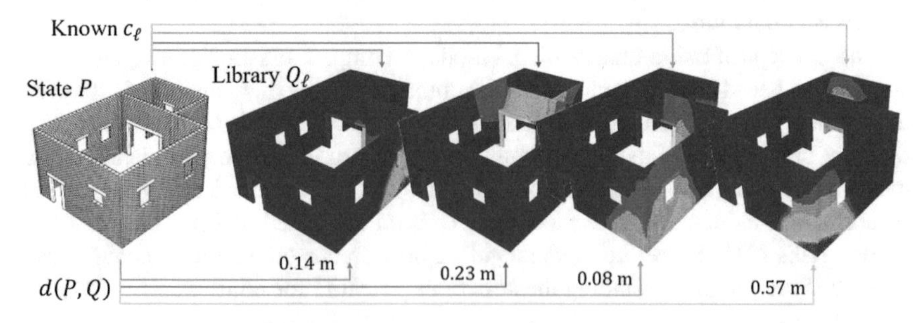

Fig. 2 Illustration of the supervised learning concept. A reference model is compared to a library of deformed structures with known damage conditions to determine the most likely damage mechanism for the observed case

is generated by applying known damage conditions to the original structure using DEM simulation. A point cloud is extracted from each of these damage simulations and clustered into discrete segments. The shape of these segments is compared to equivalent segments of the native structure by computing a dissimilarity metric. For each such segment, the likelihood of the damage being caused in the hypothesized way can be evaluated quantitatively as the reciprocal of the dissimilarity metric. Then the exact damage condition can be inferred by interpolating with Gaussian process within the library of simulated states. In this way, an estimate for the likelihood of every possible hypothesis is obtained in a smooth and continuous parameter space. Here a point cloud P representing the surface of the existing structure is considered to be an input, with the assumption that it has been deformed compared to its native, as-built state. The point cloud could be obtained from one of a variety of standard methods such as photogrammetry and laser scanning [22–24].

In the experimental case discussed in Sect. 4.1, P corresponds to the centroids of the bricks extracted from an orthorectified image captured during laboratory work (details provided in [25]). Next an entirely synthetic case is discussed in Sect. 4.2 so that known conditions can be used as ground truth and compared to the results of the method. In that example, P corresponds to the centroid of each simulated brick after a settlement event has occurred. In present methods, found in literature, damage states are not commonly separated out and linked to individual causes [26, 27]. In efforts to optimize preservation, rehabilitation, and monitoring efforts, each damage state should be diagnosed individually.

This chapter outlines an automated method for separating large models (from here on point clouds) which encompass multiple damage states into smaller ones which capture individual damage locations. By doing this, each segment of P_i can be analyzed independently which thereby decreases the computational cost by reducing the $\mathcal{O}(N^3)$ calculation associated with solving the optimal transport problem into several smaller problems, $\mathcal{O}(n \times N_i^3)$. Here N is the number of points from the original point cloud, whereas n and N_i are the number of segments and number of points per segment, respectively. This reduction of a cubic complexity enables significant cost savings for $N_i < N$. While the larger point cloud is being segmented, the following mapping is generated $m(P) \rightarrow \{P_i\}$. After computing this mapping, it is used to compare between native point cloud P and damaged point cloud Q such that segments P_i and Q_i reference the same local geometry.

By imposing this mapping system, a correspondence table between P and Q can be established when they share 1:1 correspondence or bounding volumes can be used when the point clouds may have different numbers of points. In this method, the Hungarian algorithm is utilized to obtain the correspondence since P and Q from the brick centroids were generated with the same number of bricks. Mini-Batch K-Means (MBKM) [28] was used to segment the point clouds. The only input parameter was the number of segments which is a user preference since the segmentation is only needed to separate out individual diagnoses. As detailed later in this chapter, the workflow was found to be robust to various choices of segmentation when both number of segments and algorithm applied were varied.

For this work, the individual bricks were assumed to behave as rigid bodies. This should enable a good approximation to the behavior of the masonry walls since cracking at the masonry joints frequently dominates any fracturing of the individual brick elements. A hypothetical undamaged state, U, composed of a triangulated mesh, and the parameterized damage conditions, $c = (x, y, a)$, were used as input to the DEM simulation, where x and y define the coordinate of the settlement event and a defines the amplitude of the settlement. By comparing each damaged state, Q, to the observed structure, P, the true damage c^0 can be identified which describes how the structure came to be damaged. To understand how well each hypothesis, Q, matches the true conditions, P, the earth mover's distance (EMD) is used which measures the distance d between probability distributions [29–31] (see Fig. 2). To expedite the EMD calculation over the parameter space, Gaussian process regression (GPR), a Bayesian method for model selection and updating, to approximate the distance field, is applied [32]; this approach is commonly found in the literature [33–35], the scikit-learn implementation [36] is used here. Thus, in this method GPR interpolates for continuously varying damage conditions from a discrete set of damage conditions to mitigate the computational burden of EMD. To create a weighting system which would strongly illustrate regions where the distance between P and Q is low, and thus the guess is good, $L = d^{-1}$ is interpolated. By taking this inversion, the GPR estimated field corresponds to the likelihood (measured in units of m^{-1}) that the guessed damage condition is the true one.

2.2 Unsupervised Manifold Learning

The starting point for the manifold learning strategy (illustrated in Fig. 3) is the stored image of the displaced structure and a feature vector which conveys information about how the structure was damaged. For this work, the feature vector consisted of the length of different settlement regions, the depth of different settlement regions, earthquake magnitude, earthquake orientation, point load, point load application range, and the total sum of all settlement depths. The first step in the process is the hashing of the stored image into a 2-D discrete cosine transform (DCT) [37]. Once each image has been hashed this way, the Euclidean distance between each image can be calculated.

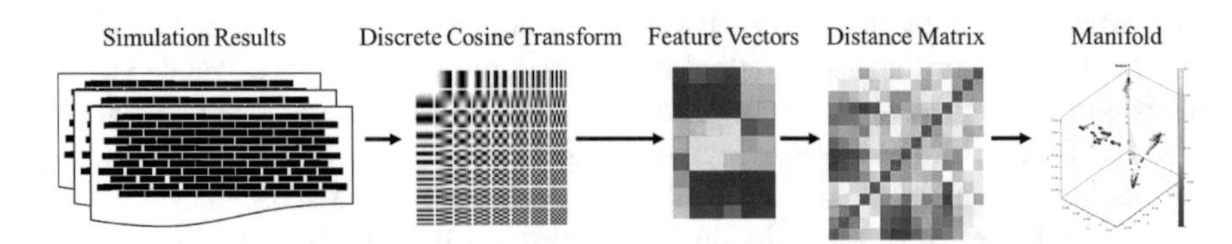

Fig. 3 Schematic illustration of the unsupervised manifold learning procedure

Having computed the distances, the coordinates of each simulation in the low-dimensional manifold can be computed. This is achieved by non-linear manifold learning using the diffusion maps approach [38]. In short, this approach uses a spectral decomposition of an affinity matrix (i.e., the reciprocal of the pairwise distances) to determine collective variables. Then, the dominant eigenvectors can be used to visualize a low-dimensional space (manifold) which contains the most significant collective variables within the data.

2.3 Experiments

To test if the proposed workflow, along with presented methods, could be used to diagnose existing crack patterns in masonry structures, a single-leaf, dry-joint masonry test wall was constructed, and cracking was induced in a controlled manner, as shown in Fig. 4 The test wall had a fixed base (cinder blocks) and region of settlement (jack). Geometry and definition of variables used in experimental setup are shown in Fig. 5. The dimensions and parameters for each brick are shown in Table 1. The settlement width was slightly smaller than the width of two bricks (190% brick length), and the depth was 0.021 ± 0.001 m. Before and after the settlement

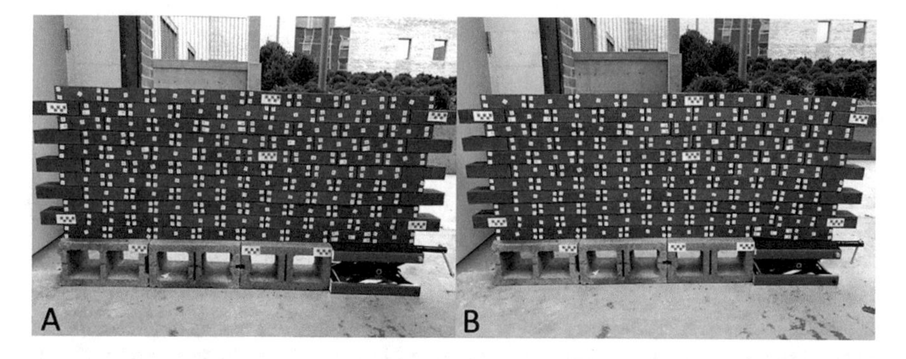

Fig. 4 Photographs of masonry walls used for experimental case studies. **a** Undeformed wall and **b** Deformed wall, after settlement is applied in lower right corner

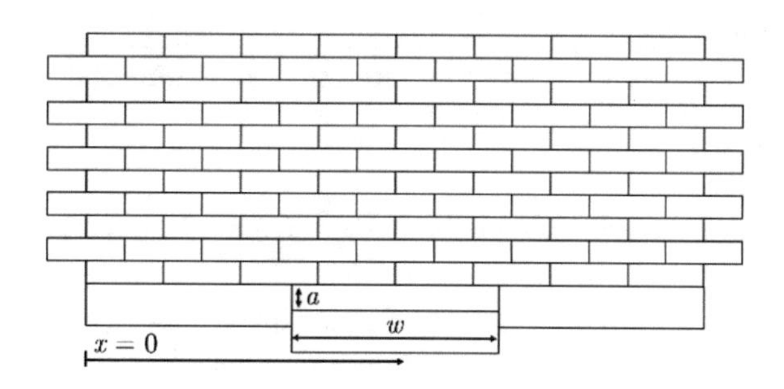

Fig. 5 Geometry and definition of variables used in experimental setup

Table 1 Measured dimensions of bricks used in experimental walls

Parameter	Value
Length (m)	0.193 m $\pm$ 0.001
Width (m)	0.090 $\pm$ 0.001
Height (m)	0.055 $\pm$ 0.001
ρ (kg/m^3)	2508.67 $\pm$ 10

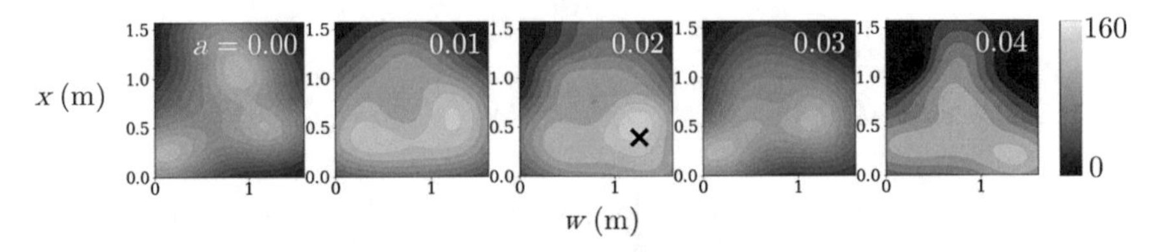

Fig. 6 Estimate of the likelihood field (c) obtained from the experimental wall. The black cross indicates ground truth settlement condition

events, photogrammetry models were generated using best practices [39–41] and orthographic photographs were taken.

3 Results and Discussion

3.1 Experimental Results

A library of 256 $Q_{\updownarrow}$ was generated by selecting $c_{\updownarrow}$ from a uniform nonrandom distribution on the domain $x \in [0, 1.54]$, $w \in [0, 1.4]$, $a \in [0, 0.04]$ (see Fig. 6). The settlements were imposed in the shape of a Gaussian kernel with dispersion $\sigma = 0.5$m. The EMD between the full point clouds was computed as $d(P, Q_{\updownarrow})$; it generated a continuous estimate of the likelihood field by applying GPR to the inverted distances L. The results are illustrated in Fig. 6. Since the likelihood field is a continuous quantity in (x, w, a) the values of the field were rendered at five discrete slices in a. The brighter values (yellow) correspond to (x, y) settlement centers, which most probably caused the damage on the target structure. The ground truth damage, as realized in test, is illustrated as a black cross. In this case, the highest likelihood is obtained very near to this experimental ground truth.

3.2 Demonstration with Synthetic Structure

Based on the success of the simple experimental case above, a more complex synthetic structure (Fig. 7) was considered. Synthetic data is useful for demonstrating the

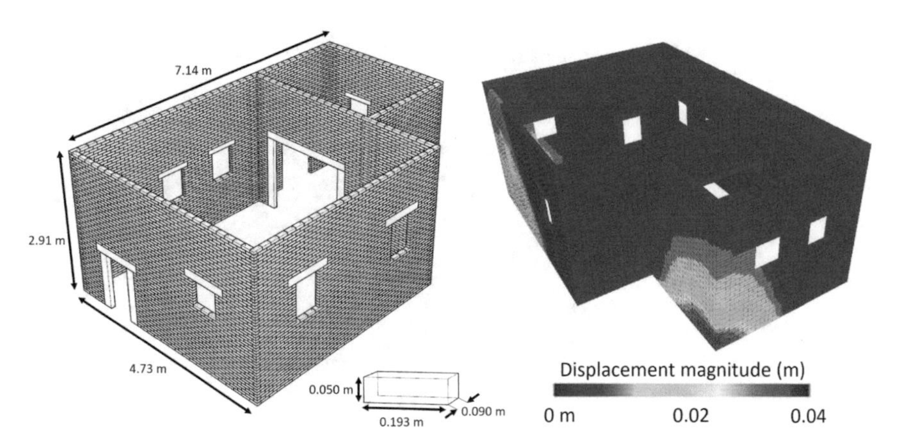

Fig. 7 (left) Geometry of synthetic masonry structure including individual brick dimensions. (right) Displacement magnitude of each brick in the synthetic structure due to two different settlement events

efficacy of the method. On the right of Fig. 7, the damaged synthetic structure with 9633 distinct elements can be seen having incurred two Gaussian settlement loads (one (x, y, a) at $(1.63, -1.73, 0.157)$ and a second at $(0.62, 3.22, 0.157)$. Since two independent damage events were considered for this synthetic case, this demonstrates the applicability of the method to more realistic scenarios where multiple damage states are present. A library of 256 damage states was generated by selecting damage cases from a uniform random distribution where the domain was $x \in [-0.39, 4.94]$, $y \in [-2.80, 5.13]$, and $a \in [0.00, 0.30]$. EMD was calculated between the ground truth model and each simulated damage state without segmentation; then, a continuous likelihood field was generated by applying GPR to the inverted distances (Fig. 8).

Considering the two bright, yellow areas which indicate a high likelihood of being the location and depth of settlement, the highest likelihood of 68.7 m^{-1} was found at $c = (0.51, 3.28, 0.198)$. When evaluating the likelihood at the true locations of settlement, it is found that they have values of 58.6 and 64.6 m^{-1} and an average

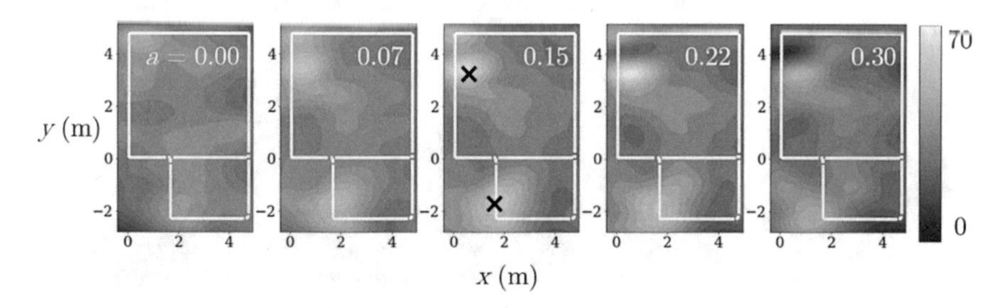

Fig. 8 Estimate of the likelihood field (c) obtained from full point cloud P. Field rendered over (x, y) slices for discrete amplitudes a as listed in each panel. Yellow color indicates greater likelihood (in units of m^{-1}). Black crosses indicate the two components of c°. White lines are centroids of bricks to illustrate the position of the structure relative to the position of settlement centers

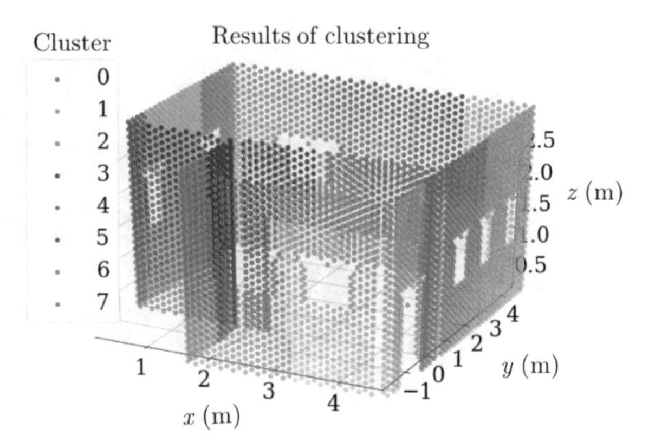

Fig. 9 Segmented point cloud for 3D structure

field value of 33.3 m^{-1}. This means that when comparing the average to the true locations, the true locations are twice as likely to be the cause of the damage. So, while there is good qualitative agreement in Fig. 8, the quantitative specificity of the result is relatively weak. To bolster the strength of the method, the damaged structure and the library of potential damages were segmented into eight different point clouds (shown in Fig. 9) using MBKM where each likelihood field corresponds to a single point cloud segment. It can be seen in a subsequent section that the method is not sensitive to the number of clusters selected in this step.

Segment 0, which is blue in Fig. 9, corresponds to the segment which covers the parts of the structure around one of the settlement events. Figure 10 presents the likelihood plot at discrete depths to understand what the possible cause of the damage in this region of the structure could be. While the true settlement occurs with a likelihood of 558 m^{-1} at $c = (1.63, -1.73, 0.157)$, the highest likelihood is 575 m^{-1} at (1.79, − 1.65, 0.156). Since the average likelihood is 105 m^{-1}, this indicates that the results are more decisive than what was previously shown in Fig. 8.

The likelihood field for Segment 7, which is gray in Fig. 9, is shown in Fig. 11. This segment lies directly over the second settlement event (and is separated from the first settlement event). The algorithm predicts the most likely settlement parameters to be (0.67, 3.28, 0.204), with a likelihood of 679 m^{-1} and the average of 94 m^{-1}.

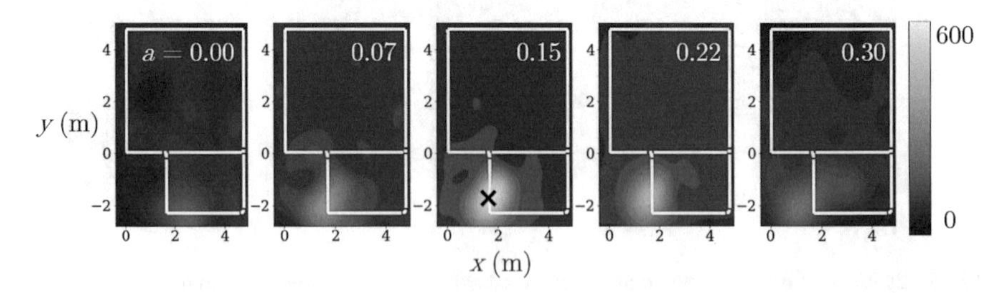

Fig. 10 Estimate of the likelihood field (c) obtained from P_0 which is the blue section of the point cloud in Fig. 9

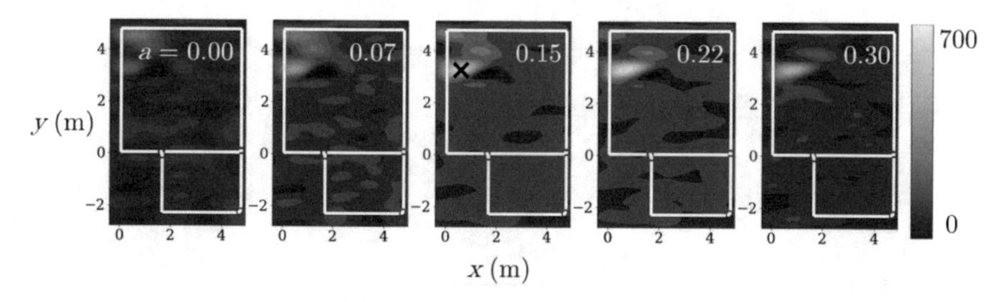

Fig. 11 Estimate of the likelihood field (c) obtained from P_7 which is the gray section of the point cloud in Fig. 9

Compared to the result for Segment 0, the likelihood is more concentrated in the $x - y$ plane and more diffuse in a. This is probably due to the delocalized nature of the settlement compared to the wall, such that certain combinations of (x, y, a) parameters induce equivalent damage states in the wall itself.

3.3 Sensitivity to Segmentation

To understand how the number of segments could affect the results of this method, segmentation was performed for the complex, synthetic structure, where the user selected 7, 8, and 9 clusters. This was repeated five times using random seeds to understand if relationships were robust. Figure 12 illustrates all 15 of the different segmentations which were carried out. Based on the broad range of high likelihood values, there is significant uncertainty in the optimal value for this parameter. This is not surprising as the peak likelihood values normally are within a concentrated region of other relatively high likelihood values compared to other locations. Since all 15 of the segmentations selected damage conditions corresponded well with the ground truth, the results are assumed to be reasonably insensitive to the number of segments selected.

3.4 Sensitivity to Noise

Since the complex example was carried out on a synthetic dataset where no noise was present, this section sought to understand how robust the method would be if the point clouds were more realistic and noise was included. Thus, noise on the order of millimeters was introduced to the synthetic dataset; random noise was added to every point in P from a uniform distribution over $[-\delta, +\delta]$ in each of (x, y, z). The analysis was repeated for Segment 7 (i.e., P_7) as outlined above. For each increase in noise, the Z-score was recorded, $z = (L(c^{\circ}) - \mu)/\sigma$, where μ and σ are the mean and standard deviation of $L(c)$ over all c in the investigated space, and error was

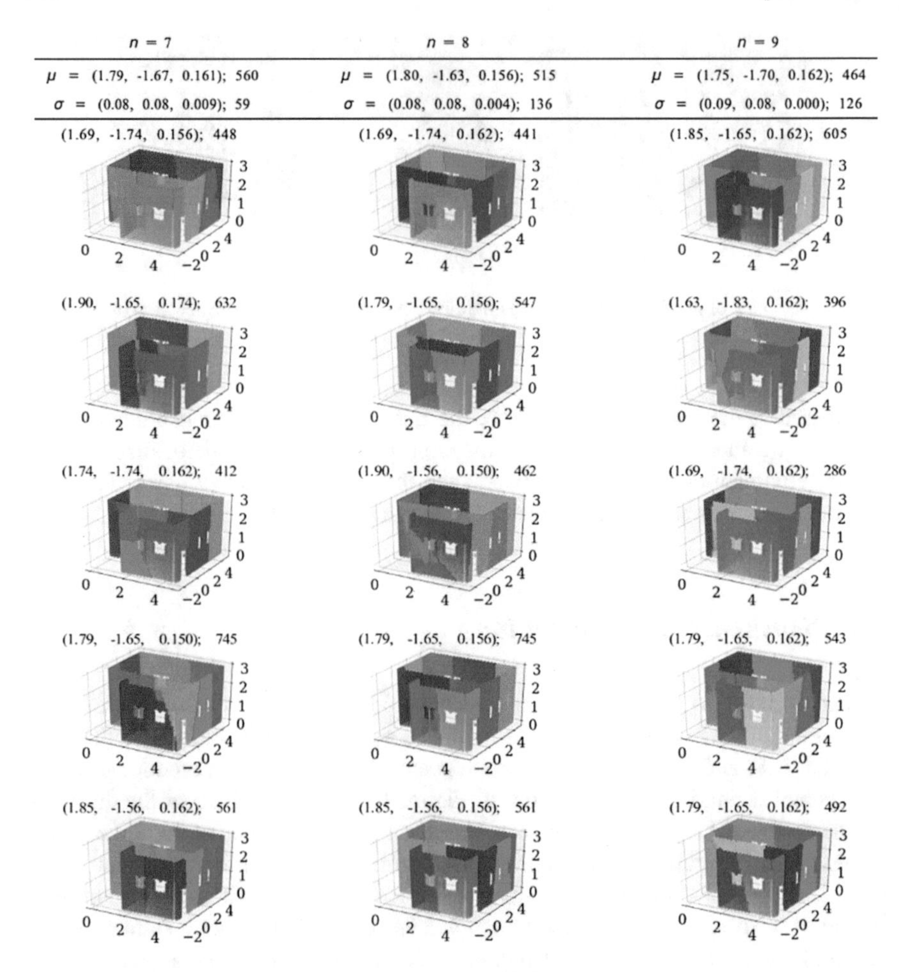

Fig. 12 Illustration of random segmentations from the MBKM segmentation while varying number of segments and random number seed. Mean and standard deviation are reported for each case. The computed maximum likelihood condition from the corresponding surrogate model is reported alongside each image

measured as the Euclidean distance between the location of peak likelihood and true cause of damage. Calculated Z-score and error are presented in Fig. 13. As one would expect, by adding noise to the point cloud, the Z-score decreased, and the prediction error increased. An encouraging result is that the magnitude of noise expected from typical point cloud acquisition methods (on the order of 10^{-3} m) does not appear to impede the method; the prediction error did not increase between 10^{-4} and 10^{-3} m.

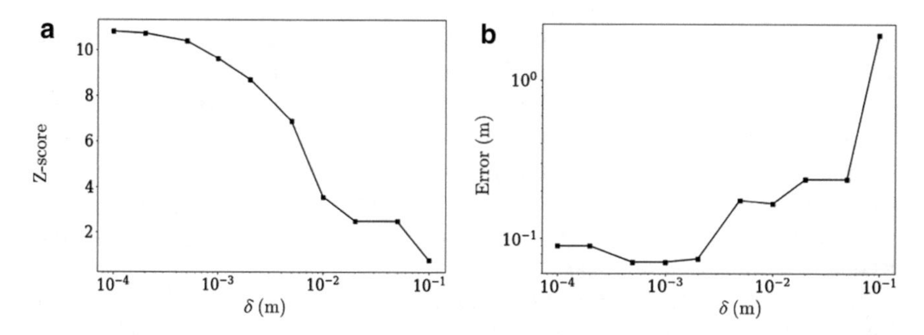

Fig. 13 Effects of noise magnitude δ (logarithmic scale in graphs) on **a** the Z-score and **b** error between peak likelihood and true value

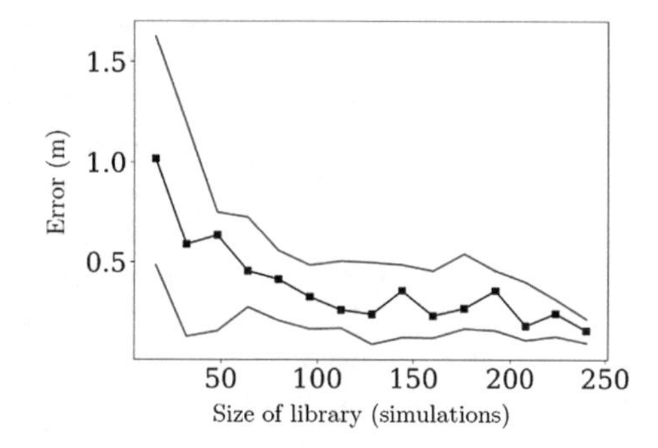

Fig. 14 Effects of the library size (in number of simulations) on the error between peak likelihood and true value. Black line indicates median value, while upper and lower blue lines indicate 75th and 25th percentiles, respectively (from 10 samples at each size)

3.5 Sensitivity to Library Size

To evaluate how the number of available simulations (hypotheses) affects the quality of the surrogate model, a sensitivity study to the library size was also performed. Simulations were randomly selected from the full library of 256 simulations, and a new model was constructed based on this sub-sample. The accuracy of the peak likelihood value versus the ground truth value was computed and reported in Fig. 14 as a function of the sample size. The accuracy clearly improves with more simulations available, but the improvement is most noticeable for libraries of less than 100 simulations.

3.6 Manifold Learning

A key limitation of the GPR-based approach is interpolation in a high-dimensional parameter space. For instance, in the cases presented above only a few discrete settlement events parameterized by a single location and depth were considered. If additional parameters such as the shape of the Gaussians (i.e., covariance matrix) and extended the analysis to include many Gaussians (i.e., representing an arbitrary surface as a Gaussian mixture model) were added, the system may have hundreds of parameters instead of only a handful. Each additional dimension in the parameter space makes the space sparser and mandates additional simulations to effectively interpolate using the Gaussian process. This makes it unlikely for the GPR approach to be tractable for complex problems. Here a method for combining numerical methods and manifold learning to reduce the number of physics-based models needed for diagnosis is outlined.

The starting point for the manifold learning strategy is the stored image of the displaced structure and a feature vector which conveys information about how the structure was damaged. For this work, the feature vector consisted of 12 total numerical features including the length of different settlement regions, the depth of different settlement regions, earthquake magnitude, earthquake orientation, point load, point load application range, and the total sum of all settlement depths. The first step in the process is the hashing of the stored image into a 2-D discrete cosine transform [37]. Once each image has been converted to a discrete cosine transform, the Euclidean distance between each image can be calculated.

For convenience of visualization, a two-dimensional manifold was constructed by taking the dominant eigenvectors ψ_2 and ψ_3 (comprised of one scalar value per observation) as shown in Fig. 15. It is important to note that the algorithm never sees the raw input feature vectors; collective variables are derived from only the distances between observed feature vectors. However, knowing what the input features can help a user to understand physical meanings behind the different eigenvectors and start to see how the latent space is related to the physical input parameters.

To see how each of the 12 input features was related to the manifold, the points on the manifold were colored by that feature. For instance, in the case of earthquake magnitude, feature 7, dark blue indicates that there was no earthquake while yellow indicates an earthquake with a magnitude of 0.3 g. As shown in Fig. 15, there is a

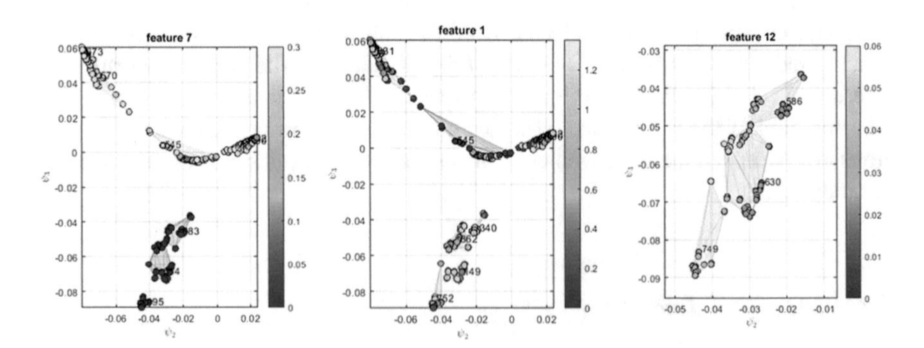

Fig. 15 Diffusion map colored by specific features. The first is colored by earthquake magnitude (gs), the second is colored by settlement width (m), and third by settlement depth (m). Third panel is zoomed in to show detail of bottom cluster only

correlation between feature 7 and where the points fall on the manifold. The simulations with lower ψ_3 values seem to not have experienced an earthquake at all, so those could be purely point load or purely settlement. And across ψ_2 there is a gradient from the very high-magnitude earthquakes at a low ψ_2, to the lesser earthquakes at higher values of ψ_2.

Not all features show these types of correlations with the eigenvectors, however. For instance, in the case of feature 1 (settlement amplitude on the left side of the wall), there does not seem to be any robust clustering possible. This indicates that this input feature did not dominate the manifold space (at least in the first two eigenvectors). Based on known information about feature 7, the bottom cluster of dark blue points in Fig. 15 is the point that did not experience any earthquake. If only the overall settlement experienced by the wall, feature 12, is considered, there are clear clusters developing. The sections where there is a lot of settlement have the highest ψ_3 and the lowest ψ_2, then the middle settlement cases have middle range ψ_2, and lastly the cases where there was little to no settlement have the highest values of ψ_2.

Based on a systematic analysis of the manifold topology, the features which exhibited the strongest correlations to the datasets were: (1) feature 7 depicted in red dominant colors representing earthquake magnitude, (2) feature 8 in green dominant colors representing earthquake orientation, and (3) feature 12 in blue dominant colors representing the total amount of settlement. These features could then be used to interpret the manifold; Figure 16 shows the annotated manifold based on these results. As ψ_3 increases, the earthquake magnitude increases, as ψ_2 increases the earthquake orientation changes, and as ψ_2 decreases, the overall settlement in the system gets larger. Thus, it would seem that ψ_2 is a latent variable encompassing information about both settlement and earthquake orientation. The samples of simulation images on the right are colored by the same conventions. Wall 630 experienced less settlement than wall 749 which is why it has less blue (RGB = 0, 0.5, 0.5) than 749 (RGB = 0, 0.5, 0.8).

Once a user has created a manifold of possible damage mechanisms, the damaged wall in question can be added to the set of inputs and tracked to see where in the manifold it ends up. Based on patterns intuited through clustering the feature-colored manifold, a user can evaluate where their damaged wall fits in with the rest of the manifold to understand what could have caused existing damages. In future work, this method should be validated against the experimental testing data such as those presented in preceding sections.

4 Conclusions and Future Directions

4.1 Summary

Here a workflow to diagnose observed damage states in masonry structures through statistical inference has been described. The approach is powered by a hybrid

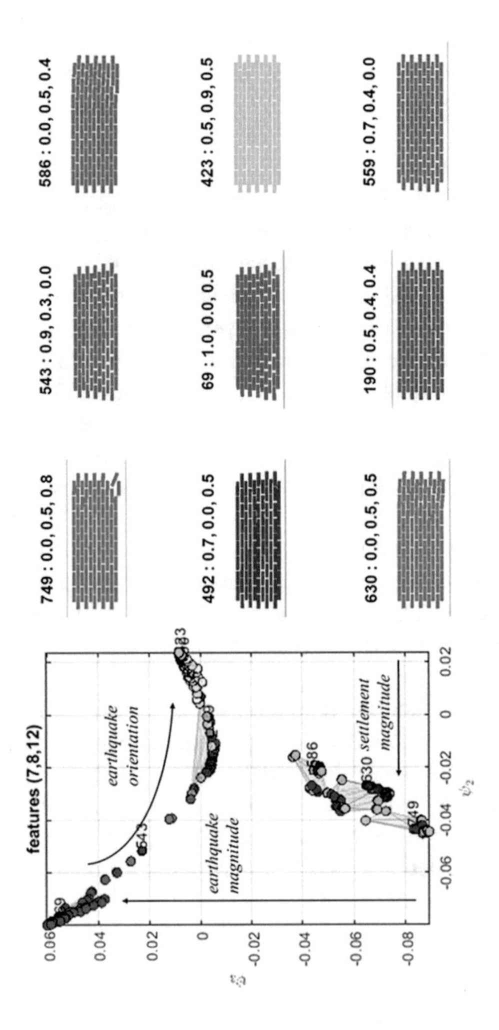

Fig. 16 (left) Annotated manifold with coloration from three features. (right) Sample walls with colors indicating location in the manifold and showing the crack pattern

modeling strategy which incorporates both physics-based and data-driven techniques. Discrete element modeling (DEM) was employed to simulate the deformation of masonry under hypothesized loads and settlements. The MBKM clustering approach was used to split the entire structure into discrete segments to support the detection of independent, localized damage mechanisms. EMD was used to determine a quantitative similarity score between hypothesized and observed damage states, which allowed interpolation in a continuous hypothesis space using GPR. By testing the method on synthetic data, for which a ground truth was known, the performance of the method in inferring the cause of two different settlements affecting the masonry structure at the same time was validated.

Furthermore, the same method was applied to experimental data through a case study on a laboratory-scale masonry wall. Settlements were systematically introduced to the wall to induce cracking across the structure which could be captured with conventional photography (or photogrammetry). To evaluate the robustness of the method in the presence of real-world variance, controlled sensitivity studies were performed on the synthetic data (again, because ground truth was known). These studies showed quantitatively that the method is relatively insensitive to noisy point clouds and randomness associated with the segmentation procedure. They furthermore quantified the number of hypotheses needed to obtain accurate results. Compared to prior work with similar objectives, this new method performed up to 10^5 times faster and required substantially less human input.

With many DEM results in hand, nonlinear manifold learning was employed to reduce the dimensionality of the damage feature space. While this manifold did not know the unique features, which caused the damages seen in the simulations, it intuited its own variables based on a spectral decomposition of an affinity matrix. This is a powerful approach because a user can see if there are trends in the damage patterns which are similar despite having disparate origins. Once a manifold has been created based on all the hypotheses for how the damage could have occurred, a user can see where their existing structure falls in the space. This will enable them to see the most probable origins of their damage patterns. This approach can be generalized to other physics-based modeling methods so long as an informative feature vector can be described.

4.2 *Future Directions*

The proof of concept shown here uses a static library of simulations drawn from a single hypothesized native state. There are several ways in which this could be extended to make it more robust. First, more sophisticated approaches are possible using iterative Bayesian optimization rather than a predetermined library of damage hypotheses. In this case, simulations would be generated one at a time or in small batches based on an acquisition function computed from prior observations. This enables intelligent navigation of the parameter space and allocates simulation resources to only those hypotheses which provide the maximal amount of new information. Second, the hypothesis space could be extended to include different native states. This would be motivated by the reality that the native state of a structure is not always known. Instead, several different native states could be postulated, and the likelihood calculations could determine the most plausible one.

As already described, while the GPR approach provides high accuracy in determining damage conditions, it is unlikely to scale efficiently to very large parameter spaces. Conversely, the manifold learning strategy described here provides a qualitative understanding of many variables without yielding reliable quantitative information about the damage conditions. The combination of these approaches may provide the advantages of each while mitigating their shortcomings. Another promising approach is topology optimization based on adjoint simulations. While this has historically been intractable for complex problems such as DEM, modern hardware and software developed for training deep learning models can provide gradients for very large and sophisticated transformations of data. Thus, a combination of data-driven and physics-based approaches will unlock new opportunities in structural health monitoring.

Acknowledgements This paper is based on work in part supported by the National Science Foundation Graduate Research Fellowship Program under Grant no. DGE-1656466. Any opinions, findings, and conclusions or recommendations expressed in this material are those of the authors and do not necessarily reflect the view of the National Science Foundation. Additional support was provided by the Dean's Fund for Innovation at Princeton and the Department of Civil and Environmental Engineering. The present research was completed as part of the Itasca Educational Partnership under the mentorship of Dr. Jim Hazzard.

References

1. Glisic B, Inaudi D (2008) Fibre optic methods for structural health monitoring. Wiley, Hoboken
2. National Trust for Historic Preservation T.: The Greenest Building (2017)
3. National Trust for Historic Preservation T.: Atlas of ReUrbanism (2017)
4. Olsen MJ (2013) In situ change analysis and monitoring through terrestrial laser scanning. J Comput Civ Eng 29:4014040
5. ISO (2005) C.S.N.: 13822 Bases for design of structures—Assessment of existing structures. CEN Brussels
6. Roca P (2004) Considerations on the significance of history for the structural analysis of ancient constructions. Structural analysis of masonry historical constructions, Balkema, Amsterdam, pp 63–73
7. Bayraktar A, Altunicsik AC, Sevim B, Türker T (2011) Seismic response of a historical masonry minaret using a finite element model updated with operational modal testing. J Vib Control 17:129–149
8. Russo S (2013) Testing and modelling of dynamic out-of-plane behaviour of the historic masonry façade of Palazzo Ducale in Venice, Italy. Eng Struct 46:130–139
9. Anderson DR, Burnham K (2004) Model selection and multi-model inference, 2nd edn. Springer, NY
10. Watt DS (2009) Building pathology: principles and practice. Wiley, Hoboken
11. Chamberlin TC (1965) The method of multiple working hypotheses. Science 80:754–759

12. Harris SY (2001) Building pathology: deterioration, diagnostics, and intervention. Wiley, Hoboken
13. Anzani A, Binda L, Carpinteri A, Invernizzi S, Lacidogna G (2010) A multilevel approach for the damage assessment of historic masonry towers. J Cult Herit 11:459–470
14. Alessandri C, Garutti M, Mallardo V, Milani G (2015) Crack patterns induced by foundation settlements: Integrated analysis on a renaissance masonry palace in Italy. Int J Archit Herit 9:111–129
15. Milani G, Valente M, Alessandri C (2017) The narthex of the Church of the Nativity in Bethlehem: a non-linear finite element approach to predict the structural damage. Comput Struct
16. Wood RL, Mohammadi ME, Barbosa AR, Abdulrahman L, Soti R, Kawan CK, Shakya M, Olsen MJ (2017) Damage assessment and modeling of the five-tiered pagoda-style Nyatapola temple. Earthq Spectra 33:S377–S384
17. Ramos LF, Lourenço PB (2004) Modeling and vulnerability of historical city centers in seismic areas: a case study in Lisbon. Eng Struct 26:1295–1310
18. Michiels T, Napolitano R, Adriaenssens S, Glisic B (2017) Comparison of thrust line analysis, limit state analysis and distinct element modeling to predict the collapse load and collapse mechanism of a rammed earth arch. Eng Struct 148:145–156
19. Douglas I, Napolitano R, Garlock M, Glisic B (2019) Reconsidering the vaulted forms of Cuba's National School of Ballet. In: Structural analysis of historical constructions, pp 2150–2158. Springer, Berlin
20. Russo V, Lignola GP, Vassallo E, Zinno A (2010) Second World War damages of the architectural heritage: St. Maria Del Popolo Agli Incurabili Church in Naples. In: Advanced Materials Research. pp 1137–1142
21. Taylor J (2005) An integrated approach to risk assessments and condition surveys. J Am Inst Conserv 44:127–141
22. Napolitano RK, Glisic B (2018) Minimizing the adverse effects of bias and low repeatability precision in photogrammetry software through statistical analysis. J Cult Herit 31:46–52
23. Patias P, Kaimaris D, Georgiadis C, Stamnas A, Antoniadis D, Papadimitrakis D (2013) 3D mapping of cultural heritage: special problems and best practices in extreme case-studies. ISPRS Ann Photogramm Remote Sens Spat Inf Sci II-5, pp 223–228
24. Barsanti SG, Remondino F, Fenández-Palacios BJ, Visintini D (2014) Critical factors and guidelines for 3D surveying and modelling in cultural heritage. Int J Herit Digit Era 3:141–158
25. Napolitano R, Glisic B (2019) Methodology for diagnosing crack patterns in masonry structures using photogrammetry and distinct element modeling (accepted). Eng Struct
26. Binda L, Gambarotta L, Lagomarsino S, Modena C (1999) A multilevel approach to the damage assessment and seismic improvement of masonry buildings in Italy. Seism. damage to Mason. Build. Balkema, Rotterdam, pp 170–195
27. Napolitano R, Hess M, Glisic B (2019) Quantifying the differences in documentation and modeling levels for building pathology and diagnostics. Arch Comput Methods Eng 1–18
28. Béjar Alonso J (2013) K-means vs Mini Batch K-means: a comparison
29. Yilmaz U (2009) The Earth Mover's Distance
30. Rubner Y, Tomasi C, Guibas LJ (2000) The earth mover's distance as a metric for image retrieval. Int J Comput Vis 40:99–121
31. Lupu N, Selios L, Warner Z (2017) A new measure of congruence: the earth mover's distance. Polit Anal 25:95–113
32. Williams CKI, Rasmussen CE (1996) Gaussian processes for regression. In: Advances in neural information processing systems, pp 514–520
33. Zhou L, Yan G, Ou J (2013) Response surface method based on radial basis functions for modeling large-scale structures in model updating. Comput Civ Infrastruct Eng 28:210–226
34. Xia Z, Tang J (2013) Characterization of dynamic response of structures with uncertainty by using Gaussian processes. J Vib Acoust 135:51006
35. Wan H-P, Ren W-X (2014) Parameter selection in finite-element-model updating by global sensitivity analysis using Gaussian process metamodel. J Struct Eng 141:4014164

36. Pedregosa F, Varoquaux G, Gramfort A, Michel V, Thirion B, Grisel O, Blondel M, Prettenhofer P, Weiss R, Dubourg V, Vanderplas J, Passos A, Cournapeau D, Brucher M, Perrot M, Duchesnay E (2011) Scikit-learn: machine learning in python. J Mach Learn Res 12:2825–2830
37. Gonzalez RC, Woods RE (2008) Digital image processing. Pearson/Prentice Hall, Upper Saddle River
38. Nadler B, Lafon S, Coifman R, Kevrekidis IG (2008) Diffusion maps—a probabilistic interpretation for spectral embedding and clustering algorithms. Lect Notes Comput Sci Eng 58:238–260
39. Letellier R, Eppich R (2015) Recording, documentation and information management for the conservation of heritage places. Routledge, Milton Park
40. Agisoft LLC (2014) Agisoft PhotoScan user manual: professional edition
41. Sapirstein P (2016) Accurate measurement with photogrammetry at large sites. J Archaeol Sci 66:137–145

Applications of Deep Learning in Intelligent Construction

Yang Zhang and Ka-Veng Yuen

Abstract Smart construction site is the concrete embodiment of the concept of smart city in the construction industry. It provides all-round, three-dimensional, and real-time supervision of construction sites with intelligent systems of controllability, data, and visualization. In particular, it is common to install many cameras in smart construction sites. These cameras only play the role of visualization, and further analysis is necessary to extract information from the images/videos and to provide safety warning signals. With the development of deep learning in the field of image processing, automatic deep feature extraction of images is possible for construction safety monitoring. This chapter summarizes the development and application of deep learning in construction safety, such as bolt loosening damage, structural displacement, and worker behavior. Finally, the application scenarios of deep learning in smart construction sites are further discussed.

Keywords Smart construction site · Deep learning · Construction safety · Computer vision

1 Introduction

Construction safety is an important issue for the society. There are many hidden dangers in construction sites, such as various types of workers, large construction machinery and complex environment. In the process of dynamic construction, the load of building structure is time-varying, and various support structures interact among themselves, leading to the stability risk of the entire construction. Buildings

Y. Zhang · K.-V. Yuen (✉)
State Key Laboratory of Internet of Things for Smart City, Department of Civil and Environmental Engineering, University of Macau, Macau 999078, China
e-mail: kvyuen@um.edu.mo

Guangdong-Hong Kong-Macau Joint Laboratory for Smart Cities, University of Macau, Macau 999078, China

Y. Zhang
e-mail: yangzhang@um.edu.mo

A. Cury et al. (eds.), *Structural Health Monitoring Based on Data Science Techniques*, Structural Integrity 21, https://doi.org/10.1007/978-3-030-81716-9_11

and structures under construction are prone to collapse, overturning and other accidents due to the influence of natural conditions, construction level, and construction quality. In addition, workers suffer from injuries caused by construction machinery, such as blow or collision. Therefore, the construction safety monitoring is necessary.

In order to improve the construction level of engineering projects and reduce the incidence of accidents, the concept of smart construction site was proposed [1]. Smart construction site is the concrete embodiment of the concept of smart city in the construction industry. It provides all-round, three-dimensional, and real-time supervision of construction sites with intelligent systems of controllability, data, and visualization. At present, the development of smart construction site is still in the initial stage, i.e., the perception stage. It uses advanced sensing technologies to monitor workers, machineries, and structures, and then identifies and locates the potential hazards. Building information modeling (BIM) takes the three-dimensional graphics of buildings as the carrier to further integrate all types of building information. The parameterized model can be used to realize construction simulation, collision detection, and other applications. Hu et al. used network analysis to improve collision detection. A component network centered on conflicting objects was constructed to represent the dependencies of components. This method can effectively identify the irrelevant conflicts and reduce the number of irrelevant conflicts by 17% [2]. Mirzaei et al. developed a novel 4D-BIM dynamic conflict detection and quantification system for the identification of spatiotemporal conflicts [3]. 3D laser scanning technology uses the principle of laser ranging to scan objects and quickly obtain 3D models. There have been many attempts in building inspection, cultural relics protection and other methods. Yang et al. [4] used terrestrial laser scanning to detect the deformation of the arch structure. The surface approximation method was used to cover the blank of measurement area, and the uncertainty of surface of different order was studied. Valenca et al. [5] proposed an automatic crack assessment method based on image processing and terrestrial laser scanning (TLS) technology. The geometric information measured by TLS was used to correct the captured image. It improved the identification accuracy of structural cracks. The combination of 3D laser scanning technology and BIM can realize the detection of structural quality and deformation. Ham et al. [6] proposed a structural safety diagnosis method based on laser scanning and BIM. The laser scanning data and BIM model were compared and analyzed to determine the deformation degree of pipe support. Chen et al. [7] proposed a point-to-point comparison method for deviation detection between automatic scanning and BIM. When there is a deviation between BIM and point clouds, it will be highlighted to remind users for further investigation.

Sensor networks can detect the deformation of high formwork support, tower crane, and bridge by different sensor nodes. Kifouche et al. [8] developed and deployed a sensor network that could collect data from multiple types of sensors. The data was transferred to a server for visualization and real-time processing. Kuang et al. [9] used fiber-optic sensors to monitor the deflection and cracks of beams. The results showed that it was possible to detect fine crack and final failure crack by optical fiber. Casciati et al. [10] used GPS to detect the displacement of steel structures in real time. The accuracy was of the order of subcentimeters. With the continuous upgrading of camera equipment, camera measurement technology has also been substantially developed. Cameras can be used to detect the deformation and vibration of structures in a close distance. Feng et al. [11] demonstrated the potential of low-cost visual displacement sensors for structural health monitoring. Meanwhile, experimental results showed that vision sensors have high precision in the full-field displacement measurement. Harvey et al. [12] used visual sensors to measure interlayer drift in real time. The dynamic characteristics were extracted to detect structural damage.

Artificial intelligence has accelerated its development, presenting new features such as deep learning, cross-border integration, human–machine collaboration, open intelligence, and autonomous control [13]. It has great potential in the field of construction safety monitoring. Deep convolutional neural networks have achieved remarkable results in the field of image processing. Compared with traditional machine learning methods, deep learning does not need to manually extract features and has a strong ability of autonomous feature extraction. Therefore, deep learning has been widely applied in various research fields [14–16].

With the rapid development of smart construction site, many cameras have been installed on construction sites. These cameras provide a lot of real-time data for construction safety detection. However, it is still an important research topic to deeply understand the image and to provide more accurate and timely monitoring results for construction safety. As one of the non-contact detection techniques, camera measurement has attracted much attention in the field of structural health monitoring. Some researchers have tried to combine deep learning with machine vision for construction safety monitoring. According to the construction characteristics, the detection content can be roughly divided into structure, worker, and mechanical operation safety. Section 2 introduces several kinds of deep learning algorithms commonly used in the field of construction safety monitoring. Section 3 presents the structural safety monitoring using deep learning. Section 4 focuses on worker safety management based on deep learning. Section 5 describes safety management of construction machinery using deep learning in the construction process. Section 6 summarizes the current situation and development of deep learning in the field of construction safety monitoring.

2 Related Deep Learning Algorithms

Convolutional neural network is the cornerstone for deep learning to achieve breakthrough achievements in the field of computer vision in recent years. Convolutional neural network uses convolutional layers to replace fully connected (FC) layers for effective feature extraction. A convolutional neural network usually consists of multiple convolutional modules. The front convolution layer has a small receptive field, which can capture local and detailed information of an image. The receptive field of the latter convolution layer is gradually enlarged to capture more complex and abstract information in the image. At first, deep convolutional neural network

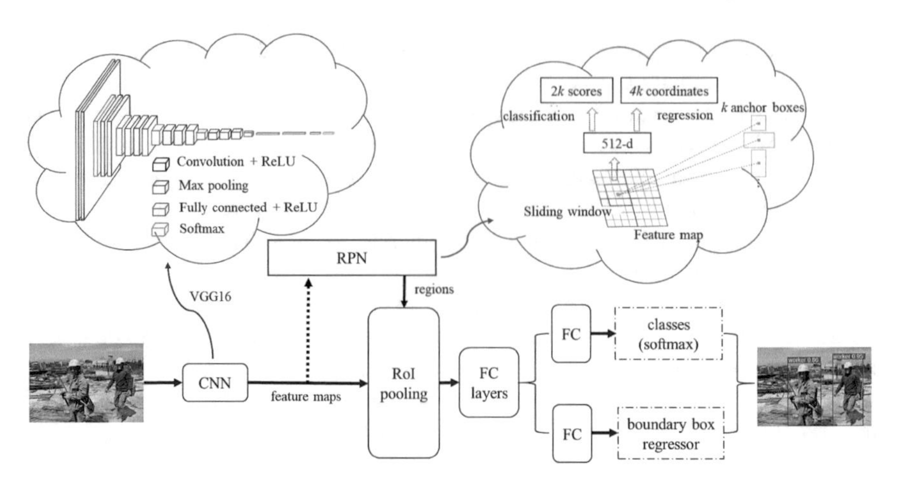

Fig. 1 Faster R-CNN

was used to perform image classification tasks and determine object categories in the image. After AlexNet was concerned by researchers, some image classification networks were produced, such as VGGNet, GoogLeNet, and ResNet. The recognition accuracy of these networks in the image classification task has almost reached the human level. However, image classification is a very crude task. Objects in an image need not only to be classified, but also to be positioned. Some algorithms combine candidate regions with image classification networks to achieve target location based on bounding boxes, such as region-based convolutional neural networks (R-CNN), Fast R-CNN, single-shot multibox detector (SSD), you only look once (YOLO), Faster R-CNN, and so on. The overall architecture of Faster R-CNN is shown in Fig. 1. Image classification network is used to extract features, and region proposal network (RPN) is used to generate candidate regions [17]. It further improves the accuracy and positioning accuracy of object recognition.

Although bounding boxes can locate objects in an image, the bounding box contains not only the target object, but also the background and other content. In order to achieve more precise recognition of objects in images, semantic segmentation and instance segmentation are proposed. These two types of algorithms belong to the pixel-level object recognition algorithm. Segmentation algorithms mostly adopt code-decoding structures [18–20], such as fully convolutional networks (FCN), SegNet, U-Net, and DeepLab. The overall architecture of FCN is shown in Fig. 2. Multiple convolutional layers are used as encoding structures for down-sampling, and multiple deconvolutional layers are used as decoding structures for up-sampling.

Human pose recognition is a special object detection task. Key point detection of human body is very important to describe human posture and predict human behavior. Action classification, abnormal behavior detection, and other tasks can be accomplished by estimating the key points of human body. Mask R-CNN, an instance segmentation algorithm, can not only identify targets at pixel level, but also estimate the key points of human body. In addition, network structures such as DensePose,

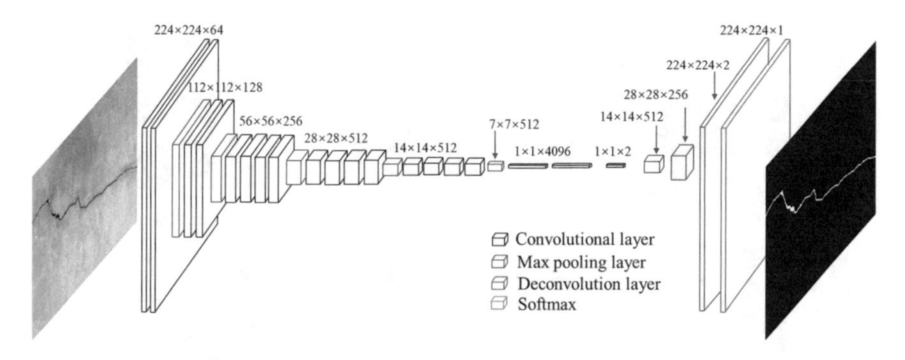

Fig. 2 FCN

OpenPose, AlphaPose, and DeepPose have also achieved remarkable effects in the field of human pose estimation.

3 Structural Safety Monitoring

3.1 Bolted Joints

In the process of dynamic construction, structural stability is poor, which can easily cause accidents. Therefore, it is necessary to monitor the status of structures under construction. Steel structures are often used in construction sites, such as steel trusses, steel columns, and support structures. Bolted joints have the advantages of simple structure, convenient installation, and strong reliability. These reasons have made bolts the preferred fastener. Under the influence of dynamic and static loads, bolts may be loosened and fallen off. In order to avoid interference with construction, non-contact monitoring methods can be preferred. The combination of machine vision and deep learning can quickly detect and count the state of bolts in steel structures. A camera or smartphone can quickly capture images of bolts. Faster R-CNN was used to identify and locate bolts in images. The image of each bolt is extracted on the basis of the localization information. Then, the edge lines of bolts in the image are extracted by using binarization and Hough transform. The detection result is shown in Fig. 3a. According to the change of edge information, the looseness angle could be recognized [21]. The head of the shoulder bolt is a regular hexagon, so this method could detect looseness angle within 60°. Grades of fastener appear on the head of each bolt. The "bolt" and "num" were identified and located simultaneously by SSD, as shown in Fig. 3b. According to the localization information, the looseness angle within 180° could be recognized [22]. When the bolt looseness angle is greater than 180°, the above method cannot effectively detect bolt looseness angle. A bolt to be loosen when the loosening is evident in a geometric change, which may even be visible to naked eyes. Therefore, Faster R-CNN can be used to directly detect large looseness

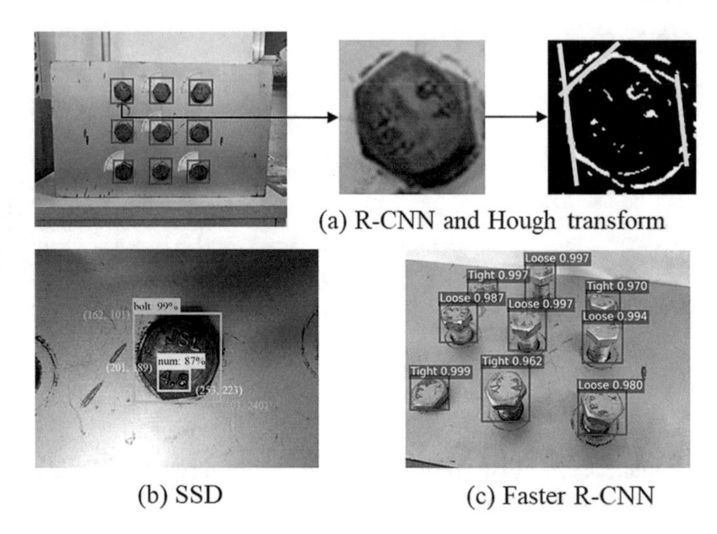

Fig. 3 Bolt damage detection

damage of bolts, as shown in Fig. 3c. This method has strong robustness even under the influence of structural vibration, illumination, and other environmental factors [23]. Deep learning is a data-driven recognition algorithm, and a large amount of data is the basic condition to ensure the recognition accuracy and generalization ability of model. 3D simulation software can quickly produce many bolt images. The detection model based on the simulation image still can accurately identify and locate bolts in a real image [24]. Although the simulation software cannot produce bolt images in a complex scene, this method provides a new idea for data augmentation.

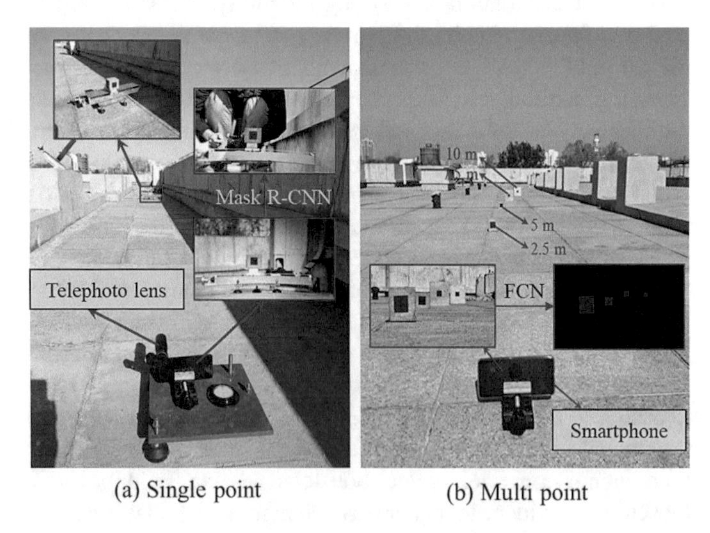

Fig. 4 Displacement monitoring based on semantic segmentation

3.2 *Structural Displacement*

During the construction process, the deformation and displacement of structures should be strictly controlled within the allowable range. With the development of high-rise and super high-rise buildings, high-support formwork and deep foundation pit appear more frequently in the construction process of building structures. These two types of structures, which have more potential hazards and high accident frequency, are the core parts of construction safety monitoring. The height of high-support formwork is more than five meters. Once it collapses, the impact and economic loss will be very serious. During the procedure of pouring concrete, the load borne by high-support formworks increase sharply, causing large deformation or even partial collapse and overall overturning. The foundation pit, located in the urban areas area, is not surrounded by sufficient space, so it is generally constructed by vertical excavation. Meanwhile, the depth of deep foundation pit is more than five meters, so it is very easy to collapse the side wall of the foundation pit. Therefore, it is necessary to monitor the displacement of high-support formwork and deep foundation pit. Deep learning can identify and locate target in an image at the pixel level. It provides the possibility for displacement monitoring. Portable devices such as smartphones are used to photograph artificial targets, which are identified and located by Mask R-CNN. Some parameters of target can be easily extracted from segmentation results. These parameters can be used to calculate the target displacement. Zhang et al. verified the feasibility of this method through static and dynamic experiments, respectively [25]. However, the proposed method can only monitor the short-range displacements. To achieve remote displacement monitoring, they installed a 22× optical zoom lens on a smartphone, as shown in Fig. 4a. The lens improves the ability of smartphones to photograph long-range targets. However, this method is limited by the lens, and it can only monitor the longitudinal one-point displacement. In order to monitor the displacement of deep foundation pit, Zhang et al. [26] proposed a longitudinal multipoint displacement monitoring method using FCN and smartphone. They used Huawei P30 Pro as the acquisition equipment, which has a high-performance camera with a 50× optical zoom. The detection result is shown in Fig. 4b. It can be used to detect the displacement of four longitudinal points within 10 m. This method does not require an external lens and is more portable.

Compared with the above method, the optical flow estimation-based method detects small motions without the use of paints or markers on the structure surface. However, the complex calculation process limits its real-time inference capacity. Deep learning can estimate optical flow with fewer parameters. With the help of GPU-accelerated computation, the optical flow method based on deep learning can be used for real-time monitoring [27]. The full-field optical flow and homography matrix were used to obtain the full-field structural displacement. Theoretically, the displacement of any point of structures can be obtained according to the full-field structural displacement diagram. Nevertheless, the optical flow estimation is affected by the background clutter. Dong et al. [28] proposed a full-field optical flow estimation algorithm based on deep learning. It reduces manual manipulation and provides more accurate measurements with less computation time. Subpixel subdivision technology can make up for the shortage of hardware and improve image resolution. Luan et al. propose a deep learning approach based on CNN to extract full-field high-resolution displacements at subpixel levels [29]. The results showed that the trained network can identify the pixels with sufficient texture contrast as well as their subpixel motions.

3.3 Structural Surface Quality

Construction quality evaluation is an important part of construction quality management. Once forms are removed, concrete surfaces may appear void, pockmarked surface, crack, and other damage. Deep convolutional neural network can identify and extract structural surface damage, it can improve the efficiency and accuracy of manual inspection. To locate surface damage in an image, sliding window algorithm cropped the image into several small images, which are fed into convolutional neural networks for classification [30]. In order to improve the speed of sliding window detector, various object recognition algorithms based on candidate regions are proposed, such as Faster R-CNN and YOLO. The identification result is shown in Fig. 5. Deng et al. [31] used YOLO to identify and locate cracks in images and compared the recognition effect with Faster R-CNN.

Object detection algorithms can locate surface damages through bounding box, which cannot be used to extract damage area, width, length, and other parameters. Semantic segmentation algorithm can identify and locate targets in an image at pixel level. For example, SegNet and FCN can be used for pixel-level recognition of cracks in images [32, 33]. Compared with object detection, the result of semantic segmentation is more precise. The identification result using semantic segmentation is shown in Fig. 6. Lee et al. [34] proposed crack width estimation method based on shape-sensitive kernels and semantic segmentation. Firstly, SegNet was used to identify cracks at pixel level. Then, the maximum width of the crack is measured based on

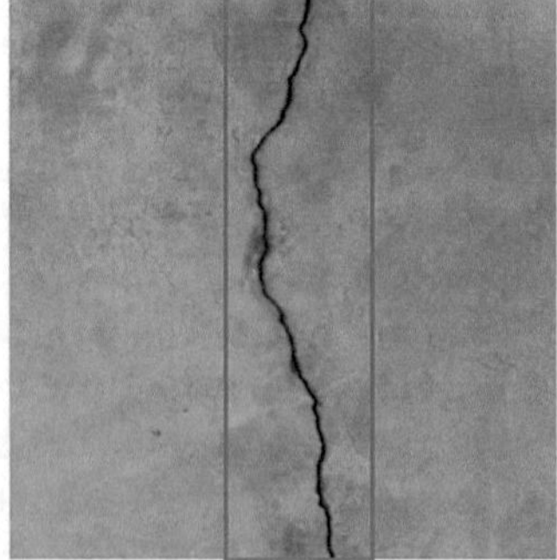

Fig. 5 Crack identification based on bounding box

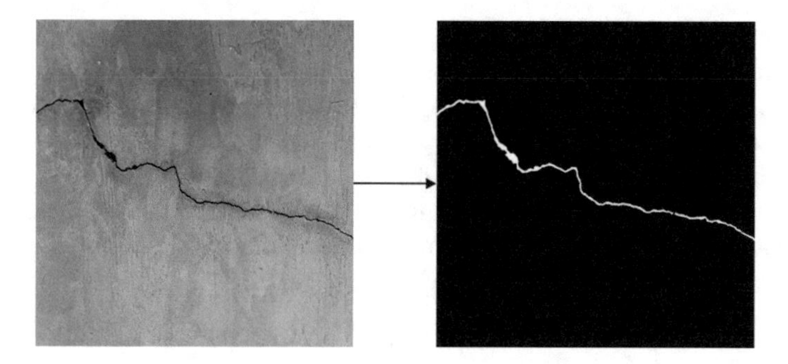

Fig. 6 Crack identification based on semantic segmentation

the identification results. A practical detection method should consider not only the detection accuracy, but also the detection speed. Cha et al. [35] proposed a real-time crack segmentation network, which has a great improvement in detection accuracy and speed. The model processes in real-time images at 1025 × 512 pixels, which is 46 times faster than in a recent work. According to the results of semantic segmentation, the parameters such as length, width, and area of damages can be obtained more conveniently. However, semantic segmentation requires pixel-level annotation data, and the annotation cost is very large. Meanwhile, semantic segmentation model is difficult to train and requires high computational performance.

To reduce the detection costs, object detection and traditional methods are combined to achieve crack refined detection. Object detection can identify and locate cracks accurately and quickly by bounding box. The positioning method based on the bounding box cannot quantitatively analyze the parameters of cracks. Some image processing methods such as binarization and filtering can be used to extract cracks in the bounding box at the pixel level. The identification result is shown in Fig. 7. Jiang et al. [36] proposed a real-time crack evaluation system based on deep learning and wall-climbing UAV. Firstly, SSD was used to identify and locate cracks in the

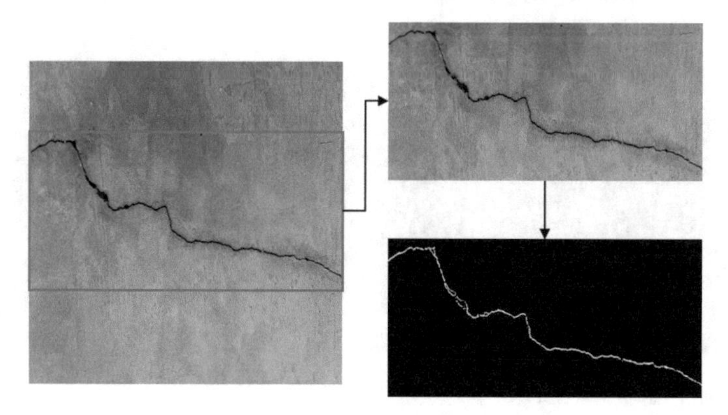

Fig. 7 Crack identification based on bounding box and image processing

images transmitted by UAV. Then, the images in bounding box were converted into binary image. The grayscale images were transformed into binary image through Otsu threshold segmentation, and morphological processing was applied to ensure continuity of cracks. These are image-based identification methods, which can only obtain two-dimensional spatial parameters of damage. Deep convolutional neural networks have also achieved remarkable results in the field of 3D point cloud segmentation. Beckman et al. [37] combined convolutional neural network with depth camera to detect the volume of concrete surface damages.

4 Worker Safety Management

A large number of workers on construction sites and the crosscutting of jobs are the main reasons for the higher casualty rate in construction industry than in any other industry. There are many potential hazards in the construction sites and workers should be vigilant of their surroundings. Therefore, it is desirable to track the number and status of workers in the construction site in real time. Object detection algorithms such as Faster R-CNN can identify and locate workers in different scenes and postures [38], as shown in Fig. 8. It can quickly count the number and location of workers in a large scene, providing a new possibility for construction worker detection.

Heavy equipment may lead to serious worker injuries, and equipment operators' view is easily obscured. Thus, it is necessary to detect workers near the equipment to provide safety warning for operators. In order to improve visibility, heavy equipment manufacturers install cameras on each side of the equipment (i.e., front, right, left, and rear) to provide a comprehensive view of the area around the equipment. However, this monitoring system cannot automatically extract information from images. Son et al. [39] proposed a real-time warning system using visual data and Faster R-CNN. This system used monocular cameras to estimate worker's position in three dimensions. In addition, some support structures without guardrails are also one of the important hazards. During the construction of an engineering

Fig. 8 Identification of worker

structure, workers tend to take shortcuts by crossing supports to perform daily activities and save time. However, crossing the support is very dangerous and forbidden. Fang et al. [40] used instance segmentation to detect workers crossing structural supports during the construction of a project. First, workers and structural supports are identified at the pixel level by the Mask R-CNN. Then, the positioning relationship between the structural support and the workers could be used to determine whether the worker is passing through the support. The proposed method could obtain the distance between workers and hazard sources, thus providing safety warning for managers and performing the correct behavior. At construction sites, workers are always in dynamic walking positions. Predicting workers' trajectories has great potential to improve workplace safety. Object identification-based tracking methods only rely on entity operation information and do not make full use of context information. Cai et al. [41] proposed a context-augmented long short-term memory approach for worker trajectory prediction. Compared with the traditional one-step prediction method, the proposed method could predict multistep trajectory to avoid error accumulation and effectively reduce final displacement error by 70%.

Workers should be vigilant of potential hazards. However, personal protection of workers is also very necessary. Safety helmets play an important role in protecting construction workers from accidents. Nevertheless, workers sometimes do not wear safety helmets for convenience. For the safety of construction workers, a high-precision and strong robustness helmet detection algorithm is urgently needed. Object detection algorithms such as Faster R-CNN, SSD, and YOLO can detect safety helmet in an image [42–44], as shown in Fig. 9. The recognition accuracy of detection algorithms based on deep learning depends on a large number of sample data. To evaluate the performance of Faster R-CNN, more than 100,000 image frames of construction workers were randomly selected from surveillance videos of 25 construction sites over a period of more than one year. The experimental results showed that this method has high accuracy, high recall rate, and fast speed, and could effectively detect non-helmet-use in different construction sites. It is conducive to improving the safety inspection and supervision level. Besides safety helmet and protective clothing, safety harness is the main protective equipment for workers working at height. Workers often forget or deliberately do not wear seat harness. This is a very

Fig. 9 Identification of safety helmet

dangerous behavior and is one of the main causes of falling from heights. Fang et al. [45] developed a visual-based automated approach that uses two convolutional neural networks to determine whether a worker is wearing a safety harness. First, Faster R-CNN was used to detect workers in an image, and the image in bounding box was cropped. Then, these cropped images were fed into a multilayer convolutional neural network to determine whether the workers wear seat harness. The accuracy rate and recall rate of the Faster R-CNN model were 99 and 95%, respectively, and the accuracy rate and recall rate of the CNN model were 80 and 98%, respectively.

Human body is very flexible, and posture can be used to identify worker behavior. Roberts et al. [46] proposed an activity analysis framework based on vision and deep learning that could estimate and track the posture of worker. Firstly, YOLO was used to identify and locate workers in an image, and the image in bounding box was cropped. Then, the cropped images were fed into a 2D skeleton detection network to estimate each joint coordinates of the worker. Posture estimation of worker is shown in Fig. 10. Finally, the pose of the same construction worker in different video frames was tracked. The proposed method was used to identify different worker activities (bricklaying and plastering), and the results showed that this method had the potential to evaluate individual worker activities. A 3D skeleton estimation network can directly determine the spatial position of human. The distance between workers can be used to estimate severity of crowding in working environment. High crowding may lead to dangerous working conditions and negative worker behavior. Yan et al. [47] proposed a vision-based crowd detection technology. First, Faster R-CNN was used to identify and locate workers in complex environment, and the image in bounding box was cropped. Then, the cropped images were fed into the 3D skeleton detection network to estimate each joint coordinates of the worker. Finally, according to the result of human skeleton estimation, the spatial position of each worker was obtained. Experimental results showed that this method could estimate the distance between two workers with an error of 0.45 m in three-dimensional space.

At present, worker status monitoring methods based on the fixed camera have some limitations, such as the perceptual mixing, occlusion, and illumination. Meanwhile, there is no camera inside the buildings under construction. Some mobile robots autonomously inspect indoor work sites and find potentially dangerous anomalies.

Fig. 10 Posture estimation of worker

Lee et al. [48] applied the perception module based on deep learning and simultaneous localization and mapping (SLAM) to target recognition and navigation of mobile robots. The proposed method could identify abnormal behaviors of some workers, such as not wearing safety helmets, standing on top of ladders, or falling. Identification of danger is only the first step of construction safety management. It is necessary to timely and accurately convey risk information to managers. Tang et al. [49] proposed a language-image framework that aims at understanding and detecting semantic roles of activities mentioned in safety rules. This framework includes semantic parsing of safety rules, construction object detectors using SSD and Faster R-CNN, and semantic role detectors. The experimental results showed that this framework can preliminarily describe the dangerous scene in an image in the form of language.

5 Construction Machinery Management

In order to improve construction efficiency, various machineries are used in the construction process, such as excavator, dump truck, bulldozer, scraper, crane, etc. However, safety accidents often occur with these construction machineries. Therefore, it is necessary to identify the position and status of construction machinery in real time. Object detection algorithms such as Faster R-CNN and R-FCN can identify and locate construction machinery in an image [50], as shown in Fig. 11. Kim et al. used R-FCN to identify five types of vehicles: dump truck, excavator, loader, cement mixer, and road roller [51]. The experimental results showed that the average accuracy of this method was 96.33%. Meanwhile, the synthesized image

Fig. 11 Identification of construction machinery

data could be used for data augmentation, which can improve the vehicle detection performance [52]. The identification and location information of construction machinery can provide managers with the type, quantity, and distribution of construction machinery. Identifying and tracking construction machines also can avoid potential collisions and other accidents. Firstly, convolutional neural network was used to detect and track vehicles. Then, the tracking trajectory was fed into a hidden Markov model (HMM) that automatically discovers and assigns activity labels to objects. Roberts et al. [53] performed activity analysis of machineries based on the object detection results. The accuracy of activity analysis was found to be 86.8% for excavators and 88.5% for dump trucks. Of all construction machinery, tower cranes are the largest in size. The tower crane operator is far from the ground and cannot clearly observe the surrounding environment of lifting objects. High-definition cameras built into Unmanned Aerial Vehicles (UAVs) could help solve this problem. Roberts et al. [54] used SSD to identify and locate tower cranes in an image, which is photographed by UAVs.

All the aforementioned identified bounding box-based construction machinery recognition methods use horizontal detection algorithms for detection. However, construction machinery can be in any direction and position, and they are not necessarily horizontal or vertical. When construction machinery is densely parked together, rotated bounding box can be more accurately fitted to the machinery region in terms of the orientation. Guo et al. [55] proposed a construction machinery identification method based on orientation-aware feature (OAF) fusion convolutional neural network. The proposed OAF-SSD could be applied not only to the construction vehicle detection, but also to the detection problem of dense multiple objects in civil engineering, which can also identify the orientation of target objects (more useful for motion tracking and estimation).

In the similar fashion as human body, the posture and movements of construction machinery need to be estimated automatically. This has a significant impact on construction safety and the use of the machinery itself. Excavator is one of the most used construction equipment. Its operation is complicated, and its posture changes more. Hence, it is more difficult to evaluate the full body posture of excavators than other equipment. Liang et al. [56] proposed a vision-based marker-less pose estimation system for estimating the joints and components of excavators. This system adopted and improved the advanced convolutional network, namely the stack hourglass network for pose estimation. The results showed that this system could estimate excavator boom, stick, and bucket joint positions but had higher estimation error for the bucket location due to the occlusion issue. Luo et al. [57] built an integrated model based on stacked hourglass and stacked pyramid network to estimate the posture of excavators in an image. This model evaluated the posture of excavators by identifying six key points of excavators, as shown in Fig. 12. Occlusion can significantly interfere with the detection results of this method. Compared to the activity identification of construction workers, research on activity identification of construction machinery is very limited, mainly because of the lack of construction machinery actions datasets. Zhang et al. [58] produced a comprehensive video dataset of 2064 clips, which included five actions (digging, swinging, dumping, moving forward,

Fig. 12 Posture estimation of construction machinery

and moving backward) of excavators and dump trucks. CNN is used to extract image features, and long short-term memory (LSTM) is used to extract time characteristics from video frame sequences. These two types of features were used to identify these five actions.

6 Conclusion

Deep learning has been growing rapidly in the field of image processing. Some new network structures promote the development of construction safety monitoring using machine vision. On construction sites, structure, worker, and machinery are the three potential hazard sources, which are the most likely to cause accidents. Machine vision-based detection method is a non-contact detection technology, which can detect hazard sources without affecting the normal construction. At present, deep learning networks applied in the field of construction safety monitoring can be roughly divided into three categories: object detection, semantic segmentation, and pose estimation.

(1) Object detection using neural network is the simplest and most used detection method. Faster R-CNN, SSD, YOLO, R-FCN, and other object detection networks were used to identify and locate workers, machineries, and structural components. This type of detection method mainly relies on bounding box to identify and locate hazard sources. It can only complete detection tasks with low positioning accuracy.

(2) Compared with object detection, semantic segmentation has higher positioning accuracy and can achieve pixel-level object recognition. FCN, SegNet, U-Net, and other networks are used to locate targets, components, and workers with high precision. However, the disadvantages of this approach are also obvious. The training cost of semantic segmentation is high, and the detection speed is slow.

(3) Pose estimation networks recognize human posture through key point estimation. It can be used to detect different types of workers and assess worker activities. Meanwhile, it can also be used to locate the key points of construction machinery and identify the working status of machinery.

Many researchers have studied construction safety with deep learning and proposed many effective detection methods. However, from the perspective of smart construction site, the research on construction safety monitoring using deep learning is still in the initial stage. These detection techniques are far from being intelligent, and only identify and locate multiple targets on construction sites. In addition, the current detection technology also has some common problems. Deep learning is a data-driven algorithm. There is a lack of large image datasets for different construction scenes. Therefore, how to train a robust and high-precision detection model is still a difficult problem. Most importantly, visual detection method is easily affected by illumination, background, occlusion, and other factors. Now more and more complex construction site environment, some targets are often blocked, which is fatal to visual detection methods. A single type of sensor data often has great limitations, so the fusion of multitype sensor data may be a more feasible scheme for construction safety monitoring.

Acknowledgements This research has been supported by the Guangdong-Hong Kong-Macau Joint Laboratory Program under grant number: 2020B1212030009 and the Science and Technology Development Fund of the Macau SAR under research grant SKL-IOTSC-2021-2023. These generous supports are gratefully acknowledged.

References

1. Hammad A, Vahdatikhaki F, Zhang C, Mawlana M, Doriani A (2012) Towards the smart construction site: improving productivity and safety of construction projects using multi-agent systems, real-time simulation and automated machine control. In: Proceedings of the 2012 winter simulation conference, pp 1–12. IEEE, Germany
2. Hu Y, Castro-Lacouture D, Eastman CM (2019) Holistic clash detection improvement using a component dependent network in BIM projects. Autom Constr 105:102832
3. Mirzaei A, Nasirzadeh F, Parchami Jalal M, Zamani Y (2018) 4D-BIM dynamic time–space conflict detection and quantification system for building construction projects. J Constr Eng Manag 144(7):04018056
4. Yang H, Omidalizarandi M, Xu X, Neumann I (2017) Terrestrial laser scanning technology for deformation monitoring and surface modeling of arch structures. Compos Struct 169:173–179

5. Valença J, Puente I, Júlio E, González-Jorge H, Arias-Sánchez P (2017) Assessment of cracks on concrete bridges using image processing supported by laser scanning survey. Constr Build Mater 146:668–678
6. Ham N, Lee SH (2018) Empirical study on structural safety diagnosis of large-scale civil infrastructure using laser scanning and BIM. Sustainability 10(11):4024
7. Chen J, Cho YK (2018) Point-to-point comparison method for automated scan-vs-bim deviation detection. In: 17th international conference on computing in civil and building engineering, pp 1–8. Springer, Finland
8. Kifouche A, Baudoin G, Hamouche R, Kocik R (2017) Generic sensor network for building monitoring: design, issues, and methodology. In: 2017 IEEE conference on wireless sensors, pp 1–6. IEEE, Malaysia
9. Kuang KSC, Cantwell WJ, Thomas C (2003) Crack detection and vertical deflection monitoring in concrete beams using plastic optical fiber sensors. Meas Sci Technol 14(2):205–216
10. Casciati F, Fuggini C (2011) Monitoring a steel building using GPS sensors. Smart Struct Syst 7(5):349–363
11. Feng D, Feng MQ (2017) Experimental validation of cost-effective vision-based structural health monitoring. Mech Syst Signal Process 88:199–211
12. Harvey Jr, PS, Elisha G (2018) Vision-based vibration monitoring using existing cameras installed within a building. Struct Control Health Monit 25(11):e2235
13. Editorial Committee (Editor-in-chief: Xin Zhao) (2017). Report on informatization of building construction industry (2017): Application and development of smart construction sites. China Building Material Industry Press, (in Chinese)
14. Hashemi H, Abdelghany K (2018) End-to-end deep learning methodology for real-time traffic network management. Comput-Aided Civil Infrastruct Eng 33(10):849–863
15. Jia Y, Johnson M, Macherey W, Weiss RJ, Cao Y, Chiu CC, Wu Y (2019) Leveraging weakly supervised data to improve end-to-end speech-to-text translation. In: IEEE international conference on acoustics, pp 7180–7184. IEEE, UK
16. De Fauw J, Ledsam JR, Romera-Paredes B, Nikolov S, Tomasev N, Blackwell S, Ronneberger O (2018) Clinically applicable deep learning for diagnosis and referral in retinal disease. Nat Med 24(9):1342–1350
17. Ren S, He K, Girshick R, Sun J (2016) Faster R-CNN: Towards real-time object detection with region proposal networks. IEEE Trans Pattern Anal Mach Intell 39(6):1137–1149
18. Long J, Shelhamer E, Darrell T (2015) Fully convolutional networks for semantic segmentation. In: Proceedings of the IEEE conference on computer vision and pattern recognition, pp 3431–3440. IEEE, USA
19. Ronneberger O, Fischer P, Brox T (2015) U-net: Convolutional networks for biomedical image segmentation. In: international conference on medical image computing and computer-assisted intervention. Springer, Germany, pp 234–241
20. Chen LC, Papandreou G, Kokkinos I, Murphy K, Yuille AL (2017) DeepLab: semantic image segmentation with deep convolutional nets, atrous convolution, and fully connected crfs. IEEE Trans Pattern Anal Mach Intell 40(4):834–848
21. Huynh TC, Park JH, Jung HJ, Kim JT (2019) Quasi-autonomous bolt-loosening detection method using vision-based deep learning and image processing. Autom Constr 105:102844
22. Zhao X, Zhang Y, Wang N (2019) Bolt loosening angle detection technology using deep learning. Struct Control Health Monit 26(1):e2292
23. Zhang Y, Sun X, Loh KJ, Su W, Xue Z, Zhao X (2020) Autonomous bolt loosening detection using deep learning. Struct Health Monit 19(1):105–122
24. Pham HC, Ta QB, Kim JT, Ho DD, Tran XL, Huynh TC (2020) Bolt-loosening monitoring framework using an image-based deep learning and graphical model. Sensors 20(12):3382
25. Zhang Y, Liu P, Zhao X (2020) Structural displacement monitoring based on mask regions with convolutional neural network. Construct Build Mater 120923
26. Zhang Y, Zhao X, Liu P (2019) Multi-point displacement monitoring based on full convolutional neural network and smartphone. IEEE Access 7:139628–139634

27. Ilg E, Mayer N, Saikia T, Keuper M, Dosovitskiy A, Brox T (2017) Flownet 2.0: Evolution of optical flow estimation with deep networks. In: Proceedings of the IEEE conference on computer vision and pattern recognition, pp 2462–2470. IEEE, USA
28. Dong CZ, Celik O, Catbas FN, O'Brien EJ, Taylor S (2020) Structural displacement monitoring using deep learning-based full field optical flow methods. Struct Infrastruct Eng 16(1):51–71
29. Luan L, Wang ML, Yang Y, Sun H (2020) Extracting full-field subpixel structural displacements from videos via deep learning. arXiv preprint arXiv:2008.13715
30. Cha YJ, Choi W, Büyüköztürk O (2017) Deep learning-based crack damage detection using convolutional neural networks. Comput-Aided Civil Infrastr Eng 32(5):361–378
31. Deng J, Lu Y, Lee VCS (2020) Imaging-based crack detection on concrete surfaces using You Only Look Once network. Struct Health Monit 20(2):484–499
32. Zhang X, Rajan D, Story B (2019) Concrete crack detection using context-aware deep semantic segmentation network. Comput-Aided Civil Infrastruct Eng 34(11):951–971
33. Dung CV (2019) Autonomous concrete crack detection using deep fully convolutional neural network. Autom Constr 99:52–58
34. Lee JS, Hwang SH, Choi IY, Choi Y (2020) Estimation of crack width based on shape-sensitive kernels and semantic segmentation. Struct Control Health Monit 27(4):e2504
35. Choi W, Cha YJ (2019) SDDNet: Real-time crack segmentation. IEEE Trans Industr Electron 67(9):8016–8025
36. Jiang S, Zhang J (2020) Real-time crack assessment using deep neural networks with wall-climbing unmanned aerial system. Comput-Aided Civil Infrastruct Eng 35(6):549–564
37. Beckman GH, Polyzois D, Cha YJ (2019) Deep learning-based automatic volumetric damage quantification using depth camera. Autom Constr 99:114–124
38. Son H, Choi H, Seong H, Kim C (2019) Detection of construction workers under varying poses and changing background in image sequences via very deep residual networks. Autom Constr 99:27–38
39. Son H, Seong H, Choi H, Kim C (2019) Real-time vision-based warning system for prevention of collisions between workers and heavy equipment. J Comput Civ Eng 33(5):04019029
40. Fang W, Zhong B, Zhao N, Love PE, Luo H, Xue J, Xu S (2019) A deep learning-based approach for mitigating falls from height with computer vision: Convolutional neural network. Adv Eng Inform 39:170–177
41. Cai J, Zhang Y, Yang L, Cai H, Li S (2020) A context-augmented deep learning approach for worker trajectory prediction on unstructured and dynamic construction sites. Adv Eng Inform 46:101173
42. Wu J, Cai N, Chen W, Wang H., Wang G (2019) Automatic detection of hardhats worn by construction personnel: a deep learning approach and benchmark dataset. Autom Construct 106:102894
43. Zhao Y, Chen Q, Cao W, Yang J, Xiong J, Gui G (2019) Deep learning for risk detection and trajectory tracking at construction sites. IEEE Access 7:30905–30912
44. Fang Q, Li H, Luo X, Ding L, Luo H, Rose TM, An W (2018) Detecting non-hardhat-use by a deep learning method from far-field surveillance videos. Autom Constr 85:1–9
45. Fang W, Ding L, Luo H, Love PE (2018) Falls from heights: A computer vision-based approach for safety harness detection. Autom Constr 91:53–61
46. Roberts D, Torres Calderon W, Tang S, Golparvar-Fard M (2020) Vision-based construction worker activity analysis informed by body posture. J Comput Civ Eng 34(4):04020017
47. Yan X, Zhang H, Li H (2019) Estimating worker-centric 3D spatial crowdedness for construction safety management using a single 2D camera. J Comput Civ Eng 33(5):04019030
48. Lee MFR, Chien TW (2020) Intelligent robot for worker safety surveillance: deep learning perception and visual navigation. In: 2020 international conference on advanced robotics and intelligent systems, pp 1–6. IEEE, UK
49. Tang S, Golparvar-Fard M (2017) Joint reasoning of visual and text data for safety hazard recognition. In: Computing in civil engineering 2017, pp 450–457. ASCE, USA
50. Fang W, Ding L, Zhong B, Love PE, Luo H (2018) Automated detection of workers and heavy equipment on construction sites: a convolutional neural network approach. Adv Eng Inform 37:139–149

51. Kim H, Kim H, Hong YW, Byun H (2018) Detecting construction equipment using a region-based fully convolutional network and transfer learning. J Comput Civ Eng 32(2):04017082
52. Kim H, Bang S, Jeong H, Ham Y, Kim H (2018) Analyzing context and productivity of tunnel earthmoving processes using imaging and simulation. Autom Constr 92:188–198
53. Roberts D, Golparvar-Fard M (2019) End-to-end vision-based detection, tracking and activity analysis of earthmoving equipment filmed at ground level. Autom Constr 105:102811
54. Roberts D, Bretl T, Golparvar-Fard M (2017) Detecting and classifying cranes using camera-equipped UAVs for monitoring crane-related safety hazards. In: Computing in civil engineering 2017, pp 442–449. ASCE, USA
55. Guo Y, Xu Y, Li S (2020) Dense construction vehicle detection based on orientation-aware feature fusion convolutional neural network. Autom Construct 112:103124
56. Liang CJ, Lundeen KM, McGee W, Menassa CC, Lee S, Kamat VR (2018) Stacked hourglass networks for markerless pose estimation of articulated construction robots. In: 35th international symposium on automation and robotics in construction, pp 843–849. Curran Associates, Germany
57. Luo H, Wang M, Wong PKY, Cheng JC (2020) Full body pose estimation of construction equipment using computer vision and deep learning techniques. Autom Construct 110:103016
58. Zhang J, Zi L, Hou Y, Wang M, Jiang W, Deng D (2020) A deep learning-based approach to enable action recognition for construction equipment. Adv Civil Eng 1–14

Integrated SHM Systems: Damage Detection Through Unsupervised Learning and Data Fusion

Enrique García-Macías and Filippo Ubertini

Abstract One of the most daunting challenges of modern structural engineering concerns the management and maintenance of ageing infrastructure. The technical response to this challenge falls within the framework of structural health monitoring (SHM), which pursues the automated diagnosis and prognosis of structures from continuously acquired sensor data. In the last years, particular attention has been devoted in the literature to ambient vibration-based SHM owing to its minimal intrusiveness and global damage identification capabilities. Nevertheless, the sheer variety of failure mechanisms that large-scale civil engineering structures may experience, some of which may be of local nature, compels the use of integrated SHM systems and data fusion for comprehensive damage identification. As a result, such systems must deal with extensive databases of heterogeneous monitoring data, being the selection of critical features a key step to link signals to decisions. This chapter presents an overview of some of the most recent statistical pattern recognition, data fusion, feature extraction and damage detection techniques for integrated SHM systems. Under an application-oriented philosophy, the theoretical basis and implementation details of these techniques are illustrated through real case studies of Italian historic buildings.

Keywords Modal identification · Historic buildings · Preventive conservation · Damage identification · Structural health monitoring · Unsupervised learning

E. García-Macías (✉) · F. Ubertini
Department of Civil and Environmental Engineering, University of Perugia, Via G. Duranti 93, 06125 Perugia, Italy
e-mail: enrique.garciamacias@unipg.it

F. Ubertini
e-mail: filippo.ubertini@unipg.it

E. García-Macías
ETS de Ingeniería de Caminos, Canales y Puertos, Av. Fuentenueva sn 18002, Granada, Spain

A. Cury et al. (eds.), *Structural Health Monitoring Based on Data Science Techniques*, Structural Integrity 21, https://doi.org/10.1007/978-3-030-81716-9_12

1 Introduction

The vast socio-economic impacts stemming from the retrofitting, replacement or failure of structurally deficient constructions have recently fostered outstanding developments in the field of SHM with a considerable number of infield applications. In the broadest sense, SHM exploits long-term monitoring data to track anomalies in the structural performance caused by damage and, desirably, to predict damage evolution and structural life expectancy [1]. Ambient vibration-based SHM systems have become particularly widespread owing to their non-destructive nature and minimum intrusiveness, without requiring heavy and costly excitation devices. Such techniques have proven proficiency to correctly assess global damage, although their ability to detect local defects with minimal effect upon the overall stiffness is rather limited (e.g. freezing/thawing cycles, chemical attack, corrosion) [2]. In this regard, static monitoring such as the assessment of crack widths, displacements or tilts is typically more effective [3]. Moreover, it is often convenient to also monitor the environmental and operational conditions (EOC) of structures to facilitate the discrimination of damage from normal operational conditions. In this light, it follows that the use of integrated monitoring systems encompassing diverse sensing solutions results crucial to attain effective local/global damage detection capabilities. The implementation of such systems comprising dense sensor networks is becoming economically viable in engineering practice thanks to the major advances and progressive cheapening of sensor technology in recent years. Under these circumstances, the integration of all these monitoring data turns the damage identification task into a big data problem where the use of data science and machine learning becomes increasingly necessary.

The comprehensive approach to SHM through the pattern recognition paradigm has been particularly successful. This framework was first formalized by Farrar et al. [4] around the idea that one can learn how to assign damage states or classes to monitoring data through the study of relations or patterns in the response of the monitored structure. The general methodology comprises four stages as sketched in Fig. 1:

(1) *Operational Evaluation*: The first stage sets boundaries to the problem and aids at tailoring the subsequent damage identification process by addressing four questions. These include the motivation and economic justification for the implementation of an SHM system, the definition of the different sources of damage that may arise in the structure to be monitored, EOC in which

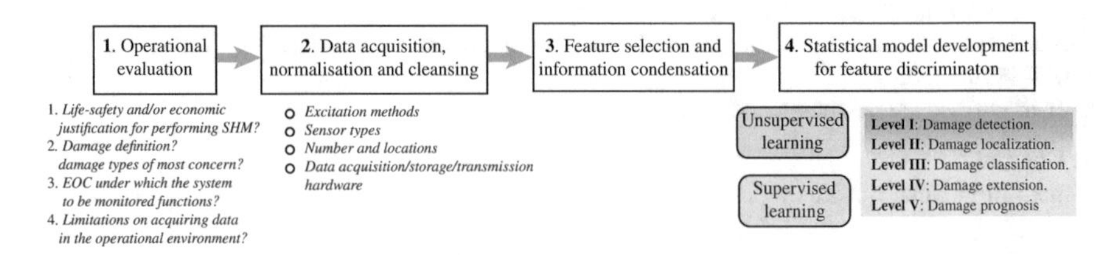

Fig. 1 Statistical pattern recognition paradigm for SHM

the system may operate, and the data acquisition limitations under operating conditions.

(2) *Data acquisition/normalization and cleansing*: The data acquisition process concerns the selection of suitable excitation methods, sensor types and coverage (number and location), and data acquisition/storage/transmission. Data normalization regards the ability of separating changes in sensor readings caused by damage from those caused by varying EOC. Finally, data cleansing regards the process of phasing out corrupted or uninformative data from the feature selection process.

(3) *Feature extraction and information condensation*: This stage relates the identification of data features that allow one to distinguish between the undamaged and damaged conditions. Inherent in the feature selection process, the information condensation step deals with the minimization of redundant features and dimensionality reduction.

(4) *Statistical model development*: This last stage concerns the implementation of statistical models mapping between the extracted features and the diagnosis, that is to say, the class or tag assignment. Such models are thus intended to enable the discrimination between features related to undamaged and damaged classes. Generally, statistical models are classified in three categories: *unsupervised learning*, *supervised learning* and *semi-supervised learning* as an intermediate solution. A pattern recognition model can be trained in a *supervised learning* fashion when data are available from both the undamaged and damaged structures. Conversely, when no information is available on the damaged structure, the algorithms fall into the category of *unsupervised learning*. In semi-supervised learning, a minimum amount of tagged training data (data labelled as "damaged") is available, but not sufficient for a full supervised learning. In this regard, *transfer learning* is a popular alternative to *supervised learning*, where tagged data are generated through a digital twin of the structure (virtual damage scenarios). In either way, statistical pattern recognition models analyse statistical distributions of the measured or derived features with the aim of performing damage identification. The damage identification process is commonly organized in a hierarchical structure of increasing complexity discussed by Rytter [5]: *Level I: Detection*; *Level II: Localization; Level III: Classification*; *Level IV: Extension*; and *Level IV: Prognosis.* All things considered, the final outcome of the paradigm of statistical pattern recognition is a set of classes from which decisions can be made, closing the data to decision (D2D) chain.

Within the paradigm of SHM as a statistical pattern recognition problem, this chapter presents a journey through the different stages involved in the damage detection process by means of integrated SHM and unsupervised learning. Motivated by an application-oriented philosophy, the presented theoretical concepts are illustrated with real experimental data acquired from three different case studies of Italian historic constructions: the Consoli Palace in Gubbio, the San Pietro bell-tower in

Perugia and the Basilica of Santa Maria degli Angeli in Assisi. These three constructions are being currently monitored with long-term integrated SHM systems, and the monitoring data are continuously processed for damage detection purposes in real time and remotely in the Laboratory of Structural Dynamics of the University of Perugia. For a detailed description of the monitoring sites, sensor layouts and monitoring conditions, readers may refer to [6–9].

The reason for focusing on unsupervised learning techniques is their more extensive application and development in the realm of historic constructions. Note that data from every conceivable damage scenario must be available to effectively apply supervised learning. Collecting such data is always a challenging task, through either modelling or experiments. This is particularly critical when dealing with historic constructions, where making copies of the structure to induce controlled damage is simply unfeasible. Modelling approaches such as finite element modelling also represent a formidable problem, since historic constructions are usually characterized by geometrical and material complexities. The modelling of damage itself is even more intricate, being possible to find multiple damage mechanisms with highly nonlinear behaviours, not to mention the frequent existence of previous damage with uncertain origin, extension and evolution. A major drawback of unsupervised learning is that it usually limits to damage detection, and damage can only be located to some extent in some particular cases. Nonetheless, level 1 diagnostic often suffices for the maintenance of historic constructions, whose invaluable architectural and economic value justifies the execution of in situ inspections every time any fault is detected. In this light, the statistical pattern recognition paradigm previously depicted in Fig. 1 is particularized for the integrated SHM of historic constructions in Fig. 2. The general workflow is organized into hardware and software components:

- **Hardware component**: The hardware part is generally composed of a heterogeneous sensor network, a data transmission system (wired or wireless) and a data acquisition (DAQ) system that permanently collects the monitoring data. Computer files containing records of certain time duration are locally processed or sent through the Internet or another transmission system to a server or the cloud.
- **Software component**: This component comprises a sequence of pattern recognition algorithms designed to translate the signals acquired by the sensing interface into condition classes (damaged or undamaged). The computer implementation follows the statistical pattern recognition paradigm previously sketched in Fig. 1: (i) *signal processing*; (ii) *feature Extraction*; (iii) *statistical pattern recognition*; and (iv) *damage detection.*

The remaining of the chapter focuses on the software component from *feature extraction* to *damage detection.* Note that the software structure furnished in Fig. 2 can be readily scripted in a sequential fashion, enabling real-time damage assessment and minimizing the need for the intervention of qualified technicians. Readers interested in more specific details on the computational implementation are referred to [10]. There, the authors reported about the architecture and algorithms included in two proprietary software programs named MOVA and MOSS. These software codes are currently being used in the Laboratory of Structural Dynamics of the University of

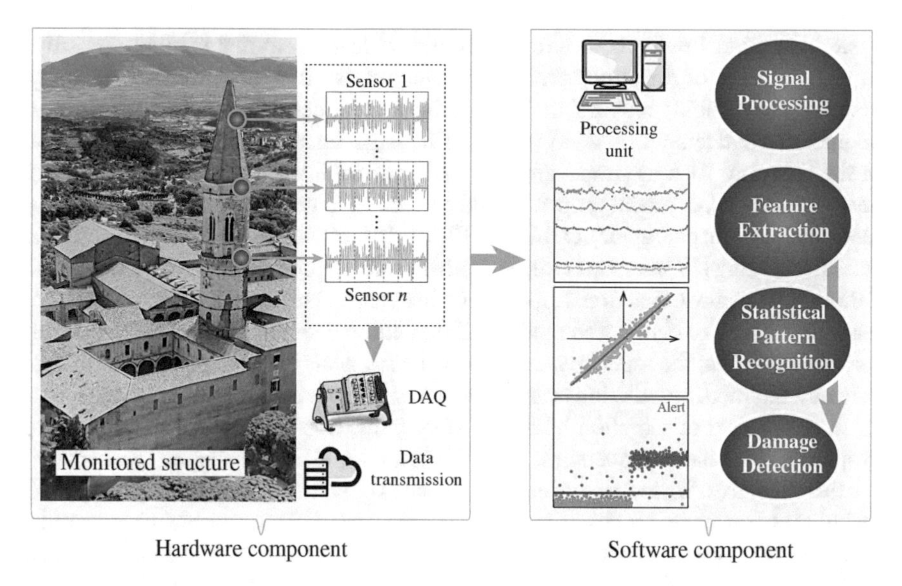

Fig. 2 General workflow of the SHM of historic constructions under the paradigm of statistical pattern recognition

Perugia for the SHM of several historic constructions, including the case studies presented herein.

2 Feature Extraction

In the fields of statistics and machine learning, feature selection and extraction are two approaches for dimension reduction aimed at producing the possible most informative, distinctive and compact set of features to improve the success of data storage and processing applications [11]. On the one hand, feature selection relates the process of selecting the most relevant attributes of a database, overlooking redundant or non-informative features. On the other hand, feature extraction creates low-dimensional data sets by functional mapping of the original features. In the realm of SHM, feature extraction relates the identification process of certain physical signatures or variables that are sensitive to the appearance of damage from the raw measurements [12]. The effectiveness of the adopted feature extraction approach largely determines the success of the damage assessment and the subsequent decision-making process, since it is a fundamental axiom of SHM that sensors cannot directly measure damage [13]. The design of the feature extraction strategy is conditioned by the operational evaluation. This includes aspects such as the employed sensing technology, accessibility limitations, the specific structural typology and engineering materials and, most importantly, the damage mechanisms of concern in the particular monitored system. The ideal outcome is a low-dimensional feature set that is highly sensitive to the condition of the structure. Such a selection is often not a trivial task, being necessary to combine robust signal processing techniques with a priori statistics and engineering judgement. A wide variety of damage-sensitive features can be found in the literature such as simple signal statistics, frequency spectra, time–frequency analysis features, or modal properties and derived signatures (e.g. mode shape curvature, modal strain energy or flexibility) [12, 14, 15]. Model updating [1, 16] and time series modelling [17] methods represent other useful approaches for model-based and data-driven feature extraction, respectively. In essence, these techniques construct (or learn) a physics- or data-based model of the monitored structure by exploiting monitoring data. Then, the model parameters or the residuals between new experimental data and the model predictions can be used as damage-sensitive features [12].

In the context of vibration-based SHM systems, it is common to employ modal properties (resonant frequencies, damping ratios and mode shapes) as damage-sensitive features. To do so, a variety of automated operational modal analysis (OMA) techniques has been proposed in the literature [18, 19]. These techniques allow to automatically extract estimates of modal signatures from continuously recorded ambient vibrations of sufficient time duration. Subsequently, the time series of modal features are obtained by so-called frequency tracking techniques. A common practice is to define a reference list or baseline modal properties extracted from a separate ambient vibration test (AVT). Once defined, the time series of modal features are traced by exploiting similarities between the reference modes and all the identified sets of modal characteristics. To illustrate this, Fig. 3a furnishes the natural frequencies identified by automated covariance-driven stochastic subspace identification (COV-SSI) of the Consoli Palace from July to December 2020. The adopted automated OMA procedure analyses sets of stabilization diagrams obtained by varying the order of the underlying state-space model and the number of blocks in the associated correlation Toeplitz matrix [10]. Afterwards, a sequential procedure involving noise mode elimination and hierarchical clustering is applied to discriminate between spurious and physical modes [18]. On this basis, a frequency tracking approach was

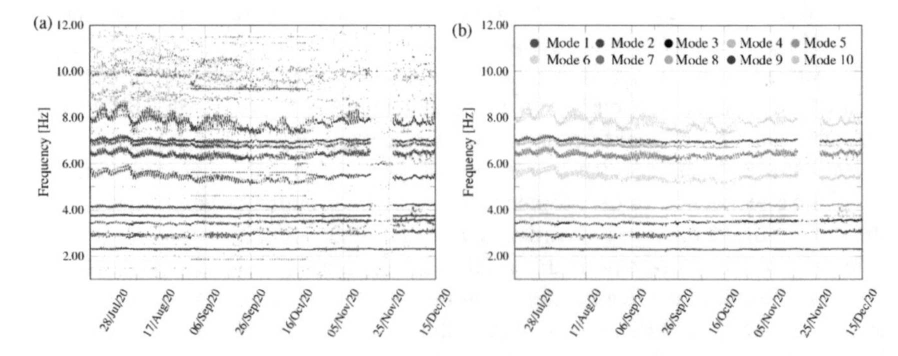

Fig. 3 Identified resonant frequencies by automated OMA of the Consoli Palace (**a**) and time series of meaningful resonant frequencies obtained by frequency tracking (**b**)

adopted to extract the time series of physically meaningful and consistently identifiable modes as shown in Fig. 3b. Specifically, a set of ten natural frequencies in the frequency broadband of 2–10 Hz were identified in a separate AVT and selected as baseline modal characteristics (see [6] for further details). Then, the frequency tracking method seeks for the most similar modes amongst all the identified ones in terms of resonant frequencies and modal assurance criterion (MAC) values. To do so, similarity thresholds in terms of resonant frequencies and MAC values are defined to pair modes from different data sets. In this particular case, maximum relative differences of 5% (modes 1–9) and 10% (mode 10) are defined in terms of resonant frequencies, while minimum MAC values of 0.8 are set up for all the modes.

It may be also useful to utilize mode shapes as damage-sensitive features. These are usually less affected by environmental conditions compared to resonant frequencies, although the number of sensors required for their correct characterization is usually larger. This motivates the use of mode shape components for damage detection, particularly during the first months of monitoring when acquired data are not enough for cleansing time series of natural frequencies from environmental effects. Mode shapes can be characterized using several metrics, although the use of MAC values between the identified modes and the reference ones is the most common approach. This statistical indicator appraises the coherence or similarity between two arbitrary complex mode shapes $\{\varphi_r\}$ and $\{\varphi_q\}$ as:

$$MAC(r, q) = \frac{\left|\{\varphi_r\}^{\mathrm{T}}\{\varphi_q\}^*\right|^2}{\left(\{\varphi_r\}^{\mathrm{T}}\{\varphi_r\}^*\right)\left(\{\varphi_q\}^{\mathrm{T}}\{\varphi_q\}^*\right)}, \tag{1}$$

where the asterisk denotes the complex conjugate. A MAC value of 1 indicates perfect correlation between the mode shapes, while a value of 0 evidences no correlation. The assessment of MAC values may be also used as a diagnostic tool to detect faulty sensors, whose effect is often not so evident in terms of resonant frequencies. To illustrate this, Fig. 4a furnishes the time series of MAC values of the modes of the Consoli Palace previously reported in Fig. 3b. It is noted in this figure that a sudden drop in the MAC values was found between 30 August and 20 October 2020. During in situ inspections, it was found out that the slot of channel 6 in the acquisition module of accelerometers had been damaged during a storm event, although it was working in an apparently normal way. This was confirmed by a statistical analysis of the mode shapes as shown in Fig. 4b. In this figure, the modal components (normalized to a maximum amplitude of 1) of the 12 channels of the SHM system are analysed during three different stages; Stage 1 17 July–29 August 2020 (before fault), Stage 2 29 August–20 October 2020 (faulty period) and Stage 3 20 October–16 November 2020 (after fault). Substantial differences in the modal components of channel 6 are noted between Sets 1 and 2. Once the slot was repaired in Set 3, the modal component of channel 6 went back to its correct value. This analysis allowed us to disregard the acceleration records of channel 6 during the faulty period as a feature selection procedure for the subsequent pattern recognition steps.

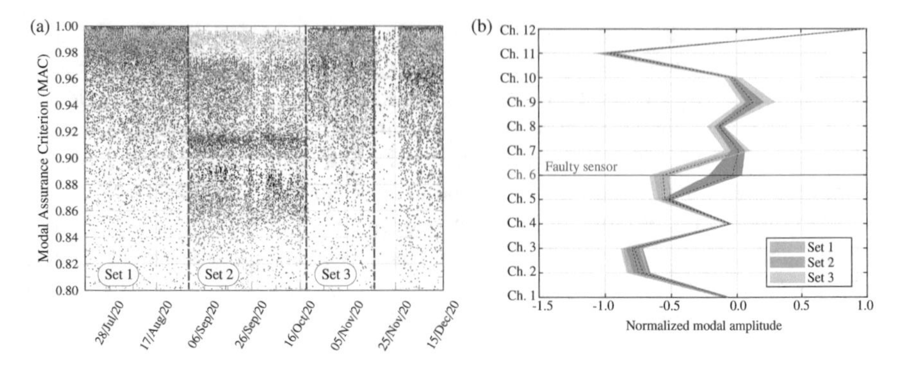

Fig. 4 Time series of MAC values (**a**) and analysis of the modal components of the fundamental mode (**b**) in the Consoli Palace. The shaded areas in (**b**) represent the 95% confidence level of the modal amplitudes

3 Data Normalization, Cleansing and Fusion

3.1 Statistical Models for Data Normalization

Data normalization constitutes the process of eliminating the variability in the selected features due to variations in EOC [12]. Such variations may attain a leading role in the variance of the selected features, exceeding in many cases the effects caused by damage, especially at initial stages. For instance, Peeters and De Roeck [20] found variations up to 18% in the first four resonant frequencies of the well-known case study of the Z24 bridge. In the context of historic structures, positive correlations between environmental temperature and resonant frequencies are often observed in masonry structures, which is usually ascribed to thermal-induced crack closure phenomena [9]. This is, for instance, the case of the Basilica of Santa Maria degli Angeli as shown in Fig. 5a. It is noted in this figure that the resonant frequencies of the first three modes exhibit both daily and seasonal fluctuations. Specifically, mean variation ranges of 11.0, 14.3 and 17.3% are found per year for Modes 1, 2 and 3, respectively. A similar behaviour can be also observed in Fig. 5b in terms of displacements measured by two linear variable displacement transducers (LVDTs) bridging the crack faces of two major cracks of the Basilica. Both magnitudes exhibit a similar positive correlation with the environmental temperature as shown in Fig. 6. However, completely different correlations can be found in practice as shown in the work by Gentile et al. [21] on the SHM of the Milan Cathedral in Italy. Their results showed a negative correlation between resonant frequencies and temperature, which was ascribed to the actions exerted by metallic tie rods in the building. Such negative correlations are also common in reinforced concrete and steel structures due to decreases in the material Young's modulus with increasing temperature [22, 23], such as asphalt concrete pavements in roadway bridges. In general, many different situations can be found depending on the specific structural typology, solar radiation

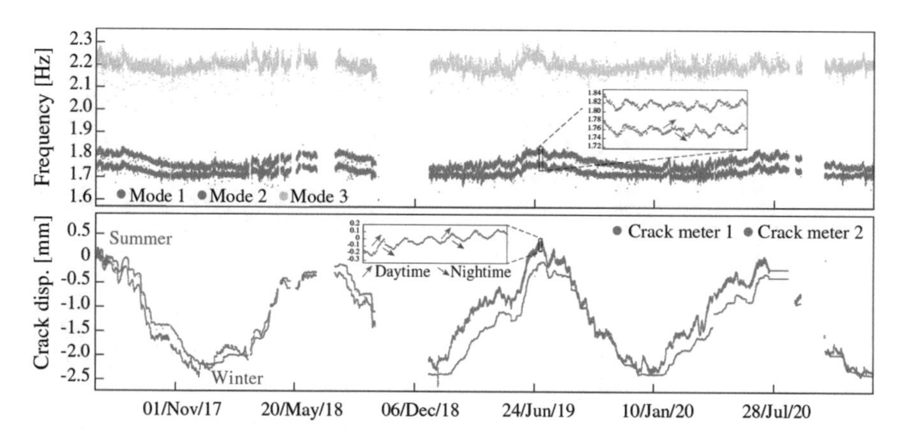

Fig. 5 Time series of resonant frequencies and crack displacements of the Basilica of Santa Maria degli Angeli from 2017 until 2020

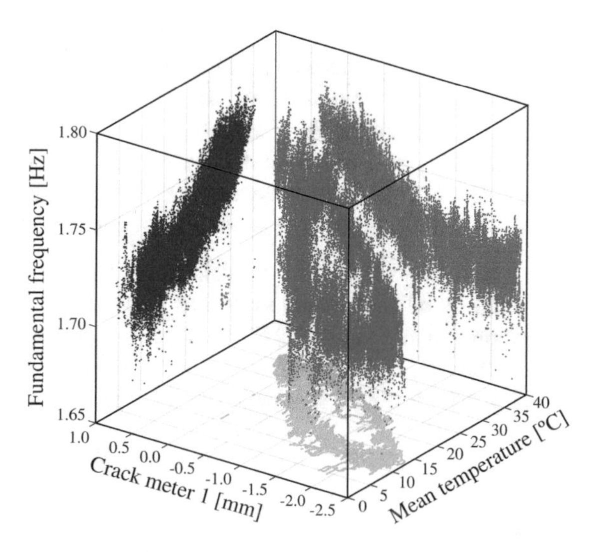

Fig. 6 Correlation analysis of the time series of ambient temperature, crack displacements and fundamental frequency of the Basilica of Santa Maria degli Angeli from 2017 until 2020

and thermal capacitance, as well as other environmental factors such as humidity, wind, snow or rain, or operational actions like traffic or human–structure interactions. These effects translate into daily and seasonal fluctuations in the selected features that may mask the appearance of damage, whereby their elimination represents a crucial step to ensure the effectiveness of the damage identification.

Generally, two different strategies can be employed for data normalization: when monitoring data of EOC are available (*input–output*) and when not (*output only*). A wide variety of statistical models can be found for each category, amongst which some of the most commonly used ones are listed below:

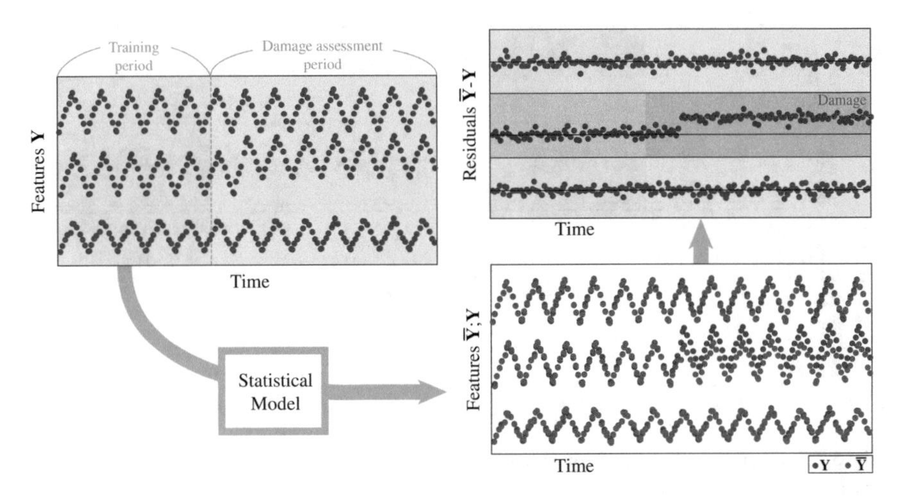

Fig. 7 Flow chart of data normalization through statistical pattern recognition

- **Input–output regression models**: Multiple linear regression (MLR) [9], autoregressive with exogenous input model (ARX) [24], artificial neural networks [25], support vector regression [26].
- **Output only regression models**: Principal component analysis (PCA) [9], kernel PCA (KPCA) [27], factor analysis (FA) [28], autoassociative neural networks (AANNs) [29], time series models [30], cointegration (CI) [31].

The general workflow of the data normalization process is sketched in Fig. 7. In general terms, statistical models for data normalization attempt to reproduce an observation matrix $\mathbf{Y} \in \mathbb{R}^{N \times n}$ containing n features and N observations. Once trained, the predictions of the model $\overline{\mathbf{Y}}$ can be used to phase out the variance due to EOC from $\mathbf{Y}$ forming the so-called residual error matrix $\mathbf{E} \in \mathbb{R}^{N \times n}$, that is:

$$\mathbf{E} = \mathbf{Y} - \overline{\mathbf{Y}}. \tag{2}$$

The statistical model must be trained using a set of t_p feature samples defining a baseline in-control population, often referred to as the *training period*. This baseline data set must statistically represent the healthy state of the structure under all the possible EOC, being a one-year period often adopted. When the system remains healthy, matrix $\overline{\mathbf{Y}}$ reproduces the part of the variance of the features driven by EOC, while $\mathbf{E}$ only contains the residual variance stemming from modelling errors. Conversely, if a certain damage develops, this only affects the data contained in $\mathbf{Y}$, while matrix $\overline{\mathbf{Y}}$ remains unaltered. Therefore, matrix $\mathbf{E}$ concentrates the damage-induced variance and is apt for being used for damage identification.

Given the widespread use of MLR and PCA as *input–output* and *output only* statistical models for the elimination of EOC in engineering structures, respectively, a concise overview and some exemplary numerical results are reported hereafter.

Additionally, an overview of the use of clustering techniques for nonlinear data normalization is also presented.

Multiple linear regression: MLR models exploit linear correlations between n features (*estimators*) and a set of p independent exploratory variables (*predictors*), which are typically taken from monitoring data of EOC. The predictions by MLR of matrix $\overline{\mathbf{Y}}$ are obtained as:

$$\overline{\mathbf{Y}} = \mathbf{Z}\boldsymbol{\beta}, \tag{3}$$

where $\mathbf{Z} \in \mathbb{R}^{N\times(p+1)}$ is an observation matrix composed of an $N \times 1$ vector of ones and an $N \times p$ matrix containing the time series of the p selected predictors, while $\boldsymbol{\beta} \in \mathbb{R}^{(p+1)\times n}$ is a matrix of regression weights composed of intercept terms in the first row and linear regression coefficients in the remaining p rows. The ordinary least squares estimate of $\boldsymbol{\beta}$ reads:

$$\boldsymbol{\beta} = \left(\mathbf{Z}^{\mathrm{T}}\mathbf{Z}\right)^{-1}\mathbf{Z}^{\mathrm{T}}\mathbf{Y}. \tag{4}$$

As an example, Fig. 8 shows the results of MLR analysis of the resonant frequency of the fourth mode of vibration (second flexural mode) of the San Pietro bell-tower from 2018 until 2020. In the analysis, the ambient temperature measurements by two thermocouples installed at the base of the cusp of the tower (T_1 indoor and T_2 outdoor) are used as predictors. Note that the model must be linear in its coefficients contained in matrix $\boldsymbol{\beta}$, while arbitrary nonlinear transformations such as exponential or logarithmic functions can be applied to the predictors. As an example, the results of

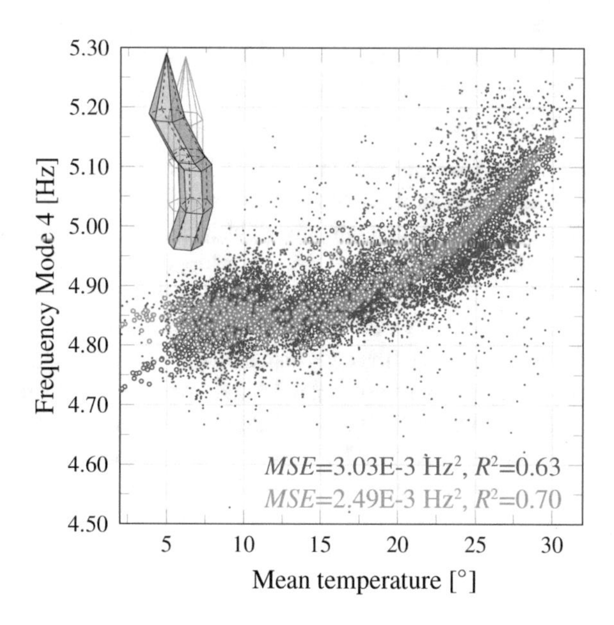

Fig. 8 Multiple linear regression analysis of the resonant frequency of the fourth mode of vibration of the San Pietro bell-tower from 2018 until 2020. *Filled blue circle* experimental data, *open red circle* MLR, $y = 4.91 - 0.44T_1 + 1.20T_2$, *open orange circle* MLR, $y = 4.91 - 1.91T_1 + 1.05T_2 + 1.65(T_1)^2 + 0.02(T_2)^2$

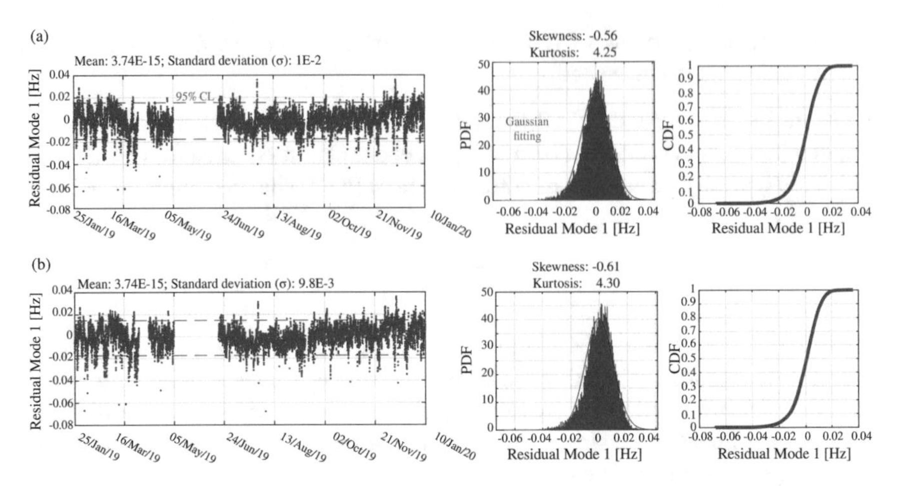

Fig. 9 Analysis of residuals of the statistical predictions by MLR of the fundamental frequency of the San Pietro bell-tower from 2018 until 2020. Models accounting for linear (**a**) and quadratic (**b**) temperature values

the MLR model using both linear and quadratic temperature values are also depicted in Fig. 8. Ideally, the obtained residuals should be normally distributed; thereby, statistical tests assessing the Gaussianity of the residuals can be used to evaluate the quality of a statistical model for data normalization. To this aim, a variety of statistical metrics can be used from simple statistics to dedicated normality tests such as the Kolmogorov–Smirnov or the Shapiro–Wilk tests. For illustrative purposes, Fig. 9 shows the analysis of the residuals obtained from the results in Fig. 8. It is noted that, although the standard deviation of the residuals decreases when quadratic terms are involved in the MLR, the kurtosis of the distribution of the residuals moves away from 3 (theoretical value for an ideal Gaussian distribution) compared to the model with linear terms, which indicates some overfitting degree.

Principal Component Analysis: PCA is a dimensionality reduction technique used to transform databases into lower-dimensional subspaces without significant losses in data variance. Mathematically, PCA is defined as an orthogonal linear transformation that converts the data to a new coordinate system where the greatest variance of the data lies on the first coordinate (called the first principal component), the second greatest variance on the second coordinate and so on. Principal components (PCs) are the eigenvectors of the covariance matrix of the original data; thereby, PCs constitute an orthogonal basis of uncorrelated components. Denoting by $\mathbf{Y}_{\mathrm{n}}$ the normalized version of matrix $\mathbf{Y}$ (i.e. the feature time series are transformed to have zero mean and unit variance) and $\mathbf{\Sigma}_{\mathbf{Y}} \in \mathbb{R}^{n \times n}$ its covariance matrix, the PCs can be obtained by the eigenvalue–eigenvector decomposition of $\mathbf{\Sigma}_{\mathbf{Y}} = \mathbf{Y}_n^{\mathrm{T}}\mathbf{Y}_{\mathrm{n}}/(N-1)$ as:

$$\mathbf{\Sigma}_{\mathbf{Y}}\mathbf{U} = \mathbf{U}\mathbf{S}^2, \tag{5}$$

where the eigenvectors of $\mathbf{\Sigma_Y}$ are the columns of $\mathbf{U}$ (loading matrix) and represent the PCs, and the eigenvalues are the diagonal terms of $\mathbf{S}^2$. The PCs are sorted in descending order according to the diagonal terms of $\mathbf{S}^2$, which represent the proportion of total variance in $\mathbf{Y}$ (i.e. $\text{tr}(\mathbf{\Sigma_Y}) = n$) explained by the PCs. The transformed data matrix $\mathbf{T} \in \mathbb{R}^{N \times n}$ (scores matrix) is the projection of the original data in $\mathbf{Y_n}$ over the space spanned by the PCs in $\mathbf{U}$:

$$\mathbf{T} = \mathbf{Y_n U}. \tag{6}$$

In the realm of SHM, PCs providing the largest contributions to the variance are assumed to encapsulate the effects of EOC on the features matrix $\mathbf{Y}$. In this light, matrix $\overline{\mathbf{Y}}$ can be estimated by mapping back the reduced subset of PCs onto the original data space. Specifically, if only the first l columns of matrix $\mathbf{U}$ are collected into a reduced matrix $\widehat{\mathbf{U}} \in \mathbb{R}^{n \times l}$, matrix $\overline{\mathbf{Y}}_n$ (normalized) can be obtained as:

$$\overline{\mathbf{Y}}_n = \mathbf{Y_n}\left(\widehat{\mathbf{U}}\widehat{\mathbf{U}}^{\mathrm{T}}\right). \tag{7}$$

The number l of PCs to be retained must be chosen according to their relative contributions to the variance in the data. If this dimension is too small, part of the EOC will not be properly captured, while an excessively large number of retained PCs will make the model suffer from overfitting with the subsequent loss of generality. As an example, Fig. 10 shows the results of the PCA of six resonant frequencies of the San Pietro bell-tower identified in the frequency broadband of 1–12 Hz. It is noted that three PCs suffice to explain more than 90% of the variance in the resonant frequencies. The comparison of the experimental identification results of the first two resonant frequencies of the tower and the predictions by PCA using three PCs is depicted in Fig. 11. It is noted that the PCA model can reproduce both the seasonal and the daily fluctuations in the resonant frequencies induced by EOC.

Clustering for local statistical pattern recognition: Results such as those previously reported in Figs. 6 and 8 evidence that EOC may induce some nonlinear

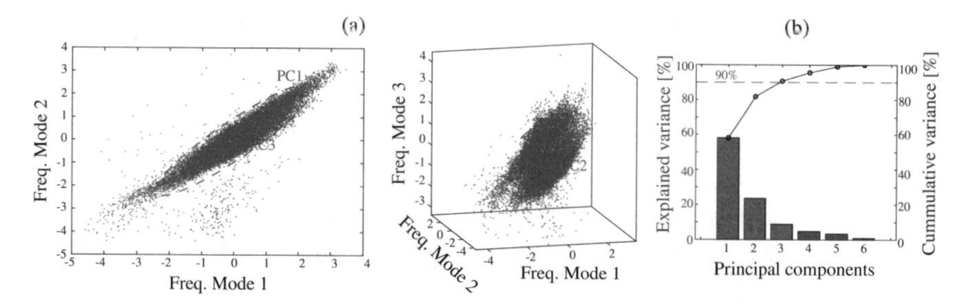

Fig. 10 Principal component analysis of the resonant frequencies of the San Pietro bell-tower. Principal components (**a**) and variance analysis (**b**). Resonant frequencies are reported in normalized values

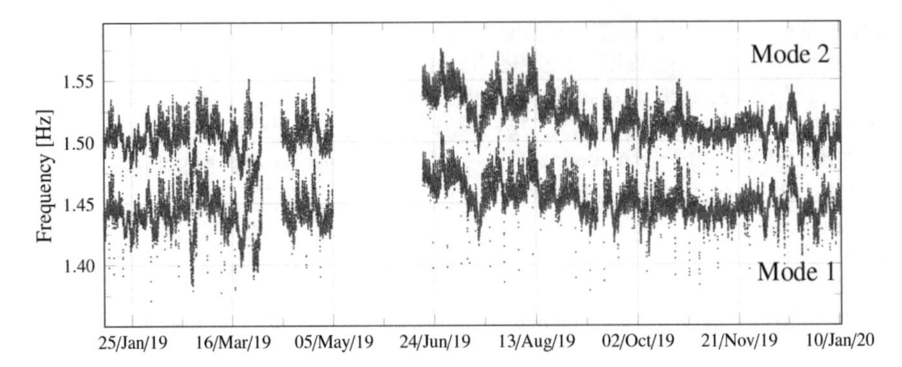

Fig. 11 Comparison between experimental data and predictions by PCA of the first two resonant frequencies of the San Pietro bell-tower. *Filled red circle* experimental data and *filled blue circle* PCA

effects upon the selected features. A remarkable example of this circumstance is when air temperature falls below zero, which leads to abrupt changes in the static and dynamic response of structures due to the formation of ice crystals in the material microporosity. A common approach to tackle such nonlinear correlations is the use of nonlinear PCA through the implementation of AANNs. An alternative and more general approach consists of the use of clustering techniques. Clustering is a technique from the field of data mining used to divide data sets into groups or clusters in such a manner that the data points that are similar lie together in one cluster. In the context of SHM, the idea is to separate the selected features into clusters exhibiting differentiated EOC effects, in such a way that the overall nonlinear correlation can be assumed to be cluster-wise linear. On this basis, piecewise extensions of any of the previously mentioned statistical models can be constructed to handle nonlinear correlations by leveraging the separation of features into clusters. A wide variety of clustering techniques for big data analytics can be found in the literature (see, e.g., [32]), although Gaussian mixture models (GMMs) have proved particularly efficient in the field of SHM. This approach assumes that the probability density function $p(\mathbf{Y})$ of the data set in the training period $\mathbf{Y}$ (in general, non-normally distributed) can be represented as a linear superposition of K Gaussian components as:

$$p(\mathbf{Y}) = \sum_{k=1}^{K} \pi_k \mathcal{N}\big(\mathbf{Y}|\boldsymbol{\mu}_k, \boldsymbol{\Sigma}_k\big). \tag{8}$$

Each component of the mixture is defined as a Gaussian distribution $\mathcal{N}\big(\mathbf{Y}|\boldsymbol{\mu}_k, \boldsymbol{\Sigma}_k\big)$ with mean and covariance matrix denoted by $\boldsymbol{\mu}_k$ and $\boldsymbol{\Sigma}_k$, respectively. The parameters $\boldsymbol{\pi} = [\pi_1, \ldots, \pi_K]^{\mathrm{T}}$ in Eq. (3) are called the mixing coefficients, and they range between 0 and 1 ($0 \leq \pi_k \leq 1$) and sum to one ($\sum_{k=1}^{K} \pi_k = 1$). The model parameters, $\boldsymbol{\mu}$, $\boldsymbol{\Sigma}$ and $\boldsymbol{\pi}$, are fitted by minimizing the log-likelihood function:

$$\ln p(\mathbf{Y}|\boldsymbol{\pi}, \boldsymbol{\mu}, \boldsymbol{\Sigma}) = \sum_{n=1}^{t_p} \ln\left\{\sum_{k=1}^{K} \pi_k \mathcal{N}\left(\mathbf{Y}|\boldsymbol{\mu}_k, \boldsymbol{\Sigma}_k\right)\right\}. \tag{9}$$

The maximum likelihood solution for the parameters ($\boldsymbol{\mu}$, $\boldsymbol{\Sigma}$ and $\boldsymbol{\pi}$) is estimated using the iterative expectation–maximization (EM) algorithm. In the expectation (E) step, the parameters (initial guess at the beginning) are held fixed and the posterior probability of assigning $\mathbf{x}_n$ to the k's cluster is given by the so-called responsibilities $\gamma(z_{nk})$ as:

$$\gamma(z_{nk}) = \frac{\pi_k \mathcal{N}\left(\mathbf{Y}_n|\boldsymbol{\mu}_k, \boldsymbol{\Sigma}_k\right)}{\sum_{j=1}^{K} \pi_j \mathcal{N}\left(\mathbf{Y}_n|\boldsymbol{\mu}_j, \boldsymbol{\Sigma}_j\right)}, \tag{10}$$

where z_{nk} is an element of a K-dimensional binary random variable $\mathbf{z}$ with a 1-of-K representation. Only one element in $\mathbf{z}$ is equal to 1, and all other elements are 0. Then, in the maximization (M) step, the parameters are re-estimated using the posterior probability calculated in the previous E step as follows:

$$\boldsymbol{\mu}_k^{\text{new}} = \sum_{n=1}^{t_p} \gamma(z_{nk})\mathbf{Y}_n,\ \boldsymbol{\Sigma}_k^{\text{new}} = \sum_{n=1}^{t_p} \gamma(z_{nk})\left(\mathbf{Y}_n - \boldsymbol{\mu}_k^{\text{new}}\right)\left(\mathbf{Y}_n - \boldsymbol{\mu}_k^{\text{new}}\right)^{\mathrm{T}} \tag{11}$$

$$\pi_k^{\text{new}} = \frac{N_k}{N},\ N_k = \sum_{n=1}^{t_p} \gamma(z_{nk}). \tag{12}$$

The log-likelihood in Eq. (3) can then be evaluated. Convergence of either the parameters of the log-likelihood is checked, and if the criteria are not satisfied, the process is iterated using the updated data values until the criteria are met. Once the K clusters have been defined in the training period, new data samples can be assigned to the cluster with the least Mahalanobis distance.

3.2 *Data Cleansing and Residual Analysis*

Outliers are always present to a certain degree in every feature set in SHM, being the result of manifold sources such as noise, identification errors, faulty sensors, imperfect mounting of sensors and more. The presence of such outliers in the training data set has two major pernicious effects. On the one hand, outliers affect the effectiveness of statistical models for pattern recognition, biasing the model parameters. On the other hand, the ability of damage classification methods based upon novelty analysis to detect early-stage damage is also affected by outliers. As it will be discussed in Sect. 5, novelty analysis techniques heavily rely on the determination of statistical moments of the damage-sensitive features in the training population. The presence of outliers will bias such moments, reducing the damage sensitivity of the classification.

An extensive variety of methodologies for outlier elimination can be found in the literature, including statistical, clustering, graph, ensemble and learning methods (refer to [33] for a comprehensive overview). In this chapter, we focus on a simple but effective methodology based upon the minimum covariance determinant (MCD) method. The MCD method [34] is a common technique in statistical analysis used to obtain robust estimates of the sample covariance of a data population. The MCD method seeks a sample subset within a multivariate data set that minimizes the covariance matrix. Consider a training population of features $\mathbf{Y}$ containing t_p samples, and let $H_1 \subset \{1, \ldots, t_p\}$ be an h-subset, that is, $|H_1| = h$. Being $\boldsymbol{\mu}_1$ and $\boldsymbol{\Sigma}_1$ the empirical mean and covariance matrix of the data in H_1, the Mahalobis distances of all the data samples in the training population read ($\det(\boldsymbol{\Sigma}_1) \neq 0$):

$$d_1(i) = \sqrt{(\mathbf{Y}_i - \boldsymbol{\mu}_1)^{\mathrm{T}} \boldsymbol{\Sigma}_1^{-1} (\mathbf{Y}_i - \boldsymbol{\mu}_1)} \text{ for } i = 1, \ldots, \mathrm{d} \tag{13}$$

Now take H_2 another h-subset such that $\{d_1(i); i \in H_2\} = \{(d_1)_{1:d}, \ldots, (d_1)_{h:d}\}$ where $(d_1)_{1:d} \leq (d_1)_{2:d} \leq \cdots \leq (d_1)_{d:d}$ are the ordered distances, and compute $\boldsymbol{\mu}_2$ and $\boldsymbol{\Sigma}_2$ based on H_2. Then, $\det(\boldsymbol{\Sigma}_2) \leq \det(\boldsymbol{\Sigma}_1)$ holds with equality if and only if $\boldsymbol{\mu}_2 = \boldsymbol{\mu}_1$ and $\boldsymbol{\Sigma}_2 = \boldsymbol{\Sigma}_1$. This process, also known as C-step, can be iteratively repeated. If $\det(\boldsymbol{\Sigma}_2) = 0$ or $\det(\boldsymbol{\Sigma}_2) = \det(\boldsymbol{\Sigma}_1)$, the algorithm stops; otherwise, another C-step is run yielding $\det(\boldsymbol{\Sigma}_3)$ and so on. The sequence $\det(\boldsymbol{\Sigma}_1) \geq \det(\boldsymbol{\Sigma}_3) \geq \det(\boldsymbol{\Sigma}_4) \geq \ldots$ is nonnegative and hence convergent, so there must be an index s such that $\det(\boldsymbol{\Sigma}_s) = 0$ or $\det(\boldsymbol{\Sigma}_s) = \det(\boldsymbol{\Sigma}_{s-1})$. An application example is shown in Fig. 13 for the resonant frequencies of the San Pietro bell-tower between 2019 and 2020. The dimension h of the subsets has been selected as $0.8d$, and once the algorithm converged, one per cent of the samples with the largest Mahalanobis distances with respect to the converged sample subset is considered as outliers.

3.3 Data Fusion

Data fusion is the integration of information from multiple sources into a new database to enhance the observability and identifiability of a system [12]. This new database is intended to be more informative, reducing uncertainties in the damage identification by increasing the information completeness and enhancing the decision-making. In general, data fusion can be achieved in three levels, namely data level, feature level and decision level [35]. In data-level fusion, the raw measurements from multiple sensors are directly combined before further processing. Feature-level fusion directly operates on statistical features or signatures of heterogeneous nature. Finally, in decision-level fusion, decision-making is performed by integrating the decisions achieved from different data sources through particular combination rules. Some common data fusion techniques are data registration, Bayesian probabilistic approaches, Dempster–Shafer (DS) evidential approach, fuzzy reasoning, state estimation, machine learning algorithms and weighted combinations. For a comprehensive review on data fusion techniques applied to SHM, readers may consult reference [36]. Within the framework of unsupervised SHM, feature-level fusion is particularly well suited. Note that the statistical models presented in Sect. 3.1. work on arbitrary sets of estimators and predictors, irrespectively of the nature of the sensor system used for their extraction. Therefore, it is possible to use feature sets combining data of very diverse nature and fuse them within the novelty analysis presented in the next section, constructing hybrid control charts. For instance, it is possible to use static data to eliminate the effect of EOC over dynamic features and vice versa.

4 Damage Detection Using Control Charts

Damage detection through unsupervised statistical pattern recognition is commonly performed by novelty analysis. The main idea is that, since the data normalization algorithm is trained using data characterizing the healthy condition of the structure, it is possible to establish the structural diagnosis based on the analysis of deviations between the predictions of the algorithm and newly acquired data. If an anomaly is detected, this implies that the system has deviated from its normal condition, possibly due to the appearance of damage. As anticipated when discussing the implications of unsupervised learning, novelty analysis is a two-class problem; that is to say, features are only classified as damaged or undamaged. Methods for novelty detection include: outlier analysis, kernel density methods, AANNs, Kohonen networks, growing radial basis function networks and control charts [37]. The use of statistical process control charts is particularly popular in unsupervised SHM due to their relative simplicity and direct automation. As sketched in Fig. 14a, control charts furnish in time a certain statistical distance accounting for nonconformities in the distribution of the residuals in $\mathbf{E}$ from Eq. (2) with respect to the healthy baseline [12]. On this basis, out-of-control processes, possibly associated with damage, are detected in the shape of data points violating an in-control region. A wide variety of control charts is available in the literature, amongst which the most popular ones are the Shewhart, cumulative sum (CUSUM), exponentially weighted moving average (WEMA) and Hotelling's T^2 control charts.

Hotelling's T^2 control chart has proved proficient in a number of vibration-based SHM applications. The plotted statistic T^2 (squared Mahalanobis distance) is defined as [38]:

$$T_i^2 = r\left(\overline{\mathbf{E}} - \overline{\overline{\mathbf{E}}}\right)^{\mathrm{T}} \boldsymbol{\Sigma}_0^{-1} \left(\overline{\mathbf{E}} - \overline{\overline{\mathbf{E}}}\right), i = 1, 2, \ldots, N/r \tag{14}$$

where r is an integer parameter referred to as subgroup size, $\overline{\mathbf{E}}$ is the mean of the residuals in the subgroup or the last r observations, while $\overline{\overline{\mathbf{E}}}$ and $\boldsymbol{\Sigma}_0$ are the mean vector and covariance matrix of the residuals in the in-control training period.

Considering sample estimations of $\bar{\bar{\mathbf{E}}}$ and $\mathbf{\Sigma}_0$, the upper control limit (*UCL*) can be derived as [38]:

$$UCL = \frac{n(k+1)(r-1)}{k(r-1)+1-n} F_{1-\alpha;n,kr-k-n+1} \tag{15}$$

where the term $F_{1-\alpha;n,kr-k-n+1}$ denotes $(1-\alpha)100\%$ confidence level of the F distribution with n and $kr-k-n+1$ degrees of freedom. The definition of *UCL* has a strong influence on the sensitivity of a control chart to detect damage; a too low value will lead to an excessive number of false alarms, while a too high value may not be able to detect damage. In order to optimally select *UCL*, as well as to compare the effectiveness of different statistical models, the analysis of simulated or synthetic damage scenarios is particularly helpful. For instance, a common approach is to include shifts in the mean values of the time series of estimators (e.g. resonant frequency decays obtained through a nonlinear modal analysis of a finite element model) and simulate control charts under different damage scenarios. As an example, Fig. 15a shows Hotelling's control charts of the residuals obtained using MLR and PCA for the data normalization of the first eight resonant frequencies of the Basilica of Santa Maria degli Angeli. Ambient temperature measurements (outdoor and indoor) are used as statistical predictions for the MLR model, while 2 PCs explaining more than 90% of the total variance in the resonant frequencies were used for the PCA model. To evaluate the effectiveness of the statistical models for damage detection, an artificial damage scenario was introduced after 1 January 2020. The artificial damage consists in a constant shift in the time series of the fundamental frequency of 0.5% of its mean value in the training period. Then, the quality of the damage classification can be appraised through the analysis of the confusion matrix, including receiver operating characteristic (ROC) curves, *precision–recall curves*, *Youden's indexes* or *F1-scores*. Figure 12 shows an example of a ROC curve and the determination of the optimal *UCL* as the cut-off threshold yielding the maximum vertical distance between the ROC curve and the diagonal curve (line of no discrimination). The area under the ROC curves (AUC) is a simple metric often used to evaluate the quality of the classification and compare different statistical pattern recognition approaches. From the particular examples shown in Fig. 15a, the resulting ROC curves are shown in Fig. 15b. It is noted that the MLR model slightly outperforms the PCA model, with AUC values of 0.87 and 0.85, respectively. These differences are also noticeable in the control charts in Fig. 15a, where two UCLs corresponding to confidence levels of 99% (*danger*) and 95% (*warning*) are represented. Text inserts indicate the proportion of data points between the 95 and 99% UCLs and above the 99% UCL. For instance, it is noted that the proportion of outliers in the assessment period varies from 4 to 9.48% and to 7.65% for the PCA and the MLR models, respectively. Interestingly, although at first glance the PCA model may seem to provide better damage identification reporting a higher proportion of outliers in the damaged period, the number of false alarms in the undamaged period is larger compared to those obtained using MLR. Global metrics such as the AUC of the ROC curves englobe the performance of a statistical pattern recognition model for minimizing the number of false alarms while offering high discrimination capabilities of true out-of-control processes.

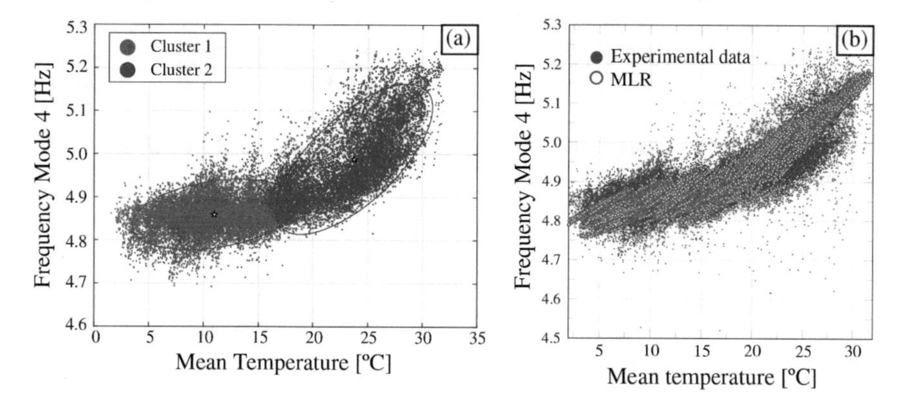

Fig. 12 Clustering results using a GMM (**a**) and statistical predictions using a cluster-wise extension of the MLR model of the fourth resonant frequency of the San Pietro bell-tower (**b**)

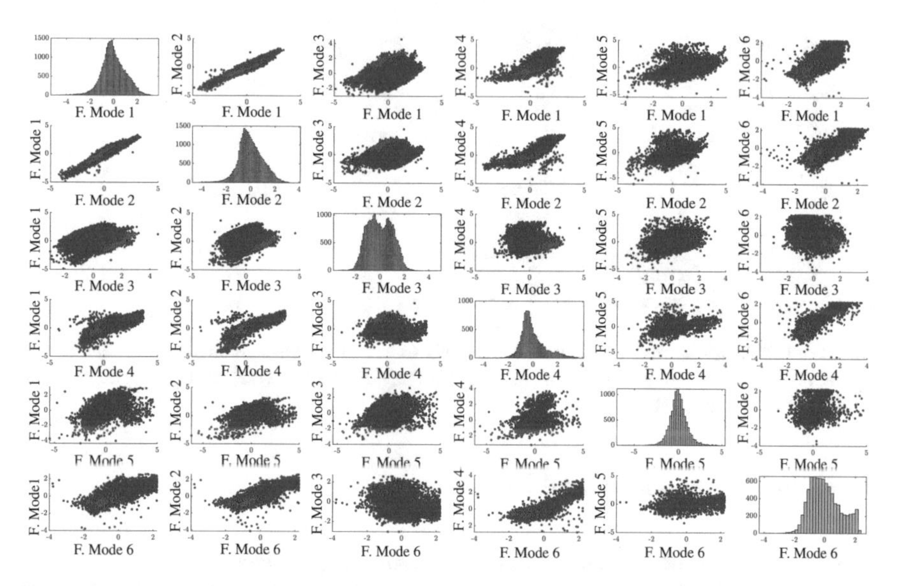

Fig. 13 Outlier identification in the database of the first six resonant frequencies in the San Pietro bell-tower between 2019 and 2020 using the MCD method. *Filled blue circle* experimental data and *filled red circle* outliers

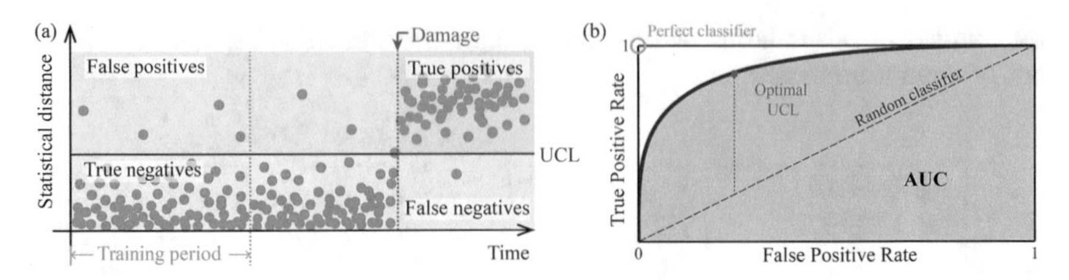

Fig. 14 Schematic representation of damage detection using control charts (**a**) and quality assessment through ROC curves (**b**)

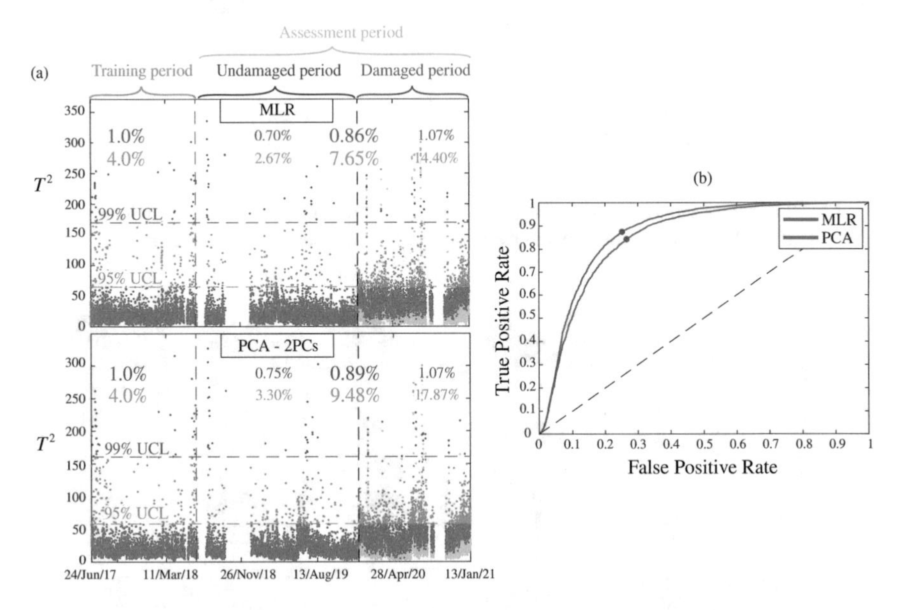

Fig. 15 Hotelling's control charts of the residuals of the first eight resonant frequencies of the Basilica of Santa Maria degli Angeli considering MLR and PCA (2 PCs) (**a**), and quality assessment in terms of ROC curves (**b**)

5 Conclusions and Final Remarks

This chapter has gone through the main stages in damage detection through unsupervised integrated SHM according to the most recent literature, and the effectiveness of the presented techniques has been illustrated with real field applications. Overall, the advanced state of development of damage detection through novelty analysis has been evidenced, being possible to completely automate the whole process. Clustering techniques have been presented as a general and effective approach for accommodating nonlinear environmental effects, and the use of outlier removal techniques along with the analysis of ROC curves has been shown suitable for the optimal definition of statistical process control charts. Most pressing open problems in the field relate to the extension of these techniques to higher diagnostic levels, specifically to local/global multiclass damage identification. Researchers have the challenge of effectively blending the injection of engineering/physics knowledge into the learning and decision phases with advanced artificial intelligence techniques. In this regard, the combination of integrated long-term SHM with digital twins within a transfer learning framework renders a promising direction for future developments.

References

1. Chen HP, Ni YQ (2018) Structural health monitoring of large civil engineering structures. Wiley Blackwell, Hoboken
2. Ramos LF, Marques L, Lourenço PB, De Roeck G, Campos-Costa A, Roque J (2010) Monitoring historical masonry structures with operational modal analysis: two case studies. Mech Syst Signal Process 24(5):1291–1305
3. Ottoni F, Blasi C (2015) Results of a 60-year monitoring system for Santa Maria del Fiore Dome in Florence. Int J Architect Heritage 9(1):7–24
4. Farrar CR, Doebling SW, Nix DA (2001) Vibration–based structural damage identification. Philos Trans R Soc Lond. Ser A: Math Phys Eng Sci 359(1778):131–149
5. Rytter A (1993) Vibrational based inspection of civil engineering structures, PhD thesis. Aalborg University
6. Kita A, Venanzi I, Cavalagli N, García-Macías E, Ubertini F (2020) Enhanced continuous dynamic monitoring of a complex monumental palace through a larger sensor network. In: XI International conference on structural dynamics, EURODYN 2020
7. Cavalagli N, Gusella V (2015) Dome of the Basilica of Santa Maria Degli Angeli in Assisi: static and dynamic assessment. Int J Architect Heritage 9(2):157–175
8. Cavalagli N, Botticelli L, Gioffrè M, Gusella V, Ubertini F (2017) Dynamic monitoring and nonlinear analysis of the dome of the basilica of S. Maria degli Angeli in Assisi. In: XI COMPDYN 2017, 1, pp 2542–2553
9. Ubertini F, Cavalagli N, Kita A, Comanducci G (2018) Assessment of a monumental masonry bell-tower after 2016 Central Italy seismic sequence by long-term SHM. Bull Earthq Eng 16(2):775–801
10. García-Macías E, Ubertini F (2020) MOVA/MOSS: two integrated software solutions for comprehensive structural health monitoring of structures. Mechan Syst Signal Process 143:106830
11. Entezami A, Sarmadi H, Behkamal B, Mariani S (2020) Big data analytics and structural health monitoring: a statistical pattern recognition-based approach. Sensors 20(8):2328
12. Farrar CR, Worden K (2012) Structural health monitoring: a machine learning perspective. Wiley, Hoboken
13. Worden K, Farrar CR, Manson G, Park G (2007) The fundamental axioms of structural health monitoring. Proc R Soc A: Math Phys Eng Sci 463(2082):1639–1664
14. Hou R, Xia Y (2020) Review on the new development of vibration-based damage identification for civil engineering structures: 2010–2019. J Sound Vib 115741
15. Avci O, Abdeljaber O, Kiranyaz S, Hussein M, Gabbouj M, Inman DJ (2021) A Review of vibration-based damage detection in civil structures: from traditional methods to machine learning and deep learning applications. Mechan Syst Signal Process 147:107077
16. García-Macías E, Venanzi I, Ubertini F (2020) Metamodel-based pattern recognition approach for real-time identification of earthquake-induced damage in historic masonry structures. Autom Construct 120:103389
17. Fassois SD, Sakellariou JS (2007) Time-series methods for fault detection and identification in vibrating structures. Philos Trans R Soc A: Math Phys Eng Sci 365(1851):411–448

18. Ubertini F, Gentile C, Materazzi AL (2013) Automated modal identification in operational conditions and its application to bridges. Eng Struct 46:264–278
19. Reynders E, Houbrechts J, De Roeck G (2012) Fully automated (operational) modal analysis. Mech Syst Signal Process 29:228–250
20. Peeters B, De Roeck G (2001) One-year monitoring of the Z24-Bridge: environmental effects versus damage events. Earthquake Eng Struct Dynam 30(2):149–171
21. Gentile C, Ruccolo A, Canali F (2019) Long-term monitoring for the condition-based structural maintenance of the Milan Cathedral. Construct Build Mater 228:117101
22. Xia Y, Chen B, Weng S, Ni YQ, Xu YL (2012) Temperature effect on vibration properties of civil structures: a literature review and case studies. J Civ Struct Heal Monit 2(1):29–46
23. Magalhães F, Cunha Á, Caetano E (2012) Vibration based structural health monitoring of an arch bridge: from automated OMA to damage detection. Mech Syst Signal Process 28:212–228
24. Sohn H, Worden K, Farrar CR (2002) Statistical damage classification under changing environmental and operational conditions. J Intell Mater Syst Struct 13(9):561–574
25. Flood I, Kartam N (1994) Neural networks in civil engineering. II: systems and application. J Comput Civil Eng **8**(2):149–162
26. Kromanis R, Kripakaran P (2013) Support vector regression for anomaly detection from measurement histories. Adv Eng Inform 27(4):486–495
27. Ghoulem K, Kormi T, Bel Hadj Ali N (2020) Damage detection in nonlinear civil structures using kernel principal component analysis. Adv Struct Eng 23(11):2414–2430
28. Deraemaeker A, Worden K (2018) A comparison of linear approaches to filter out environmental effects in structural health monitoring. Mech Syst Signal Process 105:1–15
29. Zhou HF, Ni YQ, Ko JM (2011) Eliminating temperature effect in vibration-based structural damage detection. J Eng Mech 137(12):785–796
30. Figueiredo E, Figueiras J, Park G, Farrar CR, Worden K (2011) Influence of the autoregressive model order on damage detection. Comput-Aided Civil Infrastruct Eng 26(3):225–238
31. Tomé ES, Pimentel M, Figueiras J (2020) Damage detection under environmental and operational effects using cointegration analysis–application to experimental data from a cable-stayed bridge. Mechan Syst Signal Process 135:106386
32. Gan G, Ma C, Wu J (2020) Data clustering: theory, algorithms, and applications. Soc Indus Appl Math
33. Wang H, Bah MJ, Hammad M (2019) Progress in outlier detection techniques: a survey. IEEE Access 7:107964–108000
34. Hubert M, Debruyne M, Rousseeuw PJ (2018) Minimum covariance determinant and extensions. Wiley Interdisc Rev Comput Stat 10(3):e1421
35. Klein LA (2004) Sensor and data fusion: a tool for information assessment and decision making, vol 138. SPIE Press
36. Wu RT, Jahanshahi MR (2020) Data fusion approaches for structural health monitoring and system identification: past, present, and future. Struct Health Monit 19(2):552–586
37. Markou M, Singh S (2003) Novelty detection: a review—part 1: statistical approaches. Signal Process 83(12):2481–2497
38. Mason RL, Young JC (2002) Multivariate statistical process control with industrial applications. Soc Indus Appl Math

Environmental Influence on Modal Parameters: Linear and Nonlinear Methods for Its Compensation in the Context of Structural Health Monitoring

Carlo Rainieri

Abstract Modal-based structural health monitoring (SHM) detects damage and degradation phenomena from the variations of the modal parameters over time. However, the modal parameter estimates are also influenced by environmental and operational variables (EOVs) whose effects have to be compensated. Modeling the influence of EOVs on modal parameters is very challenging, so black-box models, such as regression models, are often adopted as an alternative. However, in many applications, the set of measured EOVs is incomplete or the factors influencing the estimates cannot be identified or measured. In these conditions, output-only techniques for compensation of environmental and operational effects are an attractive alternative. Different linear as well as nonlinear methods for the compensation of the environmental and operational influence on modal parameters in the context of modal-based SHM are reviewed in the present paper. Real datasets collected from vibration-based monitoring systems are analyzed, and the results are presented and discussed to illustrate the applicative perspectives and possible drawbacks of the selected methods.

Keywords Modal-based SHM · Environmental influence · Regression model · Principal component analysis · Kernel PCA · Second-order blind identification

1 Introduction

The safety of civil engineering structures may be affected to a large extent by degradation phenomena due to aging, severe environment or fatigue, as well as by damage induced by hazardous events, such as earthquakes, fire or explosions. Thus, periodic structural assessment and maintenance become critical, above all in the case of strategic structures, such as bridges and hospitals, or densely occupied structures. Structural survey as well as destructive and non-destructive investigations are the conventional approaches usually applied to support the structural safety assessment.

C. Rainieri (✉)
Construction Technologies Institute, National Research Council of Italy, 80146 Naples, Italy
e-mail: rainieri@itc.cnr.it

A. Cury et al. (eds.), *Structural Health Monitoring Based on Data Science Techniques*, Structural Integrity 21, https://doi.org/10.1007/978-3-030-81716-9_13

However, the local nature of tests, issues related to the subjectivity of the expert judgment and limited frequency of inspections and the costs of detailed structural assessment have raised the interest toward alternative approaches that are able to automatically provide information about the structural health and performance in near real-time to a remote user. In this context, the modern structural health monitoring (SHM) technologies are very appealing because of the opportunity to carry out timely damage detection. Modal-based SHM, in particular, is currently a very active field of research even if the original idea dates back to a few decades ago. SHM basically assumes that any change of the structure that adversely affects its functionality or load bearing capacity can be referred to as damage [1]. In modal-based SHM, the relationship between the physical properties of the structure (i.e., mass, stiffness, damping) and its modal parameters is exploited for damage detection. In fact, changes in the structural response associated to variations of stiffness due to cracking in concrete or masonry structures of external (soil settlements) and/or internal restraints (loosening of tightening force in bolts of steel structures) or of mass may suggest an anomalous structural behavior. Taking into account the relationship between mass and stiffness on one hand, and modal properties on the other hand, the latter are used as damage-sensitive features in the context of modal-based SHM, and remote damage detection is based on the analysis of the variations of damage-sensitive features defined in terms of modal parameters. In spite of the increasing interest about this technology in civil engineering, practical applications to civil structures are still quite limited because of some shortcomings affecting the general applicability and reliability of the technology.

Two major approaches can be identified in the field of modal-based SHM: model-based and data-driven techniques [1]. Model-based techniques require the solution of an inverse problem to identify damage by updating a numerical model [2, 3]; as a consequence, a high degree of engineering knowledge and heavy hardware and software resources are usually needed. The main drawback is related to the uncertainties associated to the setting and refinement of the numerical model, while the main advantage with these techniques is the potential to cover all levels of SHM, from damage detection to damage prognosis [4]. Data-driven techniques, on the contrary, set data-driven models to characterize the operational response of the structure under healthy conditions so that anomalies are identified from changes in the damage-sensitive features with respect to the reference (baseline) condition [5]. They are characterized by lower requirements in terms of engineering knowledge and hardware and software resources, so they better fit the needs of continuous monitoring and near real-time damage detection. However, their effectiveness is limited to damage detection and damage location [1]. Nevertheless, these techniques currently appear as fairly mature to make the transition from research to engineering practice.

The main obstacles to the wide application of data-driven SHM systems are the availability of automated procedures for automatic feature extraction and robust anomaly detectors [4]. The first element still represents a challenge in research, even if a number of automated data processing procedures have been developed in recent years to extract the most widely-used features (eigen frequencies and mode shapes) from vibration measurements (the interested reader can refer to [6, 7] for an extensive review). The second element, instead, usually exploits approaches borrowed from statistics and machine learning for anomaly detection (the interested reader can refer to [1] for an extensive review).

Different damage-sensitive features based on modal parameters have been defined and tested over the years [1, 8], but the possibility of getting accurate estimates even in the presence of a few installed sensors [9] makes natural frequencies very appealing as damage sensitive features. The most significant drawback with the use of natural frequencies for damage detection is their high sensitivity to damage as well as to the influence of environmental and operational variables (EOVs) [10–15]. Since EOVs yield changes in natural frequency estimates that are often of the same order of magnitude of those due to damage, an effective compensation of the influence of EOVs on natural frequency estimates plays a crucial role in the development of reliable modal-based SHM systems that are able to minimize the occurrence of false or missed alarms.

The nonlinear relationship with the mechanical properties of materials and the boundary conditions and the usually large thermal inertia of structures make direct modeling of the influence of EOVs (and, in particular, temperature) on the modal properties of a structure very challenging. This is the reason why black-box models are usually adopted as the preferred alternative. If experimental measurements of the EOVs influencing the selected damage features are available, a model able to map the changes of the features with the EOVs can be set [14, 16–18]. As an example, a linear regression model can be developed to this aim [14], even if more sophisticated modeling techniques, such as neural networks and support-vector machines [17, 19], have been also applied to this aim. While these alternative approaches show high prediction capabilities, they are more computational demanding than the simple regression models [14]. In any case, it is worth mentioning that since the developed black-box model does not refer to physical laws, its applicability is restricted to the structure whose monitoring data have been used for its setting. Another common drawback with the above methods concerns the identification of the EOVs to measure and the selection of the positions of the corresponding sensors. The latter might be very challenging in the case of full-scale structures because the EOVs are often not uniform and time dependent [9, 20]. If there is a single relevant environmental variable, it might be not constant along the structure. This is the case, for instance, of the temperature. In fact, differences occur among temperature of air, indoor surfaces and outdoor surfaces of the monitored structure as a result of solar radiation and thermal inertia [5, 9, 20]. When multiple EOVs [9, 14] are responsible for the changes of the dynamic properties, multicollinearity problems associated to the selection of regressors may also occur [14].

In order to circumvent all the above-mentioned issues, methods not requiring EOV measurements [5, 10, 13, 15] are often adopted as an alternative to the previous ones. The basic idea behind those methods is that the EOV influence on the selected damage feature lies in a subspace which can be identified from the time histories of the damage feature itself and that the effect of damage lies in a different subspace. As a result, the effects of EOVs on one hand and damage on the other hand can be effectively discriminated as long as the variations of the modal properties due to damage are in some way orthogonal or uncorrelated to those caused by the EOVs [10]. However, the effectiveness of such methods often depends on the dimension of the vector of monitored features that must be large enough to make possible the identification of the subspace which the effects of EOVs belong to. In other words, multivariate data with enough redundancy to remove the unwanted effects using the data correlation structure must be available, and the training data in the healthy condition of the structure should include measurements under a wide range of environmental or operational conditions [21].

Most of the models assume linear correlation between the measured variables or features. However, the environmental or operational variations often cause nonlinear effects. For example, as the temperature falls below zero, its influence on the natural frequencies can change abruptly. This often results also in nonlinear correlation between the features, especially if the data dimensionality is low [22]. On the other hand, a linear model may be sufficient with a large data dimensionality, because the correlation structure may become linear [21]. As an alternative, when the dimension of the feature vector is insufficient, nonlinear methods can be applied. They rely on the identification of a nonlinear manifold instead of a linear subspace [23] and, as such, they are usually more complex and computational demanding. The potential of nonlinear approaches for the effective compensation of EOV influence on modal parameters is confirmed by the increasing number of studies on this topic reported in the literature. Different approaches have been tested to this aim. The use of autoassociative neural network is described in [24], while the application of Gaussian mixture model (GMM) for the compensation of nonlinear environmental or operational effects without the measurement of the underlying variables is illustrated in [22]; in a similar way, kernel principal component analysis (kPCA) has been applied for the compensation of EOV influence on natural frequencies [25] and mode shapes [26].

The primary role of techniques for compensation of environmental and operational influence on damage-sensitive features in the context of modal-based SHM motivates the present paper, aimed at illustrating the applicative perspectives and possible drawbacks of selected linear as well as nonlinear methods for the compensation of the influence of EOVs on modal parameters. The theoretical background of the selected methods is summarized in Sect. 2, while Sect. 3 illustrates and discusses relevant results obtained from their application to real datasets collected from vibration-based monitoring systems.

2 Methods for EOV Influence Compensation

The literature review reported in the previous section remarks how the compensation of the influence of EOVs on damage-sensitive features (and, in particular, natural frequency estimates) has been carried out by applying a variety of methods, depending also on the application. While promising results have been obtained in many cases, large computational efforts characterize some of the applied procedures.

This might represent a limitation to their extensive application in the context of data-driven SHM, when large volumes of data become available and need to be processed in a relatively short time for timely damage detection. As a result, in this paper, attention is focused on selected linear as well as nonlinear procedures for compensation of EOV effects on modal parameter estimates: only methods characterized by limited computational burden even in the presence of relatively large amounts of data (typically, some months of monitoring data) are herein considered.

This section briefly presents the theoretical background of the methods used in this study, pointing out relevant details specifically related to removal of the EOV influence on natural frequency estimates. Among the linear methods for compensation of EOV effects, multiple linear regression (MLR), principal component analysis (PCA), and second-order blind identification (SOBI) are considered, while kernel PCA (kPCA) is applied as nonlinear compensation method.

2.1 Multiple Linear Regression

MLR [5] can be referred to as an input–output technique for compensation of the influence of EOVs on modal parameter estimates. When applied to this aim, MLR requires measurements of the EOVs in order to formulate the mathematical model establishing how the modal properties vary with the selected EOVs. Univariate as well as multivariate MLR models can be defined. With univariate models the relation between a single dependent variable and several independent variables, the predictors, is set to predict future values of the dependent variable when only the predictors are known.

However, the variability of the natural frequencies of different modes often depends on the same predictors $\boldsymbol{R}$, so multivariate MLR can be applied:

$$\boldsymbol{f} = \boldsymbol{R}\boldsymbol{B} + \boldsymbol{E} \tag{1}$$

where $\boldsymbol{f}$, $\boldsymbol{B}$, and $\boldsymbol{E}$ hold the natural frequencies of the monitored modes, the coefficients of the MLR model, and the prediction errors associated to the analyzed modes, respectively.

The objective of regression analysis is the estimation of the model coefficients so that the best fit between model predictions and observations is obtained. They are usually obtained through the least squares method. Regression models can be also classified as static or dynamic. The former assume that the value of the output variable at a given time instant depends only on the values of the predictors at the same time instant. The latter, on the contrary, assume that the output variable at the current time instant also depends on the values of the predictors at previous time instants. Dynamic regression is often used to model phenomena characterized by delayed effects, such as heating and cooling processes in the presence of relevant thermal inertia.

In the context of modal-based SHM, the natural frequencies predicted by the regression model at the measured EOV values and those experimentally estimated are compared in order to compensate the environmental and operational influence. In fact, assuming that the regression model completely describes the variations of the natural frequency estimates with varying environmental and operational conditions, the increase in the prediction error is an indicator of the deviation of the structure from its normal operating condition.

2.2 Principal Component Analysis

Output-only compensation methods are an attractive alternative to MLR and other methods based on direct measurements of relevant EOVs. In fact, it often happens that direct measurements of the EOVs influencing the dynamic properties of the monitored structure are not available. PCA [13] is currently a well-established output-only method for compensation of EOV effects in the context of modal-based SHM. Given a multivariate dataset, the standard PCA makes a linear projection of the data aimed at expressing the covariance structure of the original set of variables as a linear combination of the variables themselves. In the framework of modal-based SHM, the compensation of environmental effects by PCA starts by turning the observed natural frequencies $\boldsymbol{f}$ into a new set of uncorrelated variables z:

$$z = \boldsymbol{T}\boldsymbol{f} \tag{2}$$

where the transformation matrix $\boldsymbol{T}$ is obtained from the singular values decomposition of the correlation matrix of $\boldsymbol{f}$. Once the reference transformation matrix $\boldsymbol{T}_{ref}$ is computed from data referring to the healthy state of the monitored structure, Eq. (2) can be applied again to obtain the principal components for the current dataset; they can be remapped afterwards in the original space as follows:

$$\widehat{\boldsymbol{f}} = \boldsymbol{T}_{ref}{}^{T}\boldsymbol{T}\boldsymbol{f} \tag{3}$$

where the superscript ^{T} denotes transpose. If only the first principal components, accounting for most of the variance in the data, are retained, a dimensionality reduction is achieved by applying Eqs. (2) and (3). As a result, the factors responsible for most of the variability in the data (such as temperature, and other significant EOVs) are retained, while secondary effects, such as those due to random errors in the identification of the natural frequencies, are neglected. On the analogy with MLR, the prediction error between the observed and the remapped frequencies $\boldsymbol{E}$ can be computed:

$$\boldsymbol{E} = \boldsymbol{f} - \widehat{\boldsymbol{f}} \tag{4}$$

Such residues are insensitive to the environmental and operational factors modeled by the retained principal components, so they can vary only as a result of damage and, as such, they can be used as damage sensitive features.

2.3 Second Order Blind Identification

The use of SOBI for compensation of environmental and operational effects on natural frequencies in the context of modal-based SHM has been recently proposed [27]. SOBI is a Blind Source Separation technique [28] and, as such, it aims at recovering the so-called sources, that is to say, an underlying set of signals from records of their mixture only. In the context of the present application, the sources are estimates of the unknown EOVs obtained from the time series of observed natural frequencies varying under the influence of those EOVs. A significant assumption of the method is that the natural frequencies are linearly related to the unknown EOVs as follows:

$$\boldsymbol{f}(t) = \boldsymbol{A}\boldsymbol{s}(t) + \boldsymbol{e}(t) \tag{5}$$

where both the mixing matrix $\boldsymbol{A}$ and the sources $\boldsymbol{s}$ are obtained from the natural frequency time series. Some noise $\boldsymbol{e}$ can be also present in the data. Under the assumption of uncorrelation of the sources among them and with the noise, the eigenvalue decomposition of the zero-lag covariance matrix allows the estimation of the whitening matrix from the largest eigenvalues and the corresponding eigenvectors, and of the noise variance from the remaining ones. Multiplying the measured data by the whitening matrix, the whitened data are obtained and used to compute the covariance matrix at different time lags. Applying the joint approximate diagonalization [29] to a number of time-shifted matrices, the mixing matrix and the sources are extracted from the measured data by exploiting their time coherence.

In the context of modal-based SHM, the compensation of environmental and operational effects by SOBI starts by estimating the reference mixing matrix $\boldsymbol{A}_{ref}$ from a set of natural frequency estimates collected when the monitored structure is healthy. Once additional datasets are collected, the corresponding mixing matrix and sources are estimated; combining the sources associated to the current dataset with $\boldsymbol{A}_{ref}$ allows predicting how the natural frequencies vary because of the EOVs represented by the identified sources. However, when setting the model of the environmental variability of natural frequencies, multiplying the reference mixing matrix by the sources estimated from the current dataset is not sufficient. Since the blind identification problem is solved up to a scaling and a permutation, the variances of the identified sources remain unknown, and the sources are, by convention, normalized to have unit variance. This circumstance implies the application of an appropriate scaling factor to avoid bias. It can be obtained as the ratio between the first singular value of the reference mixing matrix and that of the mixing matrix estimated from the current dataset [27]. Once the model $\widehat{f}(t)$ of the environmental variability of natural frequencies is set, subtracting the predicted natural frequency time series $\widehat{f}(t)$ from the measured data $f(t)$ yields residues, which are independent of the environmental and operational factors represented by the sources and, as such, can vary only as a result of damage. As a consequence, they can be used as damage sensitive features in the context of modal-based SHM.

Two relevant aspects in the practical application of SOBI for compensation of environmental effects concern the estimation of the number of sources, which cannot be larger than the number of analyzed natural frequency time series, and the possibility of estimating also the pattern of the unknown EOVs [27].

The number of sources that are responsible for the variability of the measured natural frequencies can be estimated as the number of non-zero eigenvalues of the zero-lag covariance matrix of the measured data [27]. The influence of multiple EOVs determining the variability of natural frequency estimates can be modeled [27]. In addition, the estimation of the patterns of the unknown EOVs gives the opportunity to gain a fundamental insight in the causes of that variability. In fact, the ability of SOBI to trace the patterns of the EOVs up to a scaling and an offset provides a fundamental informative support to the analysis of the operational response of the monitored structure, and it may circumvent the problem of their measure when they cannot be identified in advance or when the most appropriate sensor location cannot be predicted. The interested reader can refer to [27] for more details.

2.4 Kernel PCA

The effect of EOVs on the modal properties of structures might be nonlinear and it can affect the monitored features in different ways. Thus, the use of nonlinear data processing procedures, such as kPCA, for modal-based SHM under changing environmental conditions may be advantageous. In the context of the present study, kPCA has been selected among other nonlinear methods for compensation of the influence of EOVs on natural frequencies because it is robust and computationally efficient; moreover, the type of nonlinearity has not to be explicitly defined. However, it requires the setting of two parameters, which influence the quality of the results, as explained in the following.

The compensation of environmental and operational effects on natural frequencies by kPCA follows the approach described in [25]. A nonlinear relationship of the form:

$$\boldsymbol{\Phi}(\boldsymbol{f}_k) = \boldsymbol{H}_0\boldsymbol{u}_k + \boldsymbol{e}_k \tag{6}$$

is assumed between the selected damage features $\boldsymbol{f}_k$ (the natural frequency estimates in the present case) and the unknown EOVs $\boldsymbol{u}_k$, where the operator $\boldsymbol{\Phi}$ denotes the nonlinear mapping of damage features onto a high-dimensional feature space, and the residue vector $\boldsymbol{e}_k$ holds the information about the misfit between measured data and model predictions; by taking advantage of the Mercer's theorem [30], the kernel function k:

$$k(\boldsymbol{f}_i, \boldsymbol{f}_j) = \boldsymbol{\Phi}(\boldsymbol{f}_i)^T \boldsymbol{\Phi}(\boldsymbol{f}_j) \tag{7}$$

can be specified instead of the nonlinear mapping $\boldsymbol{\Phi}$. Solving the following eigenvalue problem:

$$\boldsymbol{K}\boldsymbol{V} = \boldsymbol{V}\boldsymbol{\Sigma} \tag{8}$$

the eigenvector matrix $\boldsymbol{V}$ can be obtained, and it can be partitioned as follows:

$$\boldsymbol{V} = [\boldsymbol{V}_1 \boldsymbol{V}_2], \boldsymbol{V}_2 \subset R^{n_t \times (n_t - n_u)} \tag{9}$$

to evaluate the error norm in the feature space as:

$$\|\boldsymbol{e}_k\|^2 = \boldsymbol{K}_{test}^T \boldsymbol{V}_2 \boldsymbol{V}_2^T \boldsymbol{K}_{test} \tag{10}$$

where $\boldsymbol{K}$ is the $n_t \times n_t$ kernel matrix computed on the training dataset, and $\boldsymbol{K}_{test}$ is the $n_t \times n_s$ kernel matrix computed between the training samples and the entire dataset. The quantities n_t, n_s, and n_u denote the number of training samples, the total number of samples, and the number of principal components in the feature space, respectively. The superscript ^{T} denotes transpose.

The Gaussian radial basis function (RBF) is applied as kernel function in agreement with [25]. It is a general-purpose kernel mostly applied in the absence of prior knowledge [31]. Such a kernel function implicitly defines an infinite-dimensional mapped feature space by setting a single positive parameter σ:

$$k(\boldsymbol{f}_i, \boldsymbol{f}_j) = e^{-\frac{\|\boldsymbol{f}_i - \boldsymbol{f}_j\|^2}{2\sigma^2}} \tag{11}$$

The misfit in the feature space can be computed once the parameters n_u and σ have been set. Their values are set in the following in agreement with the approach reported in [25] and here summarized. The number of principal components n_u is set so that a given fraction of the normal variability (usually, 99%) is accounted for. This ensures that the resulting value of n_u is large enough to account for the normal variability of the estimates due to EOVs but also small enough to be sensitive to anomalies. The value of σ is instead selected as the one maximizing the Shannon's information entropy [32]. The interested reader can refer to [25] for more details about the theoretical background of the method.

3 Applications

3.1 Pre-stressed Steel Cable

The first case study under analysis concerns the compensation of the environmental influence on the fundamental frequencies of a pre-stressed steel cable: the analyzed natural frequency time histories were collected by a monitoring system over a period of about two months [7].

The monitored pre-stressed steel cable connected the heads of two columns supporting one of the steel arches of the structure of the University Sports Hall in Campobasso, Southern Italy. Figure 1 shows the steel arches and the monitored steel cable (on the right in the picture). The cable was equipped with a monitoring system consisting of four piezoelectric accelerometers (10 V/g sensitivity, ± 0.5 g full-scale range), a thermocouple for temperature measurements and a 24-bit data acquisition system.

The analyzed dataset holds the natural frequencies of the three fundamental modes of the cable. The reduced number of monitored modes is, in principle, unfavorable to the application of linear PCA and SOBI, because those methods require a number of monitored modes larger than the number of (unknown) factors responsible for the variations of the natural frequencies. However, previous studies have shown that the natural frequency variations were due to temperature only [7]. In addition, while the

Fig. 1 Outdoor view of the University Sports Hall in Campobasso, Southern Italy, and of the monitored pre-stressed steel cable (on the right in the picture)

relationship between the natural frequencies and the cable temperature is theoretically not linear, the experimental measurements have shown that, at least for the range of temperatures recorded during the monitoring period, this relation is approximately linear, so linear methods for compensation of EOV influence can be applied [27].

The details of application of MLR, PCA and SOBI to this dataset are reported elsewhere [27]. The most relevant results are summarized here and elaborated further to highlight advantages and drawbacks of the different approaches.

Basic statistical analyses of the data have shown that the temperature yields up to 4.5% of variation of natural frequencies with respect to the corresponding average values [27]. Thus, the large environmental variability of natural frequencies can prevent early damage detection [11], and its effective compensation is fundamental to enhance the reliability of SHM.

When setting the models for prediction of the environmental variability of natural frequencies, the dataset has been divided into two parts of equal length. The first part has been used to train the model, while the second part has been used to assess the predictive performance and, as a consequence, the capability of the different methods of compensating the temperature influence on natural frequency estimates.

MLR requires explicit temperature measurements to be applied. The recorded temperature over the considered monitoring period is plotted in Fig. 2; this shows how the temperature range in the testing period was approximately the same as in the training period, with the only exception of the data collected at the beginning of the testing period when the temperature went below 0 °C for some days due to an exceptional snowfall.

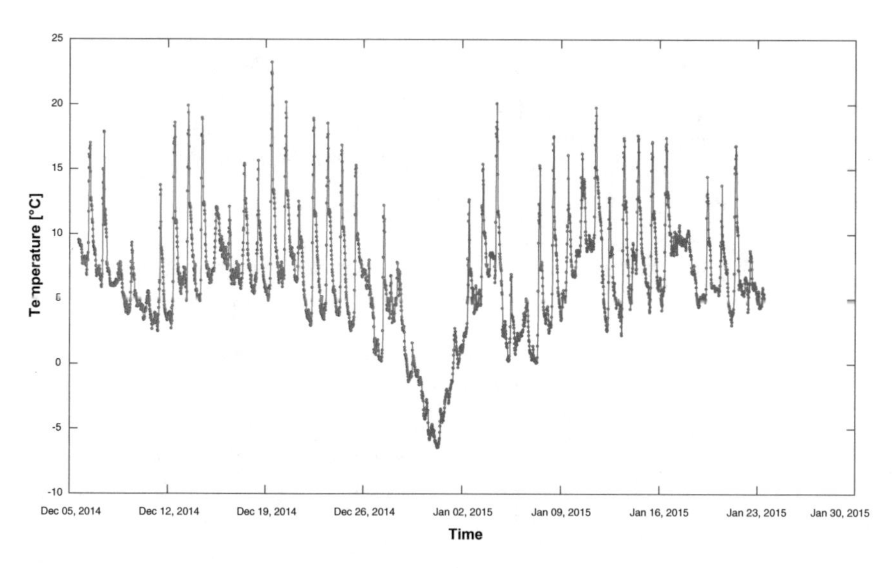

Fig. 2 Pre-stressed cable: recorded local temperature (in blue, the first half of the time series representing the measured temperature in the training period)

Figure 3 shows the residues obtained by comparing the frequencies predicted by MLR and SOBI with the measured values as per Eq. (4), and the control limits defined in the two cases. Results obtained from linear PCA are not reported, but they are similar to those obtained from SOBI. The plots in Fig. 3 remark that SOBI (and PCA) were effective in compensating the temperature influence on natural frequencies. On the contrary, some patterns are still present in the residues obtained from MLR (Fig. 3 left). The control limits are also quite different in the two cases as a result of the different performance of the methods. Moreover, in spite of the larger control limits, the results obtained from MLR are stably outside the limits at the beginning of the testing period. In a real monitoring application, this might lead to problems of false alarm.

The residues obtained from SOBI (and PCA), instead, do not show any pattern that can be correlated to EOVs (and temperature, in particular); moreover, the residues in the training phase and testing phase are very similar, with individual outliers going out of the control limits from time to time.

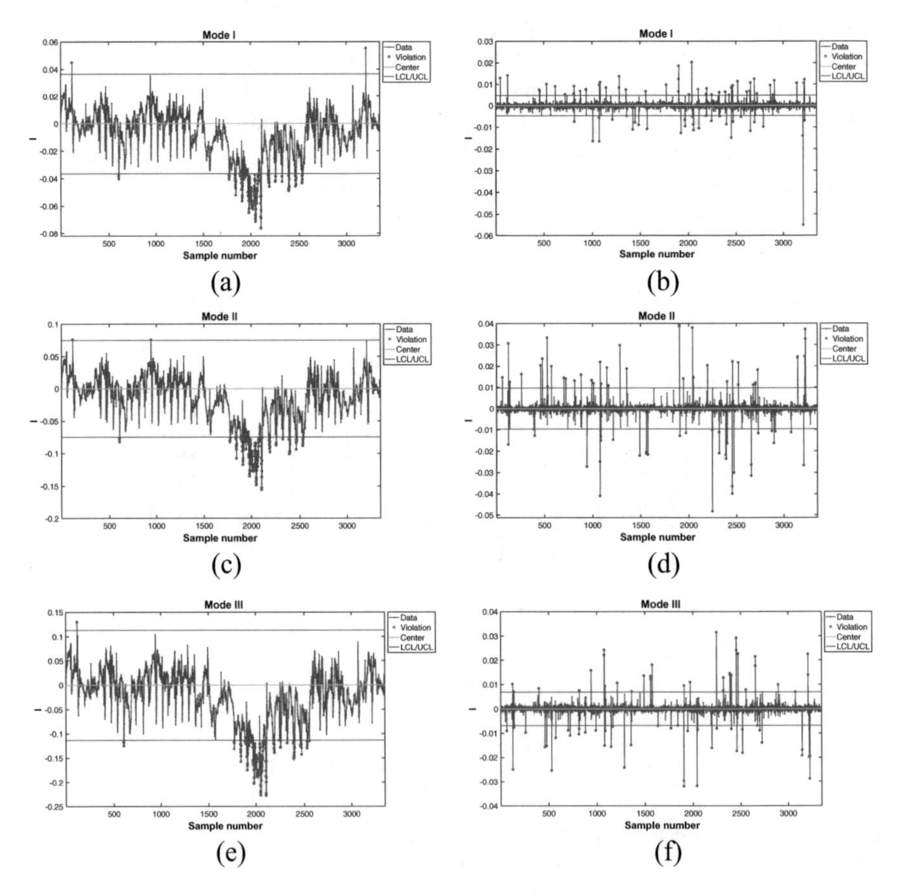

Fig. 3 Pre-stressed cable: residues from MLR (left) and SOBI (right)

The poorer performance of MLR with respect to SOBI (and PCA) can be ascribed to the well-known difficulty of measuring temperature values representative of the temperature of the cable along its full extension. In particular, direct sun radiation effects as well as persistent shadow on portions of the cable jeopardized the representativity of the temperature measurements and the predictive performance of the MLR model. When SOBI (and PCA) have been applied, on the contrary, an effective compensation of temperature influence was achieved even in the presence of just three monitored modes. This occurred because only one source (or principal component) was sufficient to describe about 99% of the variability in the data.

3.2 Hospital's Buildings

The influence of temperature on the natural frequencies of two closely spaced buildings belonging to the Campobasso's main hospital is discussed in this section starting from the experimental data collected by a monitoring system operating on the structure, albeit intermittently, for some months in between 2016 and 2017.

The main hospital in Campobasso consists of a number of high-rise as well as low-rise reinforced concrete buildings designed and built according to outdated seismic design codes. The vibration responses of two closely spaced structures hosting the inpatient wards (Fig. 4a) have been monitored [33]. The overall dimensions of the monitored structures are about 78 m $\times$ 14 m in plan and 30 m in elevation. They are separated by a narrow structural joint (Fig. 4b).

The vibration response of the structures in operational conditions was acquired by sixteen force-balance accelerometers ($\pm$0.5 g full scale range, 20 V/g sensitivity). These were installed at two upper levels along two orthogonal directions and at opposite corners of the building plans so that observability of fundamental translational as well as torsional modes was ensured. The accelerometers were wired to a centralized data acquisition system with built-in antialiasing filter and 16 bit ADC resolution. Data were sampled at 100 Hz and stored into a local MySQL database. The fundamental modal parameters were automatically extracted from the vibration records by an automated OMA procedure [7]. Sample results are shown in Fig. 5.

The first four identified modes can be described as:

- global bending of the two buildings in the transverse direction;
- global bending of the two buildings in the longitudinal direction;
- global torsion involving the two buildings;
- torsion with counter rotating buildings.

Systematic swings of the fundamental natural frequencies occurring every day can be observed, with a reduction in the night and a gradual increase in the morning until the maximum value, which is usually reached in the early afternoon. This circumstance confirms that temperature significantly affects the natural frequencies, and its effect has to be compensated in view of the development of effective modal-based SHM strategies. Plotting the (normalized) sample data of Fig. 5 in the form

(a)

(b)

Fig. 4 View of the closely spaced monitored hospital's buildings (*source* maps.google.it) (**a**) and the separation joint (**b**)

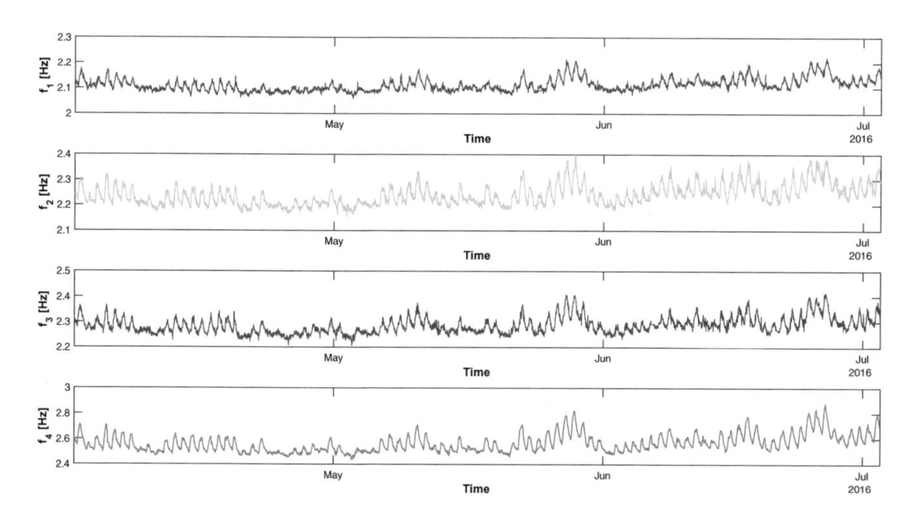

Fig. 5 Hospital's buildings—sample monitoring results: natural frequency time series of the first four modes

of histograms provides a further insight in the nature of such variations. Figure 6, indeed, shows that the distribution of data is asymmetric, with longer right tail with respect to the left one.

Taking into account that the natural frequencies increase with the temperature, the obtained monitoring results can be addressed to the influence of the very narrow joint that divides the two structures (Fig. 4b), and to direct sun radiation that mainly affects the buildings along the predominant longitudinal extension. The thermal expansion affects the infill panels as well as the relative distance between the buildings. In particular, when the temperature increases, the resulting interlocking increases the overall stiffness and, as a consequence, it yields non-symmetric distributions of the natural frequencies and their general increase with the temperature. Following these considerations, the nonlinearities in the structural response induced by temperature have to be taken into account in the compensation of the environmental influence. Nevertheless, linear (SOBI) as well as nonlinear (kPCA) methods are applied to comparatively assess their performance.

The time histories of the fundamental natural frequencies have been analyzed by SOBI first. The presence of a single dominant value among the eigenvalues of the zero-lag correlation matrix and the analysis of the source pattern [27] confirmed that a single source, the temperature, was responsible for most of the variability of the estimates. However, the residues still show some patterns (Fig. 7).

Results from application of kPCA confirm its higher effectiveness in compensating environmental effects in this case with respect to linear methods (Fig. 8): in fact, while the magnitude of misfit is similar in the training phase and testing phase, evident patterns which can push the data outside control limits cannot be detected. This is relevant to avoid possible false identification of damage.

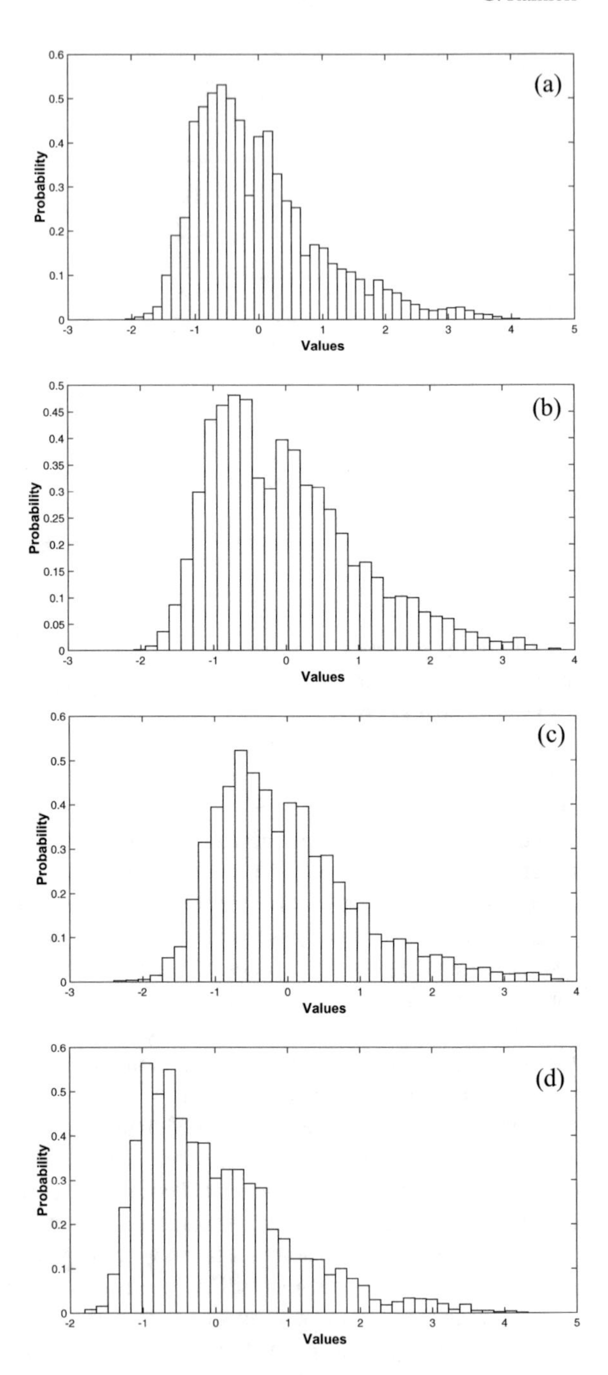

Fig. 6 Hospital's buildings—histograms of monitored frequencies: mode I (**a**), II (**b**), III (**c**), IV (**d**)

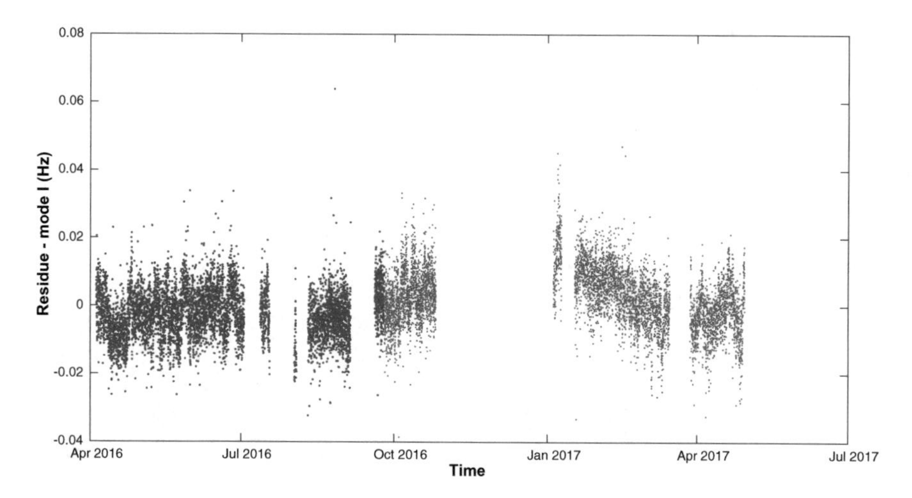

Fig. 7 Hospital's buildings: sample residues after compensation of the environmental influence by SOBI

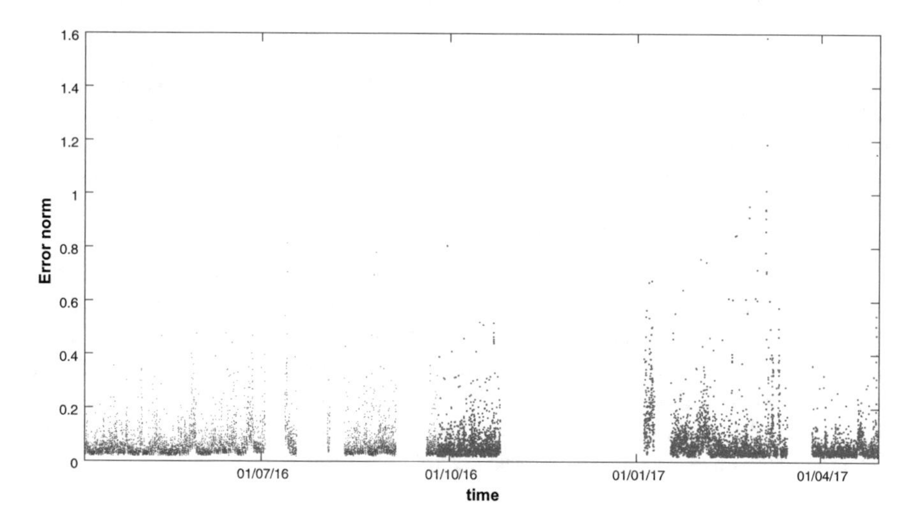

Fig. 8 Hospital's buildings: error norm from application of kPCA to the entire dataset

Repeating the analysis on a subset of the entire dataset, the influence of unmeasured conditions in the training phase on the quality of predictions can be assessed. Figure 9 shows the results obtained from application of kPCA to the data collected in the period April 1st–July 2nd, 2016. The dataset has been divided into two parts of equal number of samples, of which the first part has been used for training. Figure 9 also shows the evolution of the average local temperature, as measured by a nearby meteorological station, in the same period.

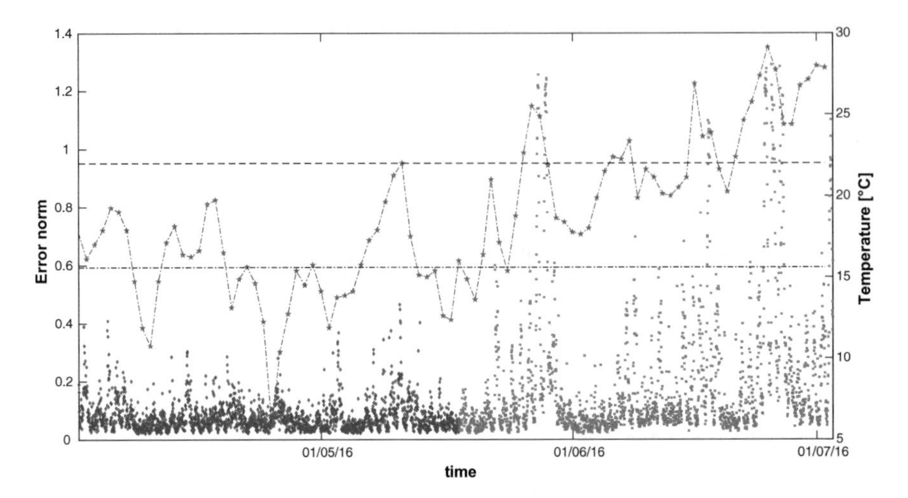

Fig. 9 Hospital's buildings: results from application of kPCA to a subset of the entire dataset; temperature evolution over time is denoted by the red dash dot line; the horizontal dash dot and dashed lines represent the average and maximum temperature in the training period, respectively

Comparing the temperature values in the training period with those in the testing period allows noting that the average daily temperature (red dash dot line in Fig. 9) is in between 5 and 22 °C—with a mean value of 15.6 °C—in the training period (from April 1st to May 17th, 2016), and it increases up to 29 °C in the testing period (from May 18th to July 2nd, 2016). This is probably the reason of the large values of the misfit that can be observed in the testing period (Fig. 9) in the days characterized by very high average daily temperature, when this goes out of the range considered in the training period. The discrete nature of the occurrence of large misfit values, on the other hand, seems to confirm the assumption that, when the temperature rises, the dynamic response of the structure abruptly changes as a result of the interlocking between the two buildings through the narrow separation joint, and if this effect was not adequately monitored in the training phase, the results in terms of prediction error are affected also by the occurrence of this phenomenon.

4 Conclusions

Different linear as well as nonlinear methods for the compensation of the influence of EOVs on natural frequencies in the context of modal-based SHM have been reviewed in the present paper and applied to real datasets collected from vibration-based monitoring systems. Based on the achieved results, the applicative perspectives and possible drawbacks of the selected methods, namely MLR, PCA, SOBI and kPCA, have been assessed. The following conclusions can be drawn.

The prediction accuracy of methods, such as MLR, requiring measures of relevant EOVs affecting the dynamic response of the monitored structure might be jeopardized by the difficulties of identifying in advance and appropriately measuring those EOVs.

Output-only methods represent an attractive alternative for compensation of environmental and operational effects on damage-sensitive features in the context of modal-based SHM since their results do not depend on the prior identification of all relevant EOVs influencing the structural response and on the effective measure of such EOVs.

Nonlinear compensation techniques might play a relevant role in enhancing the reliability of modal-based SHM in the presence of nonlinear influence of EOVs on the monitored modal parameters.

Finally, the results obtained from the analyzed case studies confirm the importance of extensive training in order to cover as much as possible the range of environmental and operational conditions the monitored structure can operate in.

References

1. Farrar CR, Worden K (2012) Structural health monitoring: a machine learning perspective. Wiley, Chichester
2. Teughels A, De Roeck G (2005) Damage detection and parameter identification by finite element model updating. Arch Comput Methods Eng 12(2):123–164
3. Simoen E, De Roeck G, Lombaert G (2015) Dealing with uncertainty in model updating for damage assessment: a review. Mech Syst Signal Process 56–57:123–149
4. Deraemaeker A, Worden K (2018) A comparison of linear approaches to filter out environmental effects in structural health monitoring. Mech Syst Signal Process 105:1–15
5. Magalhães F, Cunha A, Caetano E (2012) Vibration based structural health monitoring of an arch bridge: from automated OMA to damage detection. Mech Syst Signal Process 28:212–228
6. Rainieri C, Fabbrocino G (2010) Automated output-only dynamic identification of civil engineering structures. Mech Syst Signal Process 24(3):678–695
7. Rainieri C, Fabbrocino G (2015) Development and validation of an automated operational modal analysis algorithm for vibration-based monitoring and tensile load estimation. Mech Syst Signal Process 60–61:512–534
8. Carden EP, Fanning P (2004) Vibration based condition monitoring: a review. Struct Health Monit 3:355–377
9. Ubertini F, Comanducci G, Cavalagli N, Pisello AL, Materazzi AL, Cotana F (2017) Environ mental effects on natural frequencies of the San Pietro bell tower in Perugia, Italy, and their removal for structural performance assessment. Mech Syst Signal Process 82:307–322
10. Deraemaeker A, Reynders E, De Roeck G, Kullaa J (2008) Vibration-based structural health monitoring using output-only measurements under changing environment. Mech Syst Signal Process 22:34–56
11. Peeters B, De Roeck G (2001) One-year monitoring of the Z24-Bridge: environmental effects versus damage events. Earthquake Eng Struct Dynam 30:149–171
12. Moser P, Moaveni B (2011) Environmental effects on the identified natural frequencies of the Dowling Hall Footbridge. Mech Syst Signal Process 25:2336–2357
13. Yan A-M, Kerschen G, De Boe P, Golinval J-C (2005) Structural damage diagnosis under varying environmental conditions—Part I: A linear analysis. Mech Syst Signal Process 19(4):847–864
14. Cross EJ, Koo KY, Brownjohn JMW, Worden K (2013) Long-term monitoring and data analysis of the Tamar Bridge. Mech Syst Signal Process 35:16–34

15. Jin S-S, Cho S, Jung H-J (2015) Adaptive reference updating for vibration-based structural health monitoring under varying environmental conditions. Comput Struct 158:211–224
16. Peeters B, Maeck J, De Roeck G (2001) Vibration-based damage detection in civil engineering: excitation sources and temperature effects. Smart Mater Struct 10:518–527
17. Ni Y, Hua X, Fan K, Ko J (2005) Correlating modal properties with temperature using long-term monitoring data and support vector machine technique. Eng Struct 27:1762–1773
18. Sohn H, Dzwonczyk M, Straser E, Law K, Meng T, Kiremidjian A (1998) Adaptive modeling of environmental effects in modal parameters for damage detection in civil structures. Proc SPIE 3325:127–138
19. Ni Y, Zhou H, Ko J (2009) Generalization capability of neural network models for temperature-frequency correlation using monitoring data. J Struct Eng 135(10):1290–1300
20. Zhou HF, Ni YQ, Ko JM (2010) Constructing input to neural networks for modeling temperature-caused modal variability: mean temperatures, effective temperatures, and principal components of temperatures. Eng Struct 32:1747–1759
21. Kullaa J (2011) Distinguishing between sensor fault, structural damage, and environmental or operational effects in structural health monitoring. Mech Syst Signal Process 25(8):2976–2989
22. Kullaa J (2014) Structural health monitoring under nonlinear environmental or operational influences. Shock Vibr Article ID 863494
23. Yan A-M, Kerschen G, De Boe P, Golinval J-C (2005) Structural damage diagnosis under varying environmental conditions—Part II: Local PCA for non-linear cases. Mech Syst Signal Process 19(4):865–880
24. Sohn H, Worden K, Farrar CR (2002) Statistical damage classification under changing environmental and operational conditions. J Intell Mater Syst Struct 13(9):561–574
25. Reynders E, Wursten G, De Roeck G (2014) Output-only structural health monitoring in changing environmental conditions by means of nonlinear system identification. Struct Health Monit 13(1):82–93
26. Rizzo M, Betti M, Castelli P, Spadaccini O, Vignoli A (2019) Long-term dynamic monitoring of offshore platform for damage assessment. In: Proceedings of the 9th international conference on structural health monitoring of intelligent infrastructure: transferring research into practice—SHMII 2019, vol 1, pp 515–523. St. Louis, USA
27. Rainieri C, Magalhaes F, Gargaro D, Fabbrocino G, Cunha A (2019) Predicting the variability of natural frequencies and its causes by Second-Order Blind Identification. Struct Health Monit 18(2):486–507
28. Ans B, Hérault J, Jutten C (1985) Adaptive neural architectures: detection of primitives. In: Proceedings of COGNITIVA '85, pp 593–597. Paris, France
29. Cardoso JF, Souloumiac A (1996) Jacobi angles for simultaneous diagonalization. SIAM J Matrix Anal Appl 17:161–164
30. Mercer J (1909) Functions of positive and negative type, and their connection with the theory of integral equations. Phil Trans R Soc A 209:415–446
31. Awad M, Khanna R (2015) Efficient learning machines: theories, concepts, and applications for engineers and system designers. Apress Media
32. Shannon CE (1948) A mathematical theory of communication. Bell Syst Tech J 27:379–423
33. Gargaro D, Rainieri C, Fabbrocino G (2017) Structural and seismic monitoring of the "Cardarelli" Hospital in Campobasso. Procedia Eng 199:936–941

Vibration-Based Damage Feature for Long-Term Structural Health Monitoring Under Realistic Environmental and Operational Variability

Francescantonio Lucà, Stefano Manzoni, and Alfredo Cigada

Abstract Many vibration-based damage detection approaches proposed in the literature for civil structures rely on features related to modal parameters, since these are sensitive to structural properties variations. The influence of environmental and operational variability on modal parameters sets limits to unsupervised learning strategies in real-world applications, especially for long-time series. The chapter shows an example of unsupervised learning damage detection in a realistic environment, over a long-time period. Two damage features are compared: one from operational modal analysis and the other from autoregressive models. To start with a real though simple structure, a series of tie-rods has been considered; these are slender axially tensioned beams, widely used in both historical and modern buildings, to balance lateral forces in arches. Since the axial load is heavily influenced by temperature and eventually by other disturbances, even small changes in the environmental conditions cause dramatic changes in the dynamic tie-rod features. To investigate this problem, a set of nominally identical full-scale structures have been continuously monitored for several months under different environmental and operational conditions. It is shown how the combination of vibration-based damage features and multivariate statistics can be successfully used to detect damage in structures working under real environmental conditions.

Keywords Long-term monitoring · Damage detection · Environmental and operational variations · Vibration-based features · Multivariate statistics

1 Introduction

Civil structures are subject to degradation during their long-term service, mainly due to materials ageing or deterioration. For this reason, it is important to assess whether the safety requirement for proper operation is always met. In most cases, human inspections are still used to monitor the health of a structure: in these situations, a

F. Lucà · S. Manzoni · A. Cigada (✉)
Department of Mechanical Engineering, Politecnico Di Milano, 20158 Milano, Italy
e-mail: alfredo.cigada@polimi.it

A. Cury et al. (eds.), *Structural Health Monitoring Based on Data Science Techniques*, Structural Integrity 21, https://doi.org/10.1007/978-3-030-81716-9_14

judgement is provided by an expert after visual inspection. This approach comes with different limitations, such as the evaluation subjectivity and the qualitative content of the retrieved information. Furthermore, not all the components of a complex structure can be accessible and thus inspected. Human inspections are also characterized by being intermittent in time, with the obvious risk that if damage occurs between two timed inspections, a prompt maintenance intervention cannot always be carried out.

The development of new sensors, networks, information technology, including computing and storage capabilities, is increasing attention towards condition-based maintenance, which relies on long-term continuous monitoring. In this context, a key role is played by structural health monitoring (SHM), which is a multidisciplinary research field aiming at the definition of automatic damage detection strategies for civil, mechanical and aerospace structures [1]. All the possible approaches of SHM can be roughly divided into two families: model-based and data-driven approaches [2, 3].

In the first case, a physics-based or law-based model of the monitored structure is required. The inverse problem technique is used to calibrate numerical models, commonly finite element ones, and damage is detected by relating the measured data from the structures to the estimated data from the models. However, it is not always possible to have resources allowing one to create a complete structural model. Furthermore, an accurate model of a real structure is made difficult by the many uncertainties about materials, geometries and boundary conditions, making every structure one-of-a-kind. Moreover, in case of complex structures, it is not possible to simulate all the possible damage conditions.

For the above-mentioned reasons, the SHM community is more and more focusing on unsupervised learning data-driven approaches, where damage is detected through a statistical comparison between data referring to the structure in the current state (unknown) and a reference scenario, where the structure is assumed to be in a healthy state.

In this case, the crucial aspect is the definition of a damage-sensitive quantity or parameter that can be directly related to the health state of the structure. This quantity is commonly referred to as "damage feature". A damage feature may be defined starting from any physical quantity that can be sensitive to changes in any structural properties. Vibration-based approaches are those most commonly adopted, starting from the observation that changes in a structural system caused by damage manifest themselves as changes in mass, stiffness and energy dissipation. These changes reflect in changes of the dynamic response characteristics: for this reason, damage features extracted from vibration data are potentially sensitive to damage [2].

Modal parameters have been largely used in the literature to define damage features [4]: natural frequencies, mode shapes are those most commonly used, while modal damping is less used because of the difficulty of its identification [5]. In many cases, modal parameters can provide a physical interpretation that can help understanding the nature of damage. Low-frequency modes are more easily identified, thus very commonly adopted for damage detection algorithms. Since low-frequency modes capture the global response of the structure, they are less sensitive to local changes and this can lead to a late damage detection [6].

Another important family of approaches comes from the statistical time series methods, where time series models are fitted to the measured vibration data and model parameters are used to define damage features. Autoregressive (AR) [7, 8], autoregressive with exogenous input (ARX) [9] and autoregressive with moving average (ARMA) [10, 11] models are commonly used to fit the measured data. These models are suitable for SHM applications because they inherently account for uncertainty and do not depend on physical models; thus, they are suitable for automated damage detection. In the literature, there are several applications, where either the coefficients or the residuals of the models are used to define effective damage features [5].

When either modal-based or autoregressive-based damage features are adopted for unsupervised learning damage detection, one of the main challenges is represented by the fact that vibration-based approaches are influenced by the effect of environmental and operational conditions [12]. This happens because the structural response varies under different environmental conditions, particularly temperature, causing changes in the material properties and boundary conditions. These variations determine changes in the structural responses that could be higher than those caused by damage, at least at an early stage [13].

In the literature, there are not so many applications on real structures, under realistic environmental conditions. In this chapter, a case study has been selected to test the effectiveness of a vibration-based damage detection approach based on AR models, in a challenging realistic scenario. The case study consists of damage detection of axially loaded beam-type structures (or tie-rods) which are representative of common structural elements such as braces of truss girders, struts and ties of space trussed structure, tie-rods of arches and vaults. In operation, both the properties of the beam and those of the structure vary, due to the changes in the environmental conditions, mainly temperature. Two nominally identical tie-rods have been installed in a laboratory and continuously monitored, in a weakly controlled environment, simulating a real permanent SHM system. The environment was characterized by temperature variations and a number of working operations, such as the presence of human activities and the presence of testing machines nearby. It is worth noting that the experimental set-up allowed for the possibility of measuring the environmental and operational variables but none of them has been intentionally controlled, to create a realistic and challenging data set.

This chapter is organized as follows: in Sect. 2, the experimental set-up is described, along with the chosen instrumentation. Experimental data are presented, to show the effects of temperature on the tie-rod axial load. The adopted damage detection approach is described in Sect. 3, where a brief theoretical background of simple autoregressive models, model order selection and Mahalanobis squared distance (MSD) is provided. The experimental results are presented and discussed in Sect. 4. In the end, the conclusions are drawn in Sect. 5.

2 Case Study Description

A brief introduction to the key points of the theory on vibrations in axially loaded beams is provided to better understand the reasons behind the choice of tie-rods as an interesting SHM case study. The reader is redirected to [14] for the complete theory of vibrations on tensioned beams, while the analytical expressions for eigenfrequencies and mode shapes for different support conditions can be found in [15].

Tie-rods eigenfrequencies and mode shapes are function of the following parameters:

- Geometrical properties (cross section, momentum of inertia of the section and length of the beam): while in most cases the properties of the section can be directly measured, the length of the beam is difficult to be estimated because the side clamps may create uncertain links.
- Material properties (Young's modulus of elasticity and density): especially for ancient structures, the characteristics of the material can be roughly estimated unless some specific material tests are carried out.
- Boundary conditions (stiffness of the constraints): boundary conditions depend on different variables such as the clamping systems, the material to which the tie-rod is fixed and its deterioration.
- Loading conditions (axial load): in most cases, direct in-situ measurements are difficult and methods for an indirect tension estimation are required. Several attempts to obtain indirect estimate of the axial load can be found in the literature [16–18] showing the complexity of the problem in detail.

In a real context, detecting damage through changes in modal parameters is a complex task, since the above-mentioned parameters are subject to high uncertainty and may also change with time. A clear example is the effect of temperature, whose daily and long-term trends cause changes in the material properties that reflect in daily and long-term trends in the modal parameters. For this reason, an effective damage detection strategy must be able to separate the effects of damage from those of environmental variations.

To investigate this topic, an experimental set-up composed of two nominally identical tie-rods made of aluminium has been designed and installed in the laboratories of the Department of Mechanical Engineering at Politecnico di Milano. The tie-rods are characterized by a cross section of 15×25 mm^2 and a length $L = 4000$ mm. The geometrical characteristics are representative of tie-rods that are commonly used in civil structures to balance lateral forces of arches and vaults. The static scheme that has been reproduced is a clamped–clamped configuration, where both ends of the beam are fixed. The constraints have been realized with a structure composed of three parts (see Fig. 1a).

During the tensioning procedure, where the axial load has been applied to each of the two tie-rods, the bolted joints of clamp 1 (see Fig. 1b) were tightened-up while those of clamp 2 (see Fig. 1c) were left loose, so that the beam was not fully constrained along the axial direction. A load cell has been installed between clamp

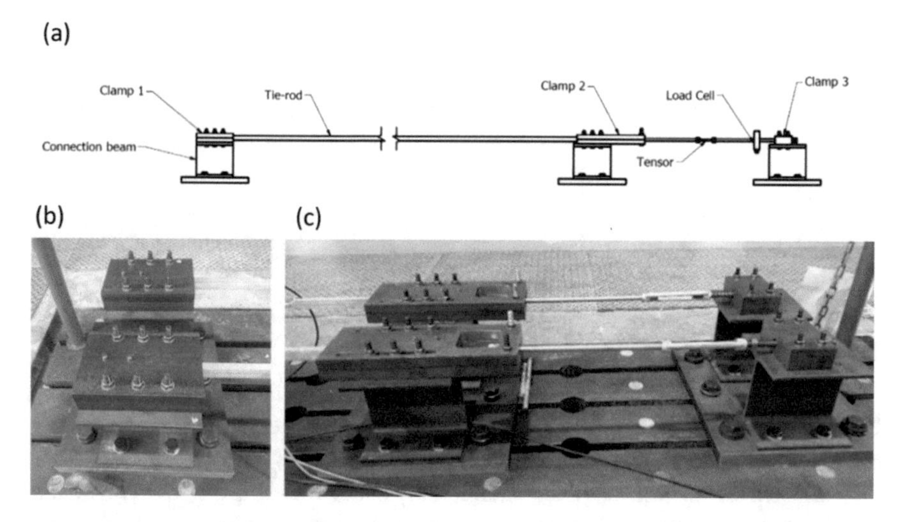

Fig. 1 Layout of the experimental set-up (**a**); detail of clamp 1 (**b**); detail of clamps 2 and 3 (**c**)

2 and clamp 3, and an axial load of 8 kN has been applied using a tensioner. Each tie-rod is instrumented with strain gauges composing a full Wheatstone bridge (thus with temperature compensation, see Fig. 2b) that has been calibrated during this phase to measure tension all through the monitoring phase.

After the strain gauge bridge calibration, the bolted joints in clamp 2 have been tightened-up and the load cell between clamp 2 and clamp 3 has been removed. Right after the tensioning procedure, a preliminary experimental modal analysis has been carried out to identify the modal parameters. The natural frequencies related to the first six vibration modes are reported in Table 1.

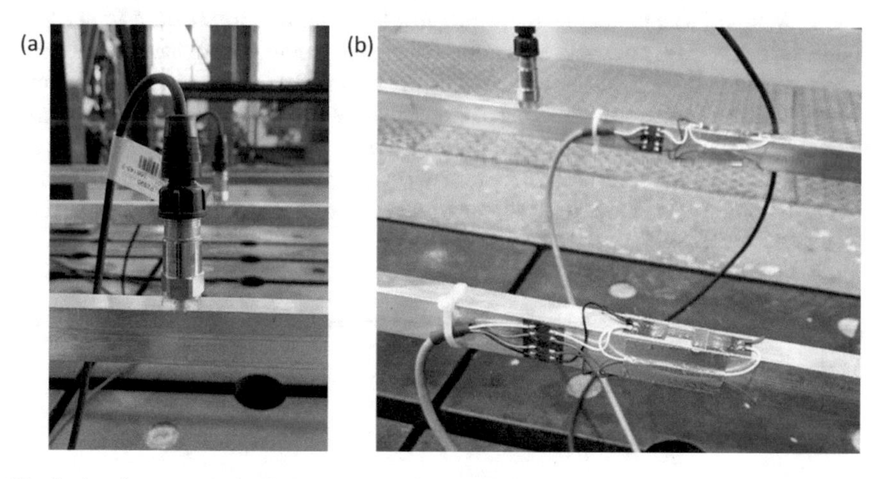

Fig. 2 Accelerometer PCB 603C01 (**a**); strain gauges (**b**)

Table 1 Eigenfrequencies of the two tie-rods, identified during the preliminary phase (October 2019)

Vibration mode	Eigenfrequencies [Hz]	
	Tie-rod 1	Tie-rod 2
1	13.89	14.17
2	30.98	31.64
3	53.36	54.46
4	81.82	83.24
5	116.55	118.66
6	157.95	160.40

Each tie-rod has been equipped with four piezoelectric industrial accelerometers PCB 603C01 (Fig. 2a). These accelerometers have a sensitivity of 100 mV/g, full-scale ±50g and spectral noise up to 10 Hz of $6.2 \times 10^3 \left(\frac{\mu m}{s^2}\right)^2/\text{Hz}$. The choice for industrial accelerometers has been intentionally made to test the proposed procedure also in case of sensors with expected lower sensitivity and signal-to-noise ratio if compared to laboratory ones, but also much cheaper, as for a real field application.

The sensor placement has been carried out taking into account the possibility to detect the first six vibration modes. The autoMAC matrix [19] has been adopted for many possible layouts, in the end selecting the one represented in Fig. 3, where dimensions are expressed as a fraction of the entire tie-rod length L (i.e., 1/20 means that the distance between the first accelerometer starting from left and the clamp 1 is $L/20 = 200$ mm).

Even if multiple sensors are available on each tie-rod, a strategy that relies only on one accelerometer has been developed in this work. This choice was made taking into account a real application, where the adoption of a high number of sensors to monitor a single tie-rod may not always be possible, for economic reasons.

The acquisition system is composed of three NI9234 modules, with anti-aliasing filters on-board. The sampling frequency has been set to 256 Hz on a total of 11 channels: eight IEPE channels to acquire the signals coming from the piezoelectric accelerometers, two channels dedicated to the acquisition of the strain gauge full-bridges, thus providing a continuous measurement of the axial load, and one channel to acquire the signal coming from a thermocouple, thus providing a measurement of the environmental temperature. The choice for the sampling frequency aims at finding a compromise between the need to detect a sufficient number of modes and the need to limit the amount of acquired data.

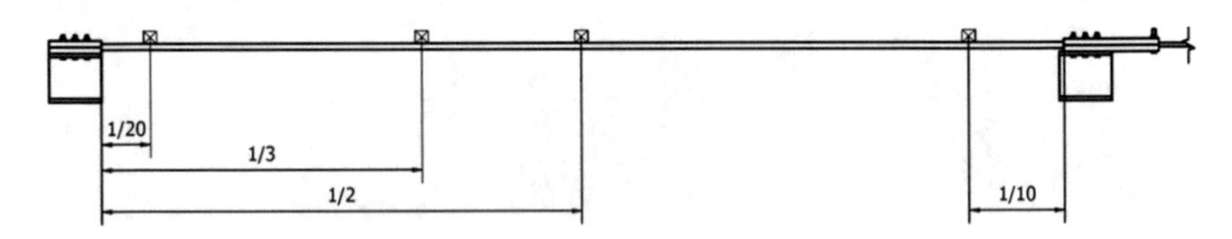

Fig. 3 Accelerometers layout

Axial load and temperature have been stored to have an additional information about the experimental conditions, but, as it will be explained in the following, these variables are not strictly required by the proposed damage detection strategy. The trends in time of the axial load for one of the two tie-rods and temperature are reported in Fig. 4a, for a period of approximately 5 months. As it is possible to see, both daily and long-term trends can be observed, with daily average axial load reaching values of 13 kN in the coldest days of the year, where the daily average temperature in the laboratory is 11 °C.

By zooming in on a shorter time window and comparing the trends of the axial load with that of temperature, it is possible to notice that the two variables are highly correlated (see Fig. 4b), with the axial load changes around 700–800 N/°C.

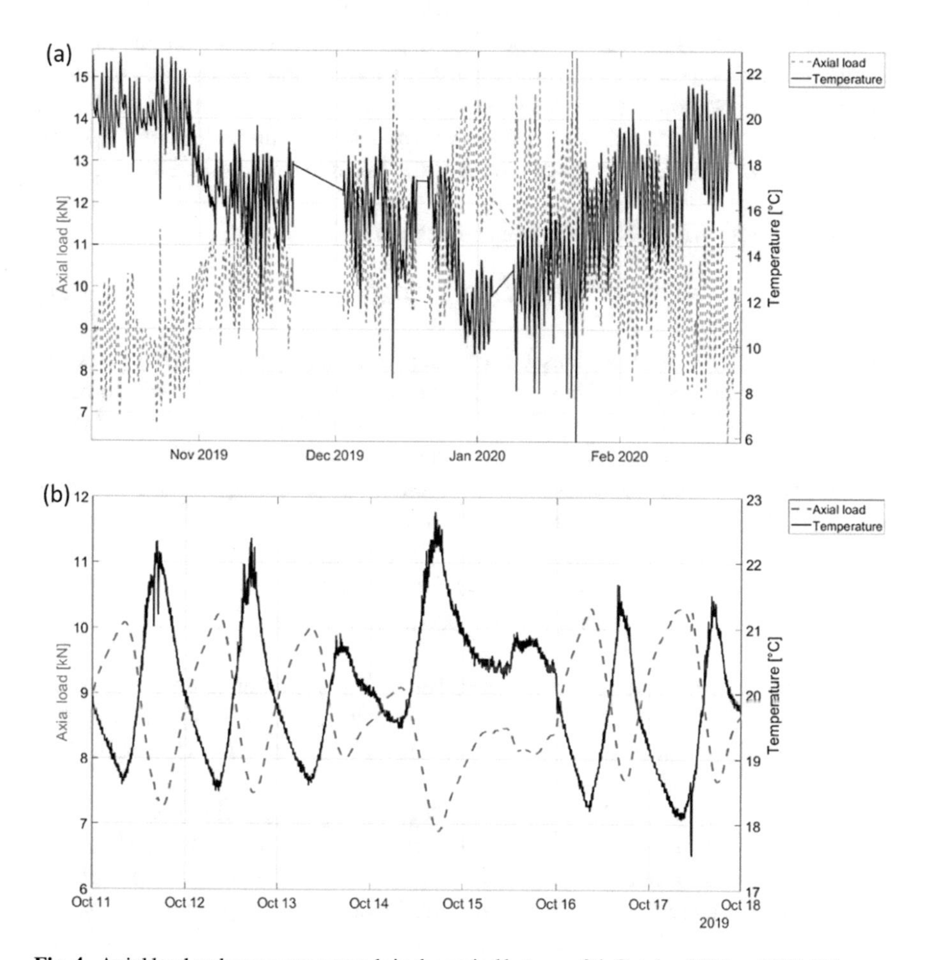

Fig. 4 Axial load and temperature trends in the period between 8th October 2019 and 27th February 2020 (**a**); axial load and temperature trends in the period 11th–18th October 2019 (**b**)

Considering the variability of the environmental conditions for the considered case study, a data-driven approach not relying on modal parameters has been developed. The damage detection strategy, which combines time series analysis (AR models) and multivariate outlier detection (Mahalanobis squared distance), is presented in the following section.

3 Methodology

The first step of the proposed damage detection strategy requires the extraction of AR coefficients from vibration data. AR models are time series models born in the field of econometrics to describe time-varying processes where the output variable linearly depends on its previous values and a stochastic term. These models can be adopted in a SHM context to represent the dynamic response of structures. In this work, autoregressive models are exploited to process only the output of a system, to accurately obtain the dynamic response of the linear system, forced by a random input.

The first order AR model AR(1) can be defined as it follows [7, 20]:

$$X_t = \phi_1 X_{t-1} + a_t \tag{1}$$

where X_t and X_{t-1} are two consecutive samples of the system output, ϕ_1 is the autoregressive coefficient, and a_t is the residual term. This latter is a Gaussian process with zero mean and variance σ_a^2. It can be demonstrated that the coefficient ϕ_1 is related to the eigenvalues of the first-order linear dynamic system [20].

For systems of order n, with $n > 1$, the autoregressive moving average model of order n, or ARMA $(n, n-1)$, must be used [20]:

$$\begin{aligned} X_t = {} & \phi_1 X_{t-1} + \phi_2 X_{t-2} + \cdots + \phi_n X_{t-n} + a_t \\ & - \theta_1 a_{t-1} - \cdots - \theta_{n-1} a_{t-n+1} \end{aligned} \tag{2}$$

where ϕ_i are the autoregressive parameters and a_t is the residual at time t using the same syntax of Eq. (1). In addition, there is the moving average (MA) part, which is composed by the residuals at previous times (from $t-1$ to $t-n+1$), weighted by the coefficients θ_i, with $i = 1, \ldots, n-1$.

Starting from the coefficients of the autoregressive part (ϕ_j with $j = 1, \ldots, n$), ARMA models may be used to obtain the natural frequencies of the system [20, 21].

A critical point is that to estimate the coefficients θ_i related to the MA part, a nonlinear least square approach is required [22] that can be subject to convergence and local minima problems, and high computational cost [23]. In order to avoid such problems, an alternative approach may be the adoption of an AR (q), with $q \gg n$, that provides and approximation of the ARMA $(n, n-1)$ model. In this case, the model is defined by the following equation:

$$X_t = \sum_{j=1}^{q} \phi_j X_{t-j} + a_t \tag{3}$$

As it is possible to see, the difference between AR (q) and ARMA $(n, n-1)$ is that the former only considers the residual at time t. In case the AR (q) model is considered, the linear least square minimization algorithm can be adopted to obtain the coefficients ϕ_j, with obvious calculation convenience. It is important to notice that by adopting an order q higher than that of the system n, to obtain a good approximation of the ARMA $(n, n-1)$ model, spurious eigenvalues will be obtained. Generally, both the order of the mechanical system and the suitable AR to describe the system output are unknown a priori. The choice of the order has consequences on the description of the system response and on the performance of damage detection too [24].

Generally speaking, higher-order models may better fit the data but may not generalize well to other data sets (overfitting), while a low-order model will not necessarily capture the underlying physical system dynamics (underfitting). In the literature, different approaches are suggested, such as Akaike's information criterion (AIC), the partial autocorrelation function (PAF) and Bayesian Information Criterion (BIC) [25]. The last in the above list has been adopted in this work; some insights on the model selection carried out for this case study are provided in the next section.

3.1 Pre-processing and Model Order Selection

Before discussing the details about the model order estimate, some pre-processing choices are briefly discussed. First of all, data continuously acquired by the monitoring system are processed every 60s, and thus the q coefficients of the AR(q) model will be estimated every 60s. This choice starts from the observation that the laboratory environment is frequently characterized by the presence of short time events (transients, harmonic excitations, impulsive forces). These can result in autoregressive coefficients which are different from those obtained when the structure is excited only by random noise. By evaluating the coefficients every 60s, a correct clustering of those records considered outliers is more easily obtained, helping in a better interpretation of results.

The time histories are initially normalized by removing their mean value and dividing them by their standard deviation. This step is carried out to allow for a comparison of AR models referring to different system conditions. Indeed the AR models are only poles functions that do not depend on the amplitude of the system input [26].

The BIC has been adopted to select the model order, which is more appropriate to describe all the possible data acquired during the tie-rods long-time monitoring. The BIC is defined as:

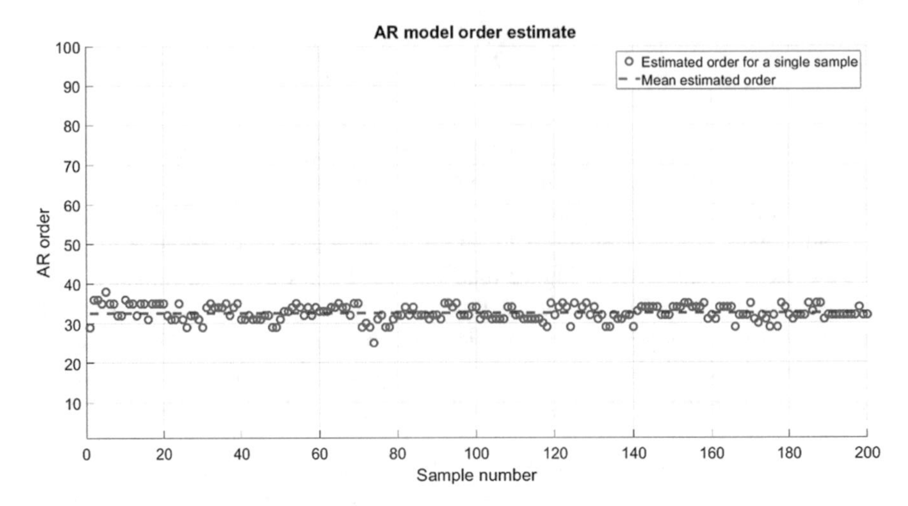

Fig. 5 AR model order selection

$$BIC(q) = q\log(N) - 2\log\left(L\left(\underline{\phi}\right)\right) \tag{4}$$

where N is the sample size, q is the number of parameters, and $\underline{\phi}$ is the set of all parameters. $L\left(\underline{\phi}\right)$ is the model likelihood, given the data, evaluated at the maximum likelihood values of $\underline{\phi}$. Comparing models with the BIC involves calculating this parameter for each model and choosing the one with the lowest BIC.

In this work, a set of 200 randomly chosen time records picked from the baseline set (see Sect. 4) have been considered. For each record, the BIC has been evaluated for $q = 1, \ldots, q_{\max}$, selecting the first order that satisfied the condition:

$$BIC(q) < toll \cdot |BIC(1) - BIC(q_{\max})| \tag{5}$$

With $q_{\max} = 100$ and $toll = 0.05$. The results of this process are reported in Fig. 5.

Finally, a model order $q = 45$ has been selected so that all the possible considered records could be correctly described (an example about the fitting quality will be presented later and discussed in Sect. 4).

3.2 Damage Feature and Outlier Detection

Every time a time history is fitted to an AR (45) model, 45 coefficients are evaluated through a linear least square minimization. Thus, the damage feature can be defined as it follows:

$$\underline{\phi}_i = \begin{bmatrix} \phi_1 \\ \phi_2 \\ \cdots \\ \phi_{\mathrm{TOT}} \end{bmatrix} \tag{6}$$

where the suffix "i" means that the coefficients refer to the time record number i.

In order to carry out unsupervised damage detection, a reference scenario must be considered. A number N_{ref} of time histories have been considered as a reference, and thus, N_{ref} observations of the damage feature may be obtained and arranged in a matrix $[\phi]_{\mathrm{ref}}$, defined as it follows:

$$[\phi]_{\mathrm{ref}} = \begin{bmatrix} \underline{\phi}_1^{\mathrm{T}} \\ \underline{\phi}_2^{\mathrm{T}} \\ \cdots \\ \underline{\phi}_i^{\mathrm{T}} \\ \cdots \\ \underline{\phi}_{N_{\mathrm{ref}}}^{\mathrm{T}} \end{bmatrix}_{N_{\mathrm{ref}} \times q} \tag{7}$$

where the symbol "T" means transpose. The matrix $[\phi]_{\mathrm{ref}}$ is characterized by multivariate mean vector $\underline{\mu}_{\mathrm{ref}}$ (size $q \times 1$) and covariance matrix $[\Sigma_{\mathrm{ref}}]$ (size $q \times q$).

Considering a new observation of data for which the health state is unknown, it is again possible to fit an AR (45) model and arrange the 45 coefficients into a new vector $\underline{\phi}_{\mathrm{new}}$.

The Mahalanobis squared distance (MSD) can be used to perform an outlier detection. The MSD between $\underline{\phi}_{\mathrm{new}}$ and the reference set $[\phi]_{\mathrm{ref}}$ can be calculated with the following expression to define the damage index DI:

$$\mathrm{DI} = \mathrm{MSD}\left(\underline{\phi}_{\mathrm{new}}, [\phi]_{\mathrm{ref}}\right) = \left(\underline{\phi}_{\mathrm{new}} - \underline{\mu}_{\mathrm{ref}}\right)[\Sigma_{\mathrm{ref}}]^{-1}\left(\underline{\phi}_{\mathrm{new}} - \underline{\mu}_{\mathrm{ref}}\right)^{\mathrm{T}} \tag{8}$$

The result of this operation provides a scalar measuring the compatibility between the new observation $\underline{\phi}_{\mathrm{new}}$ and the reference set $[\phi]_{\mathrm{ref}}$. Thus, by setting a threshold level, when DI is above the threshold, the new observation can be labelled as an outlier with respect to the reference condition, possibly pointing out an eventual damage. As from the literature, the MSD can filter out environmental effects if the training set contains a full range of environmental conditions [27].

A threshold can be set by following the procedure described in [28], once the size of the training set has been defined. The procedure is a Monte Carlo simulation made up of the following steps:

1. Build a matrix of size $N_{\mathrm{ref}} \times q$, where every element is a random number sampled from a zero mean and unit standard deviation normal distribution.
2. Calculate the MSD between each row of the matrix and the matrix itself and store the maximum MSD value.

3. Repeat the process for a high number of trials (i.e. 1000 times), obtaining an array containing all the largest Mahalanobis squared distances and then organize them in a decreasing magnitude order. The critical values for 5 and 1% tests of discordancy for a q-dimensional sample of N_{ref} observations are then given by the Mahalanobis squared distances, in the array above which 5 and 1% of the trials occur. This threshold, called t_1, refers to a situation where the baseline set also contains observations related to the outliers (i.e. damage features related to the damaged structure are in the baseline set).
4. If the baseline does not contain outliers, the threshold must be set according to the following expression:

$$t_2 = \frac{\left(N_{ref} - 1\right)\left(N_{ref} + 1\right)^2 t_1}{N_{ref}\left(N_{ref}^2 - \left(N_{ref} + 1\right)t_1\right)} \tag{9}$$

4 Results and Discussion

In this section, results obtained through the application of the proposed damage detection strategy are presented. The effect of damage has been simulated on one of the two tie-rods (named T_1) by the addition of concentrated masses, to alter the dynamic response in a reversible way, according to a strategy very commonly adopted in literature [29–31]. Two different positions have been selected: one very close to the clamped ends ($L/10$) and one at the beam mid span ($L/2$). Also, two different masses have been used: 3 and 5% of the total mass of the beam. While tie-rod T_1 was subject to the addition of concentrated masses, the second tie-rod (named T_2) was always kept in the same structural condition; i.e. no masses have been applied to T_2. In the following, the presented graphs will refer to sensors placed at $L/3$ on both tie-rods, while some final remarks regarding results obtained with the other sensors will be presented in the conclusions.

The data included in the analysis can be divided into 5 sets and labelled according to the content of Table 2. It is worth recalling that every record contains the information related to a 60s vibration data record, acquired with a sampling rate of 256 Hz.

The sets labelled as "Baseline 1" and "Baseline 2" contain data representing the reference condition, i.e. no mass added to T_1. Both T_1 and T_2 were only subject to the effect of environmental and operational variations. The two baseline sets come from two non-consecutive periods, to provide a wider range of environmental (see Fig. 6a) and operational conditions (see Fig. 6b). Into details, for every considered sample, Fig. 6a shows the corresponding average temperature and Fig. 6b shows the corresponding root mean square (RMS) of the accelerations, normalized to the maximum observed RMS. Set "V_1" is consecutive to "Baseline 1" and set "V_2" is consecutive to "Baseline 2"; "V_1" and "V_2" are used for validation; i.e. they have

Table 2 Description of the data sets

Label	# Samples	Simulated damage on T_1	Mass position
Baseline 1	10,792 (~180 h)	No simulated damage	–
V1	3597 (~60 h)	No simulated damage	–
Baseline 2	9007 (~150 h)	No simulated damage	–
V2	3002 (~50 h)	No simulated damage	–
M5	5605 (~93 h)	5% of the total mass	L/10
M3c	5898 (~98 h)	3% of the total mass	L/2
M5c	9908 (~165 h)	5% of the total mass	L/2

been recorded in the same conditions as the baseline but they are not included in $[\phi]_{\mathrm{ref}}$ (see Sect. 3.2).

The first step of the analysis consists of fitting an AR (45) model to the data. Figure 7a shows a good correspondence between the original data and the result of the autoregressive fitting on a short time window. The point-by-point difference between the experimental and fitted data is the residual, which is given in Fig. 7b.

The quality of the fitting is also confirmed by the quantile–quantile (QQ) plot of the residuals in comparison with the standard normal distribution, reported in Fig. 8. It is possible to observe a good agreement with the standard normal distribution in the range of $\pm\ 2\sigma$.

By fitting the data with the AR (45) model, it is possible to represent the response of the structure with 45 coefficients every 60s: the time trend of the coefficients for tie-rod T1 is shown in Fig. 9.

The coefficients show a daily cyclical trend that should be related to the environmental and operational conditions [32]. Furthermore, the coefficients show a different scatter in different portions of the plot. This behaviour is mainly related to the variability of the excitation sources which are part of the input to the structure: transients, impulses or harmonic forces are often present during the day, due to the working activities taking place in the surroundings, while they are reduced or nulled at night (cyclical trends in Fig. 6b). Consequently, coefficients are more scattered during working hours rather than during night hours.

At this stage, this representation does not allow for the identification of different behaviours between data sets "M5", "M3c" and "M5c" and the baseline. For this purpose, the damage index DI can be evaluated calculating the MSD between each sample and the baseline set (composed by both "Baseline 1" and "Baseline 2"), as explained in 0.

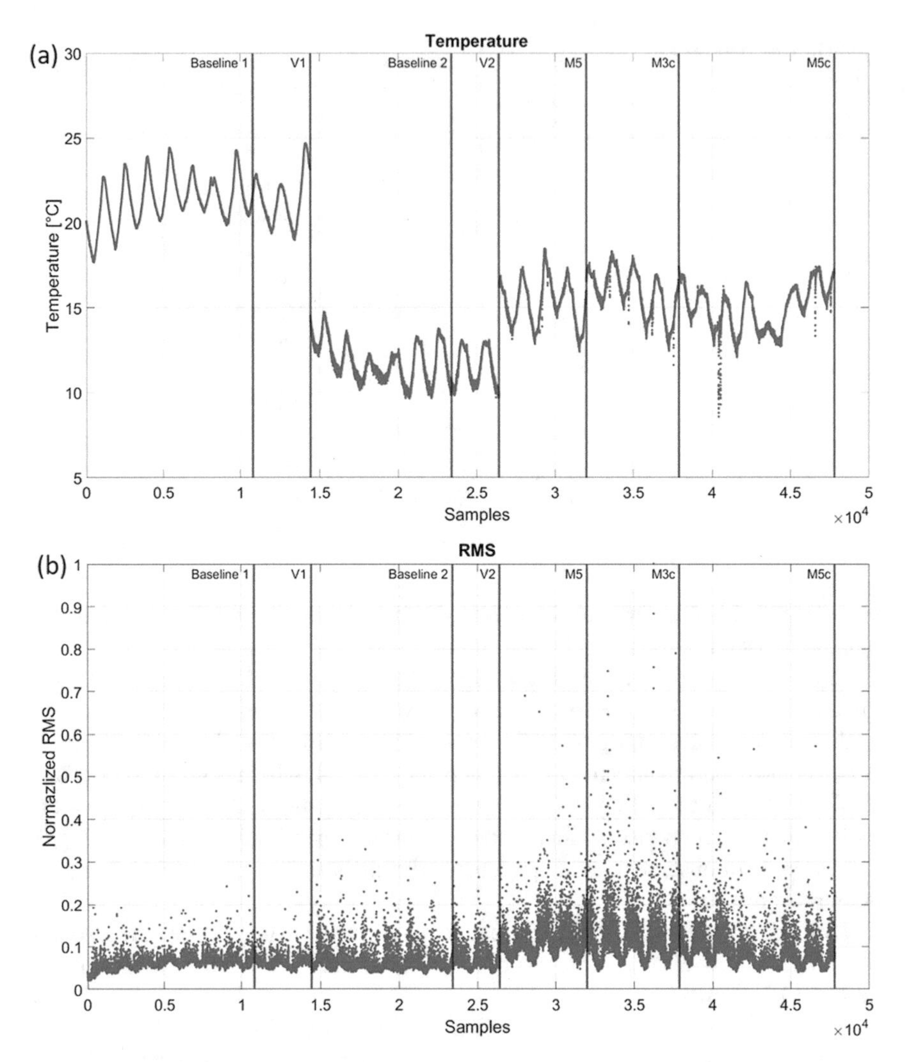

Fig. 6 Average temperature (**a**) and normalized RMS (**b**) for considered samples

In order to allow for a more clear representation, a moving average of 60 samples with overlap of 30 samples has been applied to each of the 45 autoregressive coefficients, before calculating the damage index: the result is shown in Fig. 10.

The vertical lines represent the limits between sets. As it is possible to see by comparing the red points (referring to T_1) and the blue points (referring to T_2), the method seems to be capable to detect the presence of damage. Indeed, even if both T_1 and T_2 are always subject to the same operational and environmental variations, the two trends are overlapped for data sets "Baseline 1", "V_1", "Baseline 1" and "V_2", while a separation can be observed in the samples referring to conditions "M5", "M3c" and "M5c".

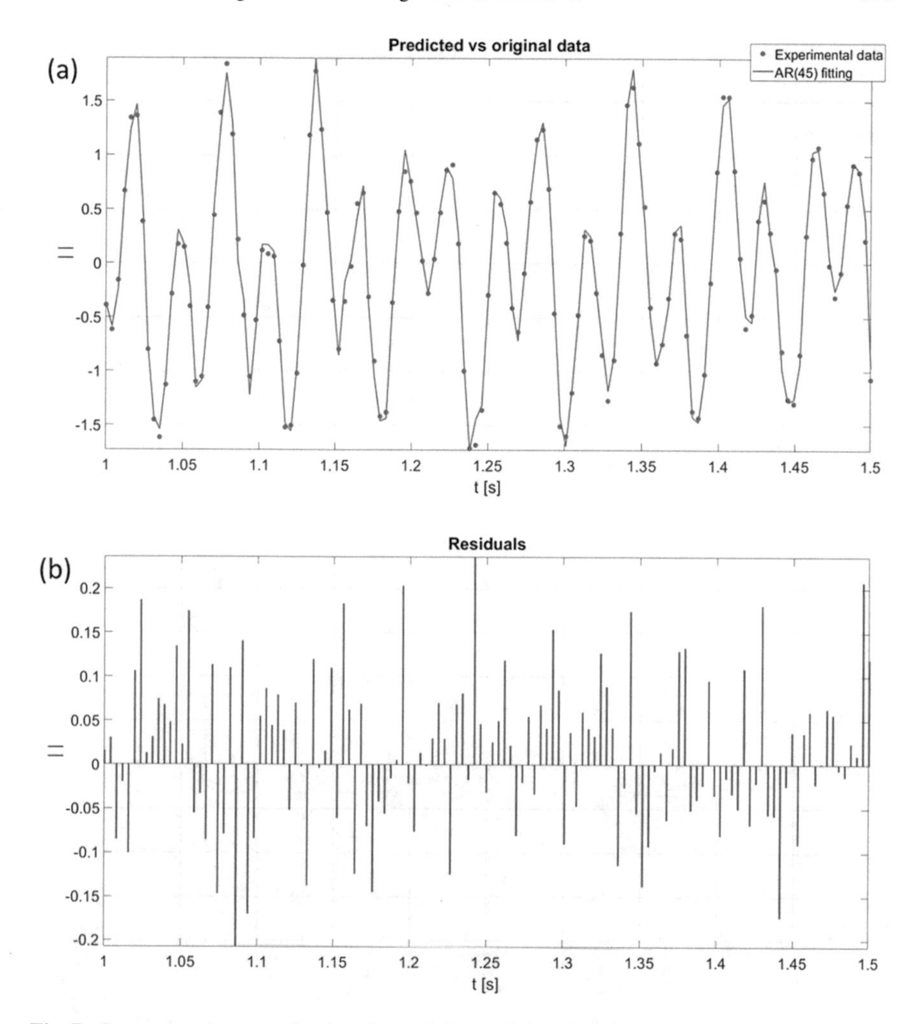

Fig. 7 Comparison between the experimental data and the AR fitting (**a**); residuals (**b**)

By only considering the points referring to T_1, it is possible to see that when the same mass is added to the structure, its effect on the damage index is dependent on the application position. Indeed, red points in data set "M5c" (mass added at the centre of the tie-rod) are higher than those of data set "m5" (mass added close to the end clamps). Furthermore, when the same position is taken into account, the damage index is sensitive to the amount of the simulated damage, as can be seen by comparing red points in "M5c" and "M3c" zones.

The threshold obtained by following the steps described in 0 is represented by a black dashed line in Fig. 10. In an automatic damage detection perspective, when the damage index is above the threshold for a given sample, an outlier is detected. In this sense, red points referring to conditions "M5", "M3c" and "M5c" are always

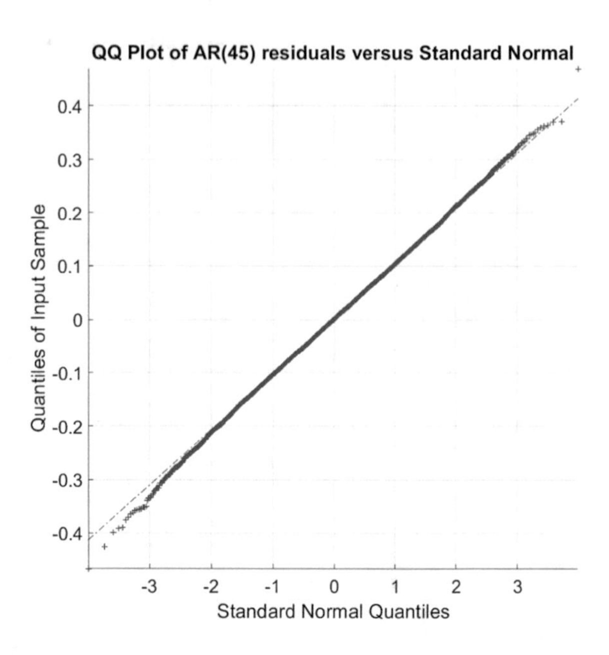

Fig. 8 QQ plot of the residuals in comparison with the standard normal distribution

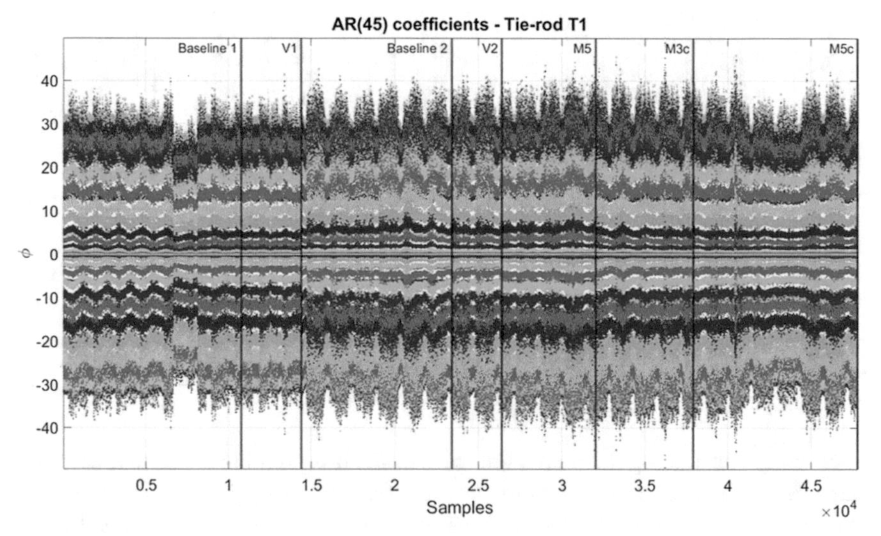

Fig. 9 Coefficients of the AR (45) model for tie-rod T_1

above the threshold; thus, no false negatives (associated with type II error, i.e., the tie-rod is damaged but the index is below the threshold) are observed.

On the contrary, some false positives (associated with type I error, i.e., the tie-rod is not damaged but the index is above the threshold) are already detected in validation data sets "V_1" and "V_2" for both tie-rods, and in data sets "M5", "M3c" and "M5c" for tie-rod T_2 with blue points that are scattered around the black dashed

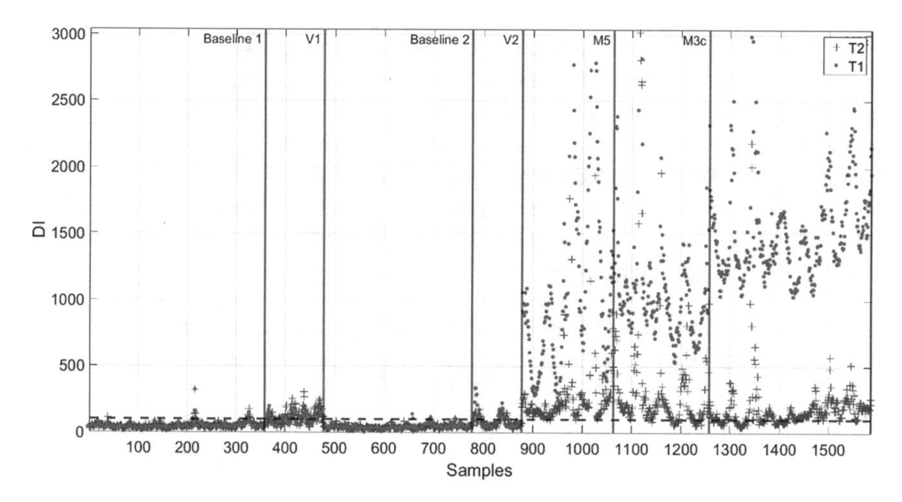

Fig. 10 Damage index calculated on AR coefficients filtered with a moving average of 60 samples with overlap of 30 samples

Table 3 Percentages of correct classifications, type I and type II errors for damage detection on T_1

Correct classifications	Type I error	Type II error
90.3%	41.1%	0%

line. Indeed, by looking at Fig. 6a and b, it is possible to notice that environmental and operational conditions of sets "M5", "M3c" and "M5c" are not totally included in those observed in "Baseline1″ and "Baseline2″. This observation suggests that the performances of the algorithm (which are reported in Table 3 in terms of percentages of correct classifications, type I and type II errors for T_1) can be improved by the adoption of a wider baseline that may include a full range of environmental and operational conditions.

5 Conclusions

This paper showed an application of a vibration-based data-driven approach to data collected in a realistic environment, under the presence of uncontrolled operational and environmental variations. A test-case consisting of two nominally identical tie-rods has been continuously monitored for a long time and damage has been simulated through the addition of masses. The environmental variations caused wide fluctuations of the axial load and consequently of dynamic properties of the structure. This aspect makes it hard to detect damage on modal parameters in a real application, where the tension is usually unknown. For this reason, a data-driven approach based on coefficients of an AR model and MSD has been adopted. The method showed to be potentially capable to separate effects of damage from those due to environmental variations, even if the performances showed a dependency on the range of environmental and operational variations observed in the baseline. Finally, the same analysis has been extended to all the sensors installed on the tie-rods. Similar performances have been observed for sensors placed at $L/2$ and $L/3$, while chances to achieve a good identification get lower for sensors close to the fixed ends, where signal-to-noise ratio is worse. A future development will be the introduction of real damage, to test for the sensitivity of the method.

Acknowledgements The research presented in this paper has been funded by the Italian National Research Program, in a project named "Life-long optimized structural assessment and proactive maintenance with pervasive sensing techniques" PRIN 2017.

Bibliography

1. Farrar CR, Worden K (2010) An introduction to structural health monitoring. CISM Int Cent Mech Sci Courses Lect 520:1–17. https://doi.org/10.1007/978-3-7091-0399-9_1
2. Farrar CR, Worden K (2013) Structural health monitoring: a machine learning perspective
3. Figueiredo E, Santos A (2018) Machine learning algorithms for damage detection. Vib Tech Damage Detect Localization Eng Struct 1–40. https://doi.org/10.1142/9781786344977_0001
4. Li H-N, Ren L, Jia Z-G, Yi T-H, Li D-S (2016) State-of-the-art in structural health monitoring of large and complex civil infrastructures. J Civ Struct Heal Monit 6:3–16. https://doi.org/10.1007/s13349-015-0108-9
5. Hou R, Xia Y (2021) Review on the new development of vibration-based damage identification for civil engineering structures: 2010–2019. J Sound Vib 491:115741. https://doi.org/10.1016/j.jsv.2020.115741
6. Doebling SW, Farrar CR, Prime MB (1998) A summary review of vibration-based damage identification methods. Shock Vib Dig 30:91–105. https://doi.org/10.1177/058310249803000201
7. Cox DD, Pandit SM (1993) Modal and spectrum analysis: data dependent systems in state space. https://doi.org/10.2307/1269677
8. Gul M, Necati Catbas F (2009) Statistical pattern recognition for structural health monitoring using time series modeling: theory and experimental verifications. Mech. Syst. Signal Process. 23:2192–2204. https://doi.org/10.1016/j.ymssp.2009.02.013
9. Roy K, Bhattacharya B, Ray-Chaudhuri S (2015) ARX model-based damage sensitive features for structural damage localization using output-only measurements. J Sound Vib 349:99–122. https://doi.org/10.1016/j.jsv.2015.03.038
10. Carden EP, Brownjohn JMW (2008) ARMA modelled time-series classification for structural health monitoring of civil infrastructure. Mech Syst Signal Process 22:295–314. https://doi.org/10.1016/j.ymssp.2007.07.003
11. Entezami A, Sarmadi H, Behkamal B, Mariani S (2020) Big data analytics and structural health monitoring: a statistical pattern recognition-based approach. Sensors (Switzerland). 20. https://doi.org/10.3390/s20082328
12. Figueiredo E, Park G, Farrar CR, Worden K, Figueiras J (n.d.) Machine learning algorithms for damage detection under operational and environmental variability, Struct. Heal. Monit
13. Ni YQ, Hua XG, Fan KQ, Ko JM (2005) Correlating modal properties with temperature using long-term monitoring data and support vector machine technique. Eng Struct 27:1762–1773. https://doi.org/10.1016/j.engstruct.2005.02.020

14. Clarkson BL (1981) Theory of vibration with applications. https://doi.org/10.1016/0022-460x(81)90474-0
15. Harris CM, Piersol AG (1962) Shock and vibration handbook. https://doi.org/10.1063/1.3058392
16. Lagomarsino S, Calderini C (2005) The dynamical identification of the tensile force in ancient tie-rods. Eng Struct 27:846–856. https://doi.org/10.1016/j.engstruct.2005.01.008
17. Campagnari S, Di Matteo F, Manzoni S, Scaccabarozzi M, Vanali M (2017) Estimation of axial load in tie-rods using experimental and operational modal analysis. J Vib Acoust Trans ASME 139. https://doi.org/10.1115/1.4036108
18. Rainieri C, Fabbrocino G (2015) Development and validation of an automated operational modal analysis algorithm for vibration-based monitoring and tensile load estimation. Mech Syst Signal Process 60:512–534. https://doi.org/10.1016/j.ymssp.2015.01.019
19. Ewins DJ (2001) Modal testing: theory, practice and application. 562. http://www.amazon.com/Modal-Testing-Application-Mechanical-Engineering/dp/0863802184
20. Pandit SM (1993) Time series and system analysis with applications
21. Datteo A, Busca G, Quattromani G, Cigada A (2018) On the use of AR models for SHM: a global sensitivity and uncertainty analysis framework. Reliab Eng Syst Saf 170:99–115. https://doi.org/10.1016/j.ress.2017.10.017
22. Cox DD, Pandit SM (1993) Modal and spectrum analysis: data dependent systems in state space. Technometrics 35:228. https://doi.org/10.2307/1269677
23. Fisher GW, Van Huffel S, Vandewalle J (1992) The total least squares problem: computational aspects and analysis. Math Comput 59:724. https://doi.org/10.2307/2153088
24. Figueiredo E, Figueiras J, Park G, Farrar CR, Worden K (2011) Influence of the autoregressive model order on damage detection. Comput Civ Infrastruct Eng 26:225–238. https://doi.org/10.1111/j.1467-8667.2010.00685.x
25. Ljung L (1998) System identification theory for the user
26. Yao R, Pakzad SN (2012) Autoregressive statistical pattern recognition algorithms for damage detection in civil structures. Mech Syst Signal Process 31:355–368. https://doi.org/10.1016/j.ymssp.2012.02.014
27. Deraemaeker A, Worden K (2018) A comparison of linear approaches to filter out environmental effects in structural health monitoring. Mech Syst Signal Process 105:1–15. https://doi.org/10.1016/j.ymssp.2017.11.045
28. Worden K, Manson G, Fieller NRJ (2000) Damage detection using outlier analysis. J Sound Vib 229:647–667. https://doi.org/10.1006/jsvi.1999.2514
29. Banerjee S, Ricci F, Monaco E, Mal A (2009) A wave propagation and vibration-based approach for damage identification in structural components. J Sound Vib 322:167–183. https://doi.org/10.1016/j.jsv.2008.11.010
30. Theodosiou A, Komodromos M, Kalli K (2018) Carbon cantilever beam health inspection using a polymer fiber Bragg Grating array. J. Light. Technol. 36:986–992. https://doi.org/10.1109/JLT.2017.2768414
31. Sakaris CS, Sakellariou JS, Fassois SD (2017) Random-vibration-based damage detection and precise localization on a lab–scale aircraft stabilizer structure via the generalized functional model based method. Struct Heal Monit 16:594–610. https://doi.org/10.1177/1475921717707903
32. Datteo A, Lucà F, Busca G (2017) Statistical pattern recognition approach for long-time monitoring of the G. Meazza stadium by means of AR models and PCA. Eng. Struct. 153:317–333. https://doi.org/10.1016/j.engstruct.2017.10.022

On Explicit and Implicit Procedures to Mitigate Environmental and Operational Variabilities in Data-Driven Structural Health Monitoring

David García Cava, Luis David Avendaño-Valencia, Artur Movsessian, Callum Roberts, and Dmitri Tcherniak

Abstract Vibration-based Structural Health Monitoring (VSHM) is becoming one of the most commonly used methods for damage diagnosis and long term monitoring. In data-driven VSHM methods, Damage Sensitive Features (DSFs) extracted from vibration responses are compared with reference models of the healthy state for long-term monitoring and damage identification of the structure of interest. However, data-driven VSHM faces a crucial problem - the DSFs are not only sensitive to damage but also to Environmental and Operational Variabilities (EOV). Machine learning and related methods, enabled by the availability of large monitoring datasets, can be used for mitigation of EOV in DSFs. EOV mitigation methods can be grouped into implicit and explicit methods. In the former, EOVs are compensated solely on the basis of the patterns identified in DSFs in the reference state. While the latter utilize measurements of the EOVs in addition to DSFs to build a cause-effect model, typically in the form of a regression. In this chapter, these two methods are discussed and illustrated via two different approaches: an artificial neural network for metric learning (implicit) and a multivariate nonlinear regression (explicit). The rationale and limitations of both methods are studied on an operating wind turbine

D. García Cava (✉) · A. Movsessian · C. Roberts
Institute for Infrastructure and Environment, School of Engineering, University of Edinburgh, Thomas Bayes Road, Edinburgh E9 3FG, UK
e-mail: david.garcia@ed.ac.uk

A. Movsessian
e-mail: artur.movsessian@ed.ac.uk

C. Roberts
e-mail: callum.roberts@ed.ac.uk

L. D. Avendaño-Valencia
Department of Mechanical and Electrical Engineering, University of Southern Denmark, Campusvej 55, 5230 Odense, Denmark
e-mail: ldav@sdu.dk

D. Tcherniak
HBK A/S, Skodsborgvej 307, 2850 Nærum, Denmark
e-mail: dmitri.tcherniak@hbkworld.com

A. Cury et al. (eds.), *Structural Health Monitoring Based on Data Science Techniques*, Structural Integrity 21, https://doi.org/10.1007/978-3-030-81716-9_15

where different damage scenarios were introduced. Additionally, the best practices of each procedure are presented through a comprehensive discussion of their potential advantages and drawbacks.

Keywords Damage detection · Environmental and operational variations · Artificial neural networks · Multivariate non-linear regression · Structural health monitoring

1 Introduction

In recent years, research on *Structural Health Monitoring* (SHM) seeks for solutions to close the loop between designing, manufacturing, building, and maintaining structures driven by continuous measurements of structural data. Integrated continuous monitoring systems allow learning from the past to decide in the present about the structural integrity, and to predict in the future the remaining useful life as well as improve new designs, all of this backed by experimental evidence [1]. *Vibration-based SHM* (VSHM) methods, namely those SHM methods employing measured structural vibration, have emerged as an attractive alternative with successful developments in the last decade across different fields of engineering. VSHM methods operate on the premise that any change in a structure, including damage, introduces a change in its vibratory behavior. These utilize *Damage Sensitive Features* (DSFs) extracted from vibration responses to monitor the status of the structure of interest [2]. This is done with the help of a baseline model, which provides a reference value for the DSFs. The baseline model can be derived either from the physics of the structure, from measured data, or a combination thereof. This chapter is focused on the data-driven VSHM approach, namely methods based solely on data measured from the structure. This does not mean that subsequent modeling is necessarily performed by disregarding the physics of the structure. Instead, physics often drives the selection of models and may also be used as constraints to build more realistic data-driven representations [1].

One of the main challenges in the development of VSHM methods relates to the sensitivity of DSFs to *Environmental and Operational Variabilities* (EOV), which hampers the algorithm's sensitivity to damage [3]. Indeed, damage in the structural components introduces local changes in its physical properties and/or boundary conditions. In turn, these changes also modify the structure's vibrational response, eventually permeating into the DSFs. However, other types of events unrelated to damage also influence the properties of the structure, introducing benign changes in DSFs, which are easily confused with damage. These events can be grouped into three main categories:

(1) *Variations imposed by the operational conditions of the structure*, which modify the way the structure operates over an extended period. Examples include set-point variables, such as the rotational speed of a rotor system [4], or the payload of a vehicle [5]. In this case, the structural dynamics exhibit

instantaneous changes that remain until the set-point is modified. Similarly, DSFs display changes of regime as the structure does. Often, the external variables related to this kind of variation are known or can be measured [6].

(2) *Variations imposed by transient events,* causing transient changes in the vibration response characteristics due to instantaneous or unexpected environmental and/or operational regimen variations [7]. This is the case of extreme meteorological conditions or earthquakes, which in short intervals introduce rare (and probably more complex) dynamic characteristics into the structural response. These events are infrequent and appear randomly due to the exposition of the structure to its natural environment. Likewise, the transition periods between operational regimes can also trigger such changes. During these transient events, DSFs will display outliers or small separate clusters.

(3) *Variations imposed by the sensitivity of the physical properties of the structure to environmental factors* [8]. For instance, the stiffness or damping properties of many materials are temperature-dependent [9]. Likewise, aerodynamic structures like wind turbine blades or bridge decks are characterized by damping properties which are a function of the incoming wind field characteristics [10]. In this case, the variations in the structural properties evolve smoothly over time, as the environmental conditions do. In turn, DSFs evidence smooth changes correlated with the external variables causing this behavior. Whereas in most cases the external variables can be measured, often noisy or partial measurements are available. For instance, in a large structure, a single temperature measurement may not be representative of the temperature gradients along the structure. Still, in those cases, it is possible to capture an overall tendency with a single or a small number of measurements.

Due to the significance of the problem of EOV in VSHM, a considerable part of the recent research efforts has been devoted in this direction. The main objective then is to minimize or control the effect of EOV in DSFs, while maximizing their sensitivity to damage. In a purely data-driven framework, enabled by the availability of large monitoring datasets, this objective can be accomplished with the help of machine and statistical learning techniques [11].

While the main scope of this chapter is not to give a complete overview of these techniques, our main objective is to postulate and analyze two fundamental methodological procedures exemplify them through respective salient techniques, and to provide a discussion on their applicability from a practical perspective. We hope that the reader can achieve a critical perspective of the different types of methods available for mitigation of EOV in damage detection problems, and so become capable of reaching opportune decisions for their own VSHM applications.

This document is organized as follows. Section 2 presents the problem statement that is discussed in this work. Section 3 introduces an approach where the mitigation of EOVs is done by DSF normalization, namely *explicit* procedure. Section 4 presents an alternative approach that uses a pattern recognition of the selected features as technique for EOVs mitigation, namely *implicit* procedure. Section 5 provides a description of the experimental campaign of the WTB monitoring system. Section 6 includes the results and discussions where both of the procedures are presented. Section 7 provides a comprehensive discussion and best practices for both methods, while Sect. 8 summarizes the main conclusions.

2 Problem Statement

Robust and reliable long-term monitoring is a must in data-driven VSHM systems. A data-driven VSHM method can be considered effective when the outputs correctly differentiate observations from undamaged and damaged states of the structure under consideration, regardless of the operational uncertainty. In practice, this is achieved by developing methodologies that consider, directly or indirectly (*i.e. explicitly* or *implicitly*), the influence of EOV in damage detection [12]. In the VSHM literature, these methods have been collectively referred to as *data normalization* methods [3, 13]. Based on this premise, the problem to be discussed in this chapter is to answer the following question: *How can data-driven VSHM methods mitigate the effect of the EOVs to build robust and reliable long-term monitoring systems?*

This challenge can be addressed by two different approaches considering (i) explicit and (ii) implicit procedures as it has been studied and investigated in recent years within the VSHM community (see Fig. 1).

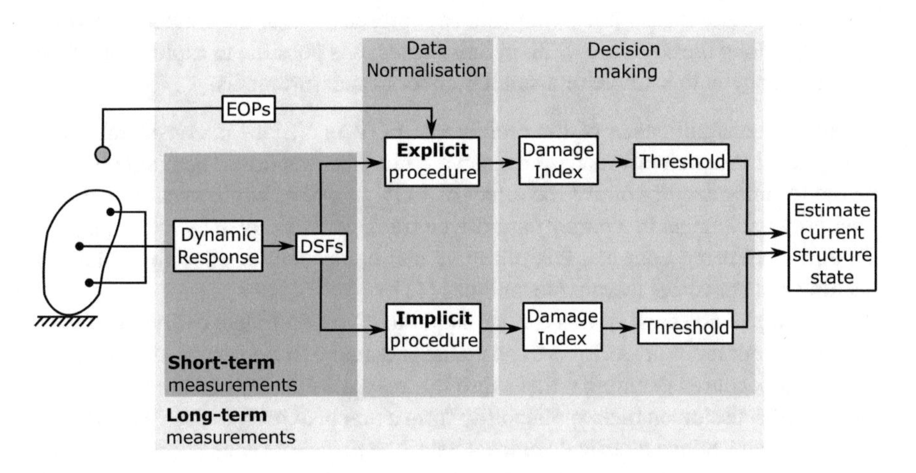

Fig. 1 Illustration of VSHM with integrated explicit and/or implicit procedure

2.1 Explicit Procedures

Explicit procedures rely on direct compensation of the trend observed in DSFs by reconstruction of the functional relationship with EOV through measurable *Environmental and Operational Parameters* (EOPs). Therefore, measurable EOPs are treated as input variables *–covariates–*, which, to a certain extent, explain the variability on DSFs. Explicit procedures are based on the fact that a *cause,* namely EOV, has an *effect* on the DSFs. Hence, these are also named *cause-effect* approaches [14]. Explicit procedures are especially effective in the case of variations originating from continuous EOPs (e.*g.* temperature or rotation speed). However, it cannot be guaranteed that the measured EOPs completely explain the variability in DSFs.

The main challenge on the application of explicit procedures is to find simple and informative manners to reconstruct these relationships. Here, the adjective informative refers to the property of a model to explain the root causes for a change in the monitored variables. Reconstruction of the underlying relationship between EOPs and DSFs can be achieved via deterministic and/or stochastic functional dependence models solved via various regression procedures. Both modeling approaches are discussed in further detail below and illustrated in Fig. 2.

Deterministic models attempt at capturing a deterministic trend in DSFs by selection of a function family, which is subsequently fit to the available observations. The model prediction error or likelihood are then optimized to achieve the best fit based on the selected function family [15]. These models are simpler to optimize and can be used for predictions compared to the stochastic models explained in the sequel. Examples include classical linear regression and different types of nonlinear regression, such as polynomial regression, neural networks, and regression trees [7, 12, 16–19]. These methods, though, are also sensitive to outliers and prone to overfitting, and may not be the most suitable for compensation of DSFs in practical applications. Regularization methods, where constraints to the size of the regression coefficients are enforced, can help alleviate these problems. Ultimately, a trade-off between the minimization of the estimation error and the compliance to the coefficient restrictions must be achieved, aided by an adjustment parameter selected by the user. Popular regularized regression schemes include ridge and lasso regressions, elastic net regularization (a combination of ridge and lasso), and dilution/drop-out in neural networks [20, Sec. 3.4]. *Support Vector Regression* (SVR) stands out as one of the most powerful methods in the deterministic class, which combines ridge regularization with kernel methods to produce a method robust to outliers and overfitting [21].

Stochastic models, beyond representing the trends in the output data, also aim at capturing the uncertainty caused by noise and model misspecification (choosing an incorrect model for the data). Many of these methods may be easily described in a Bayesian framework, where the probability of a model given observations –the *posterior*– is the quantity used to obtain a regression model [14, 22]. In turn, the posterior probability is proportional to the product of the data-fit probability –the *likelihood*– times a probability describing our original beliefs on the model –the

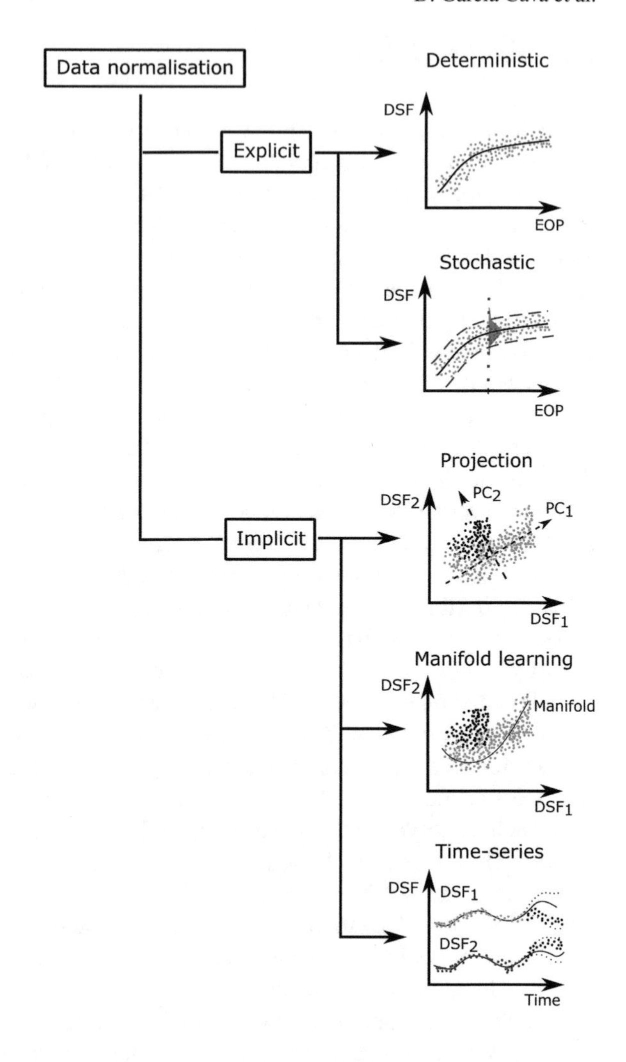

Fig. 2 Graphical representation of different data normalization methods in VSHM

prior–. As may be inferred, Bayesian regression schemes operate on a similar principle as regularization in the deterministic case, with terms penalizing modeling errors and constraining the model coefficients. In fact, in some limit cases, Bayesian regression is identical to ridge or lasso regression [20, pp. 69–73]. Nonetheless, the main difference of the Bayesian framework with traditional regularization schemes, is that it provides analytical procedures to adjust the regularization parameters, which appears as *hyperparameters* in the Bayesian context [22]. Such procedures include non-informative priors, and hierarchical and empirical Bayes methods. After hyperparameter adjustment, predictions can be calculated with the help of the *posterior predictive distribution*, which has the property of summarizing the complete uncertainty due to the model choices given the available training data. Likewise, the Bayesian framework facilitates the comparison of different competing hypotheses, which is useful for model selection but also for damage diagnosis [22]. Perhaps the most recognized method in this group corresponds to the *Relevance Vector Machine* (RVM) [23], which may be deemed as a probabilistic interpretation of SVR, while equally powerful versions of Bayesian nonlinear regressions are also available. In addition, *Gaussian Process Regressions* (GPR) comprise a complete probabilistic perspective into the regression problem, where predictions are calculated on the basis of their conditional probability given the training data, under the assumption of a joint multivariate normal probability distribution.

Another important type of stochastic model is formed by ensemble methods, where predictions are based on a pool of models obtained after adding small variations in the training data. These variations introduce some sort of randomness into the obtained models, which ultimately is reflected in the model predictions. For instance, *bagging*, or *bootstrap aggregating*, involves training a pool of models after randomly sampling different training subsets from the complete training data set. Boosting and stacking are other methods used to reach a similar effect. Models with small variations, which produce a variety of different predictions, comprise the resulting model pool. The ensemble prediction then corresponds to some sort of consensus achieved among the individual model predictions. In addition, the variations in the individual predictions also provide a measure of the uncertainty in the ensemble prediction, in a similar way as the posterior predictive distribution does for Bayesian regressions.

2.2 Implicit Procedures

Implicit procedures rely on capturing the behavior of the DSFs in a reference state of the structure, by recognizing all its different possible patterns within the target state under EOV. Once the patterns are learnt, new observations are assessed by their similarity to the reference patterns. These procedures are performed exclusively on the basis of DSFs, and thus are referred to as *effect-only* approaches. In turn, implicit procedures may follow any of the ensuing philosophies:

Projection methods. In these methods, DSFs are projected into a space where the influence of EOV might be removed or mitigated to some extent [24]. Factor analysis, *Principal Component Analysis* (PCA), Singular Spectrum Analysis and their non-linear counterparts are examples of these methods [25–29]. Upon transformation, the obtained variables are sorted according to their sensitivity to EOV. In PCA, the principal components with larger principal values are presumed to be the most affected by EOV, and thus are removed. However, there are not specific criteria that can help guiding the selection of the number of features to remove. More importantly, since these projections are calculated in an unsupervised or semi-supervised manner (with only information about the healthy state), it is difficult to guarantee the separability of EOV from damage effects in the transformed space. Therefore, it is possible that the variables judged as EOV sensitive may also carry information on damage characteristics. Hence, while these methods may reduce the sensitivity to EOV, it may also significantly diminish the damage detection accuracy.

Time-series methods. In these methods, DSFs are treated as time series, whose structure are modeled to capture non-stationary temporal patterns (trends or seasonality) taking part of the normal state. Cointegration analysis is one of the best known method for this task. It is based on the representation of DSFs as a modified vector autoregressive model referred to as *Vector Error Correction Model* (VECM). Some of the VECM properties indicate the presence of trends or seasonality patterns in the time-series data, and upon detection, the VECM can be used to project out those patterns [30, 31]. As a result, a stationary residual time series is obtained, which is expected to be free of the influence of EOV. Extensions of the original cointegration analysis algorithm have been proposed to deal with heteroscedastic variables (variables with time dependent variance) [32], and changes of regime in the DSFs [33].

Manifold learning methods. As explained in the previous section, DSFs hold functional dependencies with external variables associated with EOV. Explicit methods attempt at capturing these dependencies with the aid of measurable EOPs. Manifold learning methods, on the other hand, try to do so without the availability of EOPs and instead attempt at reconstructing the manifold in which the DSFs are embedded, solely on their multivariate characteristics. Although numerous methods can be applied for this purpose, some of the most recognizable in the VSHM context are non-linear PCA in its different versions (locally linear embeddings [34], kernel PCA [35], ANN [36], deep NN [37, 38]), and different clustering methods [39–41]. Metric learning methods follow a similar philosophy, but instead of directly learning the underlying manifold, metric learning methods attempt at capturing the patterns of the distances measured in the manifold [8, 36].

3 Multivariate Nonlinear Approach as an Explicit Procedure

3.1 Main Definition

The essence of explicit procedures consists of capturing the variations in DSFs with the help of measured EOPs. In this section, the explicit approach is illustrated by means of a multivariate nonlinear regression, which represents the DSF vector $\boldsymbol{\alpha}_p$ as a deterministic function of EOPs as shown next:

$$\boldsymbol{\alpha}_p = \mathbf{W}^T \cdot \boldsymbol{f}(\boldsymbol{\xi}_p) + \boldsymbol{u}_p, \quad \boldsymbol{u}_p \sim \mathcal{N}(0, \boldsymbol{\Sigma}_u) \tag{1}$$

$$\boldsymbol{f}(\boldsymbol{\xi}_p) = \boldsymbol{f}_1(\xi_{p,1}) \otimes \boldsymbol{f}_2(\xi_{p,2}) \otimes \ldots \otimes \boldsymbol{f}_l(\xi_{p,l}) \otimes \ldots \otimes \boldsymbol{f}_L(\xi_{p,L}) \tag{2}$$

where $\mathbf{W} \in \mathbb{R}^{L\times K}$ is the coefficient matrix of the regression model, $\boldsymbol{f}(\boldsymbol{\xi}_p)$ is a multivariate functional basis of the EOPs constructed by Kronecker products of univariate basis functions according to Eq. (2), and $\boldsymbol{u}_p$ is a zero mean normal and independently distributed random vector with covariance $\boldsymbol{\Sigma}_u$, which accounts for the variability not attributable to EOVs. In Eq. (2), $\xi_{p,l}$ represents the individual l-EOP affecting to the p-observation, so that $\boldsymbol{\xi}_p = \left[\xi_{p,1} \cdots \xi_{p,L}\right] \in \mathbb{R}^L$ denotes a vector with all the EOPs, and $\boldsymbol{f}_l(\cdot)$ represents the functional basis associated to the l-EOP with length determined by product of the number of the basis assigned to each EOP.

Given a set of input EOPs $\mathbf{X} = \left[\boldsymbol{\xi}_1\ \boldsymbol{\xi}_2 \ldots \boldsymbol{\xi}_p \ldots \boldsymbol{\xi}_{P_T}\right] \in \mathbb{R}^{L\times P_T}$ and their corresponding DSFs $\mathbf{A} = \left[\boldsymbol{\alpha}_1\ \boldsymbol{\alpha}_2 \ldots \boldsymbol{\alpha}_p \ldots \boldsymbol{\alpha}_{P_T}\right] \in \mathbb{R}^{K\times P_T}$, a standard least squares method is used to estimate the coefficient matrix of the regression model and the innovation covariance matrix, as follows:

$$\hat{\mathbf{W}} = \left[\mathbf{F}(\mathbf{X})\mathbf{F}^T(\mathbf{X})\right]^{-1}\mathbf{F}(\mathbf{X})\mathbf{A}^T \tag{3}$$

$$\hat{\Sigma}_u = \frac{1}{P_T}\left[\mathbf{A} - \left(\hat{\mathbf{W}}^T\mathbf{F}(\mathbf{X})\right)\right]\left[\mathbf{A} - \left(\hat{\mathbf{W}}^T\mathbf{F}(\mathbf{X})\right)\right]^T \tag{4}$$

where $\mathbf{F}(\mathbf{X}) = \left[\boldsymbol{f}(\xi_1)\ \boldsymbol{f}(\xi_2) \ldots \boldsymbol{f}(\xi_p) \ldots \boldsymbol{f}(\xi_{P_T})\right] \in \mathbb{R}^{L\times P_T}$ is the overall training multivariate functional basis and $(\cdot)^{\mathrm{T}}$ is the transpose.

To maximize the generalization performance and avoid overfitting, a *Leave-One-Out Cross Validation* (LOOCV) method is used to assess different model structures. The LOOCV method uses a single observation from the training set to calculate the prediction error (validation), while the remaining ones are used for model estimation (training). The process is repeated until all the instances in the training set are used for validation. In the case of the multivariate non-linear regression, the LOOCV error can be calculated efficiently with the help of the *hat* matrix, based on a single calculation on the complete training set, as explained in [20, Sec.7.10].

The compensated DSF vector. An estimate of the DSFs can be obtained based on the obtained estimates of the coefficient matrix. Then, it is possible to obtain a corrected DSF $\tilde{\boldsymbol{\alpha}}_p$ by subtracting the estimated DSF vector $\hat{\boldsymbol{\alpha}}_p = \hat{\mathbf{W}}^T \cdot \boldsymbol{f}(\boldsymbol{\xi})$ from the original DSF $\boldsymbol{\alpha}_p$ as shown in Eq. (5).

$$\tilde{\boldsymbol{\alpha}}_p = \boldsymbol{\alpha}_p - \hat{\boldsymbol{\alpha}}_p = \boldsymbol{\alpha}_p - \hat{\mathbf{W}}^T \cdot \boldsymbol{f}(\boldsymbol{\xi}) \tag{5}$$

3.2 Outlier Analysis

The corrected DSFs are used to assess the current state of the structure by a novelty index method obtained with the squared Mahalanobis distance (MD):

$$d^2\left(\tilde{\boldsymbol{\alpha}}_p, \hat{\Sigma}_u\right) = \tilde{\boldsymbol{\alpha}}_p^T \hat{\Sigma}_u^{-1} \tilde{\boldsymbol{\alpha}}_p \tag{6}$$

where $\tilde{\boldsymbol{\alpha}}_p$ is the corrected DSF vector observation from Eq. (5) and $\hat{\Sigma}_u$ corresponds to the covariance matrix estimate in Eq. (4). A basic damage diagnosis method can be made by comparing the above defined MD with a threshold $\vartheta > 0$. A new observation is then classified according to:

$$\mathbf{H}_1 : d^2\left(\tilde{\boldsymbol{\alpha}}_p, \hat{\Sigma}_u\right) \leq \vartheta \rightarrow \text{Healthy} \tag{7a}$$

$$\mathbf{H}_2 : d^2\left(\tilde{\boldsymbol{\alpha}}_p, \hat{\Sigma}_u\right) > \vartheta \rightarrow \text{Damaged} \tag{7b}$$

4 Metric Learning Approach as an Implicit Procedure

4.1 Main Definition

Due to their universal function approximation capabilities, ANNs have been frequently considered for SHM and damage detection applications. Here, ANNs are used to reconstruct the relationship between DSFs and a MD-based novelty index, thus corresponding to a metric learning configuration. To start with, an MD-based novelty index is computed from the raw DSFs, as follows:

$$\mathrm{d}^2(\boldsymbol{f}_p, \mathbf{F}) = (\boldsymbol{f}_p - \boldsymbol{\mu}_\mathrm{F})^T \boldsymbol{\Sigma}_\mathrm{F}^{-1} (\boldsymbol{f}_p - \boldsymbol{\mu}_\mathrm{F}) \tag{8}$$

where $\boldsymbol{f}_p$ is the DSF vector, with mean $\boldsymbol{\mu}_\mathrm{F}$ and covariance $\boldsymbol{\Sigma}_\mathrm{F}$.

The aim is to identify a variability pattern in the MD-based novelty index after the novelty index is computed. For this, a two-layer feedforward ANN model is adopted. This structure has been identified to approximate nonlinear relationships in several studies [42]. The 2-layer feedforward ANN architecture is defined as follows:

$$\mathrm{g}^2(f_p, \mathbf{Z}) = \sigma\left(\sum_{l=1}^{L} z_{pl}^{(2)} h\left(\sum_{r=1}^{R} z_{lr}^{(1)} f_r^p + z_{r0}^{(1)}\right) + z_{p0}^{(2)}\right) \tag{9}$$

where $\mathrm{g}^2(\boldsymbol{f}_p, \mathbf{Z})$ represents the estimated MD for DSFs $\boldsymbol{f}_p$. The output layer consists of the weights $z_{pl}^{(2)}$ and bias $z_{p0}^{(2)}$, while $z_{lr}^{(1)}$ and $z_{r0}^{(1)}$ indicate the weights and the bias for the hidden layer. A hyperbolic sigmoid activation function $\sigma(\cdot)$ is applied in the hidden layer and a linear identity transfer function $h(\cdot)$ for the output layer. The weights $\mathbf{Z}$ are trained by minimizing the least-square solution for E_D which is defined as follow:

$$E_D = \frac{1}{P_T} \sum_{p=1}^{P_T} \left(\mathrm{d}^2(\boldsymbol{f}_p, \boldsymbol{F}) - \mathrm{g}^2(\boldsymbol{f}_p, \mathbf{Z})\right)^2 \tag{10}$$

where E_D is the mean squared error between the calculated and the predicted ANN estimate of the MD-based novelty index.

4.2 *Outlier Analysis*

The ANN is trained on observations from the healthy structure (i.e. P_T observations), subsequently, damage detection is pursued upon calculation of a new novelty index. The proposed ANN-based novelty index is calculated by the relationship between MD $\mathrm{d}^2(\boldsymbol{f}_p, \mathbf{F})$ and the estimated MD $\mathrm{g}^2(\boldsymbol{f}_p, \mathbf{Z})$ by the ANN and is defined as follows:

$$\hat{d}_p^2 = \left| \log \frac{\mathrm{g}^2(\boldsymbol{f}_p, \mathbf{Z}) + \varepsilon}{\mathrm{d}^2(\boldsymbol{f}_p, \mathbf{F}) + \varepsilon} \right| \tag{11}$$

where ε is a small scalar value introduced to avoid mathematical inconsistency. In this study $\varepsilon = 10^{-2}$. The ANN learns a unique relationship between the DSFs and the MD, thus if DSFs are obtained from a structure with damage the prediction error is expected to increase. Then, a new observation is classified according to the hypotheses test outlined below.

$$\mathbf{H}_1 : \hat{d}_p^2 \leq \theta \rightarrow \text{Healthy} \tag{12a}$$

$$\mathbf{H}_2 : \hat{d}_p^2 > \theta \rightarrow \text{Damaged} \tag{12b}$$

5 Experimental Campaign of Wind Turbine Blade Monitoring

5.1 *The Test Rig, Data Collection and Artificial Damage*

The presented study is based on the data collected during the measurement campaign conducted on a Vestas V27 wind turbine during winter 2014–2015. Though its design is relatively old, this 225 kW rated power, 27 m rotor diameter, upwind, pitch regulated, horizontal axis wind turbine is representative of many modern wind turbines. In contrast to modern wind turbines, its blades are relatively stiff, and it has only two speed-regimes: 32 and 43 rpm.

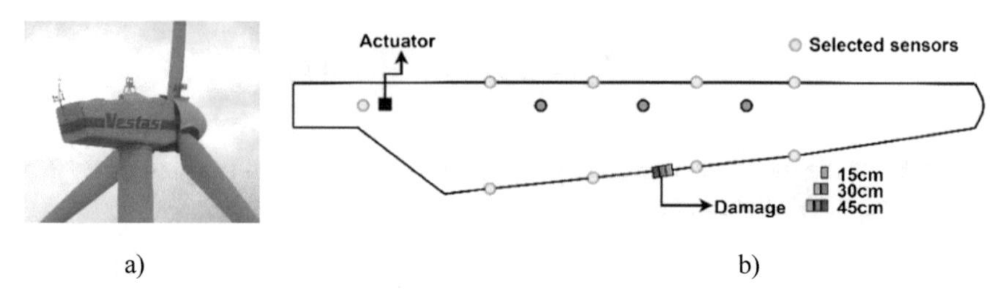

Fig. 3 a) Picture of the V27 wind turbine with the measurement equipment. b) Schematic representation of the instrumented wind turbine blade

The vibration data was collected from a single instrumented blade, which was excited by an electro-mechanical actuator. The actuator was mounted on the outer surface of the blade, about 1 m from the blade root and resembles an automatic hammer that, due to an electrical pulse, impacts the blade. The actuator was programmed to impact the blade every 5 min. The vibrational response of the blade was picked up by an array of 11 monoaxial accelerometers mounted in three rows along the blade, about 2 m from each other (Fig. 3). The synchronously sampled vibration signals were collected by the data acquisition system located inside the rotor spinner. Simultaneously, the data from the pitch sensor and two DC accelerometers were recorded. The latter allowed us to estimate the rotor azimuth angle and rotor rotational speed. From the data acquisition system, the data was transmitted wirelessly to a measurement computer located inside the wind turbine tower and stored in its hard drive as raw data time series.

Simultaneously (but not synchronously), the weather data (temperature, wind speed, wind direction, etc.) from the nearby weather mast was collected.

The measurement campaign started on Nov. 28, 2014, and finished on Mar. 12, 2015. During the campaign, the wind turbine was governed by its control system, and it was in one of its three operational modes: idle, 32 rpm, and 43 rpm. During the 104 days of the campaign, the structure of the instrumented blade was taken to five different states including: the intact blade, three artificially induced damages in the form of an opening of the trailing edge of 15, 30 and 45 cm, and the repaired blade. A more detailed description of the test campaign can be found in [22].

Data set. To illustrate how the selected explicit and implicit methods operate in a simple and comprehensive form, data corresponding only to the 43 rpm operational condition is considered. Additionally, the repaired state of the blade is selected as the reference (healthy) state, due to the longer data availability and further exposition to EOV. Also, only the accelerometers located on the leading and trailing edges are considered, leading to a number of $M = 8$ sensors. The number of observations is detailed in Table 1. The observations measured on the repaired blade (considered as healthy) are randomly divided into two groups. Then, 3456 observations are used to build the reference state (training), while 900 observations are used to evaluate the model performance (validation). Although the data set is the same in both methods,

Table 1 Number of observations on each scenario

Healthy*		Damage		
Training	*Testing*	15 cm	30 cm	45 cm
3456	900	258	194	254

*Healthy data set is built from the observations on the repaired state

the DSFs extracted from the vibration responses are different. This is explained in the sections below.

6 Results and Discussions

6.1 Application Example with an Explicit Procedure

In this section, the results of the application of an explicit procedure implemented via multivariate nonlinear approach described in Sect. 3 are presented. The vibration responses are subsequently transformed to the frequency domain by use of the discrete Fourier transform. The real and imaginary parts are concatenated in a large vector per accelerometer response. All the observations used for the training data set (see Table 1) are considered to construct a large matrix with all observations from the same accelerometer. As the vibration responses are high dimensional and highly correlated, PCA is applied to reduce the dimension of this large matrix. Finally, a DSF, $\boldsymbol{\alpha}_p$, is constructed by concatenating all the vibration responses from the accelerometers under consideration. For more detail about the DSF see [43].

Modelling the DSF dependencies with the EOPs. In this analysis only 5 EOPs were considered; temperature (°C), wind speed (ms^{-1}), azimuth angle (°), and standard deviation and maximum amplitude (ms^{-2}) from the vibration response on the accelerometer closest to the actuator. Hermite polynomials were used to build the relationships. Moreover, each EOP was evaluated from 1–5 model orders with every combination across all the 5 EOPs, so that a total of 5^5 combinations were tested. The one with the lowest LOOCV error was chosen to be the model for the corresponding feature. Figure 4 displays feature α_1 as a function of temperature (Fig. 4a) and feature α_3 as a function of the standard deviation of the actuator signal (Fig. 4b) before and after compensation from the obtained multivariate nonlinear regression model. From Fig. 4a it can be concluded that α_1 has a substantial dependency on the temperature. This can be expected as the first PCA variable represents the largest variance component. Therefore, it is expected to be highly sensitive to temperature, which in turn has a significant influence on the blade stiffness. Another example can be seen when modelling the dependence of the actuator hit and feature α_3. This is visualized in Fig. 4b where the actuator response standard deviation correlates with the variations in α_3. The regression models are created, with each individual EOP and then all combined, to model the dependencies of each individual feature variable of

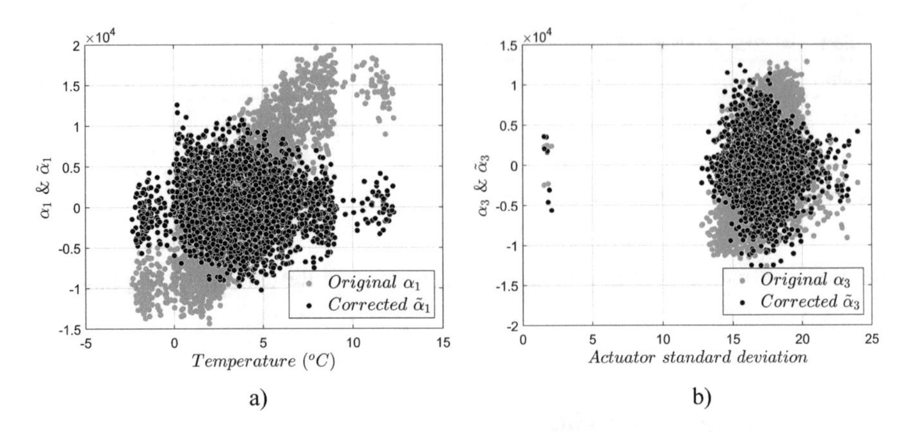

Fig. 4 Explicit method. Distribution of the feature variable a) Original α_1 and corrected $\tilde{\alpha}_1$ against temperature, b) Original α_3 and corrected $\tilde{\alpha}_3$ against actuator standard deviation

the DSF vector to all the EOPs. Once the multivariate regression models are defined, DSF vectors are easily estimated by projecting the EOPs data as shown in Eq. (5). These new estimated DSFs describe the expected dependencies with their corresponding EOP measurements. To this end, corrected DSF vectors are then calculated by subtracting the estimated DSF from the actual DSF vectors. Thus, now the new DSFs are corrected upon consideration of the EOP influence. This can be observed in Fig. 4a where the corrected feature variable $\tilde{\alpha}_1$ seems to be released from its temperature dependency. It can clearly be seen that the trend on α_1 previously observed in Fig. 4a is now removed in the new feature. Additionally, this correction can be also observed on $\tilde{\alpha}_3$ where the effects of the actuator hit is now corrected (see Fig. 4b).

Assessing the damage detectability. A comparison between the damage detectability obtained when using the original features $\boldsymbol{\alpha}_p$ (before compensation) and the corrected DSF vectors $\tilde{\boldsymbol{\alpha}}_p$ is now presented. This is conducted by implementing the outlier analysis, described in Sect. 3.2, on the DSF vectors obtained before and after compensation. The damage detection threshold is defined as the value where 98% of the training data (i.e. $P_T = 3456$ observations) lies according to a chi-squared distribution with degrees of freedom equal to the length of the corresponding DSF vector. Figure 5 presents the obtained control charts in both DSF vectors (i.e., before and after the compensation) based only on the first PC from each sensor. Therefore, the dimension of the DSF vectors is $K = 8$ with (i.e. $N_r = 1$ and $M = 8$). In both cases, the damages at different levels are detectable. However, several observations remain undetected when the uncorrected DSF vectors are used (see Fig. 5a).

Otherwise, after correction of the DSFs, an increase in the distance between the healthy and damaged cases is observed, while all the damaged cases appear outside of the selected threshold. Nonetheless, the false positive rate slightly increases within the testing set after compensation.

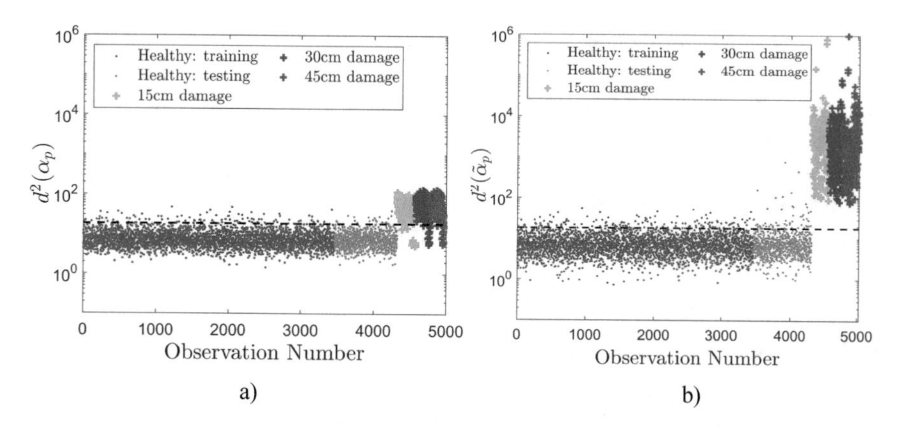

Fig. 5 Explicit method. Damage detection control chart for the wind turbine blade with only the first principal vector retained ($K = 8$) for a) non-corrected DSF vectors $\boldsymbol{\alpha}_p$ and b) corrected DSF vectors $\widetilde{\boldsymbol{\alpha}}_p$

A similar analysis is conducted using the first three principal vectors on each sensor (i.e. $K = 24$ with $N_r = 3$ and $M = 8$). Before correction (Fig. 6a), some of the damaged cases remain undetected, while after correction (Fig. 6b), all the damaged observations are clearly detected on expense of a small increment of false positives. Comparing the results obtained with $K = 8$ (Fig. 5b) and $K = 24$ (Fig. 6b), it is observed that all damages can be successfully detected in both cases, while a slight increment of false positives is observed when more principal vectors are retained ($K = 24$). In addition, it is evident that in the present case the first principal component is critical towards damage detectability.

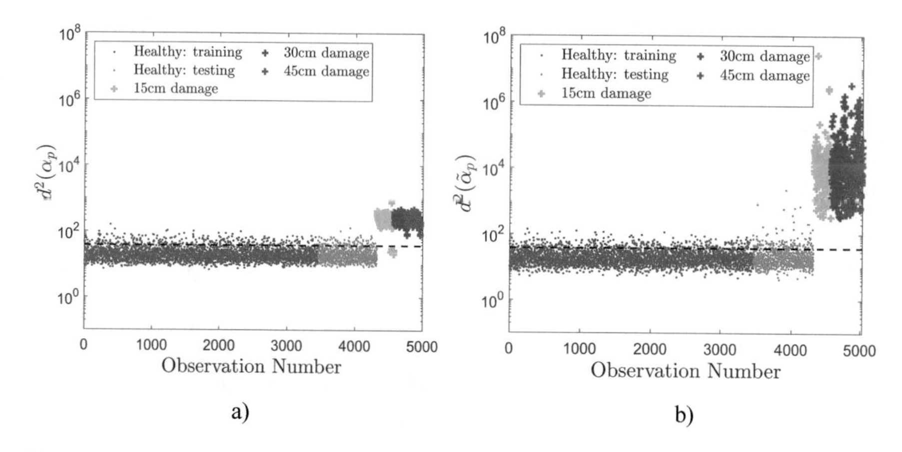

Fig. 6 Explicit method. Damage detection control chart for the wind turbine blade with only the first principal vector retained ($K = 24$) for a) non-corrected DSF vectors $\boldsymbol{\alpha}_p$ and b) corrected DSF vectors $\widetilde{\boldsymbol{\alpha}}_p$

Moreover, for both analyses, it could be seen that when the DSF vectors were compensated, some observations from the testing set move above the threshold, thus increasing the number of false positives. One reason for this could be that the EOVs experienced for these observations did not exist in the training set and, as such, are not modelled well in the multivariate regression model. However, as the distance between the damage and healthy increases, it opens the possibility to raise the threshold and reduce the number of outliers without affecting the detection rate of the damage.

6.2 Application Example with an Implicit Procedure

In this section, the results of the application of an implicit procedure implemented via the metric learning approach described in Sect. 4 are presented. As in the application of the explicit method, the data is split according to Table 1. This includes using 80% of the data for training and defining the threshold for outliers. The DSF vector in this application example relates the correlation matrix calculated among the considered accelerometers. The DSF vector $\boldsymbol{f}_p$ is then formed by organizing the elements of the upper diagonal of the acceleration correlation matrix into a single column vector. For more details about the DSF see [44].

The novelty index obtained by calculating the MD defined in Eq. (8), with mean and covariance calculated on the elements of the training set ($P_T = 3456$) is observed in Fig. 7. The threshold corresponding to a 2% false alarm rate is also displayed. As the DSFs are not pre-processed to mitigate the effects of EOPs, a large number of instances from the healthy state of the blade, both in the training and validation sets, are overlapping with the instances from the damaged state.

First, the novelty index when solely using the MD with $P_T = 3456$ is observed in Fig. 7. As the DSFs are not pre-processed to mitigate the effects of EOPs, the majority of the 2% allowed false alarms are overlapping with the 15 cm damage.

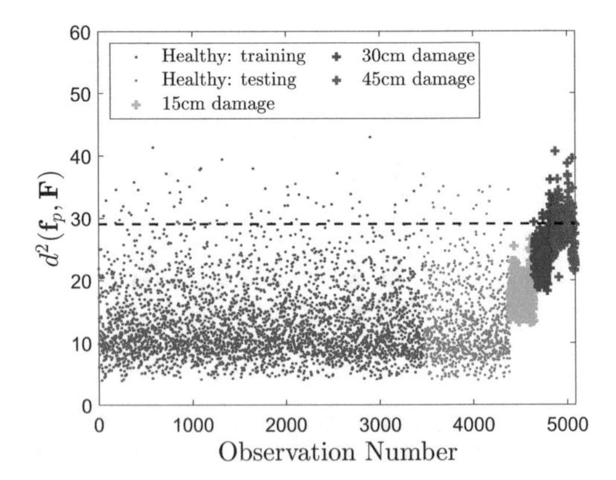

Fig. 7 Implicit method. Damage detection control chart for the wind turbine blade based on the MD of the correlation-based DSF vectors

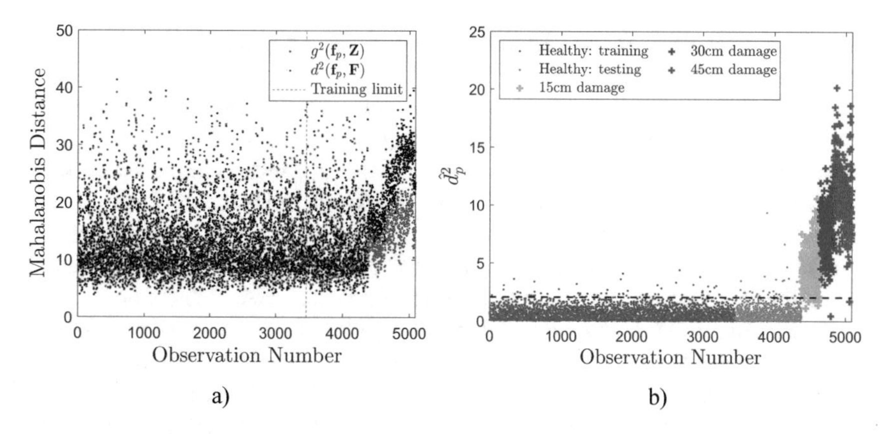

Fig. 8 Implicit method. a) Comparison of the calculated and ANN-predicted MDs of the correlation-based DSF vector. b) Damage detection control chart for the wind turbine blade based on the predicted error between the calculated and ANN-predicted MDs

Building the reference state. The ANN regression model presented in Eq. (9) is trained with $P_T = 3456$ DSFs and the corresponding $g^2(\boldsymbol{f}_p, \mathbf{Z})$ prediction target. To calculate the MD-based novelty index as the prediction target for the ANN in Eq. (9), the number of observations is reduced to 2000. The reduced number hinders pre-normalization of the novelty index and allows the ANN to learn the existing patterns. Figure 8a shows the computed and predicted MD-based novelty index. It can be observed that the model can effectively predict new observations in the validation set from the healthy state of the structure. Moreover, it is evident that the prediction error increases after the onset of damage.

The new novelty index presently corresponds to the prediction error calculated by Eq. (11), which is displayed in Fig. 8b. It is noticed that the distance between the healthy and damaged instances has increased considerably, effectively reducing the number of false positives. However, the number of missed alarms (undetected damage instances) has also increased. This can happen when the DSFs from the damaged state follow a similar pattern to that of the healthy state. This indicates that there is an overlap between the manifolds of the healthy and damaged states of the structure. As no further information regarding the EOVs is presented to algorithm, there is no way to resolve the indeterminacy.

7 Comprehensive Discussion and Best Practices

This section is devoted to present the best practices of each procedure, explicit and implicit, through a comprehensive discussion of their potential advantages and drawbacks. However, we do not aim at making a direct performance comparison between the methods discussed in the sections above. Instead, we aim to discuss how these procedures could be integrated within a VSHM methodology.

Explicit procedures. Additional EOP measurements are required to reconstruct the functional relationship with corresponding DSFs. Consequently, this type of procedures requires large datasets of different natures, which in turn calls for extra computational resources and accurate merging/synchronization of data streams. Nonetheless, in many structures of interest this information is already available. Explicit procedures introduce a controlled environment aided by EOV information, facilitating a better interpretability of the results is achieved. Thereupon, a better understanding of the monitored structure might be achieved. Also, since DSFs are associated to EOPs, then it is easy to trace back the cause for any variation in DSFs.

Regardless of their increased complexity, regularized and stochastic regressions are preferable for the EOV compensation task, due to their increased robustness towards overfitting and reduced sensitivity to outliers. Moreover, stochastic models also facilitate the calculation of uncertainty bounds and model comparison [22].

When constructing regressions, it is useful to be able to identify the variables that effectively influence the output variables, while rejecting those that are irrelevant or correlated to others. Irrelevant and correlated variables may, in the best case, add unnecessary complexity to the model, while in the worst case act as confounding variables, harming the performance of the regression model. In the present context, this problem has a special significance as the number of potentially influencing variables can be very large. This is especially true for structures equipped with Supervisory Control and Data Acquisition (SCADA) systems, which provide hundreds of different variables and descriptive statistics, many of them of a similar nature (e.g., temperatures at different locations). Hence, a good practice is to preprocess the input data to produce a set of independent variables to be used as inputs in the regression model. To this end, Principal Component, Factor, or Independent Component Analysis as well as Singular Spectrum Analysis are amenable methods. Nonetheless, such transformations come at the price of a reduced physical interpretation of the input variables. A subsequent step is to remove inputs within the model optimization loop, aided by sensitivity measures and related techniques [43].

Some types of structures will exhibit large variations in the dynamic characteristics while changing between operational regimes. For instance, wind turbines have dramatic changes on their dynamics when changing from idling condition to power production mode. In those cases, independent regression models should be built for each one of the identified operational modes.

Other aspects to consider relate to the multivariate nature of DSFs. Nonetheless, most regression methods are designed for a single output, so typically, independent regression models are fit to single DSFs. While the predictive accuracy of this approach is not harmed in any way, the efficiency of the optimization process and of the obtained model is sub-optimal. Otherwise, considering the multivariate nature of DSFs will provide a more efficient model, which at the same time, can provide further understanding into the problem. For instance, collinearity between predicted variables or a latent structure in the DSFs can be unveiled. For this purpose, polynomial regressions in a deterministic or Bayesian context can be easily upgraded to the multiple output case [22], while similar extensions for GPRs are also available [45].

In practice, explicit procedures need data from EOPs and their associated DSFs from a representative set of environmental and operational conditions to achieve accurate models. This implies large instrumentation equipment, an efficient synchronization between different measurement systems (e.g., SCADA and vibration measurement systems), and a sufficiently long initial monitoring period. Overlooking any of these aspects will most certainly impair the achievable accuracy of the final regression model.

Implicit procedures. From one hand, implicit procedures are conceptually simpler, as no extra variables beyond DSFs are required, thus eliminating the concerns for measuring, synchronization and merging of data streams. On the other hand, this lack of information is translated either into a reduced performance or higher computational complexity. Moreover, if extra variables are already available (as is often the case), why not use this information to potentially enhance the sensitivity to damage?

Implicit procedures learn patterns in DSFs without any contextual or prior knowledge. This implies a potential reduction in damage sensitivity, as the manifolds obtained in healthy and damages may overlap, especially for early damage stages. However, the overlapping sections could be resolved with the help of any EOPs.

In practice, implicit procedures need large sets of DSFs to gain as much information on the variability of DSFs as possible. In addition, when dimensionality reduction is involved, it is important to select the correct dimensionality to preserve maximum sensitivity to damages without adding unwanted confounding features. These considerations might have negative consequences if overlooked, and they are generally case-dependent based on the complexity of the problem and quality of the data measured.

In general. A large number of observations are required to learn the complex patterns of DSFs. If the dimensionality of DSFs is also large (as in the case of spectra), then an increased computational complexity is attained. Dimensionality reduction methods could be integrated to alleviate this issue, but at the same time introduce the problem of resolving the amount of information to keep/discard without affecting the sensitivity to damage. It is important to build the reference state(s) by considering a representative subset of EOVs to the best possible extent. Otherwise, there will be a higher risk of false positives stemming from unexpected events not accounted for at the training phase. While often overlooked, the selection of the model structure has a massive impact on the final performance of the method. This includes the selection of the model type, function family, structural complexity (or model order), and so on. A wrong model selection will lead to a bias in the damage indices, as the trends in the data will be misrepresented.

8 Conclusions

This chapter has been devoted to the contextualization and discussion of different procedures for mitigation of EOV in damage detection. The core of this discussion has been focused on two main philosophies: *explicit* and *implicit* procedures. A comprehensive review has been presented in both cases, aiming to explain the necessity and scope of each one of them. In short, explicit procedures consider information about EOV to reconstruct the cause-effect relationship with corresponding DSFs. Meanwhile, implicit procedures do not use a direct information of the EOVs but attempt at learning the patterns characterizing DSFs or damage indices in a baseline state. These procedures have been illustrated in the context of a wind turbine in-operation where different damage scenarios were introduced. The procedures were exemplified via two machine learning methods: a multivariate nonlinear regression to illustrate the explicit procedure; and a metric learning method assisted by an artificial neural network to illustrate the implicit procedure. A detailed discussion of the performance of each method has been presented, where their capabilities for accounting and mitigating EOVs for damage detection in wind turbine blades in-operation are described. Finally, the advantages and drawbacks of each procedure are also detailed in order to present a comprehensive best practice guideline. In general, both procedures will require a large number of observations, whether the aim is to learn the complex patterns of the DSFs or build the relationship with their corresponding EOVs for constructing a robust SHM methodology. On the other hand, both procedures provide benefits for their implementation. Implicit procedures might be simpler as they require only information from the DSFs but explicit procedures introduce a controlled environment aided by EOV information. In terms of future development, we believe that a holistic approach comprising explicit and implicit compensation procedures should be considered, as it will facilitate the mitigation of measurable and unmeasurable effects, while introducing interpretability of the outcomes in a SHM framework.

Acknowledgements Mr. Callum Roberts and Dr. David García Cava would like to acknowledge Carnegie Trust for the Caledonian Ph.D. Scholarship (grant reference number: PHD007700).

References

1. Worden K, Cross EJ, Barthorpe RJ, Wagg DJ, Gardner P (2020) On digital twins, mirrors, and virtualizations: frameworks for model verification and validation. ASCE-ASME J Risk Uncertainty Eng Syst Part B: Mech Eng 6(3)
2. Farrar CR, Worden K (2007) An introduction to structural health monitoring. Philosophical Transictions of the Royal society A. Math Phys Eng Sci 365(1851):303–15
3. Sohn H (2007) Effects of environmental and operational variability on structural health monitoring. Philosophical Transictions of the Royal society A. Math Phys Eng Sci 365(1851):539–60
4. Avendaño-Valencia LD, Chatzi EN (2020) Multivariate GP-VAR models for robust structural identification under operational variability. Probab Eng Mech 60

5. Aravanis TCI, Sakellariou JS, Fassois SD (2020) A stochastic functional model based method for random vibration based robust fault detection under variable non–measurable operating conditions with application to railway vehicle suspensions. J Sound and Vib 466
6. Zhu XQ, Law SS (2015) Structural health monitoring based on vehicle-bridge interaction: accomplishments and challenges. Adv Struct Eng 18 (12)
7. Bogoevska S, Spiridonakos M, Chatzi E, Dumova-Jovanoska E, Höffer R (2017) A data-driven diagnostic framework for wind turbine structures: a holistic approach. Sensors 17(4)
8. Movsessian A, Qadri BA, Tcherniak D, Garcia Cava D, Ulriksen MD (2021) Mitigation of environmental variabilities in damage detection: a comparative study of two semi-supervised approaches. In: Proceedings of the International Conference on Structural Dynamic 1
9. Yuen KV, Kuok SC (2010) Ambient interference in long-term monitoring of buildings. Eng Struct 32(8)
10. Hansen MH (2007) Aeroelastic instability problems for wind turbines. Wind Energy 10
11. Farrar CR, Worden K (2012) Structural health monitoring: a machine learning perspective
12. Aravanis T-CI, Sakellariou JS, Fassois SD (2020) A stochastic functional model based method for random vibration based robust fault detection under variable non–measurable operating conditions with application to railway vehicle suspensions. J Sound Vib 466
13. Sohn H, Worden K, Farrar CR (2002) Statistical damage classification under changing environmental and operational conditions. J Intell Mater Syst Struct 13(9)
14. Avendaño-Valencia LD, Chatzi EN, Koo KY, Brownjohn JMW (2017) Gaussian process time-series models for structures under operational variability. Fron Built Environ 3
15. Worden K, Sohn, H. and Farrar, CR (2002) Novelty detection in a changing environment: regression and interpolation approaches. J Sound Vib 258(4)
16. Xia Y, Hao H, Zanardo G, Deeks (2006) A long term vibration monitoring of an RC slab: temperature and humidity effect. Eng Struct 28(3)
17. Schlechtingen M, Ferreira Santos I (2011) Comparative analysis of neural network and regression based condition monitoring approaches for wind turbine fault detection. Mech Syst Signal Process 25(5):5
18. Kopsaftopoulos F, Nardari R, Li YH, Chang FK (2018) A stochastic global identification framework for aerospace structures operating under varying flight states. Mech Syst Signal Process 98
19. Spiridonakos MD, Chatzi EN (2015) Metamodeling of dynamic nonlinear structural systems through polynomial chaos NARX models. Comput Struct 157
20. Hastie T, Tibshirani R, Friedman J (2009) Elements of Statistical Learning. Springer, Second
21. Laory I, Trinh TN, Smith IFC, Brownjohn JMW (2014) Methodologies for predicting natural frequency variation of a suspension bridge. Eng Struct 80
22. Avendaño-Valencia LD, Chatzi EN, Tcherniak D (2020) Gaussian process models for mitigation of operational variability in the structural health monitoring of wind turbines. Mech Syst Signal Process 142
23. Tipping ME (2001) Sparse bayesian learning and the relevance vector machine. J Mach Learn Res 1(3)
24. Gómez González A, Fassois SD (2016) A supervised vibration-based statistical methodology for damage detection under varying environmental conditions & its laboratory assessment with a scale wind turbine blade. J Sound Vib 366
25. Deraemaeker A, Reynders E, De Roeck G, Kullaa J (2008) Vibration-based structural health monitoring using output-only measurements under changing environment. Mech Syst Signal Process 22(1)
26. Deraemaeker A, Worden K (2018) A comparison of linear approaches to filter out environmental effects in structural health monitoring. Mech Syst Signal Process 105
27. García D, Tcherniak D (2019) An experimental study on the data-driven structural health monitoring of large wind turbine blades using a single accelerometer and actuator. Mech Syst Signal Process 127
28. Yan A-M, Kerschen G, De Boe P, Golinval J-C (2005) Structural damage diagnosis under varying environmental conditions—Part I: a linear analysis. Mech Syst Signal Process 19(4)

29. Movsessian A, Garcia D, Tcherniak D (2020) Adaptive feature selection for enhancing blade damage diagnosis on an operational wind turbine. In: Proceedings of the 13th International Conference on Damage Assessment of Structures
30. Harvey DY, Todd MD (2012) Cointegration as a data normalization tool for structural health monitoring applications. In: Proceedings of SPIE - The International Society for Optical Engineering, vol 8348
31. Cross EJ, Worden K, Chen Q (2011) Cointegration: a novel approach for the removal of environmental trends in structural health monitoring data. Philosophical Transictions of the Royal society A. Math Phys Eng Sci 467(2133)
32. Zolna K, Dao PB, Staszewski WJ, Barszcz T (2015) Nonlinear cointegration approach for condition monitoring of wind turbines. Math Probl Eng 2015
33. Shi H, Worden K, Cross EJ (2018) A regime-switching cointegration approach for removing environmental and operational variations in structural health monitoring. Mech Syst Signal Process 103
34. Yan A-M, Kerschen G, De Boe P, Golinval J-C (2005) Structural damage diagnosis under varying environmental conditions—part II: local PCA for non-linear cases. Mech Syst Signal Process (19)4
35. Santos A, Figueiredo E, Silva MFM, Sales CS, Costa JCWA (2016) Machine learning algorithms for damage detection: Kernel-based approaches. J Sound Vib 363
36. Movsessian A, Garcia Cava D, Tcherniak D (2021) An artificial neural network methodology for damage detection: demonstration on an operating wind turbine blade. Mech Syst Signal Process 159
37. Zhao R, Yan R, Chen Z, Mao K, Wang P, Gao RX (2019) Deep learning and its applications to machine health monitoring. Mech Syst Signal Process 115
38. Tibaduiza D, Torres-Arredondo MÁ, Vitola J, Anaya M, Pozo F (2018) A damage classification approach for structural health monitoring using machine learning. Complexity
39. Silva M, Santos A, Santos R, Figueiredo E, Sales C, Costa JCWA (2017) Agglomerative concentric hypersphere clustering applied to structural damage detection. Mech Syst Signal Process 92
40. Langone R, Reynders E, Mehrkanoon S, Suykens JAK (2017) Automated structural health monitoring based on adaptive kernel spectral clustering. Mech Syst Signal Process 90
41. Baraldi P, Di Maio F, Rigamonti M, Zio E, Seraoui R (2016) Clustering for unsupervised fault diagnosis in nuclear turbine shut-down transients. Mech Syst Signal Process 58
42. Kostic B, Gül M (2017) Vibration-based damage detection of bridges under varying temperature effects using time-series analysis and artificial neural networks, J Bridge Eng 22(10)
43. Roberts C, Cava Garcia D, Avendaño-Valencia LD (2021) Understanding the influence of environmental and operational variability on wind turbine blade monitoring. In: Rizzo P, Milazzo A (eds) European Workshop on Structural Health Monitoring. EWSHM 2020. Lecture notes in civil engineering, vol 127. Springer, Cham. https://doi.org/10.1007/978-3-030-64594-6_12
44. Movsessian A, Garcia D, Tcherniak D (2020) Adaptive feature selection for enhancing blade damage diagnosis on an operational wind turbine. In: Proceedings of the 13th International Conference on Damage Assessment of Structures
45. Alvarez MA, Rosasco L, Neil D, Lawrence ND (2011) Kernels for vector-valued functions: a review. arXiv preprint arXiv: 1106.6251

Explainable Artificial Intelligence to Advance Structural Health Monitoring

Daniel Luckey, Henrieke Fritz, Dmitrii Legatiuk, José Joaquín Peralta Abadía, Christian Walther, and Kay Smarsly

Abstract In recent years, structural health monitoring (SHM) applications have significantly been enhanced, driven by advancements in artificial intelligence (AI) and machine learning (ML), a subcategory of AI. Although ML algorithms allow detecting patterns and features in sensor data that would otherwise remain undetected, the generally opaque inner processes and black-box character of ML algorithms are limiting the application of ML to SHM. Incomprehensible decision-making processes often result in doubts and mistrust in ML algorithms, expressed by engineers and stakeholders. In an attempt to increase trust in ML algorithms, explainable artificial intelligence (XAI) aims to provide explanations of decisions made by black-box ML algorithms. However, there is a lack of XAI approaches that meet all requirements of SHM applications. This chapter provides a review of ML and XAI approaches relevant to SHM and proposes a conceptual XAI framework pertinent to SHM applications. First, ML algorithms relevant to SHM are categorized. Next, XAI approaches,

D. Luckey (✉)
Chair of Computing in Civil Engineering, Bauhaus University Weimar, Weimar, Germany
e-mail: daniel.luckey@uni-weimar.de

H. Fritz · J. J. Peralta Abadía · K. Smarsly
Institute of Digital and Autonomous Construction, Hamburg University of Technology, Hamburg, Germany
e-mail: henrieke.fritz@tuhh.de

J. J. Peralta Abadía
e-mail: joaquin.peralta@tuhh.de

K. Smarsly
e-mail: kay.smarsly@tuhh.de

D. Legatiuk
Chair of Applied Mathematics, Bauhaus University Weimar, Weimar, Germany
e-mail: dmitrii.legatiuk@uni-weimar.de

C. Walther
Institute of Structural Mechanics, Bauhaus University Weimar, Weimar, Germany
e-mail: christian.walther@uni-weimar.de

A. Cury et al. (eds.), *Structural Health Monitoring Based on Data Science Techniques*, Structural Integrity 21, https://doi.org/10.1007/978-3-030-81716-9_16

such as transparent models and model-specific explanations, are presented and categorized to identify XAI approaches appropriate for being implemented in SHM applications. Finally, based on the categorization of ML algorithms and the presentation of XAI approaches, the conceptual XAI framework is introduced. It is expected that the proposed conceptual XAI framework will provide a basis for improving ML acceptance and transparency and therefore increase trust in ML algorithms implemented in SHM applications.

Keywords Artificial intelligence (AI) · Machine learning (ML) · Structural health monitoring (SHM) · Explainable artificial intelligence (XAI)

1 Introduction

Facilitating non-destructive condition assessment of civil infrastructure based on sensor data, structural health monitoring (SHM) has become a useful instrument to advance infrastructure maintenance, rehabilitation, and repair. Thus, SHM enhances safety and cost-efficient operation of infrastructure [1]. Unlike visual inspections that are performed periodically, SHM can be conducted continuously, in real-time, and at locations that may be difficult to physically reach [2]. Depending on type and location of the sensors, SHM provides both local and global monitoring, while visual inspections are restricted to localized areas [3]. The long-term systematic use of reliable sensors and measurement techniques, as well as appropriate data analysis methods, in SHM helps (i) to evaluate and ensure structural health, (ii) to repair damage duly, and (iii) to calculate financial expenses of structural maintenance [4].

Representing a rapidly growing and innovative research field, SHM is increasingly advanced by artificial intelligence (AI). Recent technological developments in AI have enabled SHM systems to become "smart" SHM systems, also referred to as "smart monitoring systems", allowing SHM processes to be conducted (almost) autonomously. Common AI approaches introduced in SHM for solving domain-specific problems through learning and adaption are based on machine learning (ML), a subcategory of AI. Machine learning represents the learning processes of computer systems, often described as the transformation of experience into expertise or knowledge [5]. Using ML algorithms for SHM purposes has shown to be promising, as ML algorithms can process large amounts of data and recognize patterns that may serve as a basis for condition assessment of civil infrastructure.

Concerning AI-related approaches, detection and quantification of structural damage are the main research fields of ML in SHM, applying, e.g., support vector machines [6] or convolutional neural networks [7]. Furthermore, support vector machines, artificial neural networks and Gaussian naïve Bayes techniques have been used for condition assessment using SHM systems [8]. Aiming at life-time prediction based on SHM, principal component analysis and partial least square regression have been applied [9]. An overview of the variety of ML algorithms implemented in SHM systems can be found in [10].

Because of the complex structure of ML algorithms, most ML algorithms have a black-box character, i.e. the results computed by the algorithms cannot easily be traced through the internal processes of the algorithms. As a result, engineers and stakeholders have been expressing mistrust in black-box ML algorithms and have been reluctant in the widespread adoption of ML in SHM. In recent years, the research area of so-called "explainable artificial intelligence" (XAI) has gained growing interest [11]. XAI aims to increase acceptance of and transparency in the results of ML algorithms by providing ***explanation*** and ***interpretation***, i.e. reasoning, for the algorithm output (see Sect. 3.1 for details) [12]. While explanation can be considered as a collection of interpretable features, interpretation focuses on mapping abstract concepts. Since critical infrastructure has been in operation for many decades, and has therefore been subjected to gradual changes that may require the prediction accuracy and reliability of ML algorithms to be effectively detected, it is imperative to introduce explainable ML algorithms to modern smart SHM systems. However, there is a lack of XAI approaches in SHM applications. By reviewing ML algorithms for SHM, this chapter provides an overview of ML algorithms for SHM usage suitable for XAI. Furthermore, this chapter generalizes the requirements and present possibilities for explainable ML algorithms for SHM applications.

In this chapter, a categorization of AI and ML algorithms is given to illuminate the relation between AI and ML. Subsequently, ML algorithms for SHM are reviewed to illustrate the variety of approaches relevant to SHM. The categorization of ML and the review of ML algorithms for SHM serve as a basis to highlight the possibilities of introducing XAI approaches to ML for SHM. Thereupon, an introduction to XAI and a conceptual XAI framework for SHM are presented, in an attempt to advance SHM practice. Finally, conclusions drawn from the results achieved in this study are presented and potential future work is suggested.

2 Machine Learning Algorithms for Structural Health Monitoring

This section provides the basics on AI and presents a categorization of AI and ML. Next, reviews of ML for SHM are discussed. Finally, current ML algorithms used in SHM are reviewed and summarized.

2.1 Artificial Intelligence and Machine Learning

Artificial intelligence has been a subject of constant development over the last decades. As AI covers a broad field of research and industrial applications, it does not seem possible to formulate a single, generally valid definition. Nonetheless,

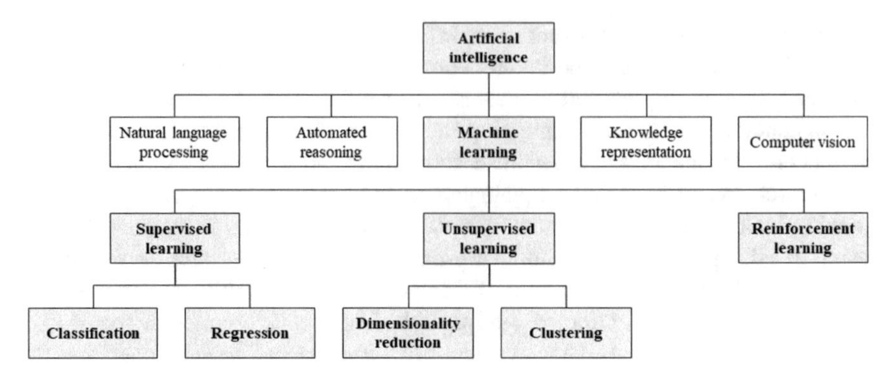

Fig. 1 Categorization of artificial intelligence and machine learning

general concepts of AI developed in each field of research and industry share similarities, identified as four characteristics: (i) perception of a complex environment, (ii) acquisition and interpretation of data, (iii) autonomous response to change, and (iv) automatic achievement of predefined goals [13].

For the purpose of automatic achievement of predefined goals, various AI algorithms have been designed to realize intelligent problem-solving strategies [14]. Furthermore, AI algorithms have been developed for applications in different fields of engineering and thus cover a wide spectrum of tasks, such as audio-visual perception of the surrounding environment [15], reproducibility of AI decisions [16], automatic data interpretation and decision making [17], and evaluation of robustness to changes in datasets [18]. AI can be broadly categorized in natural language processing, automated reasoning, ML, knowledge representation, and computer vision as shown in Fig. 1.

Machine learning is referred to as a process that improves the performance of a functional unit by new knowledge or skill acquisition or by existing knowledge or skill reorganization [19]. Designing ML algorithms requires large datasets to learn how to automate tasks and involves two steps, ***training*** and ***testing***. During ***training***, ML algorithms use datasets, ideally covering the entire possible range of values to which some meaning can be assigned, to "learn" patterns in the data. Some ML algorithms also accept incomplete or inconsistent datasets, exhibiting an acceptable level of robustness. A specific instance of a ML algorithm that is trained to solve a specific problem is referred to as a ML model. During ***testing***, the ML model obtained during the training step is given new unknown data, with which the overall performance of the ML model is tested.

In general, ML can be divided into three different learning processes:

i. supervised learning,
ii. unsupervised learning, and
iii. reinforcement learning [20].

For supervised learning, the datasets used to train a ML model contain not only the input data from which patterns will be learned but also a target feature, such

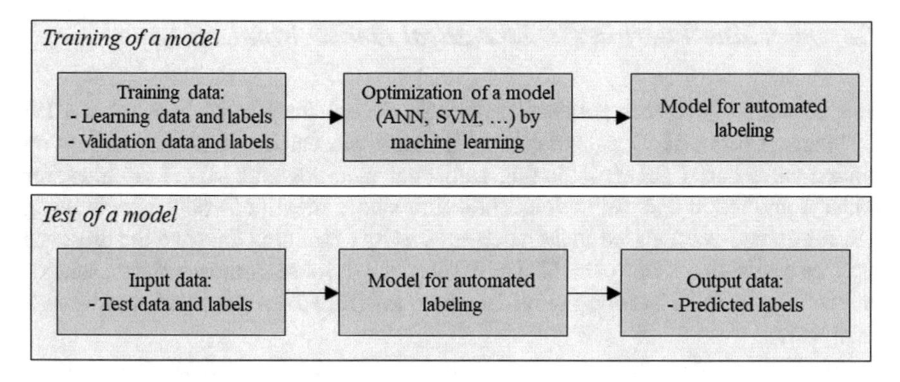

Fig. 2 Within supervised learning, the model (e.g. an ANN) is trained and tested for automatic labeling of data

as labels in classification problems and numerical continuous values in regression problems. Figure 2 presents a flow chart of the process of training a ML model. Apart from identifying patterns in the training step, some ML algorithms require to split the input data stochastically into training data and validation data. Validation data is then used to calculate the accuracy and error in the predictions of the ML model being trained, which helps to adjust and optimize the learning process. Upon completing the training step and producing the ML model adapted to the problem being solved, e.g. in the form of an artificial neural network (ANN), the ML model is tested, and if its predictions are accurate, the ML model is ready to be used in real-world applications.

For unsupervised learning, the ML algorithm is trained only on input data without a target feature. Therefore, unsupervised ML algorithms attempt to identify relationships and similarities in the input data during the training process without a predefined or expected output. Unsupervised learning methods include clustering and dimensionality reduction algorithms (Fig. 1) and are often used for anomaly detection and outlier detection. By applying clustering, automatic separation of input data into groups of similar and correlated data is performed. Dimensionality reduction usually pursues the transformation of original data from a high-dimensional space to a low-dimensional space, while maintaining the main meaning of the data.

Representing a hybrid form, in semi-supervised learning, parts of the input data are labeled, thus combining supervised and unsupervised learning. Finally, in reinforcement learning, the training dataset is not fixed and is continuously updated with new data. As such, through an iterative learning process, the ML algorithm is continuously adapted, improving the relation between input and output data.

Due to the differences in the three learning processes and the variety of ML algorithms available for use in SHM, it may be confusing to identify which learning process and ML algorithm should be used to solve a problem in SHM. Therefore, the following subsection presents a description of reviews of ML applications in SHM.

2.2 Machine Learning in Structural Health Monitoring

Several studies have been published reviewing the application of ML to solve SHM problems. Mitra and Gopalakrishnan [21] have presented a literature review on guided-wave-based SHM, including statistical methods and ML algorithms, for damage prediction and estimation. The authors have identified challenges in using ML for guided-wave-based SHM, such as scarcity of training data and the presence of noise and outliers. Bao et al. [22] have reviewed the state of the art of data science engineering in SHM by focusing on deep learning (DL) algorithms and compressive sampling.

An overview of recent research of DL approaches for SHM of civil infrastructures has been accomplished by Ye et al. [23], demonstrating history, frameworks, and datasets for DL, in addition to structural damage detection and condition assessment applications. Furthermore, the authors have indicated challenges and future trends of DL-based SHM, concluding that DL-based SHM will witness developments with new algorithms and enhanced frameworks, sufficient datasets, and computing power. Azimi et al. [24] have introduced a review of DL methods in SHM. The authors have covered vibration-based and vision-based SHM through DL, as well as applications of unmanned aerial vehicles and smartphones for DL-based SHM.

Flah et al. [25] have studied the deployment of AI approaches in SHM for damage detection of structures. The authors have introduced SHM applications utilizing AI algorithms, and have elaborated on types of structures being employed in these applications, advantages, limitations, and recommendations.

2.3 Applications of Machine Learning Algorithms in Structural Health Monitoring

Although several reviews addressing applications of ML in SHM have been proposed in recent years, a general review of ML algorithms for SHM from the point of view of XAI is lacking. Therefore, the aim of this subsection is to present a review of ML approaches in SHM keeping in mind their possible use in an XAI context. The literature collection involves studies indexed in the Web of Science Core Collection as well as conference papers indexed in the Scopus database. A representative number of papers dealing with ML and SHM published between 2016 and 2019 is selected.

The application of ML algorithms in SHM depends on the problem to be solved. As described earlier, one of the main tasks in SHM is to assess the condition of and to identify possible damage in structures. For example, Abdeljaber et al. [7] have studied structural damage detection using an one-dimensional convolutional neural network (1-D CNN), overcoming the limitation posed by the need of a large amount of measurements to train the 1-D CNN algorithms, particularly for large structures. Joshuva and Sugumaran [26] have studied the combination of ML algorithms for condition assessment of wind turbine blades using vibration signals. Das et al. [27]

have used clustering and support vector machines to classify crack modes from unlabeled acoustic emission waveform features.

The studies reviewed herein have used supervised learning (SL), unsupervised learning (UL), and semi-supervised learning (Semi-SL) algorithms. Reinforcement learning (RL) algorithms have not been used in the studies reviewed. Therefore, based on the categorization and description of learning processes presented in subsection 2.1 and omitting RL, Table 1 presents a summary of ML algorithms used in SHM, grouped by SL, UL, and Semi-SL. Furthermore, since SL algorithms may be used for classification and regression tasks, subgroups are added in the SL group. It is observed that mainly SL algorithms are being used in SHM, followed by UL algorithms. Moreover, most algorithms in the SL subgroups are classification algorithms, revealing a possible focus of XAI approaches towards SL classification algorithms.

Regarding ***SL algorithms***, support vector machine algorithms have been used for damage detection in built infrastructure [28, 29]. Moreover, Pan et al. [30] have used support vector machines for structural diagnosis and damage detection in large-scale cable-stayed bridges. Vitola et al. [31] have used *k*-nearest neighbors (*k*-NN) for damage classification, in combination with principal component analysis (PCA) for feature extraction from signals representing dynamic responses of structures. Artificial neural networks have been used for predicting accelerations to be compared with damage indices, in combination with Gaussian process algorithms, discriminating between damaged and undamaged structural conditions [32].

As for ***UL algorithms***, a damage detection methodology based on strain field pattern recognition using clustering and PCA has been presented in [33]. In addition, Diez et al. [34] have performed damage detection on bridges via clustering for grouping substructures with similar behavior.

Finally, with respect to the hybrid form of semi-SL algorithms, Sarkar et al. [35] have used a deep auto-encoder for detection and annotation of cracks on images. At the end of Table 1, a group of references containing comparative studies of different ML algorithms has been added for providing a better overview to readers interested not only in specific applications of ML in SHM, but also in comparative performance of various ML algorithms.

As shown in Table 1, SL classification algorithms have been used the most in SHM systems. Since many of these SL classification algorithms are black-box, raising trust issues regarding their outputs, the explainability and interpretability aspects of XAI may help increase the transparency and acceptance of the SL classification algorithms in SHM practice. As a result, the following section presents a conceptual XAI framework, i.e. a concept towards explainable artificial intelligence, as a solution to help explain and interpret the reasons behind the predictions of SL classification algorithms.

Table 1 Review of ML algorithms in SHM

ML in SHM	SL/UL/Semi-SL	Prob. class	Algorithm	References
	SL	Classification	Support vector machine	[6, 8, 27–30, 36–39]
			k-nearest neighbors	[31, 38–43]
			Convolutional neural network	[7, 40, 44–46]
			Artificial neural network	[37, 38, 47]
			Naïve Bayes	[8, 36, 40]
			Decision trees	[38–40]
			Long short-term memory	[40, 44]
			Classification learner toolbox from Matlab	[48]
			Nearest-mean classifier	[38]
		Regression	Artificial neural network	[32, 49–51]
			Gaussian process	[32, 52, 53]
			Partial least square regression	[9]
			Support vector regression	[54]
			Bayesian neural network	[44]
			Kernel ridge regression	[36]
	UL		Principal component analysis	[9, 31, 33, 42, 55–57]
			Clustering	[27, 34, 55]
			Gaussian mixture modeling	[27, 58]
			Deep autoencoder	[35]
			Fast independent component analysis	[56]
			Optimal baseline selection	[33]

(continued)

Table 1 (continued)

		Fuzzy cognitive map	[59]
	Semi-SL	Deep autoencoder	[60, 61]
		Manifold adaptive experimental design algorithm	[62]
	Comparative studies		[8, 9, 28, 30, 36–38, 40]

3 Design of a Conceptual XAI Framework

Explainable artificial intelligence provides a set of techniques that produce explainable ML algorithms while maintaining a high level of learning performance. By incorporating explanations, ML algorithms increase expressiveness, improve human understanding, and advance confidence in decision making [12]. As mentioned previously, XAI may be divided into two parts:

- the ***explanation part***, which can be seen as meta information added to ML algorithms, generated either by external algorithms or by a ML algorithm itself, for example features contributing to the output of ML algorithms, and
- the ***interpretation part***, which aims to find a suitable domain-specific interpretation of the meta information identified in the explanation part of XAI, for example, by linking the most relevant features to mechanical quantities from structural dynamics.

It is worth mentioning that some ML algorithms, such as decision trees and rule-based methods, are explainable by design without any need for XAI. However, as shown in Table 1, several ML algorithms employed in SHM applications are not explainable (black-box), and therefore require XAI for reliable use in engineering practice.

3.1 Overview and Existing XAI Approaches

According to Adadi et al. [63], AI can be divided into three different categories:

1. **Accurate AI**, including AI systems primarily exhibiting high accuracy in decision making,
2. **Explainable AI**, comprising advanced or newly designed AI systems exhibiting, in addition to high accuracy, a very high degree of explainability and interpretability in decision making, and
3. **Responsible AI**, encompassing future AI systems that will act and decide in a transparent, fair, and comprehensible manner, considering legal, social, and ethical aspects.

From the point of view of explainability, accurate AI represents the classical use of AI and ML algorithms in various fields of applications, and is not explainable in general, except for algorithms being transparent by design, such as decision-trees and rule-based methods. The other two categories, explainable AI and responsible AI, support transparency. Before discussing different XAI approaches proposed in recent years, it is necessary to underline the distinction between interpretation and explanation, which are defined according to [64] as follows:

Definition 1 An *interpretation* is the mapping of an abstract concept (e.g. a predicted class) into a domain of which humans can make sense.

Definition 2 An *explanation* is the collection of features of the interpretable domain that have contributed to a given problem in producing a decision (e.g. classification or regression).

From these definitions, it follows that the interpretation is problem-specific because the target domain to which the abstract concept is mapped is defined by the problem that needs to be solved using ML. In practical terms, interpretation implies that some parameters of a ML algorithm are linked to engineering quantities and are easily interpretable by engineers. By contrast, explanation is solely related to algorithms, aiming to find the dominant features contributing to the result. The problem-agnostic nature of explanation allows for flexibility in applying XAI to SHM applications, as explanation algorithms can be used for different SHM tasks, and only the interpretation part must be adapted to specific applications.

In recent years, several XAI approaches have been introduced that can be divided into two main groups [65]:

(i) **Transparent models.** Explainability is provided directly by the model or by the model structure. This group includes ML models such as decision trees, rule-based learners, general additive models, Bayesian models, models for k-nearest neighbors, and models for logistic or linear regression.
(ii) **Post-hoc explainability.** This group is divided into model-agnostic and model-specific methods, enabling subsequent explainability of ML models. Model-agnostic methods, such as visual interpretation and feature space interpretation, provide explanation by general tools and are applicable to a large number of ML algorithms. By contrast, model-specific methods are designed for providing explanations of results obtained by specific ML algorithms, and are therefore applicable only to a limited number of algorithms, such as artificial neural networks and support vector machines.

It should be mentioned that the problem formulation for using ML algorithms in civil engineering has a direct influence on applications of XAI approaches. As discussed in [10], the application of ML algorithms in civil engineering may be distinguished into (i) ML algorithms used to solve problems of data analysis, such as classification problems, and (ii) ML algorithms used for substituting conventional algorithms, i.e. for surrogate modeling. From an XAI perspective, problems addressed by surrogate models do not require explanation even if black-box ML algorithms, such as deep learning algorithms, are used. The reason is the nature of surrogate modeling: Because ML algorithms are used as substitutes of conventional algorithms, the relation between input and output is always clear, despite being nontransparent from an algorithmic point of view. By contrast, classification problems require using XAI approaches for the ML algorithms, as explained in the conceptual XAI framework proposed in the next subsection.

3.2 Conceptual XAI Framework

A conceptual XAI framework for adding transparency to the results provided by ML algorithms in SHM applications is proposed in this subsection. The framework reflects a general strategy towards using XAI approaches in SHM applications. Drawing from the previous discussion on ML algorithms and XAI approaches, as well as from ML applications in SHM, the conceptual XAI framework consists of the following steps:

Step 1. Preprocessing of the problem
The goal of this step is to clearly formulate the problem, which is intended to be solved by ML algorithms, and to analyze the problem with respect to the necessity of using XAI. Therefore, this step contains the following sub-steps:

1. Formulation of the SHM task, e.g. damage detection, to be solved using ML algorithms.
2. Analysis of the task with respect to the need of XAI. Here, it must be decided if the use of ML falls into the category of classification problems or surrogate modeling. For classification problems, using XAI is justified, otherwise there is no need for explanation, as discussed above.
3. Formulation of the SHM task as a classification problem to be solved using ML algorithms.

Step 2. Application of ML algorithms
The goal of this step is to solve the classification problem formulated in Step 1 by one or several ML algorithm(s).
Step 3. Explanation
The goal of this step is to provide the explanation of the classifier decisions from Step 2 in the sense of Definition 2. Thus, the most relevant data features influencing the classifier decision are to be identified in this step. A generic form of explanation is: The input has been classified to category A because it has the set of features Ω
Step 4. Interpretation
The final step in the framework aims at mapping the most relevant data features identified in Step 3 to the problem-specific domain of SHM. For example, in the damage identification context, the interpretation may look as follows: The set

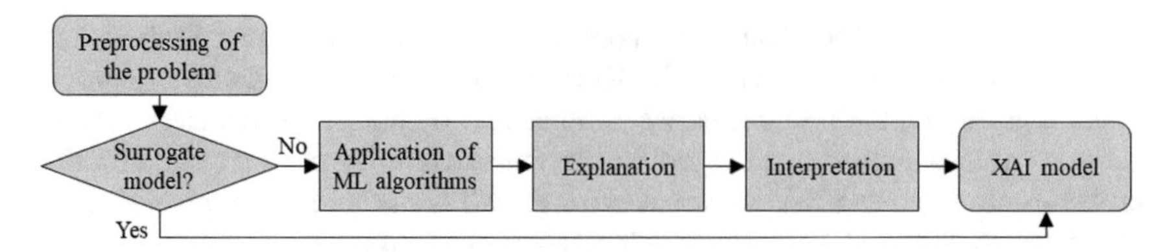

Fig. 3 XAI conceptual framework

of features Ω indicating damage contains three parameters interpreted as first eigenfrequency, peak acceleration, and peak displacement.

Upon performing the 4 steps described above, an XAI model is constructed containing a list of the most relevant features for the classification problem (explanation part), and a list of problem-specific-domain terminology linked to the most relevant features (interpretation part). The flowchart in Fig. 3 summarizes the conceptual XAI framework for SHM applications.

4 Summary, Conclusions, and Future Work

Structural health monitoring is a key component for achieving reliable, resilient, and cost-efficient – in terms of structural maintenance – infrastructure. Machine learning algorithms are frequently used in SHM to automatically analyze and detect patterns in sensor data. Despite being efficient in analyzing sensor data, the black-box nature of ML algorithms raises doubts and mistrust by engineers, thus hindering the exploitation of the ML full potential in SHM practice. Representing an emerging paradigm in data science, explainable artificial intelligence is expected to enhance the transparency of ML algorithms and, eventually, increase the acceptance of ML algorithms by SHM practitioners.

Drawing from current trends in ML applications applied to SHM, this chapter has presented a step towards introducing XAI to SHM. First, ML algorithms commonly deployed in SHM applications have been reviewed. Specifically, ML-based SHM applications, including tasks such as damage detection and damage localization, have been categorized according to the learning process, i.e. into supervised, unsupervised, and reinforcement learning approaches. From the categorization of ML-based SHM applications, it has been observed that most applications are associated with supervised learning classification problems.

Second, XAI approaches have been presented, indicating that the necessity of using XAI in SHM applications depends on the formulation of the problem that is intended to be solved using ML. In case ML algorithms substitute classical engineering algorithms, i.e. ML algorithms serve as surrogate models, XAI is not required because the reasoning behind the results obtained by ML is readily clear from the problem formulation. By contrast, classification problems solved by ML algorithms require the use of XAI because classifier decisions are generally not traceable. Given that tasks in most ML-based SHM applications are formulated as classification problems, a conceptual framework towards adopting XAI in SHM has been proposed, including four steps: (i) formulating the problem and confirming that the ML algorithms are used as classifiers, (ii) applying the ML algorithms, (iii) explain the ML algorithms outputs, and (iv) interpret the ML algorithms outputs within an SHM context.

From the results of this study, it can be concluded that introducing XAI can enhance the quality of ML-based SHM applications. Furthermore, it is concluded that ML algorithms require different levels of explanation, based on the purpose of using ML and the human individuals to which the explanation is addressed. Future work may be conducted towards an in-depth analysis of explainability and levels of explanation for ML algorithms to advance SHM and smart monitoring systems.

Acknowledgements The authors gratefully acknowledge the support offered by the German Research Foundation (DFG) under grants SM 281/9-1, SM 281/12-1, SM 281/14-1, SM 281/15-1, and LE 3955/4-1. This research is also partially supported by the Carl Zeiss Foundation. Any opinions, findings, conclusions, or recommendations expressed in this paper are those of the authors and do not necessarily reflect the views of the sponsors.

References

1. Law KH, Smarsly K, Wang Y (2014) Sensor data management technologies for infrastructure asset management. In: Wang ML, Lynch JP, Sohn H (eds) Sensor technologies for civil infrastructures. Woodhead Publishing Ltd., Sawston, UK, pp 3–32
2. Smarsly K, Lehner K, Hartmann D (2007) Structural health monitoring based on artificial intelligence techniques. In: Proceedings of the 2007 ASCE international workshop on computing in civil engineering. Pittsburgh, PA, USA, July 24, 2007
3. Smarsly K, Law KH, König M (2011).Resource-efficient wireless monitoring based on mobile agent migration. In: Proceedings of the SPIE (Vol. 7984): health monitoring of structural and biological systems 2011. San Diego, CA, USA, March 6, 2011
4. Fritz H, Smarsly K (2020) A state-of-the-art review of nature-inspired systems for smart structures. In: Proceedings of the European workshop on structural health monitoring (EWSHM). Palermo, Italy, July 6, 2020.
5. Shavlet-Shwartz S, Ben-David S (2014) Understanding machine learning. From theory to algorithms. Cambridge University Press, New York, NY, USA
6. Li R, Gu H, Hu B, She Z (2019) Multi-feature fusion and damage identification of large generator stator insulation based on Lamb wave detection and SVM method. Sensors 19(7):3733
7. Abdeljaber O, Avci O, Kiranyaz S, Boashash B, Sodano H, Inman D (2018) 1-D CNNs for structural damage detection: verification on a structural health monitoring benchmark data. Neurocomputing 275(2018):1308–1317
8. Nazarian E, Taylor T, Weifeng T, Ansari F (2018) Machine-learning-based approach for post event assessment of damage in a turn-of-the-century building structure. J Civ Struct Heal Monit 8(2):237–251
9. Sysyn M, Gerber U, Nabochenko O, Li Y, Kovalchuk V (2019) Indicators for common crossing structural health monitoring with track side inertial measurements. Acta Polytechnica 59(2):170–181

10. Luckey D, Fritz H, Legatiuk D, Dragos K, Smarsly K (2020) Artificial intelligence techniques for smart city applications. In: Proceedings of the international ICCCBE and CIB W78 joint conference on computing in civil and building engineering 2020. São Paulo, Brazil, August 18, 2020
11. Kuang C (2017) Can A.I. be taught to explain Itself? The New York Time Magazine
12. Das, A, Rad P (2020) Opportunities and challenges in explainable artificial intelligence (XAI): a survey. ArXiv abs/2006.11371
13. Samoili S, Lopez Cobo M, Gomez Gutierrez E, De Prato G, Martinez-Plumed F, Delipetrev B (2020) AI WATCH: Defining artificial intelligence. Publications Office of the European Union, Luxembourg
14. Swarnkar A, Swarnkar A (2020) Artificial intelligence based optimization techniques: a review, intelligent computing techniques for smart energy systems. Springer Singapore, 95–103
15. Tian Yh, Chen Xl, Xiong Hk et al (2017) Towards human-like and transhuman perception in AI 2.0: a review. Frontiers Inf Technol Electronic Eng 18:58–67
16. Shrestha YR, Ben-Menahem SM, von Krogh G (2019) Organizational decision-making structures in the age of artificial intelligence. Calif Manage Rev 61:66–83
17. Hofmann M, Neukart F, Bäck T (2017) Artificial intelligence and data science in the automotive industry. ArXiv abs/1709.01989.
18. Hamon R, Junklewitz H, Sanchez I (2020) Robustness and explainability of artificial intelligence—from technical to policy solutions. Publications Office of the European Union, 2020
19. ISO/IEC 2382:2015 Information technology—Vocabulary
20. Russel SJ, Norvig P (2014) Artificial intelligence: a modern approach, 3rd edn. Pearson Education Ltd., Harlow, Essex, UK
21. Mitra M, Gopalakrishnan S (2016) Guided wave based structural health monitoring: a review. Smart Mat Struct 25(5):053001
22. Bao Y, Chen Z, Wei S, Xu Y, Tang Z, Li H (2019) The state of the art of data science and engineering in structural health monitoring. Engineering 5(2):234–242
23. Ye XW, Jin T, Yun CB (2019) A review on deep learning-based structural health monitoring of civil infrastructures. Smart Struct Syst 24(5):567–585
24. Azimi M, Eslamlou AD, Pekcan G (2020) Data-driven structural health monitoring and damage detection through deep learning: State-of-the-art review. Sensors 20(10):2778
25. Flah M, Nunez I, Chaabene W, Nehdi M (2020) Machine learning algorithms in civil structural health monitoring: a systematic review. Arch Comput Methods Eng
26. Joshuva A, Sugumaran V (2018) A study of various blade fault conditions on a wind turbine using vibration signals through histogram features. J Eng Sci Technol 13(1):102–121
27. Das AK, Suthar D, Leung CK (2019) Machine learning based crack mode classification from unlabeled acoustic emission waveform features. Cem Concr Res 121:42–57
28. Gui G, Pan H, Lin Z, Li Y, Yuan Z (2017) Data-driven support vector machine with optimization techniques for structural health monitoring and damage detection. KSCE J Civ Eng 21(2):523–534
29. Ghiasi R, Torkzadeh P, Noori M (2016) A machine-learning approach for structural damage detection using least square support vector machine based on a new combinational kernel function. Struct Health Monit 15(3):302–316
30. Pan H, Azimi M, Yan F, Lin Z (2018) Time-frequency-based data-driven structural diagnosis and damage detection for cable-stayed bridges. J Bridg Eng 23(6):04018033
31. Vitola J, Pozo F, Tibaduiza DA, Anaya M (2017) A sensor data fusion system based on k-nearest neighbor pattern classification for structural health monitoring applications. Sensors 17(2):417
32. Neves AC, Gonzalez I, Leander J, Karoumi R (2017) Structural health monitoring of bridges: a model-free ANN-based approach to damage detection. J Civ Struct Heal Monit 7(5):689–702
33. Sierra-Pérez J, Torres-Arredondo MA, Alvarez-Montoya J (2017) Damage detection methodology under variable load conditions based on strain field pattern recognition using FBGs, nonlinear principal component analysis, and clustering techniques. Smart Mater Struct 27(1):015002

34. Diez A, Khoa NLD, Alamdari MM, Wang Y, Chen F, Runcie P (2016) A clustering approach for structural health monitoring on bridges. J Civ Struct Heal Monit 6(2016):1–17
35. Sarkar S, Reddy KK, Giering M, Gurvich MR (2016) Deep learning for structural health monitoring: a damage characterization application. In: Proceedings of the annual conference of the prognostics and health management society 2016, Denver, CO, USA, October 3, 2016
36. Miorelli R. Kulakovskyi A, Mesnil O, d'Almeida O (2019) Automatic defect localization and characterization through machine learning based inversion for guided wave imaging in SHM. AIP Conf Proc 2102(1):050005
37. Finotti RP, Cury AA, Barbosa FDS (2019) An SHM approach using machine learning and statistical indicators extracted from raw dynamic measurements. Latin American J Solids Struct 16(2):e165
38. Taddei T, Penn JD, Yano M, Patera AT (2018) Simulation-Based Classification; A model-order-reduction approach for structural health monitoring. Arch Computat Methods Eng 25:23–45
39. Tibaduiza D, Cerón-MH (2017) Damage classification based on machine learning applications for an unmanned aerial vehicle. The 11th international workshop on structural health monitoring. Stanford, CA, USA, September 12, 2017
40. Tripathi G, Anowarul H, Agarwal K, Prasad DK (2019) Classification of micro-damage in piezoelectric ceramics using machine learning of ultrasound signals. Sensors 19(19):4216
41. Salehi, H., Das, S., Chakrabartty, S., Biswas, S. & Burgueño, R. (2018). Damage identification in aircraft structures with self-powered sensing technology: a machine learning approach. Structural Control Health Monitoring, 25, e2262.
42. Vitola J, Pozo F, Tibaduiza D, Anaya M (2017) Distributed piezoelectric sensor system for damage identification in structures subjected to temperature changes. Sensors 17(6):1252
43. Melia T, Cooke A, Grayson S (2016) Machine learning techniques for automatic sensor fault detection in airborne SHM networks. In: Proceedings of the 8th European workshop on structural health monitoring (EWSHM), Bilbao, Spain, July 5, 2016
44. Vashisht R, Viji H, Sundararajan T, Mohankumar D, Sarada S (2018) Structural health monitoring of cantilever beam, a case study—using Bayesian neural network and deep learning. In: Proceedings of the 13th international conference on systems. Athens, Greece, April 22, 2018
45. Liu J, Yang X, Zhu M (2019) Neural network with confidence kernel for robust vibration frequency prediction. J Sensors 2019:6573513
46. Avci O, Abdeljaber O, Kiranyaz S, Boashash B, Sodano H, Inman D (2018) Efficiency validation of one dimensional convolutional neural networks for structural damage detection using a SHM benchmark data. In: Proceedings of 25th international congress on sound and vibration 2018 (ICSV 25), Hiroshima, Japan, July 8, 2018
47. Tibaduiza D, Torres-Arredondo MA, Oyaga J, Anaya M, Pozo F (2018) A damage classification approach for structural health monitoring using machine learning. Complexity 2018:5081283
48. Vitola J, Tibaduiza D, Anaya M, Pozo F (2016) Structural damage detection and classification based on machine learning algorithms. In: Proceedings of the 8th European workshop on structural health monitoring (EWSHM), Bilbao, Spain, July 5, 2016
49. Bao Y, Tang Z, Li H (2019) Compressive-sensing data reconstruction for structural health monitoring: a machine-learning approach. Struct Health Monit 19(1):293–304
50. Cadini F, Sbarufatti C, Corbetta M, Cancelliere F, Giglio M (2019) Particle filtering-based adaptive training of neural networks for real-time structural damage diagnosis and prognosis. Struct Control Health Monit (26):e2451
51. Zhao Z, Yua M, Dong S (2019) Damage location detection of the CFRP composite plate based on neural network regression. In: Proceedings of the 7th asia-pacific workshop on structural health monitoring. Hong Kong, China, November 12, 2018.
52. Wang H, Yuan S, Chen J, Ren Y (2018) Online updating Gaussian process model for fatigue crack diagnosis and prognosis. 9th European workshop on structural health monitoring July 10, 2018. Manchester, United Kingdom.
53. Gardner P, Lord C, Barthorpe RJ (2018) A Probabilistic framework for forward model-driven SHM. 9th European workshop on structural health monitoring. July 10, 2018, Manchester, United Kingdom

54. Hoang N-D, Liao K-W, Tran X-L (2018) Estimation of scour depth at bridges with complex pier foundations using support vector regression integrated with feature selection. J Civ Struct Heal Monit 8(3):431–442
55. Quaranta G, Lopez E, Abisset-Chavanne E, Duval J, Huerta A, Chinesta F (2019) Structural health monitoring by combining machine learning and dimensionality reduction techniques. Revista internacional de métodos numéricos para cálculo y diseño en ingenieria. 35(1):20
56. Ye X, Chen X, Lei Y, Fan J, Mei L (2018) An Integrated Machine Learning Algorithm for Separating the Long-Term Deflection Data of Prestressed Concrete Bridges. Sensors 18(11):4070
57. Rahim SA, Manson G, Worden K (2016) Data visualization approach for operational loading variations of an aircraft wing box using vibration based damage detection. In: Proceedings of the 8th European workshop on structural health monitoring (EWSHM). Bilbao, Spain, July 5, 2016
58. Santos A, Figueiredo E, Silva M, Santos R, Sales C, Costa J (2016) Geneticbased EM algorithm to improve the robustness of Gaussian mixture models for damage detection in bridges. Struct Control Health Monitor 24(3):e1886
59. Senniappan V, Subramanian J, Papageorgiou E, Mohan S (2016) Application of fuzzy cognitive maps for crack categorization in columns of reinforced concrete structures. Neural Comput Appl 28(1):107–117
60. Liu J, Berges M, Bielak J, Garrett J, Kovacevic J, Noh HY (2019) A damage localization and quantification algorithm for indirect structural health monitoring of bridges using multi-task learning. 45th annual review of progress in quantitative nondestructive evaluation 2102(1)
61. Karypidis D, Gil Berrocal C, Rempling R, Granath M, Simonsson P (2019) Structural Health Monitoring of RC structures using optic fiber strain measurements: a deep learning approach. The Evolving Metropolis IABSE Congress New York City 2019:114
62. Dervilis N, Papatheou E, Antoniadou I, Cross EJ, Worden K (2016) On the usage of active learning for SHM. In: Proceedings of ISMA2016. ISMA 2016, September 19, 2016, Leuven, Belgium
63. Adadi A, Berrada M (2018) 2018), Peeking Inside the Black-Box: A Survey on Explainable Artificial Intelligence (XAI. IEEE Access 6:52138–52160
64. Montavon G, Samek W, Müller K-R (2018) Methods for interpreting and understanding deep neural networks. Digital Signal Processing 73(2018):1–15
65. Barredo Arrieta A, Díaz-Rodríguez N, Del Ser J, Bennetot A, Tabik S, Barbado A, Garcia S, Gil-Lopez S, Molina D, Benjamins R, Chatila R, Herrera F (2020) Explainable Artificial Intelligence (XAI): Concepts, taxonomies, opportunities and challenges toward responsible AI. Information Fusion 58:82–115

Physics-Informed Machine Learning for Structural Health Monitoring

Elizabeth J. Cross, S. J. Gibson, M. R. Jones, D. J. Pitchforth, S. Zhang, and T. J. Rogers

Abstract The use of machine learning in structural health monitoring is becoming more common, as many of the inherent tasks (such as regression and classification) in developing condition-based assessment fall naturally into its remit. This chapter introduces the concept of physics-informed machine learning, where one adapts ML algorithms to account for the physical insight an engineer will often have of the structure they are attempting to model or assess. The chapter will demonstrate how grey-box models, that combine simple physics-based models with data-driven ones, can improve predictive capability in an SHM setting. A particular strength of the approach demonstrated here is the capacity of the models to generalize, with enhanced predictive capability in different regimes. This is a key issue when life-time assessment is a requirement, or when monitoring data do not span the operational conditions a structure will undergo. The chapter will provide an overview of physics-informed ML, introducing a number of new approaches for grey-box modelling in a Bayesian setting. The main ML tool discussed will be Gaussian process regression, and we will demonstrate how physical assumptions/models can be incorporated through constraints, through the mean function and kernel design, and finally in a state-space setting. A range of SHM applications will be demonstrated, from loads monitoring tasks for off-shore and aerospace structures, through to performance monitoring for long-span bridges.

Keywords Physics-informed machine learning · Grey-box modelling · Gaussian process regression.

E. J. Cross (✉) · S. J. Gibson · M. R. Jones · D. J. Pitchforth · S. Zhang · T. J. Rogers
Dynamics Research Group, Department of Mechanical Engineering, University of Sheffield, Mappin St, Sheffield S13JD, UK
e-mail: e.j.cross@sheffield.ac.uk

A. Cury et al. (eds.), *Structural Health Monitoring Based on Data Science Techniques*, Structural Integrity 21, https://doi.org/10.1007/978-3-030-81716-9_17

1 Introduction

As performance and monitoring data from our structures become more abundant, it is natural for researchers to turn to methods from the machine learning community to help with analysis and construction of diagnostic/prognostic algorithms. Indeed, within the SHM research field, use of neural networks, support vector machines and Gaussian processes for regression and classification problems has become common place [1]. These methods bring the opportunity to learn complex relationships directly from data, without a requirement of in-depth knowledge of the system. As an example from the authors' own work, in [2] we employed a Gaussian process (GP) regression to predict strain on a landing gear from measured accelerations across the aircraft. Use of a suitably trained GP circumvents the need to build complex physics-based models of the gear for fatigue life calculations. This kind of model is often referred to as a 'black-box' model to reflect the fact the data drives the structure of the model rather than knowledge of the physics at work.

At the other end of the spectrum the term 'white-box' model can be used to describe a model purely constructed from knowledge of physics, (e.g. differential equations and finite element models). Physics-based modelling and updating were common early themes in the structural health monitoring research field [3]. However, for large or critical engineering structures that operate in (often extreme) dynamic environments, such as wind turbines, aircraft and gas turbines, predictive modelling from a white-box perspective presents particularly difficult challenges. Loading is often unknown and unmeasured, and dynamic behaviour during operation needs to be fully captured by a computational model, but is sensitive to small changes in (or disturbances to) the structure. Validation and updating of large complex models bring their own challenges and remain active research areas [4–7].

Due to the availability of monitoring data, the inherent challenges of the physics-based approach and the promise of machine learning methods, it is fair to say that the data-driven approach to SHM has become dominant in the research field. A significant issue with the use of any machine learning method in an engineering application, however, is the availability of suitable data with which to train the algorithm. As the model learns from the data, it is only able to accurately predict behaviour present in the data on which it was trained. As an example, Fig. 1 shows a black-box model trained to predict the bending strain on an aircraft wing during different manoeuvres to inform an in-service fatigue assessment. This data set comprises of 84 flights, five of which are used for model training. The trained model is able to generalize well with a very low prediction error for the majority of the flights—Fig. 1a shows a typical strain prediction for a flight not included in the training set (normalized mean-squared error[1] (nMSE)=0.29% across the whole flight). However, for the flight shown in Fig. 1b, the model is unable to predict the strain as accurately (nMSE 4.20% across

---

1

$$\text{nMSE} = \frac{100}{n\sigma_y^2} \sum (y_i - f_i)^2 \tag{1}$$

where y_i and f_i are the measurements and predictions, respectively, $i = 1 \ldots n$.

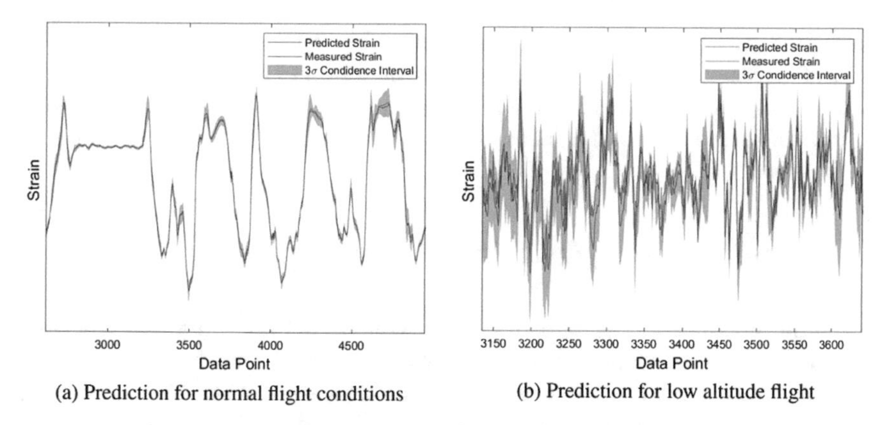

(a) Prediction for normal flight conditions (b) Prediction for low altitude flight

Fig. 1 Example of difficulty predicting behaviour outside of normal operation conditions—strain prediction on an aircraft wing during manoeuvres—see [8, 9] for more details

the whole flight)—this flight was atypical in terms of operating conditions—it was a low-altitude sortie over ground, characterized by the turbulent response one can see in the figure. These conditions are different from those included in the training set and the model is unable to generalize and predict the strain as well in this case.

In general, but especially because of the inherent flexibility in many of the machine learning models commonly used, extrapolation should not be attempted in this setting. For an SHM application, this will generally mean that training data are required from all possible operating conditions that the structure will see. For many applications, this is currently infeasible, although as data collection becomes more commonplace, the situation will improve somewhat. Where a supervised approach is needed, this problem is exacerbated by the general lack of access to data from structures in a damaged state which remains a large barrier to effective diagnosis and prognosis [10].

Currently a programme of work by the authors is pursuing a physics-informed machine learning approach to attempt to address some of these issues in a structural dynamics setting. The aim is to bring together the flexibility and power of state-of-the-art machine learning techniques with more structured and insightful physics-based models derived from domain expertise. This reflects a natural wish that any inferences over our structures will be informed by both our engineering knowledge and relevant monitoring data available.

The potential means of combining physics-based models and data-driven algorithms are many, ranging from employing ML methods for parameter estimation [11, 12], to using them as surrogates or emulators [13–15]. Of interest here are methods where the explanatory power of a model is shared between physics-based and data-driven components. We will often refer to these approaches as 'grey-box' models (a combination of white and black-box components), but the term 'hybrid modelling' is equally applicable. The philosophy followed in our work is to embed fundamental physical insight into a machine learning algorithm. In doing so, our aim is that the role of the machine learner is one of augmenting the explanatory power of the model rather than being employed to correct any potential error or bias in the physical foundation. This chapter will explore this idea in a Bayesian setting, introducing a number of different approaches and demonstrating their usefulness in an SHM setting.

2 Grey-Box Models, Overview and Literature

The term 'grey-box model' is perhaps most familiar to those from a control engineering background. Sohlberg [16, 17] provides a useful review and overview of grey-box models in this context.[2] Figure 2 attempts to capture and summarise some of the currently available modelling approaches relevant for challenges in structural health monitoring on the white to black spectrum. Note that the 'degree of greyness' of the models in the middle region will change according to implementation and application.

At the whiter end of the spectrum are modelling approaches where data are used for parameter estimation or model form selection, (with the buoyant field of equation discovery fitting in here, see [18, 19]). See also [20]. *Residual models* are those that use a data-driven approach to account for the observed difference between a physics-based model and measurements, with general form

$$y = \underbrace{f(x)}_{\text{White-box}} + \underbrace{\delta(x) + \varepsilon}_{\text{Black-box}} \tag{2}$$

where $f(x)$ is the output of the physical model, $\delta(x)$ is the model discrepancy and ε is the process noise (see for example [21–24]). The discrepancy term is often used to correct a misspecified physical model, giving rise to the term 'bias correction'. Residual-based approaches have proven effective across a range of SHM tasks including damage detection [25] and modal identification [26]. Here, we are interested in residual modelling in the context of compensation for uncaptured/missing behaviours in the physics-based model (discussed further in Sect. 3).

The term *hybrid architectures* reflects the wider possibilities for combinations of white and black models (which could include the summation form of (2)). Section 5 will demonstrate one such example of combining data-driven and physics-based models in a state-space setting.

The remainder of the spectrum contains models with structures that are data-driven/black-box in nature. Sohlberg [16]) describes *semi-physical modelling* as when features are subject to a nonlinear transformation before being used as inputs to a black-box model, we also refer to this as *input augmentation*, see [8, 27, 28]

[2] The use of grey-box models within the control community is undergoing somewhat of a revival and a good snap-shot of this can be gained by looking at the contributions to the most recent Nonlinear System Identification Benchmarks Workshop (http://www.nonlinearbenchmark.org).

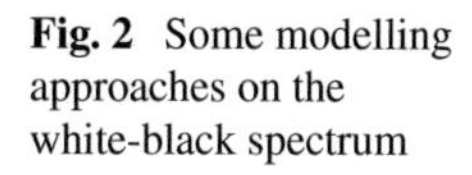

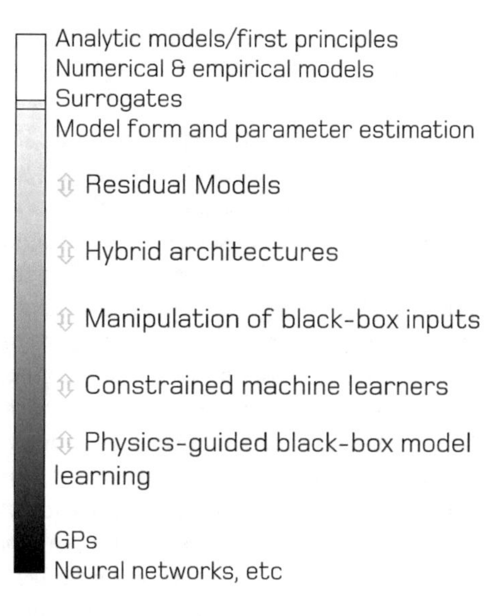

Fig. 2 Some modelling approaches on the white-black spectrum

for more examples. We place these examples under the heading of *manipulation of black-box inputs*.

Section 4 of this chapter will discuss *constraints* for machine learning algorithms—these are methods that allow one to constrain the predictions of a machine learner so that they comply with physical assumptions. Excellent examples for Gaussian process regression are [29–31] and will be discussed in greater detail later.

The final grouping of grey-box approaches mentioned here are physics-guided black-box learners. These are methods that use physical insight to attempt to improve model optimisation and include the construction of physics-guided loss functions and the use of physics-guided initialization. These will not be discussed further in this chapter but see e.g. [32–35] for more details.

2.1 Grey-Box Models for SHM

The remainder of the chapter will showcase some of the work of the authors on developing physics-informed machine learning approaches for SHM tasks. The developments here fall in the domain of residual and hybrid models (Sects. 3 and 5), and constrained machine learners (Sect. 4). In order to provide an overview, a variety of methods and results are presented, however, the implementation details given here are necessarily very brief and we refer readers to the referenced papers and our webpage[3] for specific details and more in-depth analysis. Reflecting the philosophy discussed in the introduction section, the approaches presented generally incorporate simple

[3] https://drg-greybox.github.io/.

physics-based models or assumptions and rely on the machine learner for enhanced explanatory power and flexibility (i.e. we are operating towards the blacker end of the scale).

The machine learning approach used in the work shown here will be Gaussian process (GP) regression throughout. GPs have been shown to be a powerful tool for regression tasks [36] and are becoming common in SHM applications (see for example [2, 37–40]). Their use here and throughout the work of the authors is due to their (semi)non-parametric nature, their ability to function with a small number of training points, and most importantly, the Bayesian framework within which they naturally work. The Gaussian process formulation provides a predictive *distribution* rather than a single prediction point, allowing confidence intervals to be calculated and uncertainty to be propagated forward into any following analysis (see [9] for example). As the use of GPs is now quite common, their fundamental formulation will not be introduced here, but the mathematical machinery required is briefly summarized in the Appendix—we refer unfamiliar readers to [36].

In the first examples shown here, the use of priors in the Bayesian framework is exploited as an appropriate and intuitive means of incorporating physical insight into a machine learning algorithm. In later sections, we consider the construction of constraints for GPs and, separately, their incorporation into a state-space formulation (this latter example relies more heavily on physics-based machinery than the other examples).

3 Be More Bayes

A Bayesian philosophy is one that employs evidence from data to update prior beliefs or assumptions, and has been widely adopted across disciplines, including SHM. However, most commonly, uninformative priors are utilized that do not reflect the knowledge that we have as engineers of the systems we are interested in modelling.

The formulation of a Gaussian process regression requires the selection of a mean and covariance function which form the prior process. The process is then conditioned with training data to provide a posterior mean and covariance as the model prediction. In the standard approach, no prior knowledge is assumed; a zero mean function is selected alongside a generic covariance function such as a squared-exponential or one from the Matérn class which provide a flexible process to fit to most data.

In this section, we will first employ simple physical models as prior mean functions to a GP and show how they may improve the extrapolative capability of the model. This simple means of incorporating prior knowledge is equivalent to using a GP with a zero mean prior to model the difference between the measured data and the physical model prediction and hence can be classed as a residual approach (see Sect. 2). At the end of this section, we will show how some knowledge of a system may be used to derive useful covariance functions in a regression setting.

3.1 Prior Mean Functions—Residual Modelling

3.1.1 Performance Monitoring of a Cable-Stayed Bridge

The Tamar bridge is a cable-supported suspension bridge connecting Saltash and Plymouth in the South West of England which has been monitored by the Vibration Engineering Section at the University of Exeter [41]. The interest here is in the development of a model to predict bridge deck deflections that can be used as a performance indicator (see [42, 43]). The variation in deck deflections are driven by a number of factors, including fluctuating temperature and loading from traffic (which are included as inputs to the model). Figure 3a shows the regression target considered in this example, which is a longitudinal deflection. The monitoring period shown is from September (Autumn) to January (Winter). In this figure, one can see short-term fluctuations (daily) and a longer-term trend which is seasonal and driven by the increased hogging of the bridge deck as the ambient temperature decreases into the winter months. To mimic the situation where only a limited period of monitoring data is available for the establishment of an SHM algorithm, data from the initial month of the monitoring period is used to establish a GP regression model for deflection prediction (see [44] for more details).

A GP prediction, using the standard approach of a zero mean prior, is shown in 3b. Here one can see exactly the behaviour that is expected; the model is able to predict the deck deflections well in and around the training period, but is unable to predict the deflections in colder periods towards the end of the time series. The confidence intervals widen to reflect that the inputs to the model towards the end of the period are different from those in the training set—this demonstrates the usefulness of the GP approach, as one knows to place less trust in the predictions from this period.

To formulate a grey-box model for this scenario, a physics-informed prior mean function is adopted that encodes the expected linear expansion behaviour of stay-cables with temperature [45]. Figure 3c shows the GP prediction with a linear prior mean function, where one can see a significant enhancement of predictive capability across the monitoring period. Where temperatures are at their lowest, the model predictions fall back on the prior mean function allowing some extrapolative capability. The prediction error is significantly smaller for the grey-box model in this case; the 'black-box' nMSE is 68.65, whereas the GP with the physics-informed mean function has an nMSE of 7.33.

3.1.2 Residual Modelling for Wave Loading Prediction

In this example, we follow a similar approach of adopting a physics-informed mean function, this time with a dynamic Gaussian process formulation, a GP-NARX [46], to enhance predictive capability for a wave loading assessment. The monitoring or prediction of the loads a structure experiences in service is an important ingredient

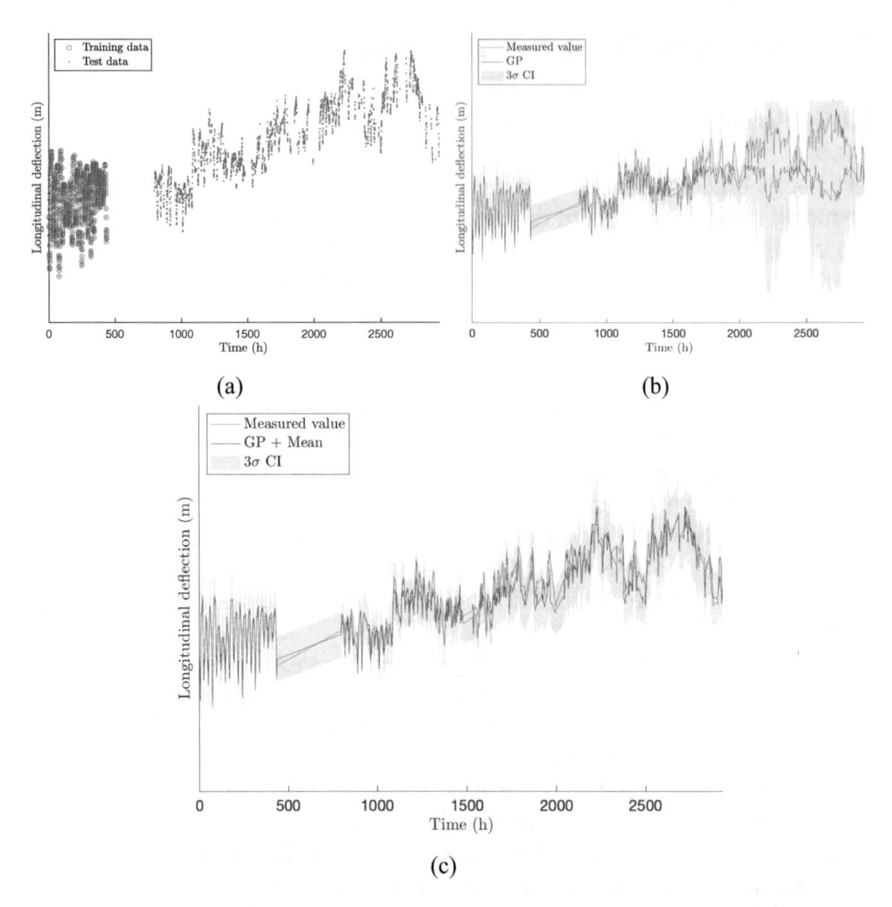

Fig. 3 Model for bridge deck deflections; **a** shows the training and test datasets for the GPs, **b** is a GP prediction with a zero mean function prior and **c** shows the prediction when a simple physics-informed mean function is incorporated. See [44] for more details

for health assessment, particularly where one wishes to infer, e.g. fatigue damage accrued.

The implementation of residual modelling is most effective where the assumptions and limitations of the white-box model are well understood. As a widely used method for wave loading prediction, Morison‘s equation [47] is employed here as a physics-informed mean function. This empirical law is known to simplify the behaviour of wave loading, not accounting for effects such as vortex shedding or other complex behaviours [48] and will typically have residual errors in the region of 20%[49]. Here we consider the addition of a data-based GP-NARX to a simplified version of Morison’s Equation in an attempt to account for these missing phenomena. The model used is:

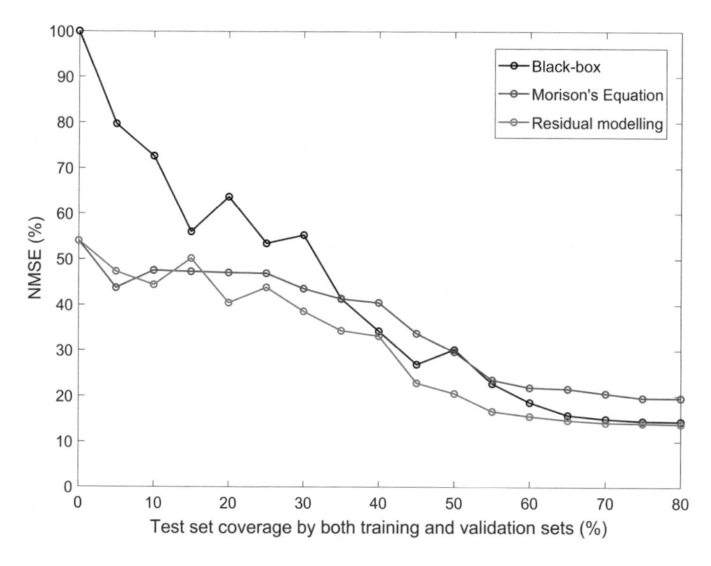

Fig. 4 A comparison of wave loading prediction model NMSEs versus test set coverage. Increasing coverage of the test set by the training and validation sets results in an increased level of model interpolation. See [50] for more details

$$y_t = \underbrace{C'_d U_t |U_t| + C'_m \dot{U}_t}_{\text{Morison's Equation}} + \underbrace{f([u_t,\ u_{t-1},\ ...,\ u_{t-l_u},\ y_{t-1},\ y_{t-2},\ ...,\ y_{t-l_y}]) + \varepsilon}_{\text{GP-NARX}} \quad (3)$$

where y_t is the wave force, C'_d is the drag coefficient, C'_m is the inertia coefficient, U is the wave velocity, $\dot{U}$ is the wave acceleration, $u_{t:t-l_u}$ are lagged exogeneous inputs and $y_{t-1:t-l_u}$ are the lagged wave force, see [50] for more details (this paper also shows an example of an input augmentation model, where Morison‘s equation is used as an additional input to the GP-NARX).

The Christchurch bay dataset is used here as an example to demonstrate the approach [51]. To explore the generalization capability with and without the physics-informed mean, different training sets for the GP-NARX are considered with increasing levels of coverage of the input space. A comparison of model errors (nMSE) with different training datasets is shown in Fig. 4. The coverage level is indicated as a percentage of the behaviour observed in the testing set that is also encountered in the training set [50].

As in the Tamar Bridge example, the model structure offers a significant improvement in extrapolation, where testing conditions are different from those in training dataset. Following this approach allows predictions to be informed by the prior mean in the absence of evidence from data. Clearly the prior specification is very important in this case and a misspecified prior could do more harm than good. Once again we advocate the use of simple and well founded physics-based models in an attempt to avoid this issue.

3.2 Physics-Derived Covariance Functions

As discussed above, in a standard approach to GP regression, a generic covariance function such as a squared-exponential or one from the Matérn class is selected as a prior. In the posterior GP, the mean is a weighted sum of observations in the training set (see Appendix), with the weightings provided by the covariance function and associated matrix. These commonly used functions encode that the covariance between points with similar inputs will be high and this allows the model to be data-driven in nature.

In the case where one has some knowledge of a process of interest, it is possible to derive a covariance function that reflects this. As an example, in [52], a composite covariance function is designed to reflect the characteristics of the guided waves being modelled.

For some stochastic processes, the (auto) covariance can be directly derived from the equation of motion of a system. An example relevant for vibration-based SHM is the single degree of freedom (SDOF) oscillator

$$m\ddot{y}(t) + c\dot{y}(t) + ky(t) = F(t) \tag{4}$$

with mass, damping and stiffness parameters, m, c, k, respectively, driven by a forcing process $F(t)$. In the case where the forcing is Gaussian white noise, the response Y is a Gaussian process with (auto)covariance

$$\phi_{Y(\tau)} = \mathbb{E}[Y(t_1)Y(t_2)] = \frac{\sigma^2}{4m^2\zeta\omega_n{}^3} e^{-\zeta\omega_n|\tau|}(\cos(\omega_d\tau) + \frac{\zeta\omega_n}{\omega_d}\sin(\omega_d|\tau|)) \tag{5}$$

where standard notation has been used; $\omega_n = \sqrt{k/m}$, the natural frequency, $\zeta = c/2\sqrt{\mathrm{km}}$, the damping ratio, $\omega_d = \omega_n\sqrt{1-\zeta^2}$, the damped natural frequency. See [53] and also [54, 55].

This covariance function can be readily used in the regression context and provides a useful prior process for oscillatory systems with a response dominated by a single frequency. This form of covariance function can be described as *expressive* [56] and proves useful even when the equation of motion of the system of interest differs from an SDOF linear assumption.

Figure 5 shows an example of a GP regression for a system with a cubic nonlinearity. Here the linear prior provides an appropriate structure for the regression and is flexible enough to incorporate the nonlinearity in the response. This is a simulated example with the GP training data shown with crosses in the figure (every 8th point)—here the nMSE is 8.09. For comparison, a GP with a squared-exponential (SE) covariance function is established with the same training data. The SE process smooths through the data as expected (nMSE = 66.7), whereas the derived covariance provides structure through the prior, resulting in good prediction during interpolation. The hyperparameters in the SDOF covariance function are physically interpretable, we are, therefore, able to guide their optimisation by providing the likely ranges for

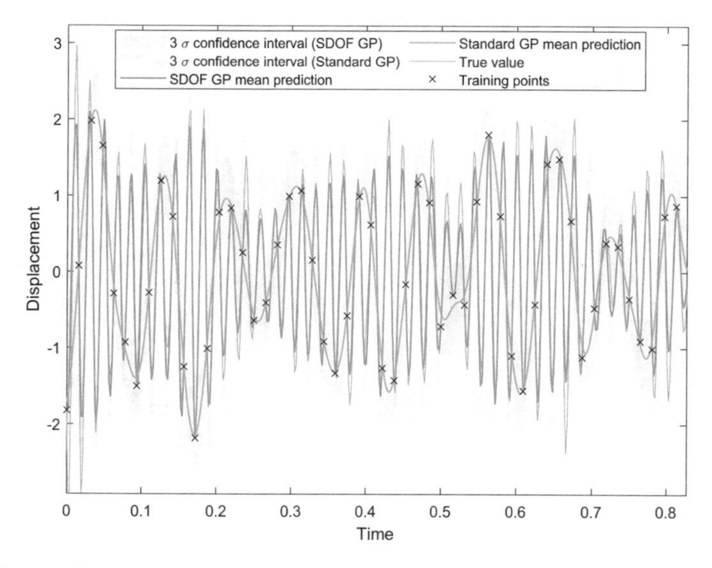

Fig. 5 A comparison of a GP regression using derived and squared-exponential covariance function. See [53] for more details

the system of interest. Here the benefit of being able to prescribe the likely frequency content of the system within the prior is clear and provides much advantage over the black-box approach—see [53] for more details.

4 Constrained Gaussian Processes

In scenarios where one lacks significant knowledge of the governing equations and solutions, grey-box methods that tend towards the black end of the spectrum can be particularly useful. An example of such approaches are constrained machine learners, which, very generally, aim to embed physical constraints into the learning procedure such that predictions made by the black-box model then adhere to these constraints

In the context of Gaussian process regression, there are a number of ways of constraining predictions, the simplest of which is to include data from boundaries within the model training. Other methods rely on applying constraints to the covariance function in a multiple output setting (see e.g. [29–31] and [57] where we employ derivative boundaries for beam deflection predictions).

Here we show an example of building known boundaries (geometry) into a GP regression via a sparse approximation of the covariance function. The approximation method relies on an eigendecomposition of the Laplace operator of a fixed domain [58]:

$$k(\mathbf{x}, \mathbf{x}') \approx \sum_{i}^{m} S(\sqrt{\lambda_i})\phi_i(\mathbf{x})\phi_i(\mathbf{x}'), \tag{6}$$

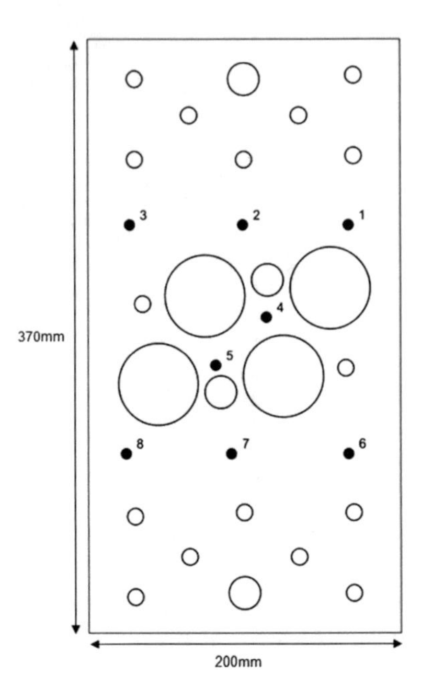

Fig. 6 Test structure schematic with sensor locations, recreated from [61]

with ϕ_i and λ_i the eigenfunctions and values, and S the spectral density of the covariance function. If one chooses the fixed domain to reflect the geometry of the problem of interest, then inference with this model is appropriately bounded (see [59, 60] for more details).

As an example here we employ constraints for a crack localisation problem via measurement of Acoustic emission (AE). The localization approach taken is to use artificial source excitations and an interpolating GP to provide a map of the differences in times of arrival (ΔT) of AE sources to fixed sensor pairings across the surface of the structure [61, 62]. Once constructed, the map can be used to assess the most likely location of any new AE sources. The bounded GP approximation allows one to build in the geometry of the structure under consideration.

To investigate the predictive capability of the constrained GP, a case study using a plate with a number of holes in is adopted. The holes, as shown in Fig. 6, provide complexity to the modelling challenge, introducing several complex phenomena such as wave mode conversion and signal reflection. Depending on the location of the source and sensor, the holes may also shield a direct propagation path to the receiver [61], adding further complication.

Neumann boundaries are imposed here around each hole and at the edge of the plate. To compare the performance of the standard and bounded GPs, differing amounts/coverage of artificial source excitations were used for model training. The initial characterization of a structure via artificial source excitation can be expensive and time consuming, for structures in operation it may also be infeasible to access all areas/components. To mimic the scenario where it is not possible to collect arti-

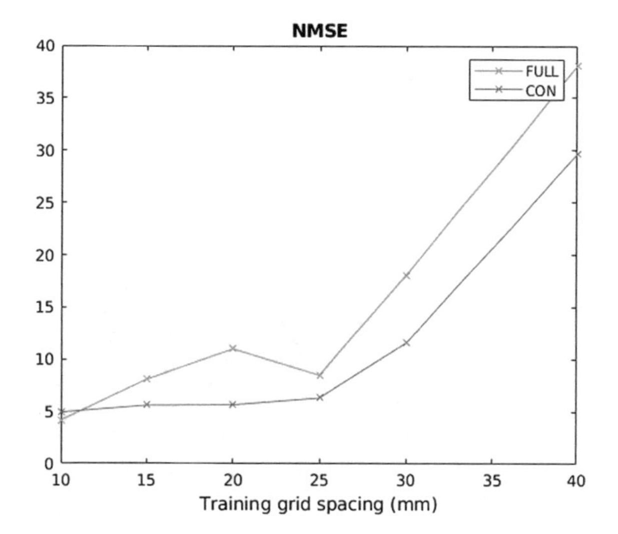

Fig. 7 Comparison between models errors for standard and bounded GPs for AE source localization study. The nMSE is averaged across all sensor pair models for each training set considered

ficial source excitations across a whole structure, here we restrict the training grid to excitation points in the middle of the plate. Figure 7 compares the performance of the standard and bounded GPs with training sets of varying grid densities. For each training set, the prediction error (nMSE) on the test set is averaged across every sensor pair (there are 8 sensors).

From Fig. 7 one can see that as the training set size reduces, the constrained GP consistently outperforms the standard full GP. This is particularly encouraging as the bounded GP remains a sparse approximation. As is consistent with our earlier observations, the inbuilt physical insight aids inference where training data are fewer. Figure 8 shows the difference in prediction error across the plate for the standard and bounded GPs for the 20 mm spacing training case and a single sensor pairing. In this case, the squared error of the full GP is subtracted from the squared error of the constrained GP, i.e. positive values indicate a larger error in the full GP, while a negative value expresses a larger error in the constrained GP.

The figure highlights the locations on the plate where the constrained GP more accurately predicts the true ΔT values. As expected, the locations at which this effect is most prominent are those that move further away from the training points, and particularly towards the extremities of the domain. At these locations, it is clear that the additional physical insight provided by the constrained GP is able to enhance the predictive performance in comparison with the pure black-box model.

5 Gaussian Processes in a State-Space Approach

One of the canonical forms for dynamic models, in structural mechanical systems and beyond, is the state-space representation of the behaviour of interest.

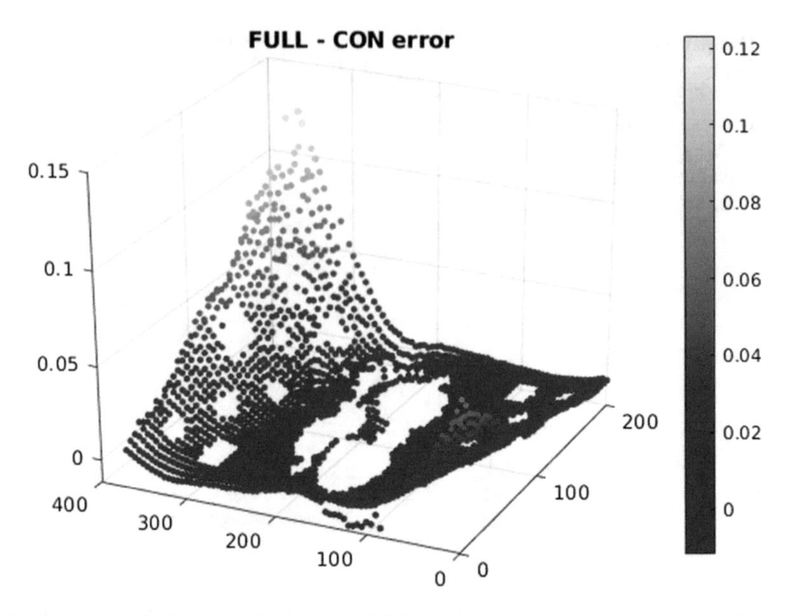

Fig. 8 Mapping of the difference between the full GP squared error and the constrained GP squared error across the test set for sensor pair 4–8

In the context of this work, the state-space model (SSM) is considered to be a probabilistic object defined by two key probability densities; a transition density $p(\mathbf{x}_{t+1} \mid \mathbf{x}_t, \mathbf{u}_t)$ and observation density $p(\mathbf{y}_t \mid \mathbf{x}_t, \mathbf{u}_t)$. The transition density relates the hidden states at a given time $\mathbf{x}_{t+1}$ to their previous possible values,[4] $\mathbf{x}_{0:t}$ and previous external inputs to the system, e.g. forcing, $\mathbf{u}_{0:t}$. The observation model relates available measurements $\mathbf{y}_t$ to the hidden states $\mathbf{x}_t$, which may also be dependent on the external inputs at that time $\mathbf{u}_t$.

The state-space formulation can be used to properly account for measurement noise (filtering and smoothing) and is commonly used for parameter estimation (the well-known Kalman filter is a closed form solution for linear and Gaussian systems). Use of the state-space models as a grey-box formulation in this setting is common within the control community [63, 64].

Here, we are interested in the case where we only have partial knowledge of a system—this could take the form of missing or incorrect physics in the equations of motion, or could be a lack of access to key measurements such as the force a system undergoes. In Sect. 3, we considered a GP-NARX formulation for wave loading prediction, with the ultimate aim of informing a fatigue assessment. The state-space formulation shown here offers an alternative means for load estimation which simultaneously provides parameter and state estimation in a Bayesian setting.

[4] The notation *subscript* $a : b$ is used to denote values in that range inclusively, e.g. $\mathbf{x}_{0:t}$ is the value of the states $\mathbf{x}$ at all times from $t = 0$ to $t = t$.

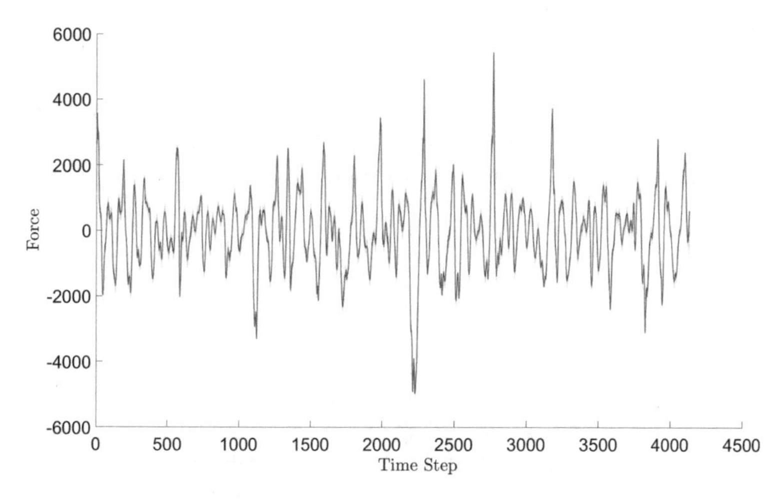

Fig. 9 Force estimation for MDOF simulation under Christchurch bay loading time history[40]

Joint input-state and input-state-parameter problems have seen growing interest in recent years, see, for example [65–69]. The approach shown here is one that considers a representation of a Gaussian process within a state-space formulation to model the unknown forcing (following [70, 71]). This is achieved by deriving the transfer function of a Matérn kernel (via its spectral density), which provides a flexible model component to account for the unmeasured behaviour. Inference over the state-space model is via Markov chain Monte Carlo to provide distributions for parameter and hyperparameter estimations.

Figure 9 shows an example of force recovery for a simulated multidegree of freedom system excited by a forcing time history from the Christchurch bay example discussed in Sect. 3. Here one can see that the force has been accurately inferred, the nMSE in this case is 1.15. For more details and analysis see [40].

The inference problem becomes significantly more challenging for nonlinear systems, and even more so if our knowledge of that nonlinearity is incomplete. Recently [72] has attempted this extension for the input-state estimation case for a known nonlinearity. The difficulty in inference is met there by employing a methodology based on Sequential Monte Carlo, specifically particle Gibbs with ancestor sampling, to allow recovery of the states and the hyperparameters of the GP.

In the face of an unknown nonlinearity, this framework may also be employed, where the GP may be used to account for missing behaviours from the assumed equations of motion. In [73], it is shown how this approach can be applied to a Duffing oscillator to learn the unknown cubic component of the model in a Bayesian manner without requiring prior knowledge of the nonlinear function. There are two particular advantages to this approach, the first is that it allows nonlinear system identification with a linear model, the Kalman filter and RTS smoother, which has significant computational advantages and arguably makes "fully Bayesian" inference industrially feasible. The second benefit is that it allows a user to apply very weak prior knowledge about the nonlinearity; in the language of grey-box models, there is a strong white box component (the second-order linear system) but the form of the nonlinearity is the very flexible nonparametric GP. Contrast this approach with the purely black-box alternative of the GP-SSM, see [74], which suffers from significant nonindentifiability and computational challenges.

6 Conclusions

This chapter has introduced and demonstrated physics-informed machine learning methods suitable for SHM problems and inference in structural dynamics more generally. The methods allow the embedding of one's physical insight of a structure or system into a data-driven assessment. The resulting models have proven to be particularly useful in situations where training data are not available across the operational envelope—a common occurrence in structural monitoring campaigns.

The Bayesian approach adopted in Sect. 3 allows predictions to fall back on a prior physical model in the absence of evidence from data. This pragmatic approach proved useful in the examples shown here but does rely on trusting the physical model in extrapolation. The ability to constrain the Gaussian process prior to known boundary conditions shown in Sect. 4 requires less physical insight and gives both an improved modelling performance, as well as providing the guarantee that predictions made adhere to known underlying physical laws of the system under consideration. At the whiter end of the spectrum, the state-space examples discussed in Sect. 5 provide a principled means of inference over structures with unknown forcing or nonlinearities. Some examples of where the presented methodology may be of benefit could include better understanding of fatigue damage accrual and parameter identification for, e.g. novelty/damage detection.

As well as providing an enhanced predictive capability, the models introduced here have the benefit of being more readily interpretable than their purely black-box counterparts. In the past, a barrier to the uptake of SHM technology has been the lack of trust owners and operators have in so-called black-box models. Perhaps naturally, there is a hesitancy to adopt algorithms not derived from physics-based models, but this may also be due, in part, to their misuse in the past. We hope that this will be ameliorated by more interpretable models which also have the benefit of being more easily optimized (Sect. 3 shows an example where the hyperparameters in the GP regression take on physical meaning).

Physics-informed machine learning is rapidly becoming a popular research field in its own right, with many promising results and avenues for investigation. This review paper [35] currently on arXiv has 300 references largely populated by papers from the last two years. It is likely that many of the emerging methods will prove useful in SHM. The work here has focussed on a Gaussian process framework, clearly the use of neural networks provide an alternative grey-box route, as these are also commonly used in our field. We look forward to seeing how these may be adopted for SHM tasks.

7 Gaussian Process Regression

Here we follow the notation used in [36]; $k(\mathbf{x}_p, \mathbf{x}_q)$ defines a covariance matrix K_{pq}, with elements evaluated at the points $\mathbf{x}_p$ and $\mathbf{x}_q$, where $\mathbf{x_i}$ may be multivariate.

Assuming a zero-mean function, the joint Gaussian distribution between measurements/observations $\mathbf{y}$ with inputs X and unknown/testing targets $\mathbf{y}^*$ with inputs X^* is

$$\begin{bmatrix} \mathbf{y} \\ \mathbf{y}^* \end{bmatrix} \sim \mathcal{N}\left(0, \begin{bmatrix} K(X,X) + \sigma_n^2 I & K(X,X^*) \\ K(X^*,X) & K(X^*,X^*) \end{bmatrix}\right) \tag{7}$$

The distribution of the testing targets $\mathbf{y}^*$ conditioned on the training data (which is what we use for prediction) is also Gaussian:

$$\begin{aligned} \mathbf{y}^* | X_*, X, \mathbf{y} \sim \mathcal{N}(&K(X^*,X)(K(X,X) + \sigma_n^2 I)^{-1}\mathbf{y}, \\ &K(X^*,X^*) - K(X^*,X)(K(X,X) + \sigma_n^2 I)^{-1} K(X,X^*)) \end{aligned} \tag{8}$$

See [36] for the derivation. The mean and covariance here are that of the posterior Gaussian process. In this work, covariance function hyperparameters are sought by maximizing the marginal likelihood of the predictions

$$\log p(\mathbf{y}|X, \boldsymbol{\theta}) = -\frac{1}{2}\mathbf{y}^T K^{-1}\mathbf{y} - \frac{1}{2}\log|K| - \frac{n}{2}\log 2\pi \tag{9}$$

via a particle swarm optimisation[5] [75].

Acknowledgements We would like to thank Keith Worden for his general support and also particularly for provision of the Christchurch Bay data and support of the state-space work. Additionally, thanks is offered to James Hensman, Mark Eaton, Robin Mills, Gareth Pierce and Keith Worden for their work in acquiring the AE data set used here. We would like to thank Ki-Young Koo and James Brownjohn in the Vibration Engineering Section at the University of Exeter for provision of the data from the Tamar Bridge. Thanks also to Steve Reed, formerly of DSTL for the provision of the Tucano data. Finally, the authors would like to acknowledge the support of the EPSRC, particularly through grant reference number EP/S001565/1, and Ramboll Energy for their support of SG and DP.

References

1. Farrar CR, Worden K (2010) An introduction to structural health monitoring. In: New trends in vibration based structural health monitoring, pp 1–17
2. Holmes Geoffrey, Sartor Pia, Reed Stephen, Southern Paul, Worden Keith, Cross Elizabeth (2016) Prediction of landing gear loads using machine learning techniques. Struct Health Monitor 15(5):568–582

[5] Yes, we could be more Bayes here.

3. Sohn H, Farrar CR, Hemez FM, Shunk DM, Stinemates DW, Nadler BR, Czarnecki JJ (2003) A review of structural health monitoring literature: 1996–2001. Los Alamos National Laboratory, USA, 1
4. Simoen Ellen, De Roeck Guido, Lombaert Geert (2015) Dealing with uncertainty in model updating for damage assessment: a review. Mech Syst Signal Process 56:123–149
5. Gardner P (2018) On novel approaches to model-based structural health monitoring. PhD thesis, University of Sheffield
6. Gardner P, Lord C, Barthorpe RJ (2019a) A unifying framework for probabilistic validation metrics. J Verificat Validation Uncertainty Quantificat 4(3)
7. Gardner P, Lord C, Barthorpe RJ (2019b) Bayesian history matching for forward model-driven structural health monitoring. In: Model validation and uncertainty quantification, vol 3. Springer, pp 175–183
8. Fuentes R, Cross E, Halfpenny A, Worden K, Barthorpe RJ (2014) Aircraft parametric structural load monitoring using gaussian process regression. In: Proceedings of the European workshop on structural health monitoring 2014. Nantes
9. Gibson S J , Rogers T J, Cross E J (2020) Data-driven strain prediction models and fatigue damage accumulation. In: Proceedings of the 29th international conference on noise and vibration engineering (ISMA 2020)
10. Barthorpe RJ (2010) On model-and data-based approaches to structural health monitoring. PhD thesis, University of Sheffield
11. Farrar CR, Worden K, Todd MD, Park G, Nichols J, Adams DE, Bement MT, Farinholt K (2007) Nonlinear system identification for damage detection. Technical report, Los Alamos National Laboratory (LANL), Los Alamos, NM
12. Kerschen G, Worden K, Vakakis AF, Golinval J-C (2006) Past, present and future of nonlinear system identification in structural dynamics. Mech Syst Signal Proces 20(3):505–592
13. Oakley J, O'Hagan A (2002) Bayesian inference for the uncertainty distribution of computer model outputs. Biometrika 89(4):769–784
14. Leser PE, Hochhalter JD, Warner JE, Newman JA, Leser WP, Wawrzynek PA Yuan F-G (2017) Probabilistic fatigue damage prognosis using surrogate models trained via three-dimensional finite element analysis. Struct Health Monitor 16(3):291–308
15. Beucler T, Pritchard M, Rasp S, Ott J, Baldi P, Gentine P (2019) Enforcing analytic constraints in neural-networks emulating physical systems. *arXiv preprint* arXiv:1909.00912
16. Sohlberg B, Jacobsen EW (2008) Grey box modelling–branches and experiences. IFAC Proc Vol 41(2):11415–11420
17. Sohlberg B (2012) Supervision and control for industrial processes: using grey box models, predictive control and fault detection methods. Springer Science & Business Media
18. Fuentes R, Nayek R, Gardner P, Dervilis N, Rogers T, Worden K, Cross EJ (2021) Equation discovery for nonlinear dynamical systems: a Bayesian viewpoint. Mech Syst Signal Process 154:107528
19. Brunton SL, Proctor JL, Kutz JN (2016) Discovering governing equations from data by sparse identification of nonlinear dynamical systems. Proc Natl Acad Sci 113(15):3932–3937
20. Noël J-P, Schoukens J (2018) Grey-box state-space identification of nonlinear mechanical vibrations. Int J Control 91(5):1118–1139
21. Kennedy MC, O'Hagan A (2001) Bayesian calibration of computer models. J Royal Statist Soc Series B (Statist Methodol) 63(3):425–464
22. Lei CL, Ghosh S, Whittaker DG, Aboelkassem Y, Beattie KA, Cantwell CD, Delhaas T, Houston C, Novaes GM, Panfilov AV et al (2020) Considering discrepancy when calibrating a mechanistic electrophysiology model. Philosoph Trans Royal Soc A 378 (2173):20190349
23. Brynjarsdottir J, Hagan AO (2014) Learning about physical parameters: the importance of model discrepancy. Inve Probl 30(11):114007
24. Worden K, Wong CX, Parlitz U, Hornstein A, Engster D, Tjahjowidodo T, Al-Bender F, Rizos DD, Fassois SD (2007) Identification of pre-sliding and sliding friction dynamics: grey box and black-box models. Mech Syst Signal Process 21(1):514–534

25. Vanli OA, Jung S (2014) Statistical updating of finite element model with lamb wave sensing data for damage detection problems. Mech Syst Signal Process 42:137–151
26. Wan H-P, Ren W-X (2015) A residual-based Gaussian process model framework for finite element model updating. Comput Struct 156:149–159
27. Rogers TJ, Holmes GR, Cross EJ, Worden K (2017) On a grey box modelling framework for nonlinear system identification. In: Special topics in structural dynamics, vol 6. Springer, pp 167–178
28. Worden K, Barthorpe RJ, Cross EJ, Dervilis N, Holmes GR, Manson G, Rogers TJ (2018a) On evolutionary system identification with applications to nonlinear benchmarks. Mech Syst Signal Process 112:194–232
29. Solin A, Kok M, Wahlström N, Schön TB, Särkkä S (2018) Modeling and interpolation of the ambient magnetic field by Gaussian processes. IEEE Trans Robot 34(4):1112–1127
30. Wahlström N, Kok M, Schön TB, Gustafsson F (2013) Modeling magnetic fields using gaussian processes. In: 2013 IEEE international conference on acoustics, speech and signal processing. IEEE, pp 3522–3526
31. Jidling C, Hendriks J, Wahlström N, Gregg A, Schön TB, Wensrich C, Wills A (2018) Probabilistic modelling and reconstruction of strain. In: Nuclear Instru Methods Phys Res Sect B: Beam Interact Mater Atoms 436:141–155
32. Karpatne A, Watkins W, Read J, Kumar V (2017a) Physics-guided neural networks (pgnn): An application in lake temperature modeling. *arXiv preprint* arXiv:1710.11431
33. Raissi M, Wang Z, Triantafyllou MS, Karniadakis GE (2019) Deep learning of vortex-induced vibrations. J Fluid Mech 861:119–137
34. Karpatne A, Atluri G, Faghmous JH, Steinbach M, Banerjee A, Ganguly A, Shekhar S, Samatova N, Kumar V (2017b) Theory-guided data science: A new paradigm for scientific discovery from data. IEEE Trans Knowl Data Eng 29(10):2318–2331
35. Willard J, Jia X, Xu S, Steinbach M, Kumar V (2020) Integrating physics-based modeling with machine learning: a survey. *arXiv preprint* arXiv:2003.04919
36. Rasmussen CE, Williams CKI (2006) Gaussian Processes for Machine Learning, vol 38. The MIT Press, Cambridge, MA, USA
37. Bull LA, Gardner P, Rogers TJ, Cross EJ, Dervilis N, Worden K (2020) Probabilistic inference for structural health monitoring: new modes of learning from data. ASCE-ASME J Risk and Uncert Eng Syst Part A Civil Eng 7(1):03120003
38. Wan H-P, Ni Y-Q (2018) Bayesian modeling approach for forecast of structural stress response using structural health monitoring data. J Struct Eng 144(9):04018130
39. Kullaa J (2011) Distinguishing between sensor fault, structural damage, and environmental or operational effects in structural health monitoring. Mech Syst Signal Process 25(8):2976–2989
40. Rogers TJ, Worden K, Cross EJ (2020a) On the application of Gaussian process latent force models for joint input-state-parameter estimation: with a view to Bayesian operational identification. Mech Syst Signal Process 140:106580
41. Ki-Young K, Brownjohn JMW, List DL, Cole R (2013) Structural health monitoring of the Tamar suspension bridge. Struct Control Health Monitor 20(4):609–625
42. Cross E, Worden K, Ki Young K, Brownjohn J (2012) Filtering environmental load effects to enhance novelty detection on cable-supported bridge performance. In: Bridge maintenance, safety and management. CRC Press, pages 745–752. 10.1201/b12352-101
43. Elizabeth C (2012) On structural health monitoring in changing environmental and operational conditions. PhD thesis, University of Sheffield
44. Zhang S, Rogers TJ, Cross EJ (2020) Gaussian process based grey-box modelling for SHM of structures under fluctuating environmental conditions. Proceedings of 10th European workshop on structural health monitoring (EWSHM 2020)
45. Robert W (2012) Environmental effects on a suspension bridge's performance. PhD thesis, University of Sheffield
46. Worden K, Becker WE, Rogers TJ, Cross EJ (2018b) On the confidence bounds of gaussian process narx models and their higher-order frequency response functions. Mech Syst Signal Process 104:188–223

47. Morison JR, Johnson JW, Schaaf SA (1950) The force exerted by surface waves on piles. Petrol Trans 189:149–157
48. Guilmineau E, Queutey P (2002) A numerical simulation of vortex shedding from an oscillating circular cylinder. J Fluids Struct 16:773–794
49. Wood AMM, Fleming CA (1981) Coastal hydraulics. Macmillan Education Limited, London
50. Grey-box models for wave loading prediction. Mech Syst Signal Process 159:107741, 2021. https://doi.org/10.1016/j.ymssp.2021. 107741 ISSN 0888-3270
51. Najafian G, Tickell RG, Burrows R, Bishop JR (2000) The UK Christchurch bay compliant cylinder project: analysis and interpretation of Morison wave force and response data. Appl Ocean Res 22(3):129–153
52. Marcus H-A, Nikolaos D, Keith W, Cross EJ, Mills RS, Rogers TJ (2021) Structured machine learning tools for modelling characteristics of guided waves. *arXiv preprint* arXiv:2101.01506
53. Cross EJ, Rogers TJ (2021) Physics-derived covariance functions for machine learning in structural dynamics. In: Submitted to SYSID 2021
54. Papoulis A (1965) Probability, random variables, and stochastic processes. McGraw-Hill Education
55. Caughey TK (1971) Nonlinear theory of random vibrations. In: Advances in applied mechanics, vol 11. Elsevier, pp 209–253
56. Wilson A, Adams R (2013) Gaussian process kernels for pattern discovery and extrapolation. In: International conference on machine learning, pp 1067–1075
57. Cross EJ, Rogers TJ, Gibbons TJ (2019) Grey-box modelling for structural health monitoring: physical constraints on machine learning algorithms. In: Structural health monitoring 2019: enabling intelligent life-cycle health management for industry internet of things (IIOT)—proceedings of the 12th international workshop on structural health monitoring, vol 2. pp 2136–2145. ISBN 9781605956015
58. Solin Arno, Särkkä Simo (2020) Hilbert space methods for reduced-rank Gaussian process regression. Statist Comput 30(2):419–446
59. Jones MR, Rogers TJ, Gardner PA, Cross EJ (2020a) Constraining Gaussian processes for grey-box acoustic emission source localisation. In: Proceedings of the 29th international conference on noise and vibration engineering (ISMA 2020)
60. Jones MR, Rogers TJ, Martinez IE, Cross EJ (2021) Bayesian localisation of acoustic emission sources for wind turbine bearings. In: Health monitoring of structural and biological systems XV, vol 11593. International Society for Optics and Photonics, page 115932D
61. Hensman J, Mills R, Pierce SG, Worden K, Eaton M (2010) Locating acoustic emission sources in complex structures using Gaussian processes. Mech Syst Signal Process 24(1):211–223
62. Jones MR, Rogers TJ, Worden K, Cross EJ (2020b) A Bayesian methodology for localising acoustic emission sources in complex structures. *arXiv preprint* arXiv:2012.11058
63. Kristensen NR, Madsen H, Jørgensen SB (2004) Parameter estimation in stochastic grey-box models. Automatica 40(2):225–237
64. Herbert JAF Tulleken (1993) Grey-box modelling and identification using physical knowledge and Bayesian techniques. Automatica 29(2):285–308
65. Lourens E, Papadimitriou C, Gillijns S, Reynders E, De Roeck G, Lombaert G (2012) Joint input-response estimation for structural systems based on reduced-order models and vibration data from a limited number of sensors. Mech Syst Signal Process 29:310–327
66. Azam SE, Chatzi E, Papadimitriou C (2015) A dual Kalman filter approach for state estimation via output-only acceleration measurements. Mech Syst Signal Process 60:866–886
67. Naets F, Croes J, Desmet W (2015) An online coupled state/input/parameter estimation approach for structural dynamics. Comput Methods Appl Mech Eng 283:1167–1188
68. Maes K, Karlsson F, Lombaert G (2019) Tracking of inputs, states and parameters of linear structural dynamic systems. Mech Syst Signal Process 130:755–775
69. Dertimanis VK, Chatzi EN, Azam SE, Papadimitriou C (2019) Input-state-parameter estimation of structural systems from limited output information. Mech Syst Signal Process 126:711–746
70. Alvarez M, Luengo D, Lawrence N (2009) Latent force models. In: Artificial intelligence and statistics, pp 9–16

71. Hartikainen J, Sarkka S (2012) Sequential inference for latent force models. *arXiv preprint* arXiv:1202.3730
72. Rogers TJ, Worden K, Cross EJ (2020b) Bayesian joint input-state estimation for nonlinear systems. Vibration 3(3):281–303
73. Friis T, Brincker R, Rogers TJ (2020) On the application of Gaussian process latent force models for Bayesian identification of the Duffing system. In: Proceedings of ISMA 2020—international conference on noise and vibration engineering and USD 2020—international conference on uncertainty in structural dynamics
74. Frigola R (2015) Bayesian time series learning with Gaussian processes. PhD thesis, University of Cambridge
75. Rogers TJ (2019) Towards Bayesian system identification: with application to SHM of offshore structures. PhD thesis, University of Sheffield

Interpretable Machine Learning for Function Approximation in Structural Health Monitoring

Jin-Song Pei, Dean F. Hougen, Sai Teja Kanneganti, Joseph P. Wright, Eric C. Mai, Andrew W. Smyth, Sami F. Masri, Armen Derkevorkian, François Gay-Balmaz, and Ludian Komini

Abstract Machine learning may complement physics-based methods for structural health monitoring (SHM), providing higher accuracy, among other benefits. However, many resulting systems are opaque, making them neither interpretable nor trustworthy. Interpretable machine learning (IML) is an active new direction intended to match algorithm accuracy with transparency, enabling users to understand their systems. This chapter overviews existing IML work and philosophy, and discusses

J.-S. Pei (✉) · D. F. Hougen · S. T. Kanneganti
University of Oklahoma, Norman, OK 73019, USA
e-mail: jspei@ou.edu

D. F. Hougen
e-mail: hougen@ou.edu

S. T. Kanneganti
e-mail: kannegantisaiteja@ou.edu

J. P. Wright
Division of Applied Science, Weidlinger Associates Inc., New York, NY 10005, USA
e-mail: jwright.wai.com@verizon.net

E. C. Mai
Lacuna Technologies, Lakewood, CO 80232, USA

A. W. Smyth
Columbia University, New York, NY 10027, USA
e-mail: smyth@civil.columbia.edu

S. F. Masri
University of Southern California, Los Angeles, CA 90089, USA
e-mail: masri@usc.edu

A. Derkevorkian
Jet Propulsion Laboratory, California Institute of Technology, Pasadena, CA 91109, USA
e-mail: Armen.Derkevorkian@jpl.nasa.gov

F. Gay-Balmaz
CNRS LMD, Ecole Normale Supérieure de Paris, 75005 Paris, France
e-mail: francois.gay-balmaz@lmd.ipsl.fr

L. Komini
Amberg Group, Trockenloostrasse 21, 8105 Regensdorf, Switzerland

A. Cury et al. (eds.), *Structural Health Monitoring Based on Data Science Techniques*,
Structural Integrity 21, https://doi.org/10.1007/978-3-030-81716-9_18

candidates from SHM to exemplify and substantiate IML. Multidisciplinary research has been making strides toward providing end users of shallow sigmoidal artificial neural networks (ANNs) with the tools and knowledge for engineering these systems. Notoriously opaque ANNs are made transparent as linear-in-the-weight parameterization tools by using domain knowledge to determine appropriate basis functions. With a small number of hidden nodes to activate these basis functions, the modeling capability of sigmoidal ANNs is systematically revealed without relying on training. The novelty is in ANN initialization theory and practical procedures that can be interpreted via domain knowledge. A rich repository of direct (non-iterative) techniques and reusable ANN prototypes can then be aggregated as the basis functions needed for specific problems, leading to interpretable ANNs as well as improved training performance and generalization as validated by simulated and real-world data.

1 Introduction

Machine learning (ML) is an approach to artificial intelligence (AI) in which algorithms develop models based on data. *Interpretable machine learning* (IML), in which machine learning models are designed to be inherently understandable [50], can make valuable contributions to structural health monitoring (SHM) by combining the accuracy of machine learning with the interpretability of physics-based modeling, as depicted in Fig. 1.

In this section, we introduce our approach to IML for SHM through the use of interpretable artificial neural networks (ANNs) for SHM function approximation problems. We introduce function approximation as a domain problem (Sect. 1.1), explain the need for IML in this domain (Sect. 1.2), outline our approach (Sect. 1.3), and present the structure of this chapter (Sect. 1.4).

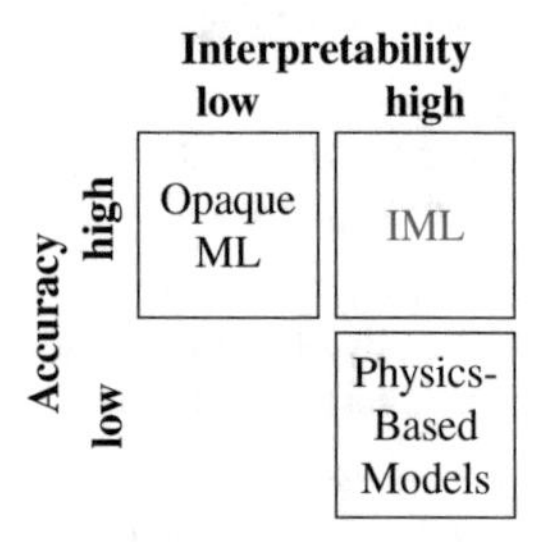

Fig. 1 Interpretable machine learning provides the benefits of both physics-based models and traditional, opaque ML nonlinear dynamics models

1.1 Function Approximation as Domain Problem

Function approximation plays an important role in AI and is essential for nonlinear control, state estimates and predictors, and more ("From Backpropagation to Brain-Like CyberInfrastructure: A Ladder of Universal Designs" on www.werbos.com). Artificial neural networks, which are variously considered components of AI or soft computing, have a long history of use in engineering mechanics and SHM, e.g., [6, 10, 12, 14, 21] including for function approximation due to their universal approximation capabilities [8, 20]. However, training sigmoidal neural networks (multilayer feedforward neural networks, FFNNs) to approximate nonlinear static functions is known to be computationally intractable [4, 24, 25]. Thus, we focus on a subset of all possible nonlinear target functions.

For example, when we use data in SHM, the input could represent the states of a nonlinear dynamical system while the output could be the underlying restoring force of this system, i.e., an internal force that restores a moving mass to its neutral position. We adopt the force-state mapping formulation [33, 38] where a single-degree-of-freedom (SDOF, as an example) restoring force is represented as $r(\mathbf{x})$ with the system's states $\mathbf{x} = [x, \dot{x}]^T$. See Fig. 2, where force-state mapping is formulated for translational motion (as an example). The nonlinear dynamics then becomes

$$\dot{\mathbf{x}}(t) = \mathbf{f}\left(\mathbf{x}, r(\mathbf{x}), u(t), t\right) \tag{1}$$

where $u(t)$ is excitation force. We use ANN to approximate $r(\mathbf{x})$.

In the work to be reviewed, we study the approximation of nonlinear static functions using a sigmoidal neural network, i.e., the universal approximator [8, 20] employing a logistic sigmoidal activation function (a feedforward neural network with one vector input, one hidden layer, and one scalar output). We focus on answering these **two key questions** concerning initialization:

1. determination of the number of hidden nodes, and
2. determination of the initial values for weights and biases.

These fundamental questions were raised at least as early as 1992 [15, 52] but still have not been adequately addressed.

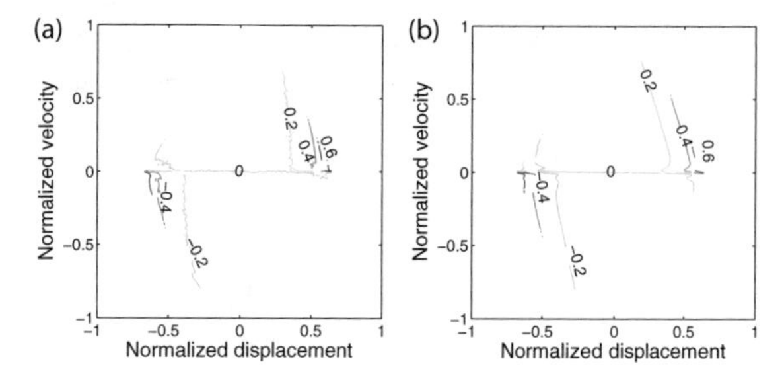

Fig. 2 **a** All normalized University of Southern California (USC) AF2 datasets including for both training and validation under the force-state mapping for translational motion, and **b** the corresponding contour plots of the outputs of the trained FFNNs using all normalized data. From [41]

The force-state mapping formulation has proven to be one of the most powerful methods to model nonlinear dynamical systems for aerospace, mechanical, and civil engineering, and many bio-mechanical systems, e.g., [11, 21, 26] thus being useful to SHM. Nonetheless, nonlinear restoring forces can be very complex especially when involving hysteresis (i.e., history/path-dependency) and when time varying (e.g., caused by deteriorating system properties due to aging or damage). Furthermore, multi-component systems involving multi-degrees-of-freedom (MDOF) can be extremely challenging. All this indicates the need of going beyond sigmoidal neural networks. The first step is to make sigmoidal neural networks interpretable for force-state mapping, and a small subset of nonlinear static function approximation, as reviewed herein.

1.2 The Need for IML

IML emphasizes using domain constraint/knowledge to develop machine learning. Engineering disciplines enjoy scientific theories, mathematical education, and quantitative practice that are as objective and repeatable as possible in a well-organized manner. This is the case in engineering mechanics where graduate-level training normally includes advanced mechanics of materials, finite element modeling and numerical methods, when engineering mechanics researchers gain experience with taking data measurements, processing them and analyzing the result. Soft computing, artificial intelligence, machine learning, and more have gradually entered engineering mechanics graduate-level training (often through multidisciplinary coursework and research practice) and are poised to become a new way of thinking with different jargons, routines, explanations, and technical challenges. Experience has shown that engineering mechanics (including SHM) researchers have the following concerns:

Concern 1 Why should AI be considered as an option in research at all? Or, what can AI offer to us that other existing options cannot offer? This concern could be first addressed practically because there are indeed significant problems that cannot be handled (effectively) by traditional approaches. Our selection of the topic of modeling nonlinear dynamics (especially hysteresis) is one such significant problem. This concern could be next addressed theoretically [2, 3] and validated numerically.

Concern 2 The reluctant attitude stems from the fact that AI does not appear to be connected with existing engineering tools. This manifests the need for IML with the goal of making the user of AI feel comfortable [17, 19, 50].

Concern 3 The use of AI itself involves subjective issues that engineering research practice, in general, tries to minimize or avoid. As reviewed elsewhere [39, 42, 45], numerous subjective issues can be encountered in the initial design and training of neural networks, and in interpretation of trained results. Among them, **the two key questions** are prominent, thus remaining the focus in our work. Variations in trained weights and biases from using different sets of training data sets can hardly be related to changes of the system [34]. Injecting interpretability into the initial design, we believe, would help alleviate this challenge for damage detection in SHM.

1.3 Overview of How We Address the Need

As explained in a classic work on interpretability in neural networks for pattern classification, "…the first layer partitions the input space into a number of cells. …The sole function of additional layers is then to group these cells into decision regions. …For nets with sigmoidal nonlinearities, one can still benefit by thinking about the problem in terms of hyperplanes in the first layer which divide the input space into cells, and that higher layers simply group these cells into different regions" [32]. This is such a powerful fundamental idea that one can find it in deep learning [29] and explainable AI (XAI) [17]. This fundamental idea is what we have been adhering to and making transparent in sigmoidal neural networks for function approximation, pioneering this approach for engineering mechanics.

We have been carrying out a series of studies that now form a small collection aimed at providing interpretability to the user of sigmoidal neural networks for a specific function approximation application in engineering mechanics [39–48]. **Techniques A to G** have been developed by us to construct sigmoidal neural network **prototypes** (see more explanation later in this section) to approximate a range of useful functions/features as summarized in Table 1.

To address Concern 1, we use theoretical work that justifies the efficiency of sigmoidal neural networks [2, 3]. When phenomenological representations (see Sect. 3 for explanation) are not available/adequate, and commonly-used fixed basis functions (e.g., polynomial fitting) are not parsimonious and hard to adapt to data, sigmoidal neural networks stand out.

To answer Concern 2, we leverage the concept of "linear parameterization." This view of basis functions is very helpful, first, because the adaptivity of the basis functions to data in sigmoidal neural networks is superb, and second, because linear parameterization connects polynomial fitting, Fourier series expansion, and wavelet decomposition to sigmoidal neural networks, making sigmoidal neural networks more interpretable.

Being a universal approximator [8, 20], sigmoidal neural networks are very powerful computational machines with the capability of accomplishing what other computational machines are devised to achieve. Finding such an equivalent solution would not only overcome the doubt to sigmoidal neural networks, but also offer a possible starting point for their adaptivity. Unfortunately, the universal approximator theorem is not constructive. Therefore, we started from scratch to approximate systematically the following basic operations using sigmoidal neural networks under **Techniques A & B**:

Table 1 Summary of techniques and nonlinear functions approximated by predetermined sigmoidal neural networks called *prototypes*

Technique ID & References	Technique name	Target nonlinear function example
A in [39, 44]	Taylor series expansion of sigmoidal function	Polynomials, summation, subtraction, and multiplication
B in [48]	Higher-order partial derivatives of sigmoidal function	Sinc, Gaussian, and Mexican hat function
C in [40]	Direct use of the three prototypes	The ten types, absolute value and reciprocal function
D in [40]	Decomposition	Harmonics, swept sine, and some hysteresis loops
E in [39, 42, 43]	Geometric capabilities of sigmoidal function	Hardening, softening, Coulomb, clearance
F in [47, 48]	Layer condensation	Velocity squared damping, division, the "Bouc-Wen" term
G in [46]	Quasi-decomposition	Unsymmetrical concrete constitutive curve, SDOF Frequency response function (FRF)

1. the four basic arithmetic operations [48], and
2. polynomial terms, p^0, p^1, p^2, p^3, $p_1 p_2$, $p_1^2 p_2$, and polynomial fitting involving these terms [39, 44].

Toward addressing Concern 3, we developed **constructive methods** to address **the two key questions** for the initialization of sigmoidal neural networks under **Techniques C to G** by building on the outcomes of **Techniques A & B**. This will be elaborated in Sect. 3. As a quick example, some existing simple damping models are given in Table 2. We establish how to set up a sigmoidal neural network to approximate each one of them because we have known how to approximate polynomials and multiplication under **Techniques A & B** as well as the signum (sgn) and absolute value function under **Technique C** (see Table 1).

IML encourages sparse learning. We develop a dictionary by asking and answering the following questions: (a) What are the useful behaviors/responses in terms of input-output? See the ten types of 1-D nonlinearities under **Technique C** illustrated in Fig. 3, and (b) What are the sigmoidal neural network prototypes that we can develop for each useful feature? See Prototypes 1–3 (and their variants a to c) illustrated elsewhere [40] that are developed to approximate these ten types of nonlinearities individually or combinatorially, which will be directly adopted as a "hint book" in Fig. 4.

Prototypes and their variants are predetermined neural networks that are not obtained from an inverse formulation of training from datasets. Instead, they are constructed in advance from a forward formulation based on either the algebraic or geometric capabilities of linear sums of sigmoidal functions to capture some dominating features of the nonlinear function to be approximated in the specified

Table 2 Damping models from [22] and recommended sigmoidal neural networks for initialization (with initial values for weights and biases obtained but not presented here)

Type	Formula	Source	Sigmoidal neural network
Linear viscous	$c \cdot \dot{x}$	Slow fluid	1 layer with 2 nodes
Air	$\alpha \cdot \text{sgn}(\dot{x})\dot{x}^2$	Fast fluid	2 layers with (3 + 4) and 8 nodes
Coulomb	$\beta \cdot \text{sgn}(\dot{x})$	Sliding friction	1 layer with 3 nodes
Displacement-squared	$d \cdot \text{sgn}(\dot{x})x^2$	Material damping	2 layers with (3 + 4) and 8 nodes
Solid, or structural	$b \cdot \text{sgn}(\dot{x})\lvert x\rvert$	Internal damping	2 layers with (3 + 4) and 8 nodes

The notation $(m + n)$ indicates an ANN layer with two separate components of m and n nodes, from two separate prototypes for multiplicand and multiplier; the 8 nodes in a layer is for multiplication (see Table 4)

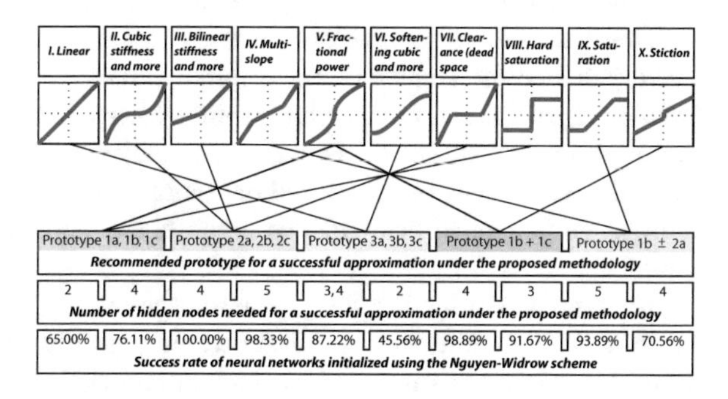

Fig. 3 Illustration of the ten types of 1-D nonlinearities, the neural network prototypes to fulfill them, and training success rates using Nguyen-Widrow initialization [37] in contrast to 100% training success rate using the initial values of weights and biases from our work, together with the number of hidden nodes

applications. Preparing prototypes and variants takes time, but can be done in a forward problem fashion by following a clear procedure. The resulting prototypes and variants are generic; i.e., they can be used for numerous individual training tasks through proper transformations leading to a high overall efficiency in addition to the built-in rationality and transparency in this proposed initialization methodology.

To complete sparse learning, we implement user-in-the-loop for feature extraction from the data, match each dominant feature with a sigmoidal neural network prototype before concatenating them within a hidden layer (and sometimes, between two hidden layers) for the backpropagation to be applied without weight freezing. This growing technique allows for iterations, especially automated iterations for future work to put in place, among other major improvements to capture more salient features and reveal more inner-workings through theoretical work and visualization.

1.4 Novelty and Structure of This Chapter

Overall, in engineering mechanics, there exists a rich body of domain knowledge—outside of ML, yet calling for new knowledge in the era of smart technologies. Intentionally and explicitly making a connection with this existing body of knowledge provides the opportunity for ML to be understood, justified, accepted, and productive. Our fundamental way of thinking concerning how to accomplish IML in terms of sigmoidal neural networks follows.

Useful basis functions and features identified according to domain knowledge are approximated by sigmoidal neural networks beforehand for reuse in initialization, where the modeling capabilities of sigmoidal neural networks are explored and exploited. An individual function approximation problem is considered an assembly of these useful basis functions and/or features leading to, often times, the growth of the width of the hidden layer of a sigmoidal neural network and realizations of possible initial values for all weights and biases in initialization.

This way of thinking enables us to use domain knowledge to directly and systematically decide the architecture and initial values for weights and biases when the training algorithm is for local minimization. Interpretability using domain knowledge shows up in two aspects: (1) Domain knowledge in engineering mechanics suggests possibly useful basis functions and/or dominant features in the data, and (2) domain knowledge in applied mathematics suggests the modeling capabilities of sigmoidal neural networks.

To complement the insights to the interpretability, the data-driven nature shows up in two different aspects: (1) Data would possibly reveal dominant features that could validate those suggested by domain knowledge in engineering mechanics or otherwise, and (2) a greater success in neural network training can be found in the balance between global and local search involved in training. Selecting a good initial point for neural network training is critical because the training process will normally result in trained values that are still in the neighborhood of their initial values. If domain knowledge could be used to influence neural network initialization, then training would more likely converge to the global minimum instead of just a local minimum, making the trained neural network more accurate and meaningful. (Note that careful weight initialization is often not considered crucial for learning in deep neural networks because their great redundancy generally allows for one of many possible low-error representations to be learned [29]. However, that approach precludes the possibility of using the interpretability method proposed herein.) This is the philosophical justification for our proposed initialization to fulfill IML.

IML is enabled by the modeling capability of sigmoidal neural networks that we have started to understand, but we are merely scratching the surface. This is shared in Sect. 2, which justifies and explains **Techniques A & B**. This section explains why it is important to systematically explore the modeling capabilities of sigmoidal neural networks in order to use them for IML, then shows how to use sigmoidal neural networks to approximate basic functions/operations. Section 3 presents the procedures of the proposed IML that underlie **Techniques C to G**, while Sect. 4

offers one case study for a sigmoidal neural network to illustrate the procedures covering USC AF2 data, and another for a nonlinear autoregressive exogenous model (NARX) to demonstrate a possible extension of the proposed IML covering NASA Jet Propulsion Laboratory (JPL) pyroshock data. Section 5 addresses justifications for this work in addition to the benefits provided by IML, while Sect. 6 discusses future work. Section 7 provides concluding remarks.

2 Modeling Capability of Sigmoidal Neural Networks

Herein, we elucidate an identified research need for the "approximation capabilities of sigmoidal neural networks." Foundational work [2, 3] proposes a complexity measure C_f for a target function f. C_f is defined as the first absolute moment of the Fourier $\mathcal{F}$ magnitude distribution of f as

$$C_f = \int\limits_{\mathbb{R}^d} |\omega| \left|\tilde{f}(\omega)\right| \mathrm{d}\omega \tag{2}$$

where we have

$$\tilde{f}(\omega) \triangleq \mathcal{F}(f(p)) \triangleq \int\limits_{-\infty}^{+\infty} f(p)e^{-i\omega p}\mathrm{d}p. \tag{3}$$

C_f plays an important role to optimize the number of hidden nodes and quantifying the approximation error bound [2, 3]. However, it has been questioned if C_f is sufficient from a computational point of view [51]. In our work, we question why C_f alone may not be a sufficient complexity measure for constructive methods by examining the set of target functions

$$f(p, w, b) = \sigma(wp + b) = \frac{1}{1 + e^{-(wp+b)}} \tag{4}$$

with $b \equiv 0$, $w > 0$, and $p \in [-a, a]$. We have shown that

$$C_f = |w| \int_{-\infty}^{+\infty} \left| \underbrace{\mathcal{F}\left(\frac{\mathrm{d}\sigma}{\mathrm{d}z}\right)}_{\text{independent of } w \text{ and } a} \star \underbrace{\frac{2\sin(wa\omega)}{\omega}}_{\text{dependent on } wa} \right| \mathrm{d}\omega \tag{5}$$

where $z = wp + b$ and $\star$ stands for convolution [47]. This means that C_f can be w-dependent. This contradicts our common sense that the actual complexity of approximating this family of target functions is the same: One logistic sigmoidal function would achieve zero approximation error. For the sake of constructive methods, C_f

perhaps tells only half of the story. The other half of the story, perhaps, is to know what is considered complex for sigmoidal neural networks and what is considered not, the so-called "modeling capabilities" of sigmoidal neural networks.

We have started to systematically explore the capabilities of sigmoidal neural networks algebraically with **Techniques A & B** and geometrically with **Techniques C to G**. Herein, we demonstrate one technique using algebra.

We utilize Taylor series expansion of a sigmoidal function to approximate polynomials. For example, two sigmoidal functions following Eq. (4) are chosen, f_1 and f_2, with $w_1 = -w_2$ and $b_1 = b_2 = 0$. The Taylor series expansion of both functions at the origin $x = 0$ to the second power can be written as

$$f_1 = \frac{1}{2} + \frac{w_1}{4} p + \frac{1}{2!} (w_1)^2 Q_2 (-w_1 \xi_1) p^2$$
$$f_2 = \frac{1}{2} - \frac{w_1}{4} p + \frac{1}{2!} (w_1)^2 Q_2 (+w_1 \xi_2) p^2.$$

The constant term can be eliminated by taking the difference of the above two functions. For a non-zero w_1, one then has

$$z = \frac{2}{w_1} (f_1 - f_2) = p + w_1 [Q_2 (-w_1 \xi_1) - Q_2 (+w_1 \xi_2)] p^2. \qquad (6)$$

One may choose z to mimic the first power of input p. It can be seen that the error bound is determined by values of w_1 and p. Using this approach as detailed elsewhere [39, 44], Table 3 gives the derived (not trained) number of hidden nodes, weights, and biases for the approximation of p^0, p^1, and p^2 terms, while their error bounds are likewise given elsewhere [44]. See Table 4 for the derived (not trained) number of hidden nodes, weights, and biases for the four basic arithmetic operations, with details given elsewhere [48].

A basic tenet of soft computing is to avoid "hard computing" methods, i.e., conventional computing methods, even for items as basic as the four arithmetic operations. Although the implementation of these basic operations draws attention in other settings such as hardware implementation required in embedded systems (e.g., [7]), the same topic has not been studied sufficiently in neural networks. When the "mapping" of basic items into neural networks cannot be understood, the acclaimed superiority of neural networks in approximating unknown functions becomes harder to accept even given some prior studies on this (e.g., [23, 28, 30, 31]). A systematic solution (not random trial-and-error) based on the governing nonlinear dynamics of neural network learning, approximation capabilities of sigmoidal neural networks and features of target functions would greatly benefit numerous applications in nonlinear system identification in addition to advancing fundamental research in neural computation.

There are other techniques for other basic target functions. Our effort in approximating a sinusoidal term [40, 48] and certain wavelet-like functions [48] is scattered and without error bounds estimated.

Table 3 Derived weights and biases in approximating polynomials from the the zeroth to second power—general results. From [44]

Target Function	Weights in Input layer, IW	Biases in Input layer, b	Weights in Output layer, LW
p^0	$\mathbf{w}_1^{[0]} = \begin{bmatrix} w_{1,1}^{[0]} \\ -w_{1,1}^{[0]} \end{bmatrix}$	$\mathbf{b}^{[0]} = \begin{bmatrix} 0 \\ 0 \end{bmatrix}$	$\mathbf{w}_2^{[0]} = \begin{bmatrix} 1 & 1 \end{bmatrix}$
p^1	$\mathbf{w}_1^{[1]} = \begin{bmatrix} w_{1,1}^{[1]} \\ -w_{1,1}^{[1]} \end{bmatrix}$	$\mathbf{b}^{[1]} = \begin{bmatrix} 0 \\ 0 \end{bmatrix}$	$\mathbf{w}_2^{[1]} = \begin{bmatrix} +\frac{2}{w_{1,1}^{[1]}} & -\frac{2}{w_{1,1}^{[1]}} \end{bmatrix}$
p^2	$\mathbf{w}_1^{[2]} = \begin{bmatrix} w_{1,1}^{[2]} \\ -w_{1,1}^{[2]} \\ w_{1,3}^{[2]} \\ -w_{1,3}^{[2]} \end{bmatrix}$	$\mathbf{b}^{[2]} = \begin{bmatrix} b_1^{[2]} \\ b_1^{[2]} \\ 0 \\ 0 \end{bmatrix}$	$\mathbf{w}_2^{[2]} = \begin{bmatrix} \frac{\left[1+e^{b_1^{[2]}}\right]^3}{\left(w_{1,1}^{[2]}\right)^2 e^{b_1^{[2]}}\left[-1+e^{b_1^{[2]}}\right]} \\ \frac{\left[1+e^{b_1^{[2]}}\right]^3}{\left(w_{1,1}^{[2]}\right)^2 e^{b_1^{[2]}}\left[-1+e^{b_1^{[2]}}\right]} \\ -\frac{2\left[1+e^{b_1^{[2]}}\right]^2}{\left(w_{1,1}^{[2]}\right)^2 e^{b_1^{[2]}}\left[-1+e^{b_1^{[2]}}\right]} \\ -\frac{2\left[1+e^{b_1^{[2]}}\right]^2}{\left(w_{1,1}^{[2]}\right)^2 e^{b_1^{[2]}}\left[-1+e^{b_1^{[2]}}\right]} \end{bmatrix}^T$

3 Procedure for IML

In engineering mechanics, a phenomenological representation of a mechanical system is a commonly used constructive method. In that representation, an engineering mechanics researcher starts a model with elements as basic as a linear spring and dashpot because these basic elements represent unambiguous static or dynamic responses under any designated cause. By connecting these basic elements in series or in parallel, and by introducing advanced versions of these basic elements (e.g., simple forms of nonlinear springs and dashpots), and more advanced elements (e.g., a slide element, and other specific forms of damping or hysteresis), the engineering mechanics researcher has a complete grasp of how the model works. In other words, the model is totally transparent and interpretable. The principle of superposition in linear system analysis, decompositions in time series analysis, and a general approach of tackling a complex problem by transforming it into an assembly of less challenging sub-problems all follow the same way of thinking.

To directly learn the spirit from this interpretable modeling methodology that is well accepted by engineering mechanics researchers, we wonder about three aspects in building a sigmoidal-neural-network-based interpretable modeling framework, which coincides with IML, that we label *interpretability elements* (IE).

Table 4 Derived weights and biases in approximating $p_1 + p_2$, $p_1 - p_2$, $p_1 \times p_2$, and $p_1 \div (100p_2)$ with $p \in [-1, +1]$ using feedforward neural networks with one hidden layer. From [48]

Target Function	Weights in Input layer, IW	Biases in Input layer, b	Weights in Output layer, LW
$p_1 + p_2$	$\begin{bmatrix} 0.1 & 0 \\ -0.1 & 0 \\ 0 & 0.1 \\ 0 & -0.1 \end{bmatrix}$	$\begin{bmatrix} 0 \\ 0 \\ 0 \\ 0 \end{bmatrix}$	$\begin{bmatrix} 20 \\ -20 \\ 20 \\ -20 \end{bmatrix}^T$
$p_1 - p_2$	$\begin{bmatrix} 0.1 & 0 \\ -0.1 & 0 \\ 0 & 0.1 \\ 0 & -0.1 \end{bmatrix}$	$\begin{bmatrix} 0 \\ 0 \\ 0 \\ 0 \end{bmatrix}$	$\begin{bmatrix} 20 \\ -20 \\ -20 \\ 20 \end{bmatrix}^T$
$p_1 \times p_2$	$\begin{bmatrix} 0.1 & 0.1 \\ -0.1 & -0.1 \\ 1 & 1 \\ -1 & -1 \\ 0.1 & -0.1 \\ -0.1 & 0.1 \\ 1 & -1 \\ -1 & 1 \end{bmatrix}$	$\begin{bmatrix} -10 \\ -10 \\ 0 \\ 0 \\ -10 \\ -10 \\ 0 \\ 0 \end{bmatrix}$	$\begin{bmatrix} 5.5076165 \times 10^5 \\ 5.5076165 \times 10^5 \\ -50.006810 \\ -50.006810 \\ -5.5076165 \times 10^5 \\ -5.5076165 \times 10^5 \\ 50.006810 \\ 50.006810 \end{bmatrix}^T$
$p_1 \div (100p_2)$	$\begin{bmatrix} 0.5 & 0.5 \\ -0.5 & -0.5 \\ 5 & 5 \\ -5 & -5 \\ 0.5 & -0.5 \\ -0.5 & 0.5 \\ 5 & -5 \\ -5 & 5 \end{bmatrix}$	$\begin{bmatrix} -10 \\ -10 \\ 0 \\ 0 \\ -10 \\ -10 \\ 0 \\ 0 \end{bmatrix}$	$\begin{bmatrix} 5.5076165 \times 10^5 \\ 5.5076165 \times 10^5 \\ -50.006810 \\ -50.006810 \\ -5.5076165 \times 10^5 \\ -5.5076165 \times 10^5 \\ 50.006810 \\ 50.006810 \end{bmatrix}^T$

IE 1 What are the behaviors/responses in mechanics that these basic modeling elements ought to capture/approximate? We introduce an ever-growing list of useful basis functions and features in our work as in Table 2 and Fig. 3.

IE 2 What are the sigmoidal neural network modeling components as basic/essential as springs and dampers? Here, we introduce neural network prototypes proposed and constructed by us.

IE 3 What are the steps that make a general initial design procedure to answer **the two key questions**? We introduce the following three steps:

Step 1 Decompose an unknown target function into a linear combination of dominant features or basis functions as in IE 1.

Step 2 Use neural network prototypes from IE 2 directly or through concatenation to construct sub-sigmoidal neural networks for each dominant feature or basis function.

Step 3 Concatenate all sub-sigmoidal neural networks to approximate the unknown target function.

IE 1, the hints, directly calls for domain knowledge. Sigmoidal neural networks are indeed used to approximate unknown functions; however, engineering mechanics researchers, given their domain knowledge, perhaps anticipate certain nonlinearities from the data. Alternatively, engineering mechanics researchers may observe some dominant features of the data under certain domain knowledge-based formulations. These hints do not need to be precise but can help significantly as shown in our work (see the loose matching of the features in Fig. 4a, b with those in the hint book in Fig. 4c). We are finding ways to identify and exploit these hints for the benefit of initializing sigmoidal neural networks—hence "what to expect when you are expecting." We examine a domain where existing knowledge has been under active development for hundreds of years. Comprehensively, systematically, and effectively representing and connecting with the rich domain knowledge indeed needs long-term study. Figure 3 illustrates a very small subset of useful features (following [53]).

IE 2, how we act on each hint in terms of answering **the two key questions** regarding (I) the number of hidden nodes, and (II) the initial weights and biases, embodies our main technical work in terms of **Techniques A to G** as in Table 1. In Fig. 3, we present the usefulness of Prototypes 1 to 3 and their combinations.

The three steps under IE 3 are simple, fast, clear, and fruitful as validated in our work [39–48].

Note that IE 1 and IE 2 are prepared outside of and before any actual data processing. IE 2 can be theoretical derivations without involving training at all, e.g., the approximation of p^1 given under Sect. 2. The quality of IE 2 could be evaluated by its reusability. Intuitively, Step 1 involves user-in-the-loop when the identification and selection of features are not automated. Step 2 echoes strongly with unsupervised pre-training that is critical to the success of deep learning [9, 18] and calls for a continuous expansion and improvement of reusability. Step 3 is a way of increasing network width but not depth. These three steps are consistent with the layered representation for classical pattern classification [32], deep learning [29], and XAI [17].

4 Case Studies

Prior work uses training examples to demonstrate the three steps under IE 3 with two carefully collected laboratory datasets [41, 47]. Here, we briefly review the dataset labeled AF2 that was previously given in Fig. 2. Figure 4 illustrates the three steps.

The Lipschitz condition for the ODE in Eq. (1) requires the approximation of $r(\mathbf{x})$ to be bounded. The specified application has further requirements (or preferences)

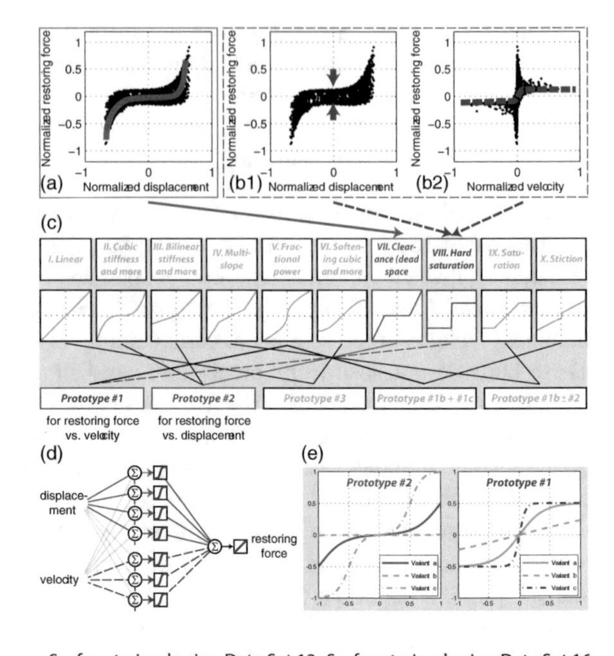

Fig. 4 A step-by-step procedure to explain the proposed neural network initialization for FFNN-5 in [41], when **the two key questions** are answered—Prototypes 2a and 1c for the restoring force vs. displacement, and restoring force vs. velocity, respectively, denoted as Prototype 2a + 1c. Slightly adapted from [47]

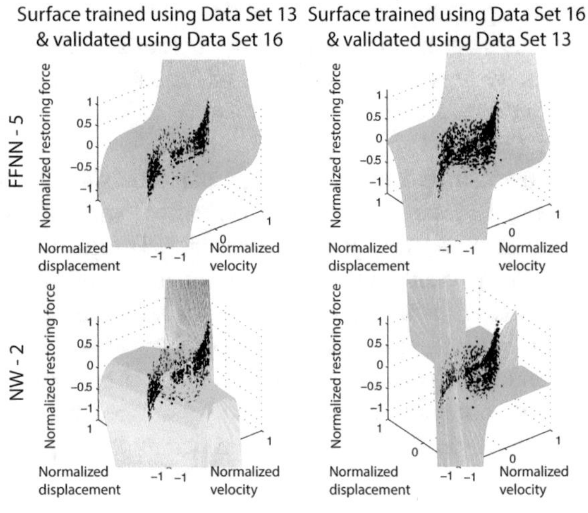

Fig. 5 Experimental data for validation (in black) compared with restoring force surfaces simulated using the trained FFNN-5 and one Nguyen-Widrow counterpart, NW-2, for Subsets 13 and 16 under AF2. From [41] with one label corrected

on sigmoidal neural networks: (i) fast training convergence, (ii) reasonable accuracy, (iii) great generalization capabilities, (iv) small or modest computational resources in terms of network size and learning time (e.g., for wireless structural health monitoring and intelligent control systems), and (v) constructive and insightful methods (i.e., not treating neural networks as “black-boxes,” a goal that we are striving for here). Prior work demonstrates all these aspects [41], while Fig. 5 contrasts the proposed with Nguyen-Widrow initialization in terms of computational stability.

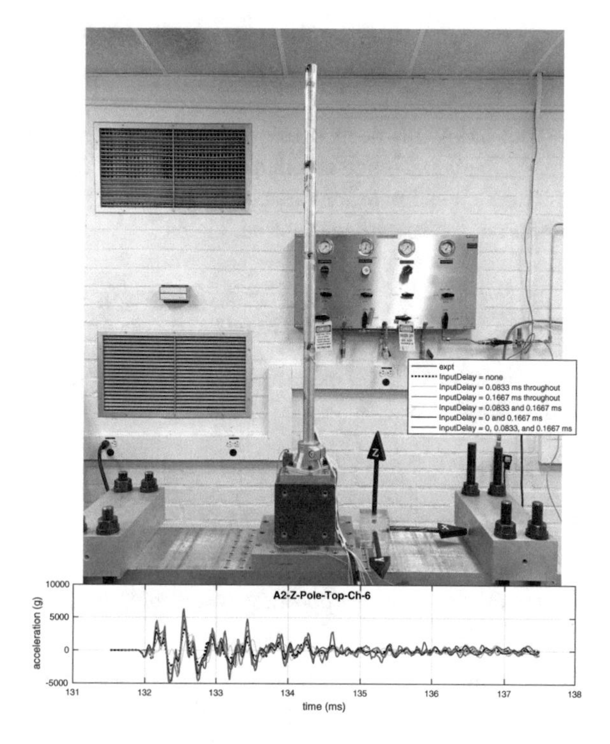

Fig. 6 Shock simulation test setup for a sample tube-like structure and predicted sample time histories of Run J by Run I

Spacecraft often subjected to high-frequency large-amplitude shock loads during flight. Shock loads are experimentally simulated using ordnance/explosive devices and shock simulation systems. These test campaigns are usually expensive, time consuming, and can potentially impose damage to sensitive electronic boxes. This is a challenge for SHM in the aerospace industry. Currently, there is a paucity of reliable computational tools to accurately model the response of complex flight systems subjected to pyroshock loads. JPL engineers have rich experiences in conducting high-quality pyroshock tests [27, 35]. Prior work launched an effort of using system identification to directly benefit pyroshock response prediction [5]. Following the same test setup, we exercised NARX using IML in a follow-up study; another effort of applying IML beyond FFNNs is to exercise tapped delay line neural networks [41].

Three pairs of datasets are in the follow-up study. Each pair contains two realizations, called *runs*, from the same test configuration. In system identification and as a starting point, we question if we can learn from the first run in order to predict the second. Similarly, we ask if we can learn from the second run in order to predict the first. See Fig. 6, where we consider the inputs as follows by incorporating input delays t_u:

$$\mathbf{u}(t - t_u) = \begin{array}{l} \{u_x(t), u_y(t), u_z(t), u_x(t - \frac{1}{2}t_u), u_y(t - \frac{1}{2}t_u), u_z(t - \frac{1}{2}t_u), \\ u_x(t - t_u), u_y(t - t_u), u_z(t - t_u)\}^T . \end{array} \tag{7}$$

Input delay has been witnessed to play a significant role for both performance improvement and inner-working interpretation. When there is no data measurement, a very rough estimate for the longest input delay using the nominal sound speed in aluminum, the specimen material, yields

$$\frac{1 \text{ meter}}{6320 \text{ m/s}} \times 10^3 = 0.1582 \text{ ms}. \tag{8}$$

Estimates using the experimental data are compatible with this number leading to the adoption of $t_u = 0.1667$ ms in Fig. 6.

5 Justifications Beyond IML

We are able to gather supporting evidence/viewpoints from sources beyond IML to justify the work reviewed here:

Optimization: Aid "local search" to balance "global search" (e.g., [36]). In the process of minimizing approximation error, we conduct a global search. Unfortunately, our training method normally does a local search. Therefore, influencing the initial point is important as has been our sole focus here. This might not be a significant issue for deep learning [29] if interpretability is not an issue.

Cognitive/Neural Theory: Have "top-down expectation" to interact with "bottom-up input" as in Adaptive Resonance Theory [16]. Clearly, we inject our expectations of types of nonlinearities, and prepare predetermined neural network prototypes to anticipate these nonlinear types before applying the training data. The injection and preparation, to realize the "top-down expectation" and to interact with "bottom-up input," are carried out through Steps 1, and 2 and 3 under IE 3, respectively.

Statistical Learning: "Purposefully introduce bias" to circumvent the bias/variance dilemma [13], which states that, "Important properties must be built-in or 'hard-wired,' perhaps to be tuned later by experience, but not learned in any statistically meaningful way. …It strikes us, however, that identifying the right 'preconditions' is the substantial problem in neural modeling." Indeed, it takes us nontrivial effort to hard-wire the preconditions in terms of deriving neural network prototypes.

Bayesian Statistical Learning: Provide model classes and prior PDF for model parameters. The initial neural networks obtained through concatenating neural network prototypes lead to the prior with interpretability. Bayesian learning could take off from here by focusing on the data-driven aspect.

6 Future Work

Following the layered representation framework in deep learning [29] and XAI [17], we may use the neural network prototypes in the layer for basic features/simple shapes. This layer is treated as pre-trained; the work reviewed here provides solutions that have been obtained in an interpretable manner. We may introduce additional layer(s) to aggregate the features. We should also introduce additional layers to devise and automate feature definition, detection, and selection, eventually leading to a deep neural network in an interpretable manner.

Neuromanifolds could be a productive path to provide new visualization ideas thus achieving interpretability in deep learning. When minimizing approximation error, state-of-the-art optimization techniques often seek insights into the underlying geometry of an error surface in a parameter space for effective numerical iterative solutions (e.g., training algorithms). Nonetheless, the geometry in this inverse problem context can be far more challenging than that in a forward problem defined by well-behaved functions. As typified by sigmoidal neural networks [1], the geometry of an error surface has a singular structure proven to exist mathematically but is obscure to end users of machine learning, who mine the data for a model but constantly encounter slow convergence without seeing the singular structure and knowing the roots for the so-called local minima. For IML, it is proposed to create a new capability for effectively visualizing the geometry of an error surface as the approximation error being generated on the fly. This is equivalent to creating and exploring an imaginary and unknown world of an often high-dimensional parameter space, whose terrains are determined and altered at every move in the numerical process of conducting an inverse problem where the data and user interact.

Philosophically, the body of knowledge has been developed, verified, and validated long before AI in the specified domain. Thus, AI could yield a very small portion of this body of knowledge. With this said, it would be rational to explain why AI makes sense to the existing body of knowledge, rather than the other way around. Along the same line, it is important to emphasize that AI should play a secondary rather than primary role when there is a significant existing body of knowledge in a specific application domain. Cross validating and merging the knowledge from these two different sources could eventually become possible in such an application domain. Physics-informed NN may be one such direction [49].

7 Summary

This review of a long-term focused effort in AI is limited to sigmoidal neural networks with a sole concentration on interpretability supported by a specific application in SHM. There is a huge existing body of knowledge in the specified domain, posing a legitimate question as to why and how the application of AI can/should be carried out. A humble and an effective approach has turned out to be taking a big step back first to study how to replicate what existing/traditional methods can do but by using sigmoidal neural networks. This backward step allows for the introduction of interpretability that can then be carried forward. For the specified function approximation application, we then identify dominant/basic features as constituents by using either domain knowledge or visualization of data, and devise sigmoidal neural networks (called prototypes) to approximate these features in predetermined settings. We finally rely on human-in-the-loop to select from the prototypes for a complete neural network design solution including the obtained number of hidden nodes and initial weight and bias values. This constructive method is transparent and interpretable to users and has the potential of directly benefiting deep learning and other IML applications.

Acknowledgements The research shown in Fig. 6 was carried out at the Jet Propulsion Laboratory, California Institute of Technology, under a contract with the National Aeronautics and Space Administration (80NM0018D0004). The first and second authors acknowledge Dr. John Antonio for enabling their collaboration.

References

1. Amari SI, Park H, Ozeki T (2006) Singularities affect dynamics of learning in neuromanifolds. Neural Comput 18:1007–1065
2. Barron AR (1993) Universal approximation bounds for superpositions of a sigmoidal function. IEEE Trans Inf Theory 39(3):930–945
3. Barron AR (1994) Approximation and estimation bounds for artificial neural networks. Mach Learn 14:115–133
4. Blum AL, Rivest RL (1995) Training a 3-node neural network is NP-complete. Neural Netw 5:117–127
5. Brewick PT, Abdelbarr M, Derkevorkian A, Kolaini AR, Masri SF, Pei JS (2018) Fusion of state-space modeling and data-driven strategies for computational shock response prediction. The American Institute of Aeronautics and Astronautics (AIAA) Journal Published online on 28 Feb 2018
6. Chassiakos AG, Masri SF (1991) Identification of the internal forces of structural systems using feedfoward multilayer networks. Comput Syst Eng 2(1):125–134
7. Choi S, Scrofano R, Parasanna V, Jang JW (2003) Energy-efficient signal processing using FPGAs. In: International symposium on field programmable gate arrays, proceedings of the 2003 ACM/SIGDA eleventh international symposium on field programmable gate arrays, Monterey, CA, pp 225–234
8. Cybenko G (1989) Approximation by superpositions of sigmoidal function. Math Control Sig Syst 2:303–314
9. Erhan YDB, Courville A, Manzagol PA, Vincent P, Bengio S (2010) Why does unsurpervised pre-training help deep learning? J Mach Learn Res 11:625–660
10. Farrar C, Worden K (2013) Structural health monitoring a machine learning perspective. Wiley, New York, http://orcid.org/10.1002/9781118443118
11. Farrar CR, Worden K, Todd MD, Park G, Nichols J, Adams DE, Bement MT, Fairnholt K (2007) Nonlinear system identification for damage detection. Tech. Rep. LA-14353, Los Alamos National Laboratory
12. Flood I (2002) Neural networks in civil engineering: A review. In: Topping BHV (ed) Civil and structural engineering computing: 2001, Saxe-Coburg Publications, pp 185–209

13. Geman S, Bienenstock E, Doursat R (1992) Neural networks and the bias/variance dilemma. Neural Comput 4:1–58
14. Ghaboussi J, Wu X (1998) Soft computing with neural networks for engineering applications: fundamental issues and adaptive approaches. Struct Eng Mech 6(8):955–969
15. Greene R, Yau ST (eds) (1993) Differential geometry: partial differential equations on manifolds, vol 54
16. Grossberg S (2013) Adaptive resonance theory. Scholarpedia 8(5):1569. http://orcid.org/10.4249/scholarpedia.1569, revision #145360
17. Gunning D, Stefik M, Choi J, Miller T, Stumpf S, Yang GZ (2019) XAI—explainable artificial intelligence. Sci Robot 4(37):eaay7120
18. Hinton GE, Salakhutdinov RR (2006) Reducing the dimensionality of data with neural networks. Science 313(5786):504–507
19. Holzinger A, Kieseberg P, Weippl E, Tjoa AM (2018) Current advances, trends and challenges of machine learning and knowledge extraction: from machine learning to explainable AI. In: Holzinger A, Kieseberg P, Tjoa AM, Weippl E (eds) Machine learning and knowledge extraction. Springer, Hamburg
20. Hornik K, Stinchcombe M, White H (1989) Multilayer feedforward networks are universal approximators. Neural Netw 2:359–366
21. Housner GW, Bergman LA, Caughey TK, Chassiakos AG, Claus RO, Masri SF, Skelton RE, Soong TT, Spencer BF, Yao JTP (1997) Structural control: past, present, and future. J Eng Mech Spec Issue 123(9):897–971
22. Inman DJ (1994) Engineering vibration. Prentice Hall
23. Jones LK (1990) Constructive approximations for neural networks by sigmoidal functions. Proc IEEE 78(10):1586–1589
24. Jones LK (1997) The computational intractability of training sigmoidal neural networks. IEEE Trans Inf Theory 43(1):167–173
25. Judd JS (1990) Neural network design and the complexity of learning. MIT Press, Cambridge
26. Kerschen G, Worden K, Vakakis AF, Golinval JC (2006) Past, present and future of nonlinear system identification in structural dynamics. Mech Syst Sig Process 20:505–592
27. Kolaini AR, Nayeri R, Kern DL (2009) Pyroshock simulation systems: are we correctly qualifying flight hardware for pyroshock environments. In: 25th Aerospace testing conference
28. Lapedes A, Farber R (1988) How neural nets work. In: Anderson D (ed) Information neural. American Institute of Physics, Processing Systems, pp 442–456
29. LeCun Y, Bengio Y, Hinton G (2015) Deep learning. Nature 521:436–444
30. Lippmann RP (1987) An introduction to computing with neural nets. IEEE ASSP Mag 4–22
31. Lippmann RP (1989) Pattern classification using neural networks. IEEE Commun Mag 47–64
32. Makhoul J, El-Jaroudi A, Schwartz R (1989) Formation of disconnected decision regions with a single hidden layer. IJCNN Int Joint Conf Neural Netw I:455–460
33. Masri SF, Caughey TK (1979) A nonparametric identification technique for nonlinear dynamic problems. J Appl Mech 46:433–447
34. Masri SF, Smyth AW, Chassiakos AG, Caughey TK, Hunter NF (2000) Application of neural networks for detection of changes in nonlinear systems. ASCE J Eng Mech 126(7):666–676
35. NASA (1999) Pyroshock test criteria NASA technical standard. Tech. Rep. NASA-STD-7003, National Aeronautics and Space Administration
36. Nelles O (2000) Nonlinear system identification: from classical approaches to neural networks and fuzzy models. Springer, Berlin
37. Nguyen D, Widrow B (1990) Improving the learning speed of 2-layer neural networks by choosing initial values of the adaptive weights. Proc IJCNN III:21–26
38. O'Donnell KJ, Crawley EF (1985) Identification of nonlinear system parameters in space structure joints using the force-state mapping technique. Tech. rep., MIT space systems lab., SSL#16-85, p 170
39. Pei JS (2001) Parametric and nonparametric identification of nonlinear systems. Ph.D. dissertation, Columbia University

40. Pei JS, Mai EC (2008) Constructing multilayer feedforward neural networks to approximate nonlinear functions in engineering mechanics applications. ASME J Appl Mech 75
41. Pei JS, Masri SF (2015) Demonstration and validation of constructive initialization method for neural network to approximate nonlinear function in engineering mechanics applications. Nonlinear Dyn 79(3):2099–2119
42. Pei JS, Smyth AW (2006a) A new approach to design multilayer feedforward neural network architecture in modeling nonlinear restoring forces: part I—formulation. ASCE J Eng Mech 132(12):1290–1300
43. Pei JS, Smyth AW (2006b) A new approach to design multilayer feedforward neural network architecture in modeling nonlinear restoring forces: part II—applications. ASCE J Eng Mech 132(12):1310–1312
44. Pei JS, Wright JP, Smyth AW (2005a) Mapping polynomial fitting into feedforward neural networks for modeling nonlinear dynamic systems and beyond. Comput Methods Appl Mech Eng 194(42–44):4481–4505
45. Pei JS, Wright JP, Smyth AW (2005) Neural network initialization with prototypes—a case study in function approximation. In: Proceedings of international joint conference on neural networks 2005 (IJCNN'05), Montreal, Canada, pp 1377–1382
46. Pei JS, Mai EC, Wright JP, Smyth AW (2007) Neural network initialization with prototypes—function approximation in engineering mechanics applications. In: Proceedings of international joint conference on neural networks 2007 (IJCNN'07), Orlando, FL, IEEE Catalog Number 07CH37922C, ISBN 0-4244-1380-X
47. Pei JS, Wright JP, Masri SF, Mai EC, Smyth AW (2011) Toward constructive methods for the universal approximator—function approximation in engineering mechanics applications. In: Proceedings of international joint conference on neural networks (2011) IJCNN, San Jose CA
48. Pei JS, Mai EC, Wright JP, Masri SF (2013) Mapping some functions and four arithmetic operations to multilayer feedforward neural networks. Nonlinear Dyn 71(1–2):371–399
49. Raissi M, Perdikaris P, Karniadakis G (2019) Physics-informed neural networks: a deep learning framework for solving forward and inverse problems involving nonlienar partial differential equations. J Comput Phys 378:686–707
50. Rudin C (2019) Stop explaining black box machine learning models for high stakes decisions and use interpretable models instead. Nat Med Intell 1:206–215
51. Scarselli F, Tsoi AC (1998) Universal approximation using feedforward neural networks: a survey of some existing methods, and some new results. Neural Netw 11(1):15–37
52. Sontag ED (1992) Feedforward nets or interpolation and classification. J Comput Syst Sci 45:20–48
53. Worden K, Tomlinson GR (2001) Nonlinearity in structural dynamics: detection, identification and modelling. Institute of Physics Pub., p 680

Partially Supervised Learning for Data-Driven Structural Health Monitoring

Lawrence A. Bull, A. J. Hughes, T. J. Rogers, Paul Gardner, Keith Worden, and Nikolaos Dervilis

Abstract The cost of labelling data by engineer inspections remains a significant issue for performance and health monitoring. In many cases, this is because the actual data annotation process is expensive (e.g. non-destructive testing) or it is simply infeasible to label all the measurements (e.g. lack of access). Often, however, it is possible to provide a small number of budget-restricted labels, to describe the measurements. In these scenarios, methods for *partially supervised learning* are proposed. *Active learning*, *semi-supervised learning*, and *transfer learning* are summarised here—demonstrated with simulated monitoring examples. Each family of algorithms is shown to significantly improve conventional methods for data-driven monitoring.

Keywords Partially supervised learning · Active learning · Semi-supervised learning · Transfer learning · Structural health monitoring · Prognostics and health management · Condition monitoring

L. A. Bull (✉) · A. J. Hughes · T. J. Rogers · P. Gardner · K. Worden · N. Dervilis
University of Sheffield, Mappin St, Sheffield, S1 SJD, UK
e-mail: l.a.bull@sheffield.ac.uk

A. J. Hughes
e-mail: ajhughes2@sheffield.ac.uk

T. J. Rogers
e-mail: tim.rogers@sheffield.ac.uk

P. Gardner
e-mail: p.gardner@sheffield.ac.uk

K. Worden
e-mail: k.worden@sheffield.ac.uk

N. Dervilis
e-mail: n.dervilis@sheffield.ac.uk

A. Cury et al. (eds.), *Structural Health Monitoring Based on Data Science Techniques*, Structural Integrity 21, https://doi.org/10.1007/978-3-030-81716-9_19

1 Limited Labels in Data-driven Performance and Health Monitoring

When monitoring engineering systems via measured signals [23], data *labelling* to describe the operational, performance, or health state can be difficult to obtain for a number of reasons; for example:

- There is some high-cost associated with the data labelling process itself—e.g. expert inspection via non-destructive testing or evaluation procedures.
- It is infeasible to inspect the structure because of difficulty in access—e.g. for offshore wind farms.
- Some high cost is associated with downtime—e.g. suspending turning operations to inspect the cutting tool in a lathe during high-value manufacturing.

In these scenarios, providing descriptive labels for all measured data is economically infeasible or impractical. This problem renders many conventional data-driven *classification* models inapplicable, as they require fully labelled or *supervised* training data; in other words, each measurement vector $\mathbf{x}_i \in \mathbb{R}^d$ must have its own descriptive (scalar) label $y_i \in \mathbb{R}$, to define a training set,

$$\mathcal{D}_l = \{\mathbf{x}_i, y_i\}_{i=1}^n$$

where n is the number of training samples used to *learn* the classifier $f(\mathbf{x}_i) = y_i$ which monitors the system.

The absence of complete labelling $\{y_i\}_{i=1}^n$ will typically force a dependence on *unsupervised* learning, i.e. data-driven models learnt from the unlabelled measurements $\tilde{\mathbf{x}}_i$ only; thus, the training data are

$$\mathcal{D}_u = \{\tilde{\mathbf{x}}_i\}_{i=1}^m$$

where m is the number of unlabelled training samples. While unsupervised techniques have proved successful in many applications, they can limit monitoring procedures to *novelty detection* [23]: in other words, an indication of *normal* or *abnormal* operation only.

In many cases, however, it is often feasible to label a *small* number of measurements, given a budget determined by the performance/health monitoring regime, for example:

- The cutting tool in a lathe could be inspected between turning operations.
- Or a wind turbine could be inspected during scheduled maintenance trips.

In these scenarios, there are two sets of data to consider in training: the supervised set $\mathcal{D}_l$ and the unsupervised set $\mathcal{D}_u$. Because of budget restrictions, the number of labelled data will generally be much smaller than the number of unlabelled data; i.e. $m \gg n$.

With both labelled and unlabelled data, it would be illogical to learn a classification algorithm given only $\mathcal{D}_l$, while ignoring information in the larger set $\mathcal{D}_u$; likewise, the converse is true. Instead, data-driven models should utilise the labelled and unlabelled data in a *combined* approach, such that the union of both sets is considered,

$$\mathcal{D} = \mathcal{D}_u \cup \mathcal{D}_l$$

Conveniently, there are statistical and machine learning tools designing for learning from partly labelled data; these are referred to as *partially supervised learning* [51].

Partially supervised learning is used here as an umbrella term, to refer to methods of *leaning from fewer labelled examples*[1] [47]. Here, a brief introduction is provided for:

- Active learning,
- Semi-supervised learning, and
- Transfer learning.

Many related techniques exist, some of which are outlined in Sect. 6. Each method is applied here to a simulated SHM data set. Results for applications to operational data (from previous publications) are also provided.

2 Inspection Management: Active Learning

The first family of methods considered here are active learners [53]. The main premise is to improve the predictive performance of the mapping $f(\mathbf{x}_i) = y_i$ while querying (i.e. requesting labels) as little as possible [47]. In most scenarios, queries are taken from the unlabelled data in $\mathcal{D}_u$ to automatically extend the labelled set $\mathcal{D}_l$.

In general, there are two main approaches: *stream-based* and *pool-based*. In stream-based methods, the data in $\mathcal{D}_u$ arrive incrementally (in real-time), and the active learner must determine whether to query or not. On the other hand, pool-based methods iteratively select the most informative datum from a static set of unlabelled examples.

Intuitively, active learning has the potential to manage the inspection budget in SHM. The learner can automatically suggest the measurements for which inspections appear necessary, to improve (or maintain) the predictive performance of f. A critical step, therefore, is determining which data should be investigated and labelled. A brief review of some probabilistic approaches to active learning is provided.

[1] It also includes methods of learning from weakly labelled data.

2.1 Simulated Data

For the first active learning examples, a simulated (vibration-based) data set is considered, based on an eight-degree-of-freedom benchmark system designed by the Los Alamos National Laboratory [23]. Following identification of the system parameters via modal analysis, time series data were simulated for six conditions. Each health state represents progressive damage, approximated via reductions in the stiffness of the system (a common assumption in the literature [23]):

- [$y_i = 1$] normal (system parameters unchanged),
- [$y_i = 2$] damage 1: spring five stiffness 97%,
- [$y_i = 3$] damage 2: spring five stiffness 93%,
- [$y_i = 4$] damage 3: spring five stiffness 88%,
- [$y_i = 5$] damage 4: spring five stiffness 82%,
- [$y_i = 6$] damage 5: spring five stiffness 70%.

From the simulated time series, eight-second windows were converted to the frequency domain (transmissibilities) to define 100 frequency-domain observations per class, according to the procedure in [10]. Of these data, two-thirds are set aside for training ($\mathcal{D} = \mathcal{D}_l \cup \mathcal{D}_u$) and one-third are held out as an independent test set $\mathcal{D}^*$. To visualise the data set and active learning models, the frequency-domain features are projected via principal component analysis onto two dimensions, as plotted in Fig. 1.

2.2 Query Schemes: Uncertainty Sampling

Perhaps, the most obvious way to query data is to select instances that appear uncertain, given the current model [7]; this procedure is known as *uncertainty sampling* [53]. That is, starting from a small number of labelled data, further points can be queried according to those that appear 'uncertain' based on various statistics.

To demonstrate, a Gaussian mixture model (GMM) is learnt for the data presented in Fig. 1, estimating the parameters in a Bayesian manner [46]. This involves fitting a Gaussian distribution to each class-conditional density[2] $p(\mathbf{x}_i|y_i = k)$ (where $k \in \{1, \ldots, 6\}$) given some small initial labelled set $\mathcal{D}_l$ (i.e. leading to six clusters of data). The GMM resulting from 3% labelled data is visualised by ellipses, one for each class-conditional, also in Fig. 1.

Given the current GMM and the unlabelled data, observations in $\mathcal{D}_u$ that appear most uncertain can be queried. Two measures of uncertainty are considered in this first example: high entropy and low-likelihood.

[2] Herein, probability distribution functions (mass or density) will be denoted $p(\cdot)$.

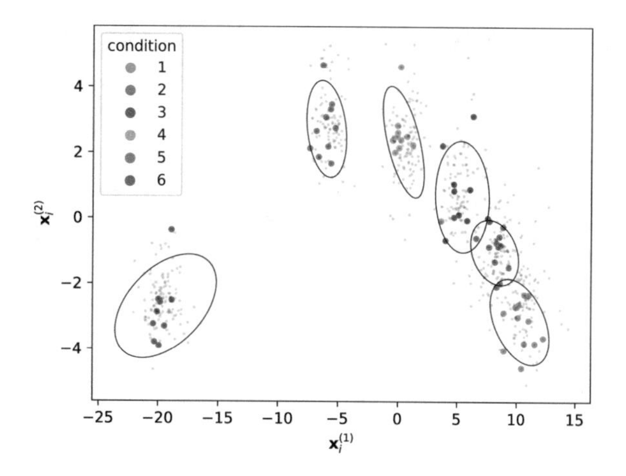

Fig. 1 Six-class, progressive damage simulated data. Colours represent labelled training data $\mathcal{D}_l$, and small black markers represent unlabelled data $\mathcal{D}_u$. Ellipses (two-sigma) visualise the (supervised) GMM learnt for 3% labelled data

2.2.1 Maximum Entropy Sampling (MES)

Entropy can be used to query observations whose *predicted labels* appear to be the most 'confused' or 'conflicted' [54]. Specifically, measurements whose classification into all states is equally likely have the highest predictive entropy [5].

Typically, the Shannon entropy of the posterior predictive distribution over the unobserved labels $\tilde{y}_i$ is used[3],

$$H(\tilde{y}_i) = -\sum_{k=1}^{K} p(\tilde{y}_i = k \mid \tilde{\mathbf{x}}_i, \mathcal{D}_l) \log p(\tilde{y}_i = k \mid \tilde{\mathbf{x}}_i, \mathcal{D}_l) \tag{1}$$

The result of querying labels with maximum entropy is to select those data that appear at the boundaries between existing classes. To visualise, a heat-map of the predictive entropy over the feature space[4] is presented in Fig. 2; unlabelled data lying in dark red regions would be queried by the learner.

2.2.2 Lowest Likelihood Sampling

An alternative measure of uncertainty considers data $\tilde{\mathbf{x}}_i$ that appear *unlikely* given the current classification model. These ideas align with that of *least confident* sampling [53], although least confident samples consider the distribution over y_i, as opposed to $\mathbf{x}_i$. In contrast to MES, lowest likelihood samples select data that appear in the extremities of the model, rather than at class boundaries. Considering the GMM example, low-likelihood measurements are those with a low (marginal) likelihood,

[3] That is, the averaged Shannon information content [42] of the possible label outcomes $k \in \{1, \ldots, K\}$, where $K = 6$ for the simulated data.

[4] Given the model and data from Fig. 1.

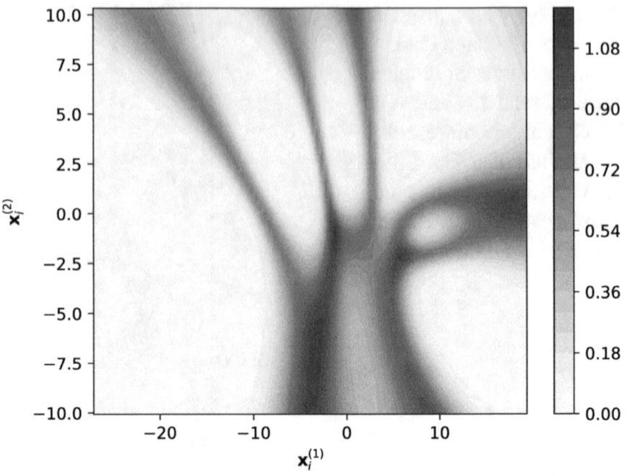

Fig. 2 Maximum entropy sampling: darker regions represent areas in which data would be queried. Queries are preferred at the boundaries between existing clusters; i.e. data that are confused given the GMM predictive distribution over y_i

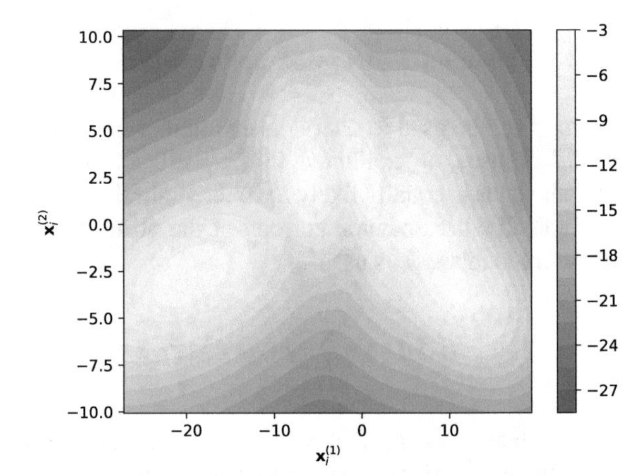

Fig. 3 Lowest likelihood sampling: darker regions represent areas in which data would be queried (colour map corresponds to the log-likelihood). Queries are preferred at the extremities of the model; i.e. data that appear unlikely given the GMM

$$p(\tilde{\mathbf{x}}_i|\mathcal{D}_l) = \sum_{k=1}^{K} p(\tilde{\mathbf{x}}|\tilde{y}_i = k, \mathcal{D}_l)\, p(\tilde{y}_i = k|\mathcal{D}_l) \tag{2}$$

Low-likelihood regions are represented by dark blue areas in Fig. 3. These queries are useful in discovering new classes of data, as they sample measurements that appear *novel* given the model (rather than confused). As such, these queries are arguably most useful in novelty detection—when applying active learning to streaming data, for example [7].

2.3 *Query Schemes: Information-Theoretic Active Learning*

Another view of active learning considers a Bayesian experimental design perspective [31]. The goal is to select data that appear to *improve the model* as quickly as possible. [41] proposed a querying scheme by selecting observations whose labels are expected to lead to the greatest reduction in entropy of the posterior distribution over the parameters. In other words, those labels that provide the most information about the model (via $\boldsymbol{\theta}$) when queried[5]. This process is typically formalised by defining a *utility* for querying a point $\tilde{\mathbf{x}}_i$,

$$U(\mathbf{x}_i) = H(p(\boldsymbol{\theta}|\mathcal{D}_l)) - \mathbb{E}_{p(\tilde{y}_i|\tilde{\mathbf{x}}_i,\mathcal{D}_l)}[\mathbb{H}(p(\boldsymbol{\theta}|\mathcal{D}, \{\tilde{\mathbf{x}}_i, \tilde{y}_i\}))] \tag{3}$$

(where $\mathbb{E}_A[p(\cdot)]$ is the expected value of the distribution $p(\cdot)$ with respect to A). Unlike expression (1), this utility is focussed on the entropy of the posterior distribution over the parameters $p(\boldsymbol{\theta}|\mathcal{D})$—rather than the predictive distribution over the labels $p(\tilde{y}_i|\tilde{\mathbf{x}}_i, \mathcal{D}_l)$. In practice, however, (3) can also be rewritten in terms of the predictive distribution, as (3) is usually intractable for complex models in the above form [32].

An advantage of the information-theoretic approach is that it takes both measurement and parameter uncertainty into account; this prevents the learner from selecting only 'the most confused' labels [46]—particularly for discriminative classifiers. Variation in the sampling procedure is useful to prevent training data from becoming unrepresentative of the underlying distribution of data—a recognised issue for active learning methods, referred to as *sampling bias* [33, 53] (Fig. 4).

2.3.1 A Note on Sampling Bias

While the reasoning behind active learning is intuitive, the performance of active learners can sometimes prove to be worse than conventional (passive) learning [20, 58]. Specifically, problems occur if sampling becomes too focussed on specific queries, such that the training data become poorly representative of the underlying distribution, i.e. *sampling bias* [33, 53]. As such, queries should not focus too much on specific regions of the feature space—the cluster boundaries, for example. To avoid sampling bias, various methods have been proposed: combining different query methods [33] or more 'adaptive' query schemes [32].

2.4 *Application Examples*

The query methods summarised here are applied to the eight-degree-of-freedom data. Initiating with a small (random) sample of labels, the unlabelled data in $\mathcal{D}$ are queried sequentially, as if each point were the last (i.e. *myopic* active learning [31]). The three

[5] For the GMM, the random vector $\boldsymbol{\theta}$ contains cluster centres and covariances.

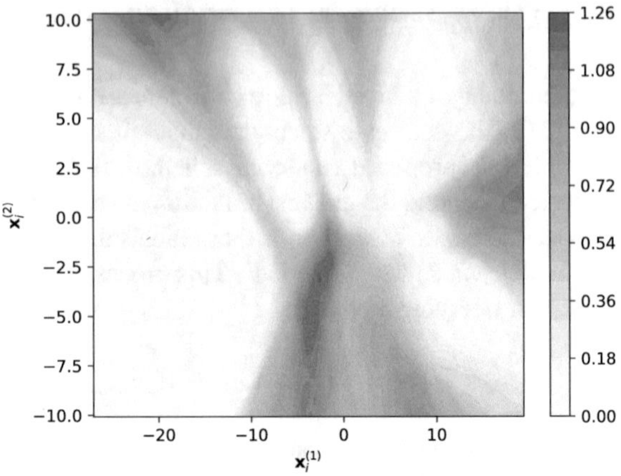

Fig. 4 Information-theoretic active learning: darker regions represent areas in which data would be queried. Similar to Fig. 2, queries are preferred at the boundaries between existing clusters; however, in this case, the parameter uncertainty for each cluster is taken into account

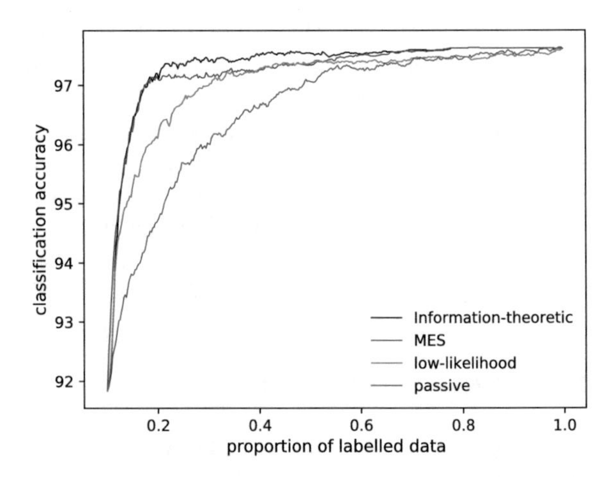

Fig. 5 Averaged classification accuracy for active learning methods compared to conventional (passive) learning, for 400 repeats

query methods—maximum entropy sampling (MES), low-likelihood sampling, and information-theoretic sampling—are compared to conventional passive learning, that is, querying data by random sampling.

The classification accuracy for an increasing number of queries is presented in Fig. 5 (until all of the data are labelled). Each of the active query schemes outperforms passive learning in this example, with information-theoretic sampling providing the best accuracy. Low-likelihood queries lead to the smallest performance increase; as aforementioned, these queries are arguably more suited to streaming data, which might explain the reduced improvements.

2.5 Query Schemes: Value of Information

It may be desirable to apply active learning to develop a classifier with a decision process, as well as the classifier. In the context of health/performance monitoring, an operations and maintenance (O&M) example might involve the selection of a *maintenance action* based on a prediction of the state of the monitored system [35]. In this setting, one can consider the *value of information* as a potential query measure [36].

In decision theory, the *expected value of perfect information* (EVPI) is understood to be the amount of utility (or resource) a decision-maker is willing to expend to gain access to the ground truth of an uncertain or unknown state. More formally, EVPI can be defined as [39]

$$\mathrm{EVPI}(d|y_i) := \mathrm{MEU}(\mathcal{I}_{y_i \to d}) - \mathrm{MEU}(\mathcal{I}) \tag{4}$$

where $\mathrm{EVPI}(d|y_i)$ is the expected value of querying an incipient point $\{\tilde{\mathbf{x}}_i, \tilde{y}_i\}$ to obtain the labelled observation $\{\mathbf{x}_i, y_i\}$, with perfect information, before making a future decision d. In Eq. (4), $\mathrm{MEU}(\mathcal{I})$ denotes the *maximum expected utility* of a decision process $\mathcal{I}$ involving the random variables $\{\tilde{\mathbf{x}}_i, \tilde{y}_i\}$ and a decision d. Similarly, $\mathrm{MEU}(\mathcal{I}_{y_i \to d})$ denotes the maximum expected utility of a modified decision process $\mathcal{I}_{y_i \to d}$, in which y_i is obtained prior to decision d being made. Note that the EVPI is strictly non-negative, and $\mathrm{EVPI}(d|y_i) = 0$ if and only if the optimal policy for d in $\mathcal{I}$ remains optimal for d in $\mathcal{I}_{y_i \to d}$. That is, value of information only arises when labelling has the potential to result in a change in optimal policy.

For stream-based active learning, the EVPI of an incipient data point can be used to form a convenient heuristic for triggering inspection. Simply, an inspection is mandated if the following criterion is satisfied,

$$\mathrm{EVPI}(d|y_i) > C_{\mathrm{insp}} \tag{5}$$

where C_{insp} can be interpreted as the 'cost' of inspection. For pool-based (myopic) active learning, candidate data can be ranked according to their EVPI and queried preferentially.

2.5.1 Application Example

A value-of-information scheme is demonstrated for a numerical data set, originally presented in [36]. The data include an undamaged state and three damaged states, each with increasing severity. Importantly, the most advanced damage state is associated with a *costly* mode of failure: it is the goal of a decision-maker to avoid this failure mode via preventative maintenance. Details of the probabilistic models and utility functions can be found in [36].

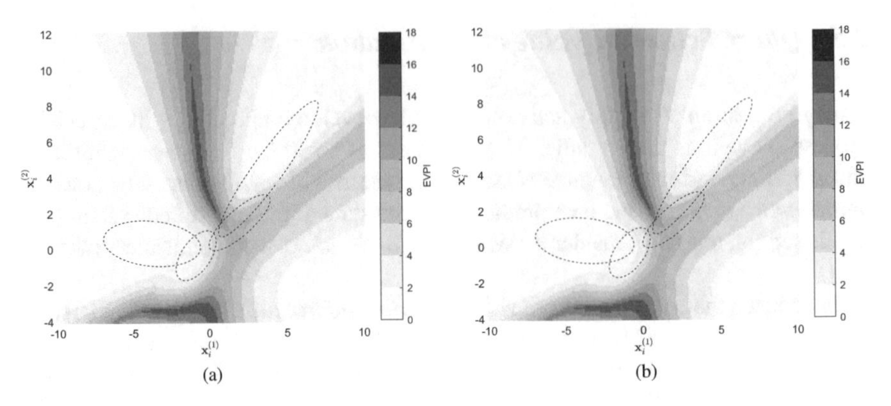

Fig. 6 Value-of-information active learning

Fig. 7 Mean decision accuracy for a given number of queries for an agent utilising a classifier trained using (i) EVPI-based querying and (ii) random sampling (passive learning). Dashed lines show ±1 standard deviation

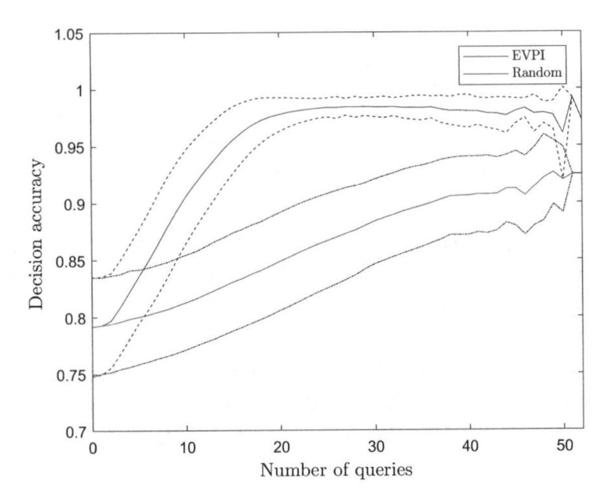

Similar to the previous examples, a GMM is used as the predictive model. Given a small sample of labelled data, a heat-map of the EVPI over the feature space is shown in Fig. 6a, with clusters corresponding to increasing damage from left to right. Figure 6b shows the EVPI once the GMM has been updated via value-of-information queries. Comparing Fig. 6a, b, the region of high EVPI becomes more pronounced between the two far-right clusters. This occurs, as queries are preferred in areas between clusters corresponding to high-severity and lower-severity health states, for which there are differing optimal decision policies. Thus, Fig. 6 highlights that the value-of-information method queries data that strengthen more 'costly' decision boundaries.

Figure 7 is provided to demonstrate that the performance of a decision-making agent can be improved via value-of-information sampling. Again, for details of the decision process, the reader is directed to [36].

It would be remiss to fail to highlight the assumptions in this EVPI-based example. Notably, to evaluate EVPI as presented here, it is necessary to assume that the number of classes (i.e. health states of interest) is known *a priori*, to attribute utilities to each. Additionally, it is assumed that perfect information of the health state can be obtained via inspection. An alternative approach to decision-theoretic active learning, which relaxes assumptions of perfect information, is summarised in [47].

3 Combining Labelled and Unlabelled Data: Semi-supervised Learning

Another partially supervised technique suited to performance/health monitoring is *semi-supervised* learning [16, 51]. The focus of these methods is to actually use the remaining unlabelled data in $\mathcal{D}_u$ to help infer the parameters of the classifier. In other words, the model $f(\mathbf{x}_i)$ is learnt from the union set of labelled and unlabelled data $\mathcal{D} = \{\mathcal{D}_u \cup \mathcal{D}_l\}$ within a *unifying* training scheme[6].

These models have potential in monitoring applications as a small set of labelled data (annotated by the engineer) can be combined with the larger sets of unlabelled measurements. Unsurprisingly, there are numerous ways to enforce semi-supervision. Arguably, the most interpretable is *self-training* (also self-labelling and pseudo-labelling) [16, 47]. In simple terms, the predicted labels for $\tilde{\mathbf{x}}_i$ are used as *pseudo-labels* to train the algorithm in subsequent learning. Returning to themes of entropy, self-labelling implicitly encourages models with low-entropy predictions (i.e. confident label predictions) [47].

3.1 Entropy Minimisation

Formally, *entropy minimisation* techniques [30] minimise the following loss function for the unlabelled data [47],

$$L = -\sum_{i=1}^{m}\sum_{k=1}^{K} p(\tilde{y}_i = k|\tilde{\mathbf{x}}_i)\log p(\tilde{y}_i = k|\tilde{\mathbf{x}}_i) \tag{6}$$

One notices similarities to the entropy expression for active learning (1), and that (6) is minimised when points are assigned to a single class of data with unit probability. In simple terms, the parameters of the model $\boldsymbol{\theta}$ are adjusted such that the unlabelled data are classified with the maximum possible confidence.

[6] This is in contrast to active methods, whereby unlabelled instances are only sampled from to extend the labelled training set.

A special case can be implemented via expectation maximisation (EM), originally proposed by [48], such that the expected joint log-likelihood is maximised [16],

$$\mathcal{L}(\boldsymbol{\theta} \mid \mathcal{D}) = \mathcal{L}(\boldsymbol{\theta} \mid \mathcal{D}_u, \mathcal{D}_l) \propto \underbrace{\sum_{i=1}^{m} \log \sum_{k=1}^{K} p\left(\tilde{\mathbf{x}}_i \mid \tilde{y}_i = k, \boldsymbol{\theta}\right) p(\tilde{y}_i = k \mid \boldsymbol{\theta})}_{\mathcal{D}_u} \dots$$

$$+ \underbrace{\sum_{i=1}^{n} \log \left[p\left(\mathbf{x}_i \mid y_i = k, \boldsymbol{\theta}\right) p(y_i = k \mid \boldsymbol{\theta})\right] + \log p(\boldsymbol{\theta})}_{\mathcal{D}_l} \tag{7}$$

For details of how (7) relates to an entropy minimisation viewpoint, refer to [1, 30]. Expression (7) implies that the full joint log-likelihood of the model is maximised, considering *both* the labelled (term one) and unlabelled data (term two).

3.1.1 Application Example

To demonstrate improvements via semi-supervised learning, a Gaussian mixture model (GMM)—learnt from the eight-degree-of-freedom data—is shown before and after semi-supervised updates in Fig. 8. These results were originally presented in [10]. Figure 8 visually highlights the improvements to density estimation when considering both labelled and unlabelled data to infer the parameters, i.e. using the information in large sets of unlabelled measurements to support learning from a small set of inspected data.

Figure 9 is provided to quantify model improvements as the proportion of labelled data is increased. In the same test scheme as Sect. 2.4, the semi-supervised model is learnt while increasing the number of labelled data until all of the training data in $\mathcal{D}$ are labelled, i.e. $\mathcal{D}_u = \emptyset$. In these tests, the model is compared to standard supervised learning, whereby inference only considers the labelled set $\mathcal{D}_l$ while ignoring the unlabelled data $\mathcal{D}_u$. For the range of label proportions, there is a notable decrease in the classification error when implementing semi-supervised updates.

3.1.2 Considerations for Semi-supervised Learning

During semi-supervised learning, incorporating unlabelled signals has the potential to decrease the predictive performance if the structure imposed by classifier proves inappropriate [16]. This can be particularly problematic for generative methods [20], such as the GMM presented here. Additionally, when self-training, if a model initially categorises the data poorly, it will (iteratively) retrain on incorrect predictions. These errors can propagate, until the classifier returns an invalid solution—a phenomenon termed *confirmation bias* [47].

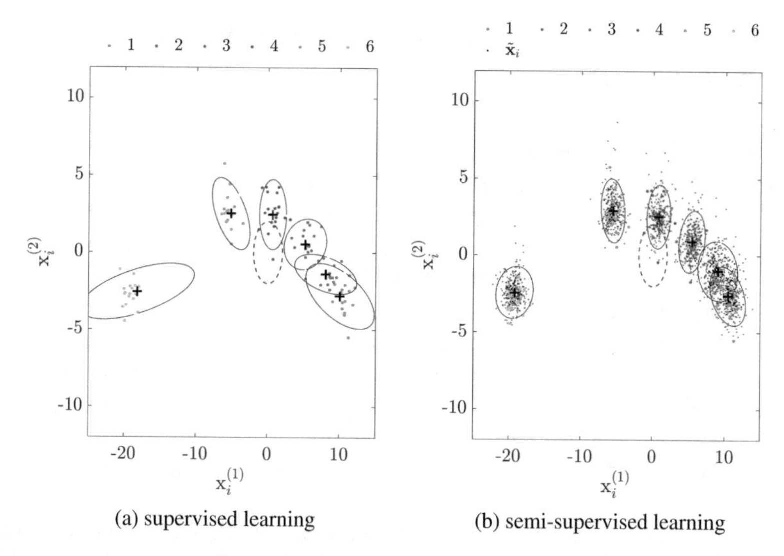

(a) supervised learning (b) semi-supervised learning

Fig. 8 Black ellipses represent the supervised and semi-supervised GMMs (the blue ellipse represents the prior of the class-conditionals). Coloured markers correspond to the labelled set, while black markers are unlabelled. Improvements to density estimation via semi-supervised learning are clear

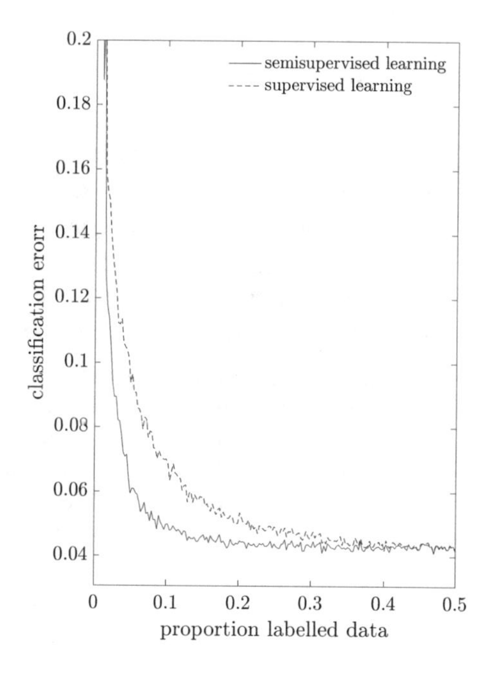

Fig. 9 Averaged classification error for semi-supervised learning, compared to conventional supervised learning, for 50 repeats [10]

4 Transfer Learning: Transferring Labels Between Similar Domains

The final family of methods summarised here are *transfer learners* [49]. In this setting, the labelled and unlabelled data are split across two or more *domains*, rather being generated by the same underlying distribution. This implies that labelled data from one domain might be used to aid a *task* in another unlabelled domain, where the data are generated by different (but similar) processes.

In the context of health and performance monitoring, the ideas of transfer imply that labelled data from one structure might help annotate the unsupervised measurements from another. Alternatively, the labelled data generated by some model or simulation might prove useful to annotate in-the-field measurements. Critically, in these examples, conventional machine learning cannot be applied, as the data are generated by different distributions for each case: either different structures or models.

When modelling data from two or more structures, these ideas align with that of population-based SHM—an emergent field in the SHM literature [9, 26, 29]. An example of the population-based framework to match labels between structures is visualised in Fig. 10.

4.1 Concepts for Transfer

Typically, one considers the transfer of information between two domains at a time[7] [49]. Some common terminology is useful in this setting:

- The **source domain** is associated with finite set of observations and corresponding labels, i.e. $\mathcal{D}_s = \{\mathbf{x}_i, y_i\}_{i=1}^n$ where $\{\mathbf{x}_i \in \mathbf{X}_s,\ y_i \in \mathbf{y}_s\}$.
- The **target domain** is associated with observations and (in this example) *unknown* labels, i.e. $\mathcal{D}_t = \{\tilde{\mathbf{x}}_i, \tilde{y}_i\}_{i=1}^m$ where $\{\tilde{\mathbf{x}}_i \in \tilde{\mathbf{X}}_t,\ \tilde{y}_i \in \tilde{\mathbf{y}}_t\}$.

At this stage, it is interesting to highlight how transfer ideas are somewhat analogous to semi-supervised learning (when target domain labels are unknown). That is, a limited set of labelled data in $\mathcal{D}_s$ are used to support a *task*—characterised by $f(\cdot)$—in an unlabelled domain $\mathcal{D}_t$. However, there is a critical difference: semi-supervised learning assumes the labelled and unlabelled data are generated by the same underlying distribution, whereas transfer learning allows for changes in this distribution.

[7] However, multi-domain methods also exist [21].

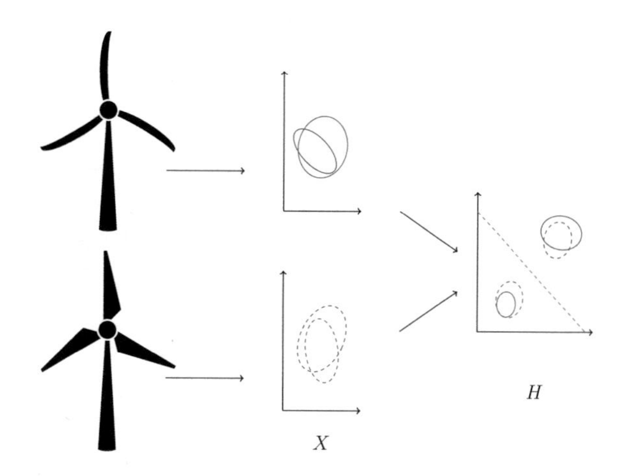

Fig. 10 Separation of normal and novel data via transfer learning. Normal data are blue, and novel data are red. Labelled and unlabelled data sets are represented by solid and dashed lines, respectively

4.2 Application Example

For demonstration, *domain adaptation* [50] is applied as a form of transductive transfer [2]. Transductive transfer is appropriate when a model $f(\cdot)$ will not generalise between the source and target domains because of differences in the distributions $p(\mathbf{x}_i)$ and $p(\tilde{\mathbf{x}}_i)$ [28]. In other words, the source and target *tasks* are the same (classification), while the *domains* (feature space and measurements) are different.

To demonstrate, consider simulated data from *two* three-degree-of-freedom systems. The pair of systems can be considered to represent a pair of multi-storey civil structures, similar to those presented in [12]. The first three natural frequencies are used as features to monitor the health/environmental states of each structure. Both of the three-degree-of-freedom systems (S1 and S2) are identical, other than an increase in the mass at the top floor of the second structure (S2). In practice, this mass change could represent operational variations or a difference in design.

For systems S1 and S2, two classes of data are simulated:

- $[y_i = 1]$ the normal condition (S1: 100 training points, S2: 50 training points).
- $[y_i = 2]$ environmental effects—increased stiffness in the beams of the structure[8] (S1: 75 training points and S2: 30 training points).

(There are 1000 test observations from each class, in each domain.) Importantly, only the data from S1 are labelled, while the data from S2 are unlabelled; this implies that S1 is associated with the source domain $\mathcal{D}_s = \{\mathbf{x}_i, y_i\}_{i=1}^{n}$, while S2 is associated with the target domain $\mathcal{D}_t = \{\tilde{\mathbf{x}}_i\}_{i=1}^{m}$. The common task across both domain structure pairs is classifying observations as normal $[y_i = 1]$ or environmental effects $[y_i = 1]$.

To visualise why domain adaptation is necessary, the data from both structures can be stacked and then projected onto two dimensions via principal component analysis (PCA)—shown in Fig. 11. Clearly, the data from systems S1 and S2 are

[8] Stiffness changes due to temperature effects are a typical assumption in the literature [23].

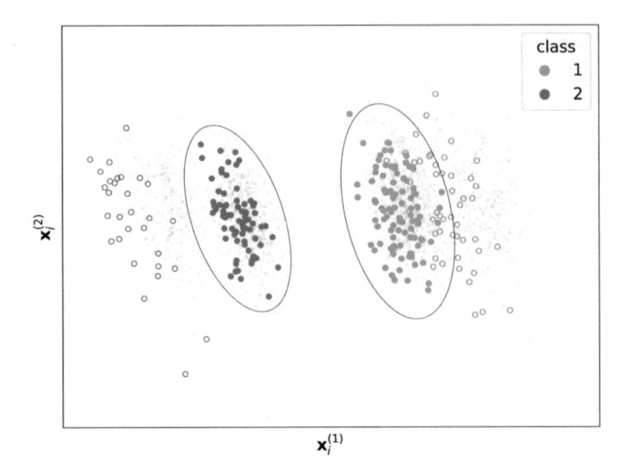

Fig. 11 Principal component subspace (X) for the source and target data (*filled circle* and *open circle* markers, respectively). The labelled source data are associated with structure S1; the unlabelled target data are associated with structure S2. The source and target data appear to be sampled from distinct (bi-modal) distributions; (three-sigma) ellipses represent the GMM learnt from the source data. Test data are shown by light · markers

sampled from different (bi-modal) distributions. As such, a two-class GMM learnt from the supervised S2 data (visualised by the ellipses) would generalise poorly to the S1 data. It would be useful, therefore, to define some formal/robust way of using the supervised data from S1 to model and label the data from S2.

4.2.1 Transfer Component Analysis

As the data in each domain have a similar structure, transfer component analysis (TCA) [50] can be applied to *match* source and target distributions in a shared latent space. Distribution matching is achieved using a projection $\psi(\cdot)$ such that the distributions can be considered *approximately* equivalent,

$$p\left(\psi\left(\mathbf{x}_i\right)\right) \approx p\left(\psi\left(\tilde{\mathbf{x}}_i\right)\right) \tag{8}$$

In brief terms, $\psi(\cdot)$ defines a nonlinear projection[9] that minimises the distance between the source and target distributions in the subspace,

$$\text{Dist}\left[p\left(\psi\left(\mathbf{X}_s\right)\right),\ p\left(\psi\left(\tilde{\mathbf{X}}_t\right)\right)\right] \tag{9}$$

[9] A linear projection on a kernel, similar to kernel PCA [46].

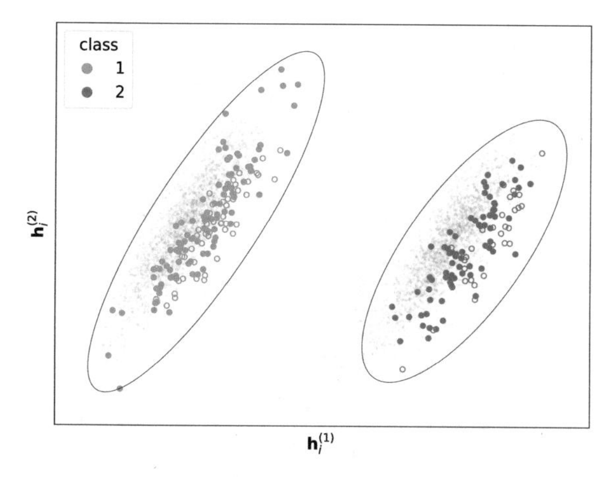

Fig. 12 Transfer component subspace (H) for the source and target data (*filled circle* and *open circle* markers, respectively). The labelled source data are associated with structure S1; the unlabelled target data are associated with structure S2. The source and target data appear to be sampled from the same (bi-modal) distributions; (three-sigma) ellipses represent the GMM learnt from the source data. Test data are shown by light · markers

Thus, in the latent space, a shared model ($f = f_s = f_t$) can be learnt from the supervised data in S1, to classify and label the S2 target data: in some sense, this step can be considered similar to *self-training* for semi-supervised learners.

The TCA subspace[10] for the data from S1 and S2 is shown in Fig. 12, where the distributions have been (approximately) matched using $\psi(\cdot)$. In consequence, the GMM classifier learnt from $\mathcal{D}_s$ (represented by ellipses) can be seen to generalise to $\mathcal{D}_t$; in other words, a single model can be shared between the structures,

$$f \approx f_s \approx f_t$$

Thus, the labels should propagate (more reliably) from the source to the target data.

4.2.2 A Note on Negative Transfer

It is critical to consider whether the application of transfer learning is appropriate before attempting to transfer information between domains. If the source and target domains are not related—measurements from a boat and an aeroplane, for example—there would be a high risk that the source domain actually decreases the performance in the target domain [49]. This phenomenon is referred to as *negative transfer*; a visual example of how classes might be confused in a monitoring context is shown in Fig. 13.

[10] To be comparable with PCA, a linear kernel is used here.

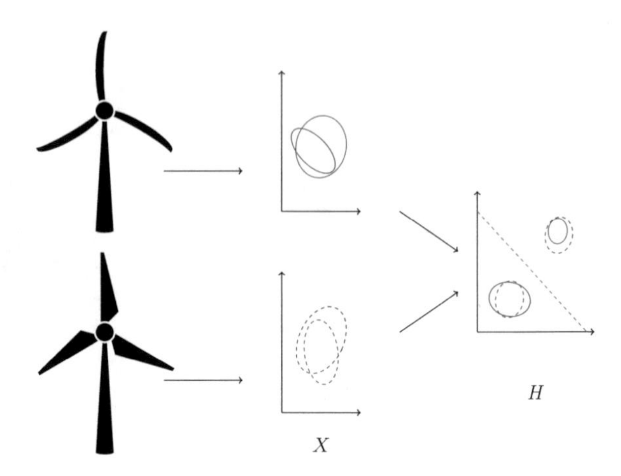

Fig. 13 A visualisation of negative transfer. Labelled and unlabelled data sets are represented by solid and dashed lines, respectively. The normal data (blue) and novel data (red) are confused between the labelled and unlabelled domains/structures

5 A Brief Review of SHM Applications

Increasingly, methods for partially supervised learning are being applied in the performance and health monitoring literature. In particular, transfer learning has been investigated quite frequently, in a range of settings. Some examples are provided for each of the topics discussed.

Transfer learning:

The majority of literature concerning transfer learning for SHM focusses on image classification using convolutional neural networks (CNNs), e.g. [13, 22, 25, 37]. Typically, applications consider crack detection [22, 25, 37] and *fine-tuning* the parameters of a CNN trained on a source domain to aid generalisation in a target domain. Generally, fine-tuning does not aim to transfer label knowledge from source to target domains; instead, it focusses on repurposing expensive-to-train 'deep' neural networks [28].

In addition to fine-tuning, neural networks have been applied for domain adaptation within SHM, for example [40, 55, 57]. These techniques use a neural network to define the mapping from the domain data into a shared latent space, where a classification model is learnt. An interesting study is presented by [45], demonstrating domain adaptation for condition monitoring, applied to a fleet of power plants.

Besides neural networks, a paper by [14] applies a probabilistic method for transfer learning by defining an objective function such that information in the source and target domains is considered jointly. This application focusses on the important issue of *sensor coverage* (for a single structure) rather than transferring information between similar systems.

Domain adaptation has been discussed in an SHM context by [27, 28], considering methods for knowledge transfer between a simulated source and target domain, as well as a simulated source and experimental target structure. Additionally, transfer component analysis has been applied to a group of tailplane structures (i.e. experimental data) to transfer damage detectors between members of a population for SHM [11].

Semi-supervised learning:

In the context of bridge monitoring, Chen *et al.* introduced a graph-based approach for label propagation [18, 19]. The objective function of a multi-resolution classifier [17, 43] is modified such that the weighting parameters are optimised over the labelled and the unlabelled data. Another graph-based application considers fault diagnosis for condition monitoring of bearings and pumps [56]. On the other hand, label propagation within hierarchical clustering has been investigated with experimental aircraft data [6] and pipe monitoring [52]. Generative mixture models have also been adapted for probabilistic applications of semi-supervised monitoring with vibration-based aircraft data [10].

Further examples consider applications of K-means [4], Dirichlet processes, and fuzzy C-means [34] clustering for semi-supervised heuristics. Huang *et al.* [34] use fuzzy C-means within an online SHM strategy; the proposed method becomes partially supervised during a *label-matching step*, where the unsupervised clusters are compared to known classes from the supervised data. Bouzenad *et al.* [4] define a similar online heuristic using K-means; in this case, new clusters are created when a distance-based threshold is broken within the unsupervised algorithm. These methods can be considered as simple form of *clustering with constraints* [16], whereby partial supervision is introduced via constraints on an *unsupervised* algorithm. (However, in the examples referenced here, partial supervision is enforced within the SHM framework, rather than the algorithm.)

Active learning:

Active learning is somewhat less explored in performance and health monitoring applications. Existing studies include generative mixture models for (batch) online clustering in tool wear [8] and bridge monitoring regimes [7]. Neural networks have been applied with uncertainty sampling to classify images of defects in a data set concerning civil structures [24]. Martinez Arellano and Ratchev [44] propose a Bayesian convolutional neural network for tool-monitoring, using maximum entropy sampling. Additionally, an adaptive probabilistic framework is proposed in [15] for active data selection to aid a particle filter based damage-progression model.

6 Concluding Remarks

Emerging technologies for learning from *fewer labelled data* in performance and health monitoring have been summarised and demonstrated. Collectively, these methods are termed *partially supervised learning*. Because of the inherent cost associated with labelling data in monitoring regimes, it is the authors' opinion that these algorithms will prove critical in developing structural health monitoring, while ultimately aiding its transition into standard practice in industry.

Active learning, semi-supervised learning, and transfer learning have been proposed as examples here:

- Active learning can be used as a technique to guide/automate system inspections.
- Semi-supervised learning can be used to combine these inspection results with large volumes of unlabelled data.
- Transfer learning is proposed to transfer label information between similar systems (or monitoring regimes).

As with most data-driven monitoring, care must be taken to ensure that any model assumptions are appropriate; the caveats for each method have been highlighted throughout. When implemented correctly, however, partially supervised learning has the potential to significantly improve performance and health monitoring frameworks.

Considering the foundations in information theory [42] (particularly active and semi-supervised learning), there are various related problems, in particular: Bayesian experimental design [38], Bayesian optimisation, and model updating [3]. There are also related tools from the machine learning literature, including: meta learning, few shot learning, and data augmentation [47], all of which indicate interesting avenues of future work.

Acknowledgements The authors gratefully acknowledge the support of the UK Engineering and Physical Sciences Research Council (EPSRC) through grant references EP/R003645/1, EP/R004900/1, EP/R006768/1.

References

1. Amini MR, Gallinari P (2002) Semi-supervised logistic regression. In: ECAI, pp 390–394
2. Arnold A, Nallapati R, Cohen WW (2007) A comparative study of methods for transductive transfer learning. In: Seventh IEEE international conference on data mining workshops (ICDMW 2007), IEEE, pp 77–82
3. Behmanesh I, Moaveni B, Lombaert G, Papadimitriou C (2015) Hierarchical Bayesian model updating for structural identification. Mechan Syst Sig Proces 64:360–376
4. Bouzenad AE, El Mountassir M, Yaacoubi S, Dahmene F, Koabaz M, Buchheit L, Ke W (2019) A semi-supervised based k-means algorithm for optimal guided waves structural health monitoring: a case study. Inventions 4(1)
5. Bull L (2019) Towards probabilistic and partially-supervised structural health monitoring. PhD thesis, University of Sheffield
6. Bull L, Worden K, Manson G, Dervilis N (2018) Active learning for semi-supervised structural health monitoring. J Sound Vibr 437:373–388
7. Bull L, Rogers T, Wickramarachchi C, Cross E, Worden K, Dervilis N (2019a) Probabilistic active learning: an online framework for structural health monitoring. Mechan Syst Signal Proces 134:106294
8. Bull L, Worden K, Rogers T, Wickramarachchi C, Cross E, McLeay T, Leahy W, Dervilis N (2019b) A probabilistic framework for online structural health monitoring: active learning from machining data streams. In: Journal of Physics: Conference Series, IOP Publishing, vol 1264, p 012028

9. Bull L, Gardner P, Gosliga J, Rogers T, Dervilis N, Cross E, Papatheou E, Maguire A, Campos C, Worden K (2020a) Foundations of population-based SHM, part I: Homogeneous populations and forms. Mechan Syst Signal Process 148:107141
10. Bull L, Worden K, Dervilis N (2020b) Towards semi-supervised and probabilistic classification in structural health monitoring. Mechan Syst Signal Proces 140:106653
11. Bull L, Gardner P, Dervilis N, Papatheou E, Haywood-Alexander M, Mills R, Worden K (2021a) On the transfer of damage detectors between structures: an experimental case study. J Sound Vibr 116072
12. Bull LA, Gardner P, Rogers TJ, Cross EJ, Dervilis N, Worden K (2021b) Probabilistic inference for structural health monitoring: new modes of learning from data. ASCE-ASME J Risk Uncertainty Eng Syst Part A: Civil Eng 7(1):03120003
13. Cao P, Zhang S, Tang J (2018) Preprocessing-free gear fault diagnosis using small datasets with deep convolutional neural network-based transfer learning. IEEE Access 6:26241–26253
14. Chakraborty D, Kovvali N, Chakraborty B, Papandreou-Suppappola A, Chattopadhyay A (2011) Structural damage detection with insufficient data using transfer learning techniques. In: Sensors and smart structures technologies for civil, mechanical, and aerospace systems 2011, International Society for Optics and Photonics, vol 7981, p 798147
15. Chakraborty D, Kovvali N, Papandreou-Suppappola A, Chattopadhyay A (2015) An adaptive learning damage estimation method for structural health monitoring. J Intell Mater Syst Structs 26(2):125–143
16. Chapelle O, Scholkopf B, Zien A (2006) Semi-supervised learning. MIT Press, Cambridge
17. Chebira A, Barbotin Y, Jackson C, Merryman T, Srinivasa G, Murphy RF, Kovavcevic J (2007) A multiresolution approach to automated classification of protein subcellular location images. BMC Bioinform 8(1):210
18. Chen S, Cerda F, Guo J, Harley JB, Shi Q, Rizzo P, Bielak J, Garrett JH, Kovacevic J (2013) Multiresolution classification with semi-supervised learning for indirect bridge structural health monitoring. In: 2013 IEEE international conference on acoustics, speech and signal processing, pp 3412–3416
19. Chen S, Cerda F, Rizzo P, Bielak J, Garrett JH, Kovacevic J (2014) Semi-supervised multiresolution classification using adaptive graph filtering with application to indirect bridge structural health monitoring. IEEE Trans Signal Process 62(11):2879–2893
20. Cozman FG, Cohen I, Cirelo MC (2003) Semi-supervised learning of mixture models. In: Proceedings of the 20th international conference on machine learning (ICML-03), pp 99–106
21. Daumé III H (2009) Frustratingly easy domain adaptation. arXiv preprint arXiv:09071815
22. Dorafshan S, Thomas RJ, Maguire M (2018) Comparison of deep convolutional neural networks and edge detectors for image-based crack detection in concrete. Constr Build Mater 186:1031–1045
23. Farrar CR, Worden K (2012) Structural health monitoring: a machine learning perspective. Wiley, Hoboken
24. Feng C, Liu MY, Kao CC, Lee TY (2017) Deep active learning for civil infrastructure defect detection and classification. Comput Civil Eng 2017:298–306
25. Gao Y, Mosalam KM (2018) Deep transfer learning for image-based structural damage recognition. Comput-Aided Civil Infrastruct Eng 33(9):748–768
26. Gardner P, Bull L, Gosliga J, Dervilis N, Worden K (2020a) Foundations of population-based SHM, part III: Heterogeneous populations-mapping and transfer. Mechan Syst Signal Process 149:107142
27. Gardner P, Fuentes R, Dervilis N, Mineo C, Pierce S, Cross E, Worden K (2020b) Machine learning at the interface of structural health monitoring and non-destructive evaluation. Philos Trans R Soc A 378(2182):20190581
28. Gardner P, Liu X, Worden K (2020c) On the application of domain adaptation in structural health monitoring. Mechan Syst Signal Process 138:106550
29. Gosliga J, Gardner P, Bull L, Dervilis N, Worden K (2020) Foundations of population-based SHM, part II: Heterogeneous populations-graphs, networks, and communities. Mechan Syst Signal Process 148:107144

30. Grandvalet Y, Bengio Y, et al (2005) Semi-supervised learning by entropy minimization. In: CAP, pp 281–296
31. Houlsby N (2014) Efficient Bayesian active learning and matrix modelling. PhD thesis, University of Cambridge
32. Houlsby N, Huszar F, Ghahramani Z, Hernández-lobato J (2012) Collaborative gaussian processes for preference learning. Adv Neural Inform Process Syst 25:2096–2104
33. Huang SJ, Jin R, Zhou ZH (2010) Active learning by querying informative and representative examples. In: Advances in neural information processing systems, pp 892–900
34. Huang Y, Gong L, Wang S, Li L (2014) A fuzzy based semi-supervised method for fault diagnosis and performance evaluation. In: 2014 IEEE/ASME international conference on advanced intelligent mechatronics, pp 1647–1651
35. Hughes A, Barthorpe R, Dervilis N, Farrar C, Worden K (2021a) A probabilistic risk-based decision framework for structural health monitoring. Mechan Syst Signal Proces 150(107339)
36. Hughes A, Bull L, Gardner P, Barthorpe R, Dervilis N, Worden K (2021b) A risk-based active learning approach to inspection scheduling. In: Proceedings of the 10th international conference on structural health monitoring of intelligent infrastructure
37. Jang K, Kim N, An YK (2019) Deep learning-based autonomous concrete crack evaluation through hybrid image scanning. Struct Health Monit 18(5–6):1722–1737
38. Kleinegesse S, Drovandi C, Gutmann MU (2021) Sequential Bayesian experimental design for implicit models via mutual information. Bayesian Anal 1(1):1–30
39. Koller D, Friedman N (2009) Probabilistic graphical models: principles and techniques. MIT Press, Cambridge
40. Li X, Zhang W, Ding Q, Sun JQ (2019) Multi-layer domain adaptation method for rolling bearing fault diagnosis. Sign Proces 157:180–197
41. Lindley DV (1956) On a measure of the information provided by an experiment. Annal Math Stat 986–1005
42. MacKay DJ, Mac Kay DJ (2003) Information theory, inference and learning algorithms. Cambridge University Press, Cambridge
43. Mallat S (2008) A wavelet tour of signal processing. The Sparse Way, 3rd edn. Academic Press, Cambridge
44. Martinez Arellano G, Ratchev S (2019) Towards an active learning approach to tool condition monitoring with bayesian deep learning
45. Michau G, Fink O (2019) Domain adaptation for one-class classification: monitoring the health of critical systems under limited information. Int J Prognost Health Manage 10:11
46. Murphy KP (2012) Machine learning: a probabilistic perspective. MIT Press, Cambridge
47. Murphy KP (2021) Probabilistic machine learning: an introduction. MIT Press, Cambridge
48. Nigam K, McCallum A, Thrun S, Mitchell T et al. (1998) Learning to classify text from labeled and unlabeled documents. AAAI/IAAI 792:6
49. Pan SJ, Yang Q (2009) A survey on transfer learning. IEEE Trans Knowl Data Eng 22(10):1345–1359
50. Pan SJ, Tsang IW, Kwok JT, Yang Q (2010) Domain adaptation via transfer component analysis. IEEE Trans Neural Netw 22(2):199–210
51. Schwenker F, Trentin E (2014) Pattern classification and clustering: a review of partially supervised learning approaches. Pattern Recogn Lett 37:4–14
52. Sen D, Aghazadeh A, Mousavi A, Nagarajaiah S, Baraniuk R, Dabak A (2019) Data-driven semi-supervised and supervised learning algorithms for health monitoring of pipes. Mechan Syst Signal Proces 131:524–537
53. Settles B (2012) Active learning. Synthesis lectures on artificial intelligence and machine learning 6(1):1–114
54. Shewry MC, Wynn HP (1987) Maximum entropy sampling. J Appl Stat 14(2):165–170
55. Wang Q, Michau G, Fink O (2019) Domain adaptive transfer learning for fault diagnosis. In: 2019 prognostics and system health management conference (PHM-Paris), IEEE, pp 279–285
56. Yuan J, Liu X (2013) Semi-supervised learning and condition fusion for fault diagnosis. Mechan Syst Signal Process 38(2):615–627

57. Zhang W, Peng G, Li C, Chen Y, Zhang Z (2017) A new deep learning model for fault diagnosis with good anti-noise and domain adaptation ability on raw vibration signals. Sensors 17(2):425
58. Zhu XJ (2005) Semi-supervised learning literature survey. Department of Computer Sciences, Technical Report, University of Wisconsin-Madison

Population-Based Structural Health Monitoring

Paul Gardner, Lawrence A. Bull, Julian Gosliga, Nikolaos Dervilis, Elizabeth J. Cross, Evangelos Papatheou, and Keith Worden

Abstract One of the dominant challenges in data-based structural health monitoring (SHM) is the scarcity of measured data corresponding to different damage states of the structures of interest. A new arsenal of advanced technologies is described here that can be used to solve this problem. This new generation of methods is able to transfer health inferences and information between structures in a population-based environment—population-based SHM (PBSHM). In the category of *homogeneous* populations (sets of nominally identical structures), the idea of a Form can be utilised, as it encodes information about the ideal or typical structure, together with information about variations across the population. In the case of sets of different structures and thus *heterogeneous* populations, technologies of transfer learning are described as a powerful tool for sharing inferences (technologies that are also applicable in the homogeneous case). In order to avoid negative transfer and assess the likelihood of a meaningful inference, an abstract representation framework for spaces of structures will be analysed as it can capture similarities between structures

P. Gardner (✉) · L. A. Bull · J. Gosliga · N. Dervilis · E. J. Cross · K. Worden
University of Sheffield, Mappin Street, Sheffield S1 3JD, UK
e-mail: p.gardner@sheffield.ac.uk

L. A. Bull
e-mail: l.a.bull@sheffield.ac.uk

J. Gosliga
e-mail: j.gosliga@sheffield.ac.uk

N. Dervilis
e-mail: n.dervilis@sheffield.ac.uk

E. J. Cross
e-mail: e.j.cross@sheffield.ac.uk

K. Worden
e-mail: k.worden@sheffield.ac.uk

E. Papatheou
University of Exeter, Exeter EX4 4QF, UK
e-mail: e.papatheou@exeter.ac.uk

A. Cury et al. (eds.), *Structural Health Monitoring Based on Data Science Techniques*, Structural Integrity 21, https://doi.org/10.1007/978-3-030-81716-9_20

via the framework of graph theory. This chapter presents and discusses all of these very recent developments and provides illustrative examples.

Keywords Population-based structural health monitoring (PBSHM) · Machine learning · Graph theory · Complex networks · Transfer learning · Forms

1 Population-Based Structural Health Monitoring

Despite significant successes in data-based approaches to structural health monitoring (SHM) [6], several limitations have prevented wide-scale adoption of these techniques in industry. One of these limitations that prevent data-based approaches from progressing beyond novelty detection is a scarcity of measured data corresponding to the damage states of interest for the structures in question. This sparsity of labelled health-state data means that supervised (and even semi-supervised) techniques are limited in their effectiveness, unable to classify observations on a structure that correspond to a health state not previously seen on the structure (unless inspections are performed for the particular observation in question). As a consequence, conventional data-based approaches that are developed for individual structures are often limited in industrial applications to performing novelty detection (in the absence of labelled health-state data), where these techniques typically are not only sensitive to damage, but detect novelty for a variety of reasons, such as due to confounding influences and other benign effects [18].

In the light of these challenges, population-based structural health monitoring (PBSHM) [4, 8, 9, 19, 21] provides a variety of tools that seek to expand the available data for performing SHM, by considering observations from a population of structures. By utilising data from multiple structures, observations of health states from across the population can be shared in diagnosing different members of the population. A population-based viewpoint therefore overcomes problems associated with a scarcity of health-state (or usually, *damage-state*) labelled data and enables diagnostic predictions from the start of an SHM campaign. This chapter introduces key concepts for PBSHM, such as population types and the tools most applicable for each type, with the focus of this chapter being on *learning* for PBSHM.

It is helpful at this stage to provide an illustration of a typical industrial setting for PBSHM. Imagine a scenario in which an asset manager of a wind farm is interested in performing SHM for each wind turbine in the farm, as shown in Fig. 1. Each wind turbine in the farm is of the same model type and can be considered nominally identical—this type of population is termed a *homogeneous population* [4, 8, 9]. During the complete operational phase, each structure may transition from its normal operating condition to a different health state; however, it is unlikely that any one turbine will observe *all* health states of interest to the asset manager (particularly as these structures are designed for low failure rates). The lack of observed labelled health states for each wind turbine means that conventional data-based SHM is limited to novelty detection. In fact, even observing the complete normal condition for any one wind turbine may not be possible for a variety of reasons, such as local differences in weather, local interactions between structures and different operational patterns. Despite the fact that some wind turbines may have limited or no labelled health-state data, the asset manager is still tasked with maintaining and monitoring the complete population. The asset manager therefore requires a population-based approach to SHM, where the information across the population is used to create a machine learner that will both generalise across the population and will allow label information to be transferred to any wind turbine in the farm, allowing robust health diagnostics for all wind turbines in the population. Figure 1 demonstrates this process, where data from across the farm are mapped into some space where a data-based model can be constructed utilising the population-level information.

The above example describes a scenario involving a *homogeneous population* where the structures are nominally identical. A second category of population also exists, termed a *heterogeneous population*, where every structure in the population is different for various reasons, broadly categorised as geometric, material and topological differences [4, 8, 9]. Staying with the wind farm illustration, imagine the asset manager is tasked with overseeing multiple wind farms situated around the world, with each farm containing wind turbines of different model types. A population-based approach can be extended to consider a larger population, covering all the wind turbines in the portfolio. However, more care must be taken in scenarios where the differences between structures in a population are large. For example, the asset manager must consider the physics of the structures that are grouped into a population; are there common failure types that each turbine will experience such that labelled data can be shared, are there sub-systems or components that are common across the population? To answer these questions, it will be necessary to demonstrate an abstract representation framework for spaces of structures, as this will allow an engineer to quantify and capture similarities between structures via the framework of graph theory.

A key component of population-based SHM concerns how learning algorithms are constructed in a population setting. Clearly, data from one structure cannot naively be used to classify data from another structure, without some mapping to harmonise datasets, as the generative distributions from each member of the population will be different. In the context of population types, this chapter outlines two key technology types for performing population-based SHM, the concept of a *Form* and *transfer learning*. Each are suited to different problem types in PBSHM, which are discussed within the chapter. Briefly, a *Form* seeks to capture the essence of a population, typically by defining a data-based model that captures the expected normal condition as well as the variability across the population. By contrast, transfer learning seeks to leverage information from source structures where label information is known and *transfer* this knowledge via some mapping to partially or unlabelled target structures. These types of learning are outlined in the context of each population type along with illustrative examples.

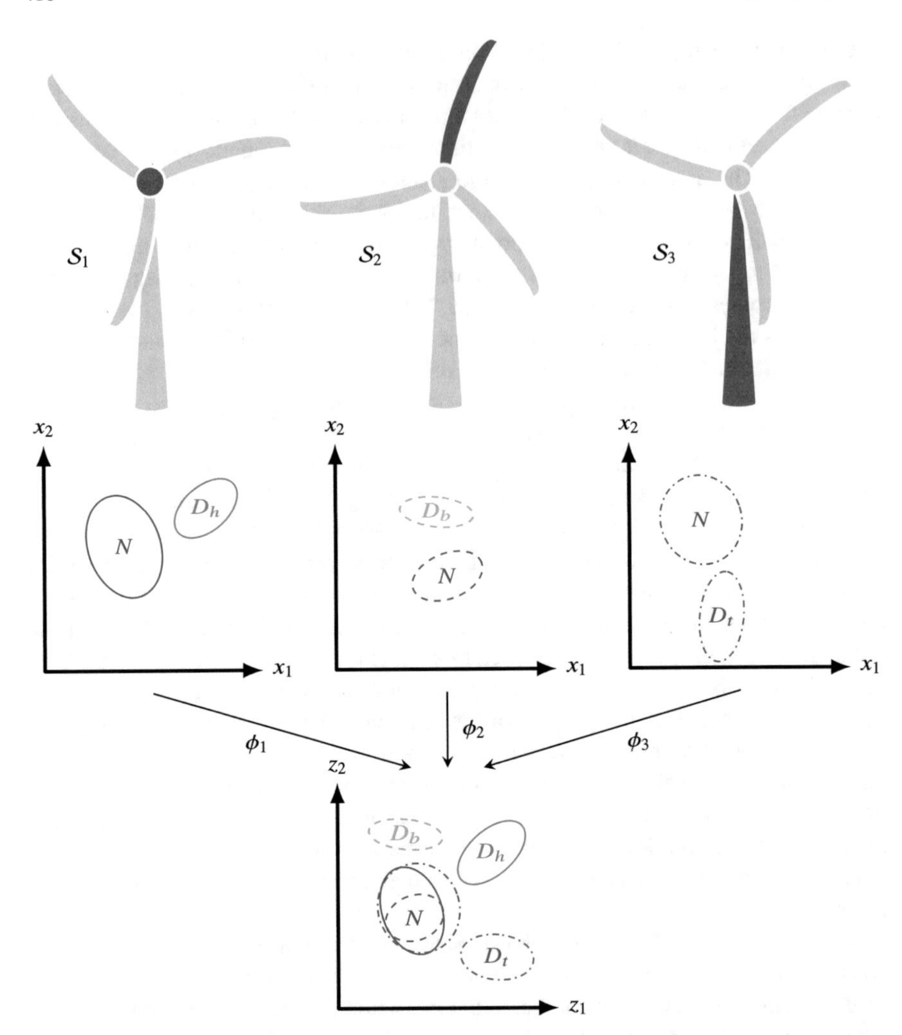

Fig. 1 A typical population-based structural health monitoring scenario. The scenario depicts a population of wind turbines ($\{\mathcal{S}_1, \mathcal{S}_2, \mathcal{S}_3\}$) where different types of damage have been observed for each wind turbine over their operational phase (N—normal condition, D_h—damage to the rotor hub, D_b—damage to a blade, D_t—damage on the tower). Population-based SHM methods seek to define some model that captures information from across the population, generally via some mapping ϕ

2 Homogeneous Populations

2.1 The Concept of a Form

As discussed, homogeneous populations are groups of structures that can be considered *nominally identical* [4]. Some examples include same model vehicle fleets, or turbines within a wind farm. In this special case, a general and *shared* representation—referred to as the *Form*—may be be used to monitor the collected group of systems.

The concept of the *Form* for PBSHM is motivated by the work of Plato. Initially in Meno [15] and later The Republic [16], Plato considers Forms to be the essence of things, existing as abstract entities: eternal, immutable and representative of the highest level of reality. Ordinary objects derive their nature and properties by 'participating' in the Forms. For example, all cats in the world are recognisable as such because they participate in the Form of *cat*.

To apply Plato's Form to a group of structures for monitoring purposes, an extension is needed: not only to capture the essence of things, but the extent of variations in their participants [4]. For example, consider a specific model of vehicle; one could argue that the essential nature of the vehicle is captured in the complete design specification; in reality, variations will occur over the production run: manufacturing tolerances, changes in operating environments, as well as other inconsistencies.

Herein, the term *form* will be used in a mathematical sense to denote a model, in some feature space, of an object (the uncapitalised *form* distinguishes the model from the conceptual Form). The model attempts to capture the two ingredients of the extended Form: the *essential nature* of the object and the *variations encountered* when the object is embodied in the real world. The *object of interest* need not be the structure itself, but rather a feature or measurement vector—which represents the structure for SHM purposes. The feature, therefore, is part of the description of the form.

2.1.1 A Motivating Example: Wind Turbine Power Curves

For a specific model of turbine, the *power curve* captures the relationship between wind speed and power output; the associated function can be used as a indicator of performance [14, 22]. For a wind farm consisting of identical turbines, this trend should be relatively consistent across the group. Variations in the power curve can exist for an array of reasons; the results of operator control, or shadowing and wake effects from other turbines [2], for example.

Intuitively, the power curve defines a convenient object (i.e. feature) to consider as the form for a wind farm. This relationship captures the *essential nature* of power production, while also the *variations* across the group. To demonstrate, operational power-curve data (SCADA [22]) from a wind farm are presented in Fig. 2a. A regression of these data should generalise to future measurements, given optimal power generation for turbines within the farm.

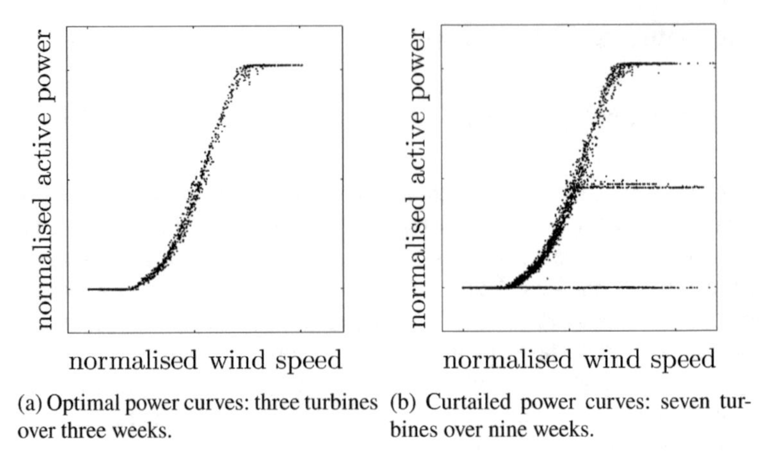

(a) Optimal power curves: three turbines over three weeks.

(b) Curtailed power curves: seven turbines over nine weeks.

Fig. 2 Wind farm power curves (normalised)

In practice, however, only a subset of measurements or turbines would be approximated by a form modelled on the data in Fig. 2a. This is because, in actuality, *power-curtailments* will appear as additional functional components; that is, further variations in the form. An example of operational data including curtailments is shown in Fig. 2b; here, three trends can be observed in the power data: (i) the ideal power curve (ii) ≈50%-limited output and (iii) zero-limited output. Curtailments usually correspond to the output being controlled (or limited) by the operator for various reasons; e.g. responding to requirements of the electrical grid [20] or the mitigation of loading/wake effects [2]. As a result, the form object is multi-valued, differing significantly from the ideal curve. However, as these data capture important *variations* that are expected in practice, they should be useful to model a more complete form for the wind farm.

2.1.2 The Power Curve Form as a Mixture of Gaussian Processes

As the functional feature (Fig. 2b) is multi-valued, conventional regression would prove inappropriate for this expression of the form. The work in [5] proposes that an overlapping mixture of probabilistic regression models (Gaussian processes) is used to approximate the power-curve relationship.

Specifically, the overlapping mixture of regression models (introduced by [11]) assumes that there are K latent functions to approximate the form,

$$y_i^{(k)} = \left\{ f^{(k)}(x_i) + \epsilon_i \right\}_{k=1}^{K} \tag{1}$$

i.e. the power y_i at each input x_i (wind speed) is found by evaluating one of K latent functions $f^{(k)}(x_i)$ with additive noise ϵ_i. From the power-curve data, it should be

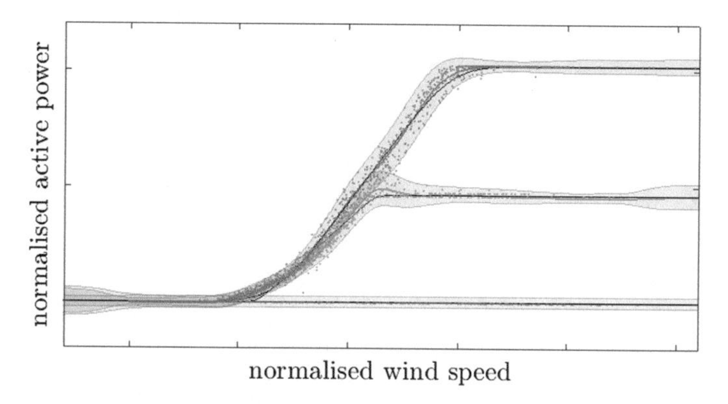

Fig. 3 Wind farm form as an overlapping mixture of Gaussian processes

clear that an appropriate number of components is $K = 3$: (i) ideal, (ii) 50% limited, and (iii) zero power.

Labels to assign each observation $\{x_i, y_i\}$ to function k are unknown, so a latent variable is introduced to the model, $\mathbf{Z}$; this is a binary indicator matrix, such that $\mathbf{Z}[i, k] \neq 0$ indicates that observation i was generated by function k. There is only one nonzero entry per row in $\mathbf{Z}$ (each observation is found by evaluating one function only). Therefore, for N data, the likelihood of the model is [11],

$$p\left(\boldsymbol{y} \mid \{\boldsymbol{f}^{(k)}\}_{k=1}^{K}, \mathbf{Z}, \boldsymbol{x}\right) = \prod_{i,k=1}^{N,K} p\left(y_i \mid f^{(k)}(x_i)\right)^{\mathbf{Z}[i,k]} \tag{2}$$

A Gaussian process (GP) is associated with each of the (three) latent functions $f^{(k)}$. Briefly, each GP can be described by its mean and kernel function [17], which can be specified for power-curve modelling; for details, refer to [5]. Unlike a conventional GP, the computation of the posterior distribution $p\left(\mathbf{Z}, \{f^{(k)}\} \mid \mathcal{D}\right)$ is intractable; thus, methods for approximate inference are implemented to infer the latent variables and functions—a variational inference and expectation maximisation (EM) approach was proposed in [1]. Additionally, input-dependent noise is approximated according to the scheme [10]. The resulting form is shown in Fig. 3.

2.1.3 The Form for Wind Farm Performance Monitoring

To demonstrate the form as a performance-monitoring (or diagnostic) tool, the model can be compared to future (test) data from *all* turbines within the population. As such, the form is treated as a general model and used to make predictions across the wind farm.

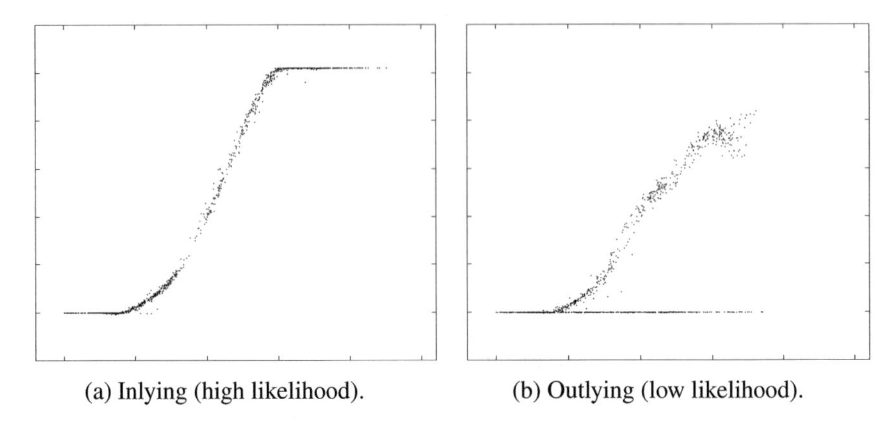

(a) Inlying (high likelihood). (b) Outlying (low likelihood).

Fig. 4 Weekly datasets, compared to the form in [4]

In the experiments presented in [4], a similar power-curve model was used to inform outlier analysis, by measuring the deviation of future data from the form (via the combined predictive-likelihood of the mixture model). Examples of (weekly) data (from across the wind farm) that appeared as inlying or outlying with respect to the form are shown in Fig. 4. In other words, data that appear *likely* or *unlikely* when compared to the currently modelled population form respectively. The examples are sampled at random from the most extreme inlying and outlying weeks of data in the test set [4].

Specifically, Fig. 4b resembles a typical *sub-optimal* power curve [14]. On the other hand, the inlying example in Fig. 4a resembles one of the permitted normal conditions associated with the form—in this case, the ideal curve.

2.1.4 Form Difficulties: Increased Population Variance

As variation across individuals increases, variation in the population data is also likely to increase; thus, it becomes progressively difficult to approximate the *form*. In particular, when the underlying distributions of data vary dramatically between individuals, more involved techniques are required to infer a *shared* model. An idea presented in [3] suggests that dissimilar population data might be projected into a shared and more consistent space, where the form can then be inferred. These concepts align closely with those of *transfer learning*.

2.2 Transfer Learning for Homogeneous Populations

Transfer learning—a branch of machine learning—provides an alternative viewpoint to the concept of a form. Rather than seeking to capture the essence and variation of a population, transfer learning seeks to leverage label information from source structures in aiding classification of unlabelled (or partially-labelled) target structures. Transfer learning is applicable across a variety of PBSHM scenarios, including homogeneous populations, where the data vary too greatly to be modelled by the form framework in the previous section, preventing a learner trained on one (or more) structure(s) from generalising to another.

Two objects are required to formally define transfer learning,

- A **domain** $\mathcal{D} = \{\mathcal{X}, p(X)\}$ consists of a feature space $\mathcal{X}$ and a marginal probability distribution $p(X)$ over the feature data $X = \{\mathbf{x}_i\}_{i=1}^{N} \in \mathcal{X}$, a finite sample from $\mathcal{X}$.
- A **task** $\mathcal{T} = \{\mathcal{Y}, f(\cdot)\}$ consists of a label space $\mathcal{Y}$ and a predictive function $f(\cdot)$ which can be inferred from training data $\{\mathbf{x}_i, y_i\}_{i=1}^{N}$ where $\mathbf{x}_i \in \mathcal{X}$ and $y_i \in \mathcal{Y}$.

Using these objects, transfer learning between a single source domain and single target domain is defined as [13],

Definition 1 **Transfer learning** is the process of improving the target prediction function $f(\cdot)$ in the target task $\mathcal{T}_t$ using knowledge from a source domain $\mathcal{D}_s$ and a source task $\mathcal{T}_s$ (and a target domain $\mathcal{D}_t$), whilst assuming $\mathcal{D}_s \neq \mathcal{D}_t$ and/or $\mathcal{T}_s \neq \mathcal{T}_t$.

Within the field of transfer learning, *domain adaptation* arguably offers one of the most useful tools for PBSHM and is defined as,

Definition 2 **Domain adaptation** is the process of improving the target prediction function $f(\cdot)$ in the target task $\mathcal{T}_t$ using knowledge from a source domain $\mathcal{D}_s$ and a source task $\mathcal{T}_s$ (and a target domain $\mathcal{D}_t$), whilst assuming $\mathcal{X}_s = \mathcal{X}_t$ and $\mathcal{Y}_s = \mathcal{Y}_t$, but that $p(X_s) \neq p(X_t)$ and typically that $p(Y_s \mid X_s) \neq p(Y_t \mid X_t)$.

Domain adaptation is appropriate for scenarios where a classifier will not generalise across domains because of differences in the underlying data distributions, such as the example outlined in Fig. 1. For this reason, the transfer learning methods demonstrated in this chapter are therefore all forms of domain adaptation.

To contextualise transfer learning for homogeneous populations, a case study is provided. The case study considers a special case of the homogeneous population type, when the source and target structure are exactly equivalent; i.e. they are the same structure. Even in this context, PBSHM provides a useful framework for over coming challenges with data-based SHM; in this particular instance, the problem of structural repairs and how they change the underlying data distribution of a system. Structural repairs introduce modifications that change (even if locally) the mass, stiffness and damping of the structure, causing shifts in the underlying generating distributions and manifest as drift in the output feature space. This dataset shift means that a data-based model trained on pre-repair labelled data will not generalise to the post-repair structure because the distributions of the dataset in training and testing are not the same. The implication of this dataset shift is that a new labelling campaign would be required every time structural repairs are made. Instead, a PBSHM viewpoint can be taken by treating the two datasets as coming from a population of two homogeneous structures. The following case study considers two datasets from the

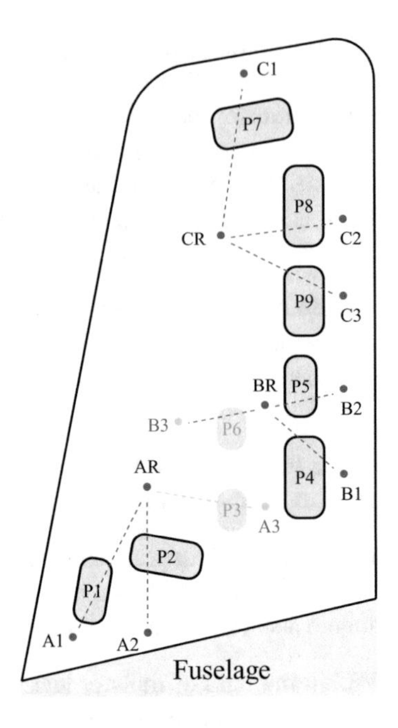

Fig. 5 A representative schematic of the Gnat aircraft starboard wing (not to scale), depicting inspection panel, accelerometer and transmissibility path locations. Recreated from [12]

same Gnat trainer aircraft, before and after the inspection panels have been removed and reattached (simulating a repair scenario).

The Gnat aircraft dataset was collected as part of an experimental campaign in which a network of uni-axial accelerometers were used to obtain transmissibility features (under white noise excitation) from the starboard wing of a Gnat trainer aircraft in situ [12]. During the experiments, psuedo-damage was introduced into the structure by removing individual inspection panels—the locations of which are represented in Fig. 5. The sensor network was designed such that each transmissibility path targeted a specific inspection panel (i.e. the transmissibility targeting panel 1 (P_1), denoted T_1, is computed from the reference accelerometer AR and response $A1$), with each transmissibility covering a frequency range of 1024–2048Hz containing 1024 spectral lines, with the magnitudes being utilised as the feature data. In the following analysis, the feature data are seven stacked transmissibilities (i.e. $\mathbb{R}^{1024\times 7}$) covering panels $\{P1, P2, P4, P5, P7, P8, P9\}$ where the label space is $\mathcal{Y} \in \{1, 2, 4, 5, 7, 8, 9\}$ (with only the panels with a large surface area being considered in this analysis). For more details about the experiments, the reader is referred to [12].

The dataset simulates a repair scenario, because the experimental sequence (removing all panels one-by-one and replacing them on the structure with the same applied torque on the fasteners) was repeated twice. In fact, the maintenance process on the Gnat aircraft wing typically involves removing and reattaching inspection panels. In order to visualise the problem caused by the repair action, Figures 6 and 7 present two depictions of the dataset. Figure 6 demonstrates the changes in the first

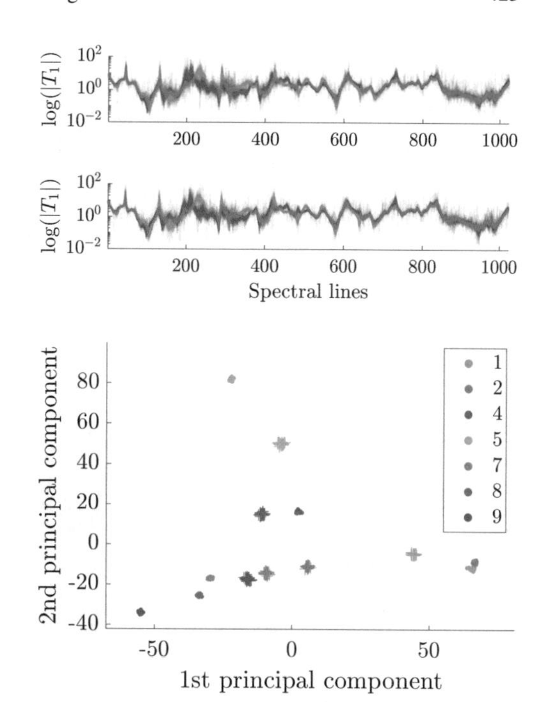

Fig. 6 A comparison of T_1 for the pre- (red) and post-repair (blue) scenarios; top and bottom sub-panels depict the normal condition and the removal of Panel One

Fig. 7 Visualisation of the pre- and post-repair datasets; the first two principal components of the pre-repair (•) and post-repair (+) datasets

transmissibility path before (show in red) and after (depicted in blue) repairs had taken place for two health states, the undamaged normal condition (top sub-panel) and when Panel One has been removed (bottom sub-panel). It is clear from this figure that there are larger changes due to the repair action than due to damage (e.g. the large differences around 200–400 spectral lines between pre- and post-repair transmissibilities), and hence, a classifier will not generalise from the pre- to post-repair scenarios. Figure 7 shows the first two principal components of the complete feature space, demonstrating that the data distributions have changed significantly between the pre- and post-repair structural states, and hence $p(Y_s, X_s) \neq p(Y_t, X_t)$ (where the pre-repair data has been denoted as the source domain, subscript $_s$, and the post-repair data is denoted as the target domain, subscript $_t$), so domain adaptation is applicable.

In order to harmonise the pre- and post-repair datasets, such that label information from the pre-repair dataset can be used to diagnose the unlabelled post-repair data, a domain adaptation algorithm has been applied. The algorithm, *metric-informed joint domain adaptation* [7], seeks to find a mapping $Z = KW \in \mathbb{R}^{(N_s+N_t \times k)}$, by learning some weight matrix $W \in \mathbb{R}^{(N_s+N_t \times k)}$ that projects a kernel matrix, formed from the joint dataset $K \in \mathbb{R}^{(N_s+N_t \times N_s+N_t)}$, onto a k-dimensional space. The weight matrix is inferred by minimising the distances between the joint distributions from the source and target data, formed as an optimisation problem that minimises the maximum mean discrepancy distance between the marginal and class conditional distributions (for more information about the algorithm, the reader is referred to [7]).

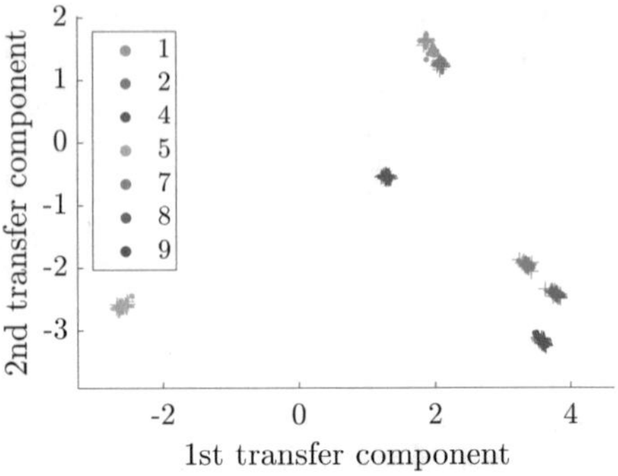

Fig. 8 Transfer components of the pre-repair (•) and post-repair (+) from the metric-informed joint domain adaptation approach

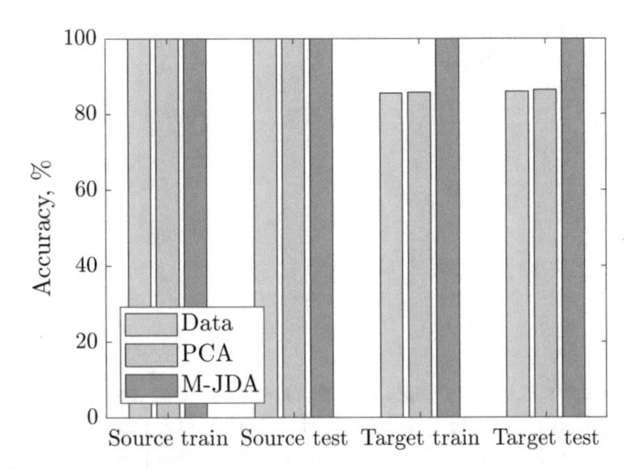

Fig. 9 Comparison of classification accuracy given feature spaces with no transfer learning (Data and PCA) and the feature space after transfer learning (M-JDA)

The projected space, defined by a set of *transfer components*, where $k = 2$, is shown in Fig. 8, where the algorithm has matched the joint distributions in the projected space. In this space, label data from the pre-repair state can be used to classify the post-repair data, transferring the label information. A k-nearest neighbour classifier was trained on the pre-repair data from three different feature spaces, the original transmissibilities (Data), the principal components of the transmissibilities (PCA) and the transfer components (M-JDA). The classification results in Fig. 9 show that domain adaptation has allowed for label information to be successfully transferred to the target domain, signified by classification accuracies in the target domain of 100%, and that, because of differences in between the pre- and post-repair datasets, a classifier trained on either the pre-repair transmissibilities or principal components, does not generalise to the post-repair data.

3 Heterogeneous Populations

3.1 Transfer Learning for Heterogeneous Populations

In the field of population-based SHM, heterogeneous populations provide a more complex set of challenges for transfer learning. The reason for this increased difficulty is that transfer learning assumes that there is some shared commonality between the source and target domains. Heterogeneous population push these assumptions towards their limits, and as structures in a population become more dissimilar, the risk of negative transfer increases. The term *negative transfer* describes the scenario where transfer learning incorrectly maps information from one domain onto another, reducing the performance of the learner (discussed in more detail in Sect. 3.2). It is therefore important in heterogeneous populations to understand and quantify the level of similarity between structures, such that transfer learning is only attempted in contexts where transfer will be successful and beneficial. Later on in this chapter, an approach for assessing the similarities between structures (before attempting transfer learning) is introduced. Briefly, the approach converts structures into an abstract representation, called irreducible element models, where a graph theory framework can be used to quantity similarities. The remainder of this section looks at illustrating transfer learning in the context of heterogeneous populations via two example populations of n-storey buildings.

The first example considers a population of two n-storey structures, a three-storey structure where the feature data are labelled, denoted the *source* structure, and a four-storey target structure, where the feature data is unlabelled. The two structures form a heterogeneous population as they have different topologies—with their nominal geometries and material properties being the same. Each structure is modelled as a lumped-mass model, shown in Fig. 10, where the spring stiffness between each floor is modelled as four springs in parallel. The SHM problem is that of locating damage, in the form of open cracks at one of the beams at a particular floor using lateral bending natural frequencies of the whole structure as features.

In this example, the PBSHM problem is that of transferring localisation labels from the three-storey structure to the four-storey structure. An interesting challenge arises in the context of a heterogeneous population when performing a localisation task between structures with different topology, namely that the labels spaces between the two structure are not exactly equivalent, termed label inconsistency. This phenomenon means that care must be taken when transferring information between members of a heterogeneous population. In this example, both structures have an undamaged condition ($Y = 1$) and can be damaged at floors one to three ($Y = \{2, 3, 4\}$, respectively). As a result, the complete label set from the three-storey structure can be transferred to the four-storey structure, where the algorithm should not try to pair data points relating to damage at the fourth floor ($Y = 5$) of the target structure with data from the source structure. This type of PBSHM problem is termed an $L + 1$-problem, as there is one more class label in the target domain than in the source domain (i.e. $Y_s \in \{1, 2, 3, 4\}$ and $Y_t \in \{1, 2, 3, 4, 5\}$. Furthermore, most

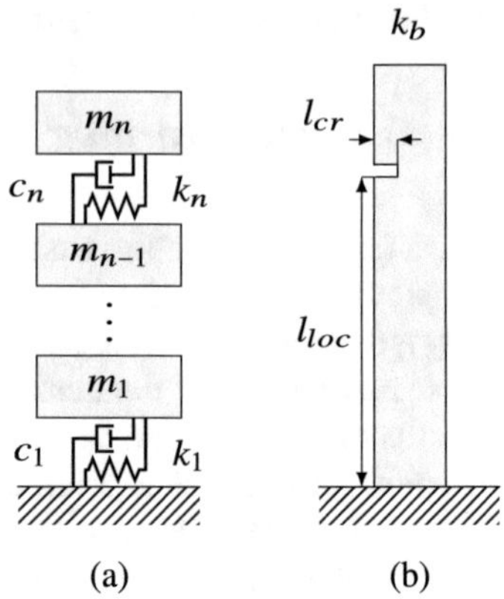

Fig. 10 Schematic of the n-storey building structures lumped-mass models, panel (**a**), and beam component, panel (**b**)

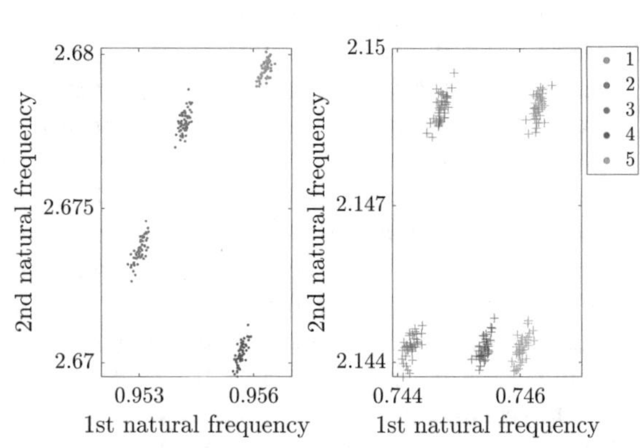

Fig. 11 A visualisation of the first two natural frequencies (in Hz) from the three- (•) and four (+) storey structures

domain adaptation techniques require that both the source and target feature spaces to be of the same dimension, such that negative transfer is minimised. In this example, the feature spaces are the first three bending natural frequencies such that both feature spaces are $\mathbb{R}^3$, with Fig. 11 presenting a visualisation of the first two natural frequencies for each structure.

The approach taken to solve this particular $L + 1$-problem is outlined next. Firstly, an unsupervised clustering method (namely a Gaussian mixture model) is used to identify and group the unlabelled target domain feature data. Once unlabelled target clusters are identified, each target cluster is removed iteratively from the domain adaptation training dataset. A mapping, in the form of $Z = KW \in \mathbb{R}^{(N_s+N_t \times k)}$, is subsequently identified from the (complete) source dataset to the particular target dataset (where one cluster has been removed), where the domain adaptation algorithm uses the maximum mean discrepancy as a cost function and $k = 2$. The algorithm then selects the mapping that produces the smallest distance between the source and target training datasets. This methodology is based on a naive form of manifold assumption; i.e. it is expected that the manifold of the source and target clusters is the 'same'. Figure 12 presents the 'optimal' mapping, where the correct target clusters were used

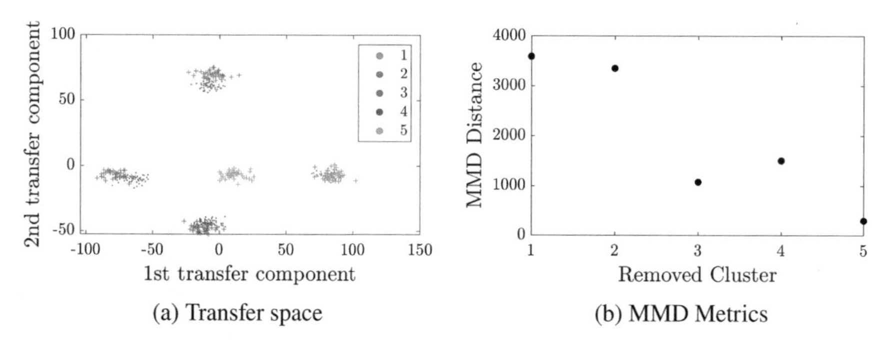

Fig. 12 Transfer learning results for the three- to four-storey example. Panel **a** displays the transfer components for the three- (•) and four-storey (+) structures, with panel **b** demonstrating the maximum mean discrepancy (MMD) distances in the transformed space when one cluster has been removed from the target domain

in training, and the distance—in the form of a maximum mean discrepancy distance (MMD)—between the mappings with different target clusters removed. It can be seen that the 'optimal' mapping is selected by this approach, as the smallest MMD distance is produced when the target cluster corresponding to $Y = 5$ (i.e. damage at the fourth storey) is removed in training. A classifier trained on the source domain data in the transfer component space in Fig. 12 can be shown to classify the target structure with 100% accuracy using a semi-supervised Gaussian mixture model. This case study demonstrates the challenges heterogeneous populations cause for transfer learning techniques, with care being needed in order to minimise negative transfer.

The second example involves a heterogeneous population comprised of a numerical physics-based model and an experimental structure, as shown in Fig. 13. The aim in this example is to transfer damage-extent label information from an unvalidated (and in this case a deliberately poor-performing) numerical model to unlabelled data from an experimental structure. This case demonstrates the flexibility of a PBSHM approach in which a variety of sources of label information can be utilised, and shows a significant advantage of the PBSHM viewpoint, namely that damage-state data can be generated in a cost-effective manner from physics-based models, even when computer model validation is challenging. The numerical model, constructed using the approach in Fig. 10 as a three degree-of-freedom lumped-mass model, was formed using the measured dimensions of the experimental structure and with typical material properties that matched those from the structure.

The SHM problem was to classify the extent of damage, in the form of open cracks from 0 to 20 mm, where label $Y = 1$ denotes the undamaged condition, $Y = 2$ refers to a 5 mm crack, $Y = 3$ for a 10 mm crack, etc. Again, it is reiterated that the numerical model has been oversimplified such that the example demonstrates the effectiveness of domain adaptation in utilising physics-based models in labelling real world structures. The simplified physics-based model therefore reflects that physics-based models are challenging to validate in SHM contexts and may not fully

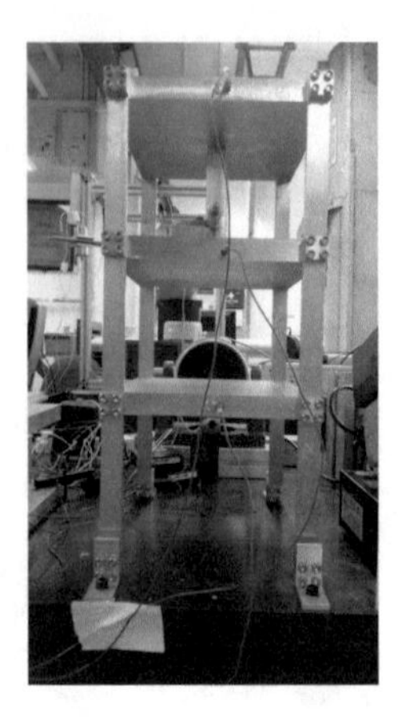

Fig. 13 Experimental three-storey building structure [8]

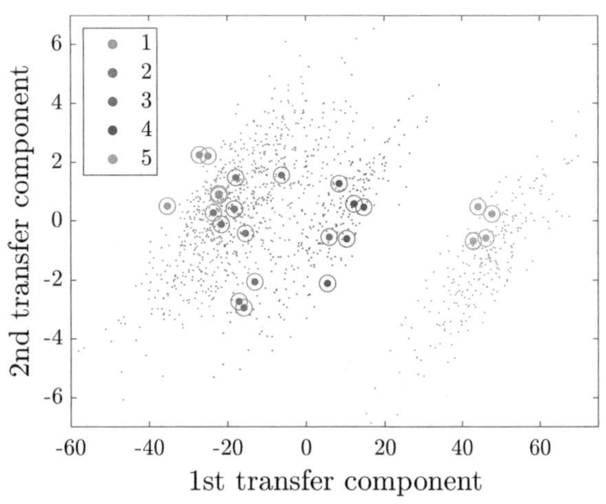

Fig. 14 The transfer components from the numerical (·) and experimental (● predicted label and ◯ true label) datasets

agree with observational data, because of model form-errors. The feature data in this example were the first three lateral bending natural frequencies of the system.

Domain adaptation was applied such that a mapping in the form of $Z = KW \in \mathbb{R}^{(N_s+N_t \times k)}$ could be identified (where $k = 2$), using an MMD-based cost function [8]. The inferred mapping is visualised in Fig. 14 where it can be seen that the numerical and experimental data have been aligned. A k-nearest neighbour classifier was trained using the numerical model data, both before and after the transfer mapping. The classifier trained on the untransformed data produced a testing accuracy in the target domain of 48%; this is compared to a testing accuracy 88% from the classifier trained on the transfer components. This result demonstrates the effectiveness and applicability of utilising both physics-based models and observational datasets within a PBSHM framework.

Both of these case studies highlight the potential of PBSHM beyond populations of nominally similar structures. Of course, there are a number of research questions that are posed by considering the full extent of heterogeneous populations. It is therefore important to explore similarities between structures and datasets, and to form groupings of PBSHM problems that address each scenario, while monitoring the potential for negative transfer.

3.2 The Problem of Negative Transfer

Negative transfer has been mentioned several times in this chapter because of the risk it poses to making inferences of health states in a PBSHM approach. As previously stated, negative transfer occurs when class data from a source domain has been incorrectly and confidently paired with class data from a target domain, i.e. Class One in the source domain is mapped onto Class Two in the target domain. Generally, negative transfer is any scenario where transfer learning reduces the performance on a classifier when compared to not performing any transfer. Negative transfer is a particular concern for a PBSHM viewpoint, as typically there are no labelled data points in the target domain that can be used to validate the inferred transfer mapping. As such, it is important to quantify how related the source and target structures are and to only perform transfer when positive transfer is likely. This is a significant and open research question, with the following section outlining one approach to identified the similarities between members of a population.

To illustrate the effect of negative transfer, the three- to four-storey example is reintroduced. The problem in this example was to transfer localisation labels from the three-storey source structure to the unlabelled four-storey target structure. The problem is an $L + 1$-problem, meaning that there is one more class label in the target domain than in the source domain, and that all other class labels exist in both domains. Figure 15 shows the identified mapping when data corresponding to classes $Y_s \in \{1, 2, 3, 4\}$ and $Y_t \in \{1, 2, 4, 5\}$ were used to train the domain adaptation mapping—this gave the second smallest distance in Fig. 15 and therefore was not selected as the optimal mapping, but will be useful for this discussion. Of particular note, negative transfer has occurred between the source data for class $Y = 3$ and the target data for class $Y = 4$. This is interesting, as source data for class $Y = 4$ were used in training, but the domain adaptation algorithm, given no label information about the target, has inferred a mapping that incorrectly pairs these classes. Negative transfer has also occurred between class $Y = 4$ in the source domain and $Y = 5$ in the target, which arises as the cost function in the domain adaptation approach only considers a mapping with the smallest distance between the two domains.

In this example, it should be clear that the risk of negative transfer was high, as the data from the source and target domains did not correspond to the same classes in training. However, it is worth noting that negative transfer can occur in scenarios where both the source and target training data do refer to the same classes, and where the two datasets are more dissimilar than can be accounted for given the assumptions and adaptation method (i.e. the type of mapping) used in the transfer learning algorithm. For PBSHM to be applied with confidence using transfer learning

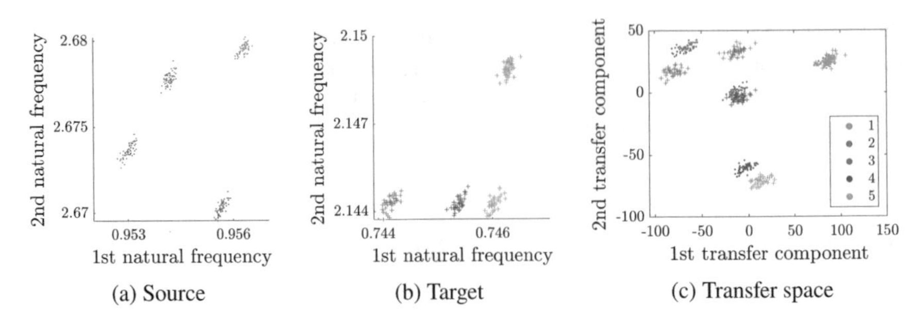

(a) Source (b) Target (c) Transfer space

Fig. 15 An example of negative transfer between the three- and four-storey structures. Panel **a** shows the first two natural frequencies of the source domain data (•) used in training, panel **b** shows the first two natural frequencies of the training target data (+), and panel **c** shows the inferred transfer components after domain adaptation, where negative transfer has occurred

approaches, it will be important to estimate the probability of negative transfer from an algorithm, such that the risk of negative transfer is always a minimum, and if possible, zero.

3.3 Abstract Representation Framework for Spaces of Structures

The abstract representation mentioned previously in this chapter is a method for describing engineering structures systematically in a way that lends itself to comparison. This abstract representation focusses on three areas believed to be important for avoiding negative transfer in an SHM context: geometry, material properties and topology. This abstract representation is known as an *irreducible element* (IE) model. The irreducible elements which give the modelling approach its name describe the constituent parts of a given structure. In a simplified example describing a wind turbine, these elements could be considered to be the blades, the hub, the nacelle, the tower and the foundations (Fig. 16). These elements possess attributes which describe their geometry and material properties.

These elements are connected by joints, labelled with numerals in Fig. 16. Joints describe the physical connections between elements. Describing the physical connections between elements requires a description of which elements are connected by a particular joint, as well as the nature of a particular connection; for example, whether a connection is welded or bolted. Combining the joints with the elements within a structure allows one to determine the topology of the structure.

Boundary conditions, describing how a structure interacts with its environment, are also included in the IE model. The boundary conditions constitute an element-joint pair (element 1 and joint 7 in Fig. 16), where a special element describes the nature of the boundary, for example, the ground, and the joint describes the nature of the connection between the structure and the boundary. Together, the element-joint

Fig. 16 A simple illustrative example of how a wind turbine (**a**) may be conceptualised as a series of elements (**b**). These elements then form the nodes in a graph representation of the structure (**c**), with the physical connections between elements represented by the edges in said graph

pair fully describes the effect of the boundary condition on the behaviour of the structure.

Isomorphic to the information within the IE model is the attributed graph (AG) representation of the physical structure; that is, the representations in panel (b) and (c) within Fig. 16 contain the same physical structural information. One could consider that the AG provides a more structured form of the data describing an IE model, and as such facilitates the storage and creation of IE models within a database. The formal data structure of an AG also facilitates the use of graph-matching algorithms to find similar physical structures within said database.

In PBSHM, IE models can be used to compare two structures to determine the overall level of similarity. If two models are sufficiently similar, then transfer between the two structures should be possible. What 'sufficiently similar' means however, is a complicated question; one way of determining this is to examine the largest substructure common to two structures. If the largest common substructure is in fact as large as the two structures in question, then the two structures can be said to be homogeneous. In which case, as discussed, not only should transfer learning be possible, but in some cases may not even be required. Of course, for the majority of comparisons between structures, this situation will not be the case, with the largest substructure only representing only a small part of the overall structure. In this case, transfer learning may be possible within the substructure. For example, in Figs. 17 and 18, which show the AGs for two different beam and slab bridges, the intermediate pier is topologically identical in both. If upon further examination, the attributes show that this pier is identical in terms of geometrical and material properties, then it would seem that some form of information transfer should be possible between

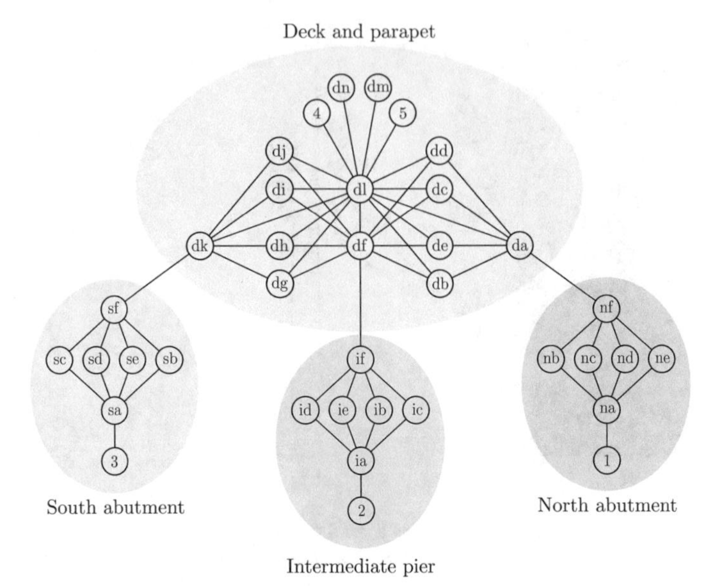

Fig. 17 A graph representation of a beam and slab bridge, located near Castledawson in Northern Ireland. This bridge features a deck supported by four longitudinal beams, as well as columns at the North and South abutments

the intermediate piers within these two bridges, making allowances for the influence from the rest of the structure. At the very least, the label spaces would be consistent between the two.

A more complex problem would be transferring between the deck and parapet sections for the two bridges, since in Fig. 18 there is an additional support beam. Here the label spaces are not consistent if one were to attempt a damage localisation problem. However, in theory, both deck sections should still exhibit some similar behaviour, since both are plate type elements supported with longitudinal beams for support and intermediate piers and so some physics should be common to the two decks. Therefore, there would be an expectation that some information could be transferred from a classifier trained on one bridge to another.

The largest common substructure found for the two bridges in fact involves not only the intermediate pier, but also the four longitudinal beams within the deck structure, the parapets, the deck itself and associated boundary conditions, as well as the cap beams linking the deck to either the supports or foundations. This situation creates a problem where there is still a label mismatch between the substructure and the original substructure, as shown in Fig. 18. How to cope with these issues and how to perform meaningful similarity comparisons within a transfer learning context remain open research questions.

A definition of similarity for PBSHM needs to be designed with transfer learning in mind and will likely vary depending on the SHM problem in question. A set of criteria informed by cases where negative transfer is likely to occur would be a

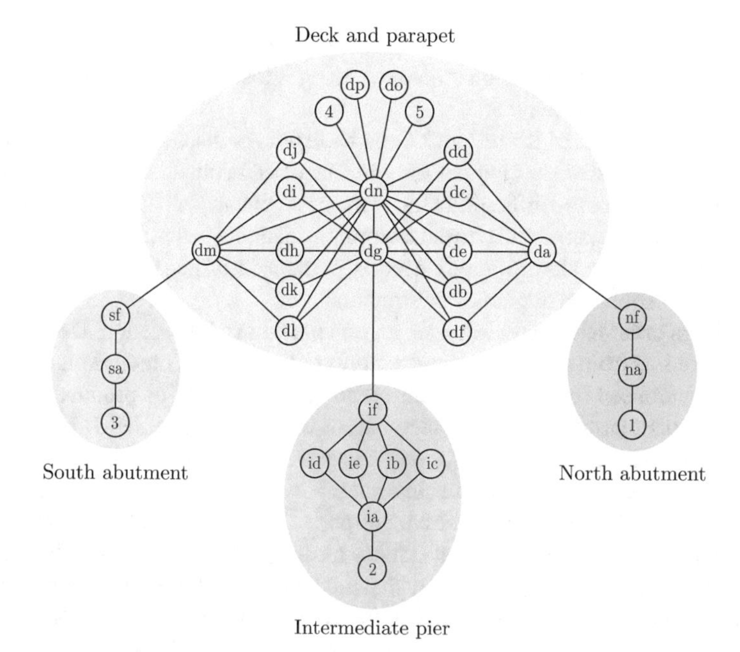

Fig. 18 A graph representation of a slightly different beam and slab bridge, located near Randalstown in Northern Ireland. This bridge features a deck supported by five longitudinal beams, and the cap beams sit directly on top of the foundations at either abutment

useful thing to have. Currently, it is hoped that by ensuring material, geometric and topological similarity, the criteria for transfer learning yielding improved classifier performance are met.

4 Conclusions and Future Directions

Clearly, the subject of PBSHM is in its infancy. While there have been scattered papers on SHM in systems-of-systems or on fleet-based maintenance, etc., the formal framework presented in this chapter has only the theoretical foundations established [4, 8, 9, 19, 21]. This situation means that the scope for 'future work' is very open indeed; however, to bring a little focus to the discussion, it will be suggested here that there are three main areas in which research is needed.

The general framework is based on knowledge transfer in populations, where the structures of the population are represented in an abstract representation space; this suggest the three areas for development. In the first case, the abstract representation of the structures requires research. There will need to be rules for building IE models with consistency and for developing the theory of the representation space in which the structures are embedded. Perhaps most important is that the representation space should have a metric (or metrics), so that structures can be compared; transfer of knowledge will be contingent on structures being 'close together' in some sense, so that negative transfer is avoided.

The second main area for research is on *transfer*. As discussed in the main body of this chapter, transfer learning has only recently been considered for PBSHM; the problems which arise are difficult. In the worst case scenario, different structures will have different label spaces and current technology cannot be applied with confidence. New transfer learning algorithms are needed, perhaps informed by physics, as a great deal is known about the dynamics of structures.

Finally, in order to transfer knowledge, one must have knowledge. Data for diagnostics can come from sensors on the structures or from model predictions; these data should be optimised for diagnostics across populations, and this produces demands and constraints that have not been seen before in conventional SHM. In particular, the role of physics-based models is very interesting in PBSHM. Because the main algorithms will come from transfer learning, it will not be necessary that models conform perfectly to their physical counterparts (although digital twin technology may offer prospects in terms of high-fidelity representation); it will only be required that transfer is possible. In fact, the way that models of structures will be embedded in populations means that they are not distinct from real structures. PBSHM therefore offers the prospect of a final convergence of model-based and data-based SHM.

Acknowledgements The authors gratefully acknowledge the support of the UK Engineering and Physical Sciences Research Council (EPSRC) via Grant references EP/R003645/1, EP/R004900/1 and EP/R006768/1. They would also like to thank their colleagues David Hester and Andrew Bunce of Queen's University Belfast, for their help and advice in building IE models of bridges and for access to the details of specific bridges.

References

1. Blei D, Kucukelbir A, McAuliffe J (2017) Variational inference: A review for statisticians. J Am Stat Assoc 112:859–877
2. Bontekoning M, Sanchez Perez-Moreno S, Ummels B, Zaaijer M (2017) Analysis of the reduced wake effect for available wind power calculation during curtailment. J Phys Conf Ser 854:012004
3. Bull L, Gardner P, Dervilis N, Papatheou E, Haywood-Alexander M, Mills R, Worden K (2021a) On the transfer of damage detectors between structures: an experimental case study. J Sound Vibr 501:116072
4. Bull L, Gardner P, Gosliga J, Dervilis N, Papatheou E, Maguire A, Campos C, Rogers T, Cross E, Worden K (2021b) Foundations of population-based structural health monitoring, Part I: Homogeneous populations and forms. Mechan Syst Sig Proces 148:107141
5. Bull L, Gardner P, Rogers T, Dervilis N, Cross E, Papatheou E, Maguire A, Campos C, Worden K (2021c) Statistical modelling of curtailments in wind turbine power curves. Submitted to: mechanical systems and signal processing
6. Farrar C, Worden K (2012) Structural health monitoring: a machine learning perspective. Wiley, Chichester
7. Gardner P, Bull L, Dervilis N, Worden K (2021a) Overcoming the problem of repair in structural health monitoring: metric-informed transfer learning. Submitted to: Jour-

nal of Sound and Vibration. 510:116245. https://www.sciencedirect.com/science/article/pii/S0022460X21003175
8. Gardner P, Bull L, Gosliga J, Dervilis N, Worden K (2021b) Foundations of population-based structural health monitoring, Part III: Heterogeneous populations, transfer and mapping. Mechan Syst Sig Process 149:107142
9. Gosliga J, Gardner P, Bull L, Dervilis N, Worden K (2021) Foundations of population-based structural health monitoring, Part II: Heterogeneous populations and structures as graphs, networks, and communities. Mechan Syst Sig Proces 148:107144
10. Kersting K, Plagemann C, Pfaff P, Burgard W (2007) Most likely heteroscedastic Gaussian process regression. In: Proceedings of the 24th international conference on machine learning, pp 393–400
11. Lázaro-Gredilla M, Van Vaerenbergh S, Lawrence N (2012) Overlapping mixtures of Gaussian processes for the data association problem. Pattern Recogn 45:1386–1395
12. Manson G, Worden K, Allman D (2003) Experimental validation of a structural health monitoring methodology: Part III. Damage location on an aircraft wing. J Sound Vibr 259:365–385
13. Pan S, Yang Q (2010) A survey on transfer learning. IEEE Trans Knowl Data Eng 22:1345–1359
14. Papatheou E, Dervilis N, Maguire A, Antoniadou I, Worden K (2015) A performance monitoring approach for the novel Lillgrund offshore wind farm. IEEE Trans Indus Electron 62:6636–6644
15. Plato (2005) Protagoras and Meno. Penguin Classics
16. Plato (2007) The Republic. Penguin Classics
17. Rasmussen C, Williams C (2005) Gaussian processes for machine learning. The MIT Press
18. Sohn H (2007) Effects of environmental and operational variability on structural health monitoring. Philos Trans R Soc Ser A 365:539–560
19. Tsialiamanis G, Mylonas C, Chatzi E, Dervilis N, Wagg D, Worden K (2021) Foundations of population-based structural health monitoring, Part IV: The geometry of spaces of structures and their feature spaces. Mechan Syst Signal Proces 157:107142
20. Waite M, Modi V (2016) Modeling wind power curtailment with increased capacity in a regional electricity grid supplying a dense urban demand. Appl Energy 183:299–317
21. Worden K, Bull L, Gardner P, Gosliga J, Rogers T, Cross E, Papatheou E, Lin W, Dervilis N (2020) A brief introduction to recent developments in population-based structural health monitoring. Front Built Environ 6:146
22. Yang W, Court R, Jiang J (2013) Wind turbine condition monitoring by the approach of SCADA data analysis. Renew Energy 53:365–376

Machine Learning-Based Structural Damage Identification Within Three-Dimensional Point Clouds

Mohammad Ebrahim Mohammadi and Richard L. Wood

Abstract Damage identification via remotely sensed data amid routine structural inspections or assessments in the aftermath of extreme events has been the subject of intensive research in recent years. By analyzing remotely sensed data, these tasks can be performed rapidly, safely, and economically while maintaining high accuracy and objectivity. Therefore, many methods have been proposed to detect damage from remotely sensed point clouds of the civil structures. These methods use various feature extraction techniques based on geometry and other spectral information and classification based on supervised or unsupervised learning algorithms. However, the proposed solutions are typically optimized to detect particular types of damage or group various types of damage as a single class, limiting the application of these methods. This article proposes an advanced workflow to identify damage from point clouds by extracting semantic information using supervised and unsupervised learning methods. The supervised learning method is comprised of a deep learning model developed based on both voxel- and index-based data structures to segment the point cloud data into various objects semantically. An unsupervised learning algorithm classifies the segmented scene into instances for damage detection into various damage states and handles rare cases of damage instances.

Keywords Damage identification · Point clouds · 3D Machine learning · Imbalance dataset classification · Semantic segmentation

1 Introduction

Damage assessment from civil infrastructure following extreme events (e.g., a tornadic event) is a primary task of civil and structural engineers. While this task was traditionally conducted through site visits and visual inspection, with the rapid

M. E. Mohammadi · R. L. Wood (✉)
University of Nebraska-Lincoln, Lincoln, NE 68588, USA
e-mail: rwood@unl.edu

M. E. Mohammadi
e-mail: me.m@huskers.unl.edu

A. Cury et al. (eds.), *Structural Health Monitoring Based on Data Science Techniques*, Structural Integrity 21, https://doi.org/10.1007/978-3-030-81716-9_21

development of remote sensing technologies, various remote sensing platforms have improved and complemented these data collection and reconnaissance efforts. As one of the emerging remote sensing platforms, an unpiloted aerial system (UAS) has been increasingly utilized to collect aerial images following extreme events due to its ease of deployment, operation, and data collection efficiency. UASs enabled researchers to document perishable damage data of the affected areas, which improves and informs emergency management and recovery operations, facilitates forensic analyses, and allows forecasting and characterizing the hazard risks. Ultimately, the damage assessment results will provide a baseline or validation dataset for resiliency models and numerical simulations [1].

UAS platforms can collect a large number of images in a short time, which makes manual processing of the images inefficient. Therefore, preprogrammed processes are required to categorize and analyze the collected images for various applications. While images provide rich texture or potentially other spectral information that can be used to detect and localize the damaged areas, the damage analyses based on images can be limited due to the lack of depth information. In other words, while the damaged areas can be observed and detected in images, if the observed damage is part of an intact or damaged structure, it is not observable directly or reliably. Therefore, to evaluate the integrity of a structure and its current damaged state, more complete geometric and depth information is required. The collected aerial images can be processed through the structure-of-motion (SfM) technique to extract depth information and ultimately reconstruct the three-dimensional (3D) point cloud representation of the surveyed area. As a result, the damage analysis can be performed using 3D point clouds that incorporate the depth information and other textual information. This manuscript aims to propose an advanced workflow to analyze the UAS-SfM-derived 3D point clouds of affected areas following an extreme event for the task of damage detection and assessment. Damage detection is essentially a classification task using an imbalanced dataset, where the damaged data can be the minority class.

Moreover, the damaged data has random and unique patterns. The proposed workflow aims to semantically classify the point cloud data by using supervised and unsupervised learning algorithms. The workflow includes a combination of a 3D fully convolutional network (3DFCN), the k-means clustering algorithm, and 3D convolutional neural network (CNN). The workflow tags each vertex (or point) within the point cloud dataset into ten classes, including vehicles, debris, water bodies, roads, healthy trees, destroyed structures, severely damaged structures, moderately damaged structures, minorly damaged structures, and undamaged structures. Within this study, various damage states are obtained based on the field observations dataset conducted by an NSF-sponsored Structural Extreme Events Reconnaissance (StEER) field assessment team following the March 2020 Tornado outbreak in northern Tennesee [2].

2 Background

Many research studies have proposed workflows based on machine learning approaches to detecting damage from 2D images of civil structures. Early studies used traditional machine learning methods to detect damage, where a series of features are initially extracted from the given set of images, and then the damage areas within each image are detected through analyzing the extracted features by a learning algorithm. Vetrivel et al. [3] introduced one such method by evaluating the support vector machine (SVM) and random forest (RF) learning algorithms. To identify the damaged areas within this study, the off-nadir aerial images were initially collected by UAS from the city of Mirabello, Italy, that sustained damaged following the 2012 Emilia earthquakes. Then, the features were extracted from the images based on the histogram of gradients (HOG) and Gabor wavelet techniques. Afterward, the learning algorithm detects the damaged areas within the images independently based on each set of features, which results in an accuracy of 65 to 72%. The study reported that the proposed method for feature extraction could reach an overall accuracy of 80%. Duarte et al. [4] investigated the application of CNN models with residual connections and dilated convolutions to classify satellite, piloted, and unpiloted aerial images for the task of building damage detection. The aerial images used within this study were collected following the 2010 Haiti and 2012 Emilia earthquakes, while satellite images belonged to the 2016 central Italy earthquakes. Duarte et al. [4] proposed three network architectures that are developed based on multi-resolution images and compared the results to two benchmark networks trained only on the satellite image. The final feature maps were generated from airborne and satellite image samples. Duarte et al. reported that networks trained using multi-resolution images demonstrated a 4% overall improvement compared to models developed based on only satellite images. Spencer et al. introduced a workflow to create a color-coded 3D representation of damaged areas based on detected damage from UAS collected images in the aftermath of the 2017 Central Mexico Earthquake [5]. In this study, a model with FCN and a residual network architecture were initially used to semantically segment the images for damage detection. Afterward, the color-coded images were used to reconstruct the 3D scene. The authors have reported that the proposed workflow reached an average accuracy of 91%. Gao and Mosalam [6] introduced a model developed based on transfer learning from VGGNet introduced in 2015 [7]. Gao and Mosalam [6] investigated the application of transfer learning through fine-tuning and transfer learning through feature extractions to develop the model. The developed model can detect the type of structure that was being inspected (e.g., bridge, wall, or building) if the analyzed image contains any visible spalling, the damage level, and the type of damage. The authors reported while the model was able to identify the type of structure with an accuracy of 88.8%, it detected spalling with an accuracy of 85%, damage level with an accuracy of 77.0%, and type of damage with an accuracy of 57.7%. More recently, Tilon et al. [8] developed a damage detection model based on generative adversarial network (GAN) architecture that is trained only on pre-event images. As a result, the damaged data were detected as anomaly instances, which accounts for the imbalanced nature of the damage. The GAN model generator architecture consisted of U-net encoder-decoder model, and the discriminator architecture consisted of a series of convolutional layers [9]. To test the model, the authors used post-disaster images following a volcanic eruption took place. Tilon et al. [8] reported the developed model could detect damaged areas within images with recall and precision of 59 and 97%, respectively.

Similar to images, 3D point clouds have been investigated to address various tasks within civil infrastructure assessments. One of the popular workflows to detect damage and changes within the area of interest (ROI) is through change detection analysis. As one of the early studies, Olsen [10] introduced a workflow to detect spatial changes for a given ROI through change detection analysis. Within this study, the point cloud of an ROI at two different epochs are collected and registered through the georeferencing process. Afterward, the point clouds were segmented into a series of cell based on selected gird sizes, and the points within the corresponding gird are compared to detect changes and potentially damaged areas. Tran et al. [11] introduced a machine learning-based change detection analysis of point clouds collected by aerial laser scanners (ALS). Within this study, initially, each point cloud dataset collected at different epochs were analyzed to extract four classes of features, including point distribution, normalized height, ALS-based features per each dataset, and the localized changes between dataset per each point. Afterward, these features were analyzed by a learning algorithm to predict the changes. The developed model based on the designed features could detect changes as accurately as 90%. While change detection analysis is affected by the imbalance dataset classification issues, it relies on the availability of a baseline dataset. Its accuracy is the function of the registration of point clouds collected at two different epochs. In addition, to change detection-based workflows, a series of studies followed a traditional classification workflow where a series of features were identified through engineered features, and then the points are classified based on the extracted features through a classifier. As one of the early studies, Axia et al. [12] used the normal vector as the extracted feature and studied the normal vectors variation of a large point cloud dataset to a global reference vector to identify damaged regions. The dataset utilized within this study was collected following the 2010 Haiti earthquake and covered a 1.5 km by 1.5 km wide area of Port-au-Prince, Haiti. The damages areas were identified if the deviation of points' normal vector exceeds the predefined threshold value. While the proposed method could be programmed to analyze point clouds rapidly, the authors reported that this approach could classify partially damaged structures and undamaged areas. He et al. [13] used a series of 3D shape descriptors to detect roof damage from aerial point clouds of Haiti's National Palace, Port-au-Prince, Haiti, which is collected in the aftermath of 2010 Haiti. Within the proposed method, the scene's point cloud representation was initially processed into a digital elevation model (DEM). Then, the point cloud data and created DEM were further processed to identify the building locations, and the 3D shape descriptors were computed for each building. The proposed solution then classified the buildings' point clouds into damaged and undamaged classes by thresholding the contour shapes based on the jaggedness. The authors reported that the proposed workflow resulted in an overall accuracy of 89%, while the recall and precision values were 89 and 58%, respectively. Axel and van Aardt [14] improved the damage detection method based on normal vector variations by introducing preprocessing steps to eliminate the points representing vegetation and ground surfaces based on above the ground (AGL) height and surface roughness. To detect damaged structures, Axel and van Aardt [14] used ALS-based point cloud data of damaged areas following the 2010 Haiti earthquake and utilize region growing approach to detect and separate structures from the rest of the scene and then analyzed the structures for damage through evaluating the variation of normal vectors and height analysis. The proposed method resulted in overall detection accuracy of 93% and damage classification of 78.9%.

In addition to the studies that focused on detecting damaged areas from point clouds, multiple studies have also investigated the task of semantic segmentation from point clouds. Hackel et al. [15] introduced a fast semantic segmentation method to classify lidar point clouds to a series of semantic objects. Similar to previous studies, the proposed method started by extracting features from point clouds directly; however, Hackel et al. [15] relied on the geometric features extracted from a point and its selected neighboring points at various distances based on principal component analysis (PCA) and approximating the three-dimensional shape context features using histogram-based descriptors. The study has reported that the proposed method can be predicted and classified points with an overall accuracy of 90.3%.

Besides change detection analysis and traditional methods to detect damage or semantically classify the point clouds into various objects, various research studies have focused on investigating the application of deep learning to analyze point cloud datasets. Here, deep learning corresponds to the learning algorithms that learn features during the training process. As 3D data can be represented based on various structures, various deep learning-based solutions based on these 3D representations are introduced. The first group of studies that are reviewed here developed deep learning models based on volume element (voxel) grid representations of the 3D point clouds. Prokhorov [16] was an early study to use voxel representations to develop a model to classify 3D point cloud data. Prokhorov proposed a model with an architecture similar to that of CNN to classify various objects' point clouds. The developed model had one 3D convolutional layer, one 3D pooling layer, and two fully connected layers, which is followed by 2-class output later. The training strategy used within the study includes pretraining the parameters within the convolutional layer based on lobe component analysis. Maturana and Scherer [17] expanded the model proposed by Prokhorov [16] for the task of object recognition with a new training strategy and method to create voxel models, which is known as Voxnet. The proposed model had two tandem 3D convolutional layers, one max-pooling layer, and one fully connected layer, which was followed by the output layer. In contrast to the study conducted by Prokhorov [16], Maturana and Scherer [17] did not pre-trained the developed network while the model resulted in a performance on par or better than the network proposed by Prokhorov [16], which could be attributed to the training strategy and optimization process.

As voxel representation of point clouds requires a larger memory footprint compared to the raw point cloud representations, multiple studies have investigated other data structures that require smaller memory to store data. As one of the main studies, Qi et al. [18] developed a deep learning model using raw point cloud data as input instances, known as PointNet. Within this model, a multi-linear perceptron (MLP) was applied on each input file point, and these MLPs are combined to extract a signature for the input instance. However, the PointNet architecture did not account for the local geometric features. To address this matter, Qi et al. [19] proposed a hierarchical model that applies the PointNet model in a recursive pattern on point cloud segments that were created by sampling and grouping layers, known as PointNet++ . Recently, Can et al. [20] utilized PointNet++ to classify the UAS-SfM-derived point clouds of large urban areas and reported an overall accuracy and IOU score of 82 and 45.3%. While the PointNet++ took the local geometric information of point clouds into account, this model's input size was limited to a predefined number of points, and the model could not use color or other spectral information to improve the classification. Kolkov and Lempitsky [21] introduced a deep learning model based on kd-tree representations of point clouds, which is named kd-network. The computational graph was created using the kd-tree structure, and the model shares similarities to CNN models, including sharing learnable parameters and learning the first-order features, and combining the learned features with higher-order extracted features to analyze point clouds. Kolkov and Lempitsky [21] compared the developed model's performance on the Shape-Net-core dataset with the Voxnet and PointNet and reported that while the kd-network performed 3% better than Voxnet on average, it demonstrated similar performance to PointNet. Similarly, Riegler et al. [22] proposed a deep learning model based on the octree representation of point clouds known as OctNet. The developed model used the octree spatial partitioning method to create a computational graph, and similar to CNNs, OctNet utilizes a convolution operation to extract and combine features. Riegler et al. [22] tested the developed model to semantically segment the façade of a structure and reported an overall accuracy and IOU score of 81.5 and 59.2%, respectively. Unlink PointNet, the kd-network, and OctNet models could use the local geometric information; however, only OctNet demonstrated the potentials to accept color and other spectral information along with geometric information. More recently, Wen et al. [23] introduced a model by combining the graph-based attention layers and CNNs to analyze raw point cloud representations based on global and local geometric information, known as GACNN. The developed model contained a graph attention module and encoder-decoder network. The graph attention module accepted raw point clouds and identified the kd-tree representation of the input instance. Then, the model analyzed the data based on a series of MLPs and two attention layers, namely global and local attention layers, each of which was constructed by a series MLPs. In the encoder-decoder model with skip connections, the encoder network downsampled the input and used the graph attention module to analyze the parameters. The decoder interpolated the data and used the graph attention module to compute the parameters. Wen et al. [23] utilized the model to segment the scene into nine classes semantically, included powerlines, low vegetation, impervious surfaces, car, fence, roof, façade, shrub, and tree. The study had concluded that the developed model could semantically segment the scene with overall precision and recall values of 74 and 70%, respectively. However, it was noted that the model could only detect some of the classes (e.g., impervious surfaces that represent road or sidewalks) with high accuracy compared to models that are developed based on PointNet architecture.

While the previous studies have demonstrated various methods to detect damage from aerial point clouds, most machine learning-based solutions that were proposed to analyze the point clouds did not focus on the task of damage detection and damage state classification. This manuscript aims to introduce a new approach to analyzing the UAS-SfM-derived point cloud data of large areas after an extreme event (e.g., hurricanes or tornadic events) to identify damaged areas semantically segment the point cloud representations. As a result, a workflow based on 3D fully convolutional network (3D FCN), k-means clustering algorithm, and 3D CNN model were developed to semantically segment the UAS-SfM point clouds for the task of damage identification. The 3D FCN and 3DCNN models within this study are independently trained to learn the features based on the training instances. The developed workflow learns each class based on the 3D geometry of input instances and the color information collected during the data collection. Ultimately, this method allows the analyst to perform a damage assessment of built-up areas directly.

3 Dataset

On March 3, 2020, a tornado impacted northern Tennessee and the Nashville metro area as an EF3 with estimated peak wind speeds of approximately 265 kph (165 mph). It was reported that the tornado remained on the ground over 60 miles, making it the longest tornado path officially recorded since 1950 (the earliest documentation of path length), and resulted in five fatalities and 220 injuries [24]. Following the event, a team on behalf of NSF-StEER conducted a series of field assessments along the tornado path to document the distribution and intensity of the damage within the affected areas. The StEER team reconnaissance focus included door-to-door (d2d) assessment, aerial surveys via UASs, and vehicle-mounted street view imaging to document damaged areas and buildings' performance during the event [2].

One of the surveyed areas included the Lockland Springs neighborhood, as shown in Fig. 1, (approximately at a latitude of 36.17501 and a longitude of − 86.72643) [2]. The aerial images were collected using a DJI Mavic Pro 2 platform at an altitude

Fig. 1 Lockland Spring dataset (unit in meters)

of approximately 74 m (243 ft) AGL. A total of 11 flights were conducted, which resulted in a total of 5293 images. Due to the time and site access limitations, no ground control was available. As a result, the geolocation and scale for the UAS-SfM point cloud data of the site were approximated by the onboard GNSS on the UAS platform. The collected images were then processed using the Pix4Dmapper software to reconstruct the 3D point cloud representation of the site. The final point cloud resulted in an average ground sampling distance (GSD), or the distance between pixel centers measured on the ground, approximately 2.2 cm (0.88 in).

The resulted point cloud dataset was segmented manually into one of the following 14 classes: boats, cars, debris, fallen trees, healthy trees, water bodies, roadways, terrain, poles, destroyed structures, severely damaged structures, moderately damaged structures, minorly damaged structures, and undamaged structures. This was done to create training and testing instances. Figures 2 and 3 demonstrate an example instance for each class. The boat classification contains any marine vessel that can be propelled on water. The car classification broadly consists of anything used to transport people or goods, such as an all-terrain vehicle (ATVs), passenger car (or sedan), van, sport utility vehicle, truck, cart, recreational vehicle (RV), trailer, and construction vehicle (e.g., excavators). The debris class is comprised of both human-made and untouched debris, including tree branches and other vegetation; however, the debris classification does not include a collapsed structure. The fallen tree class is restricted to tree trunks that have been uprooted or snapped at their base. Healthy trees consist of single trunks or multiple thin trunks. The water bodies classification includes swimming pools, ponds, or creeks. Roadways are comprised of

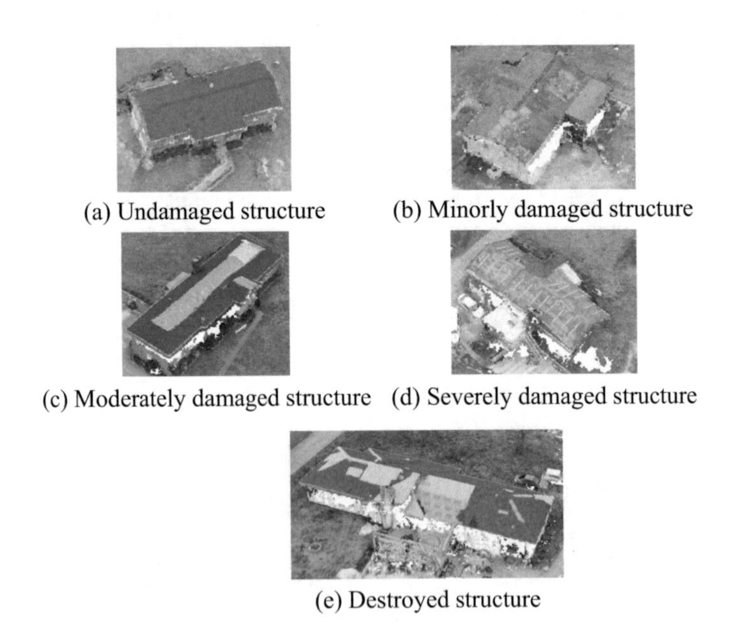

Fig. 2 Examples of various structure instances existing in the Lockland dataset

Fig. 3 Examples of various objects except structures existing in the Lockland dataset

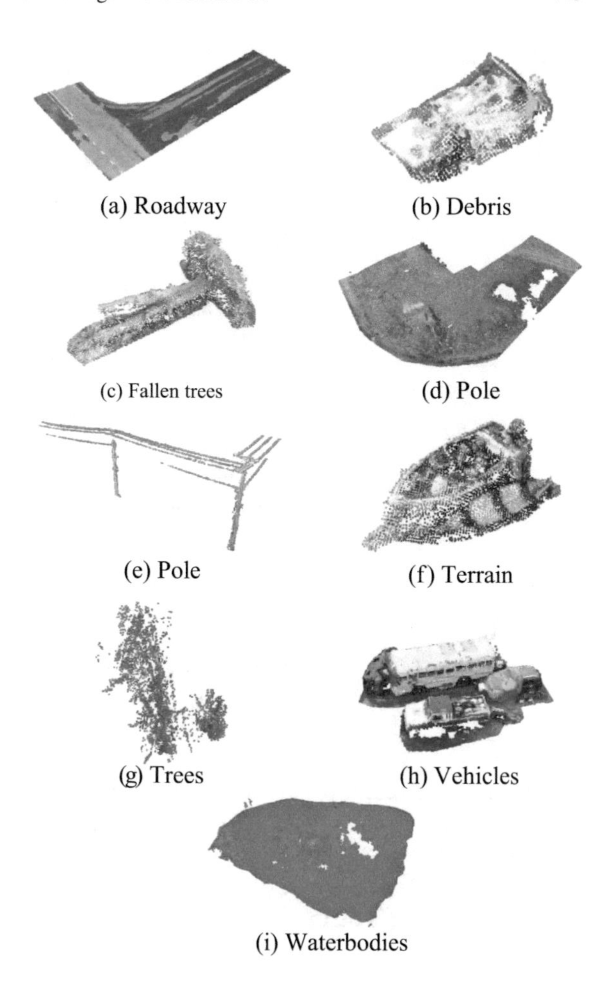

paved roads, sidewalks, bridges, parking driveways, and gravel roads or driveways that are regularly used for vehicle movements. The terrain class contains grasslands, bushes, utility boxes, postboxes, fences, chairs, and patio furniture. The poles class consists of transmission poles and wires supporting a variety of services, including power, telephone, and data.

To classify structures, five different categories are selected based on the StEER d2d ratings included destroyed structures, severely damaged structures, moderately damaged structures, minorly damaged structures, and undamaged structures [2]. A structure is considered destroyed if more than 15% or higher percentage of the structure's roof or walls is failed, or more than 25% roof deck is damaged. The severely damaged structure classification comprises structures that sustained 50% or more damage within the roof covering or walls, lost between 5 and 25% of their roof sheathing, or lost 15% or less of the roof structure. The moderately damaged structure classification comprises structures that either 15–50% of the roof covering

Table 1 Summary of number points for each class

Instance	Number of points (Thousands)	Percentage of the total (%)
Boat	31	0.01
Debris	14,432	3.54
Destroyed structure	4258	1.04
Fallen tree	49,2523	12.07
Healthy tree	54,789	13.42
Minorly damaged structure	10,171	2.49
Moderately damaged structure	7765	1.90
Pole	1362	0.33
Roadway	44,521	10.91
Severely damaged structure	5769	1.41
Terrain	196,515	48.15
Undamaged structure	14,474	3.55
Vehicle	3880	0.95
Water body	938	0.23

or walls sustained damage or less than 5% of their substructure is failed. The minorly damaged structure classification includes the structures that have been damaged at 15% or less in terms of roof cover or wall damage. Lastly, only structures that do not exhibit any visible damage are considered undamaged structures. Table 1 summarizes the number of points that were segmented for the Lockland Spring area dataset.

4 Data Classification Methodology

The classification of UAS-SfM point clouds for the task of damage detection introduces a unique set of challenges. While the color, geometry, and point density vary between objects of the same class, these variations combined with random and unique geometric damage patterns existing within the point clouds make the task of damage detection more difficult than the semantic segmentation of the scene. Mohammadi et al. [25] and Liao et al. [26] presented deep learning models based on 3DFCN to detect damage from UAS-SfM-derived point clouds. While the 3DFCN models introduced in those studies were able to learn and predict the geometry input instances (e.g., terrain class), the overall precision and recall values for damaged and undamaged instances were limited and only between 10 and 20%, respectively. This manuscript expands on this earlier work and introduces a new workflow based on the fusion of multiple deep learning approaches. This includes 3DFCN, k-means clustering algorithm, and 3DCNN to detect damaged areas from the UAS-SfM derived point clouds. The main idea behind the introduced workflow is that the damaged instances can be detected with higher accuracy within the pre-segmented point cloud data compared to classifying the entire scene for damaged areas and other classes.

The developed workflow initially classifies the point cloud data into seven general classes including vehicles (includes boats and cars), debris (includes debris and fallen trees), water bodies, roadways, healthy trees (includes health trees and poles), terrain, and structures (at various damage states, including destroyed, severely damaged, moderately damaged, minorly damaged, and undamaged structures). Once the point cloud data are segmented into these seven classes, only points that are classified as structures are analyzed by the k-means clustering algorithm to detect each individual structure, and lastly, the 3DCNN network evaluates each structure class and further classify it into five damage states (or classes) of destroyed, severely damaged, moderately damaged, minorly damaged, and undamaged (or no damage) structures. This section initially describes the data preparation and transformation process to convert raw point cloud data into 3D voxels used within this workflow, then presents the developed network architecture, and finally reviews the training strategy used to develop the two models. The 3DFCN and 3DCNN presented in this study were implemented in TensorFlow. The k-means clustering algorithm was implemented through the Scikit learn library.

4.1 Voxel Transformation

The process to create instances within this study is similar to the approach proposed by Mohammadi et al. [25], which is briefly described here. As stated, the process to create the training and testing instances are initiated by manually segmenting the UAS-SfM point clouds into the 14 classes. Afterward, a label corresponding to each segmented object was assigned to all its points. Then, all the labeled objects are compiled into a single file. Lastly, the compiled file is divided into 10 m $\times$ 10 m (32 ft $\times$ 32 ft) parts to create segments of equal dimensions consisting of multiple objects that can be used for semantic segmentation. The selected dimensions here permit the user to control each instance point-to-point spacing or resolution. In the first step, the minimum value for each instance point coordinates is computed. Then, the values are subtracted by the identified minimum values for each component (i.e., X, Y, and Z) to transfer the data into positive ordinates. In the next step, the positive coordinates are downsampled based on the selected voxel dimensions. Within this study, a voxel dimension of 64 is used, which results in a sampling of roughly 16 cm (6.2 in). The sampling rate of 16 cm results in an acceptable voxel representation for building damage assessment following windstorms [27]. In the last step, the voxel coordinate values are multiplied by the selected dimension for the voxel representations, normalized based on coordinates computed ranges, and rounded. Lastly, as each voxel model has a number of empty arrays or cells and occupied arrays, an extra label corresponding to empty space is added to the training labels known as neutral class.

4.2 3DFCN Model

The 3DFCN developed for this study is inspired by the previous work of Long et al. [28], which introduced the FCN architecture to analyze 2D images (Fig. 4). The developed model has an input layer that accepts three 3D voxels with 64^3 cells matching the three RGB channels and comprises encoding and decoding parts. The encoder part has a total of six convolutional layers with linear rectifier units, and the decoder part has a total of six transpose convolutional layers with linear rectifier units. The network architecture contains skip connections that allow the convolutional layers to copy their input into transpose convolutional layers. This allows the latter layers to learn new information but retain the initial knowledge learned by previous layers, minimizes the potential gradient vanishing issues [9]. The kernel sizes in the convolutional and transpose convolutional layers had a size of 3, and in each layer, a total of 32 kernels existed. Furthermore, the stride parameter was set to a value of 1 for these layers. In the last layer of the network, each kernel also has a self-attention module that is comprised of two multilayer perceptrons with 32 nodes. These self-attention layers enabled the model to restore the labels at a higher rate. Lastly, the output layer is a single voxel with a size similar to input size, each of which cells represent the label of the input point cloud instance. The model was trained based on the minibatch size of 64 using an Adam optimizer and an L_2 regularization with the lambda of 0.0005 to reduce the overfitting potential during training.

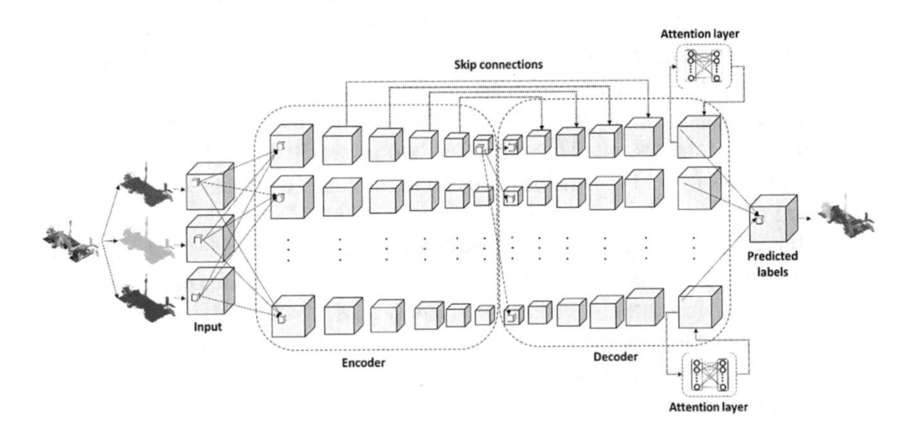

Fig. 4 Developed 3D fully convolutional network (3DFCN)

4.3 K-Means Clustering Algorithm

Once the point cloud data is classified into seven classes by the 3DFCN model, a query search initially collects all the points classified as structure, and a statistical outlier removal (SOR) minimizes the number of sparse points within building segments [29]. Afterward, the k-means clustering algorithm categorizes the points that represent structures into individual buildings. The k-means clustering algorithm here requires two input parameters, including the number of clusters and a threshold value. The first parameter, the number of clusters, in this study is determined by double the number of buildings within the ROI that are being analyzed. The second parameter, the threshold value, was estimated based on the minimum distance between two structures. Within this study, the threshold value was set to the value of 0.20 m.

4.4 3DCNN Model

The 3D CNN model developed within this study was inspired by Voxnet [17]. The developed model, similar to 3DFCN, has an input layer that accepts three 3D voxels with 64^3 cells matching to RGB channels, two convolution layers followed by a pooling layer, and two fully connected layers or multilayer perceptrons. Lastly, the network has five nodes within its classification layer, as shown in Fig. 5. The convolutional layers have linear rectifier units with stride parameters of 1 for both convolutional layers, respectively. Within this model, a max-pooling approach with a stride parameter of 2 was utilized to reduce the number of parameters and introduce invariance to translations. The network utilizes the SoftMax function to determine the class of the inputted voxel instance within its classification layer. Like 3DFCN, the model was trained based on the minibatch size of 32 using an Adam optimizer and an L_2 regularization with a lambda of 0.001 to reduce the overfitting potential during training.

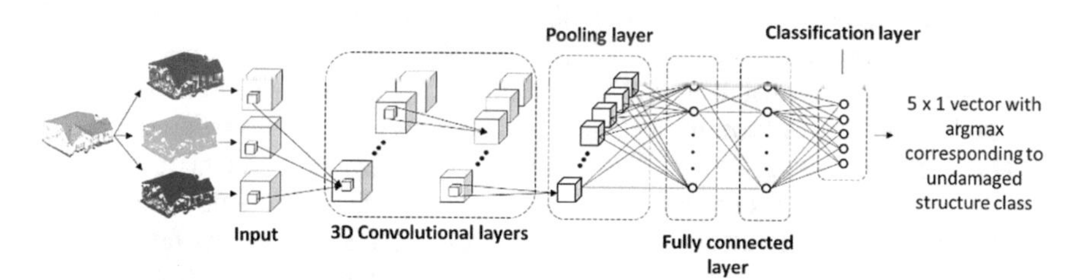

Fig. 5 Developed 3D convolutional neural network (3DCNN)

5 Results and Discussion

The hyperparameter search and the developed 3DFCN and 3DCNN models were primarily conducted on the GPU resources at the Holland Computing Center, located at the University of Nebraska-Lincoln. The testing and part of the training process of 3DCNN were conducted locally on a workstation with Nvidia RTX A6000 GPU. To evaluate the developed models, the confusion matrix and three performance measures (precision, recall, and IOU) were calculated for both training and testing phases. Both 3DFCN and 3DCNN models were trained on roughly half of the instances and tested on the rest of the data to investigate the learning transformability of the proposed workflow. To prepare the training dataset for 3DFCN and to increase the number of instances, a total of 8677 unique 10 m × 10 m segments were augmented by randomly rotating each instance two times around its vertical axis. This results in a total of approximately 17,000 training instances. As for 3DCNN, all the structure classes were augmented by randomly rotating each structure instance ten times around its vertical axis resulted in a total of approximately 9000 instances.

5.1 3DFCN Model Training and Testing

The 3DFCN model was trained for a total of approximately 1600 epochs prior to testing. To identify a set of optimized parameters, multiple networks were initially trained through a factorial design process, in which the L_1 and L_2 regularization lambdas learning rates, and the number of kernels per layer varied. Through this hyperparameter tuning study, the network with a total of 32 kernels per layer, L_1 regularization lambda of 0, L_2 regularization lambda of 0.0005, and a learning rate of 0.0015 proven to be the most successful model. The most successful model was identified through evaluating model performance measures, including precision and recall, as well as training mean square error values. As a result, this model was selected for the extended training process. Table 2 presents the performance measures for the training and testing sets, where the model was able to detect instances with average precision, recall, and IOU of 51, 52, and 38%, respectively. Moreover, Fig. 6 demonstrates the confusion matrix for the training and testing phases. As shown, the model was able to learn all instances with precision and recall value of 40% or higher except for classes water bodies, vehicles, and roadways. However, it was noted that these three classes represent the lowest number of instances within the training dataset, making it difficult for the model to learn. The testing results demonstrate a trend in performance measures where an average value of 47, 43, and 33% was observed for precision, recall, and IOU metrics, respectively. Due to a large number of learnable parameters within 3DFCN and continued performance improvement of the model on the testing dataset, it is believed that further training will result in improved testing results. However, the current model was able to learn and predict

Table 2 Performance measures for training and testing datasets of 3DFCN

Instance	Train (%)			Test (%)		
	Precision	Recall	IOU	Precision	Recall	IOU
Neutral	100	97	97	100	100	100
Vehicles	0	2	0	0	1	0
Debris	72	41	35	47	27	20
Healthy trees	55	60	41	53	45	32
Water bodies	3	37	3	0	16	0
Roadways	24	55	20	26	57	21
Terrain	82	68	59	71	59	48
Structures	76	52	45	76	35	31

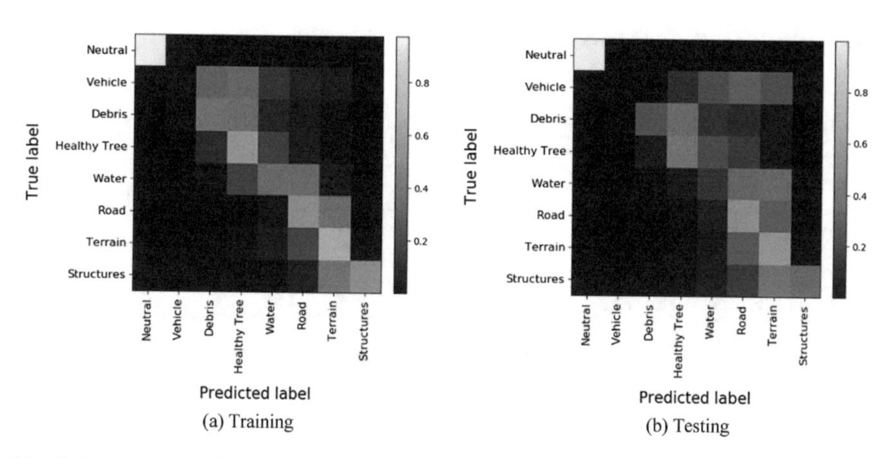

Fig. 6 3DFCN confusion matrix for **a** training and **b** testing results

the correct labels of the neutral, terrain, debris, and structure classes which were subsequently used within the k-means clustering and 3DCNN workflows (Fig. 7).

5.2 3DCNN Model Training and Testing

The 3DCNN model was trained on the created training dataset for only a total of 15 epochs prior to testing, unlike 3DFCN to avoid overfitting informed based on evaluation results of the validation set. Multiple networks were initially trained through the factorial design process to identify a set of optimized parameters similar to 3DFCN. The L_1 and L_2 regularization lambdas, learning rate values, number of kernels in convolutional layers, and number of nodes in the fully connected layers were varied. Through the factorial design process, the network with a total of 32 kernels per convolutional layer, L_1 regularization lambda of 0, L_2 regularization lambda of 0.005, the

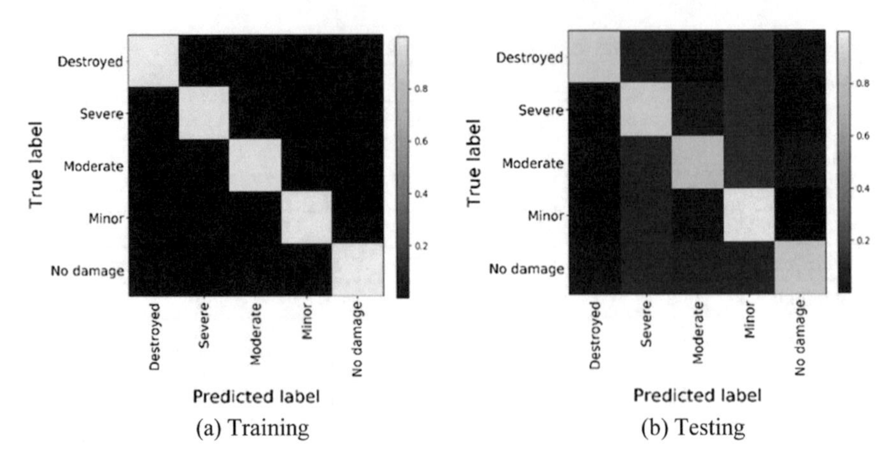

Fig. 7 3DCNN confusion matrix for **a** training and **b** testing results

Table 3 Performance measures for training and testing datasets of 3DCNN

Instance	Train (%)			Test (%)		
	Precision	Recall	IOU	Precision	Recall	IOU
Destroyed structure	96	95	91	76	67	55
Severely damaged structure	95	94	90	63	68	49
Moderately damaged structure	94	94	89	68	63	48
Minorly damaged structure	95	96	91	61	74	50
Undamaged structure	95	95	90	73	65	53

learning rate of 0.001, and 256 nodes for fully connected layer proven to be the most successful model. The best model was identified through assessing the performance measures, included precision and recall, as well as training-based mean square error values. Table 3 presents the performance measures for training and testing results where the model was able to detect instances with average accuracy and recall of 95, 94, and 90%, respectively. In addition, Fig. 8 demonstrates the confusion matrix for the training and testing processes. As shown, the model was able to learn all instances with precision, recall, and IOU score values of 68, 67, and 67%, respectively, which demonstrates significant improvement over previous studies presented by Liao et al. [25], where a single 3DFCN model was only able predicted damaged areas with precision and recall value of 34 and 32%, respectively.

The performance of the proposed workflow is further demonstrated by analyzing a 70 by 70-m point cloud segment. Figure 8 depicts the final semantic segmentation results for the 3DFCN model. As illustrated, the 3DFCN model learned and classified the points pertained to the terrain and roadways classification with an acceptable level of accuracy. In addition, while the points corresponding to the structure class were detected within the test segment, it was noted that not all the points representing

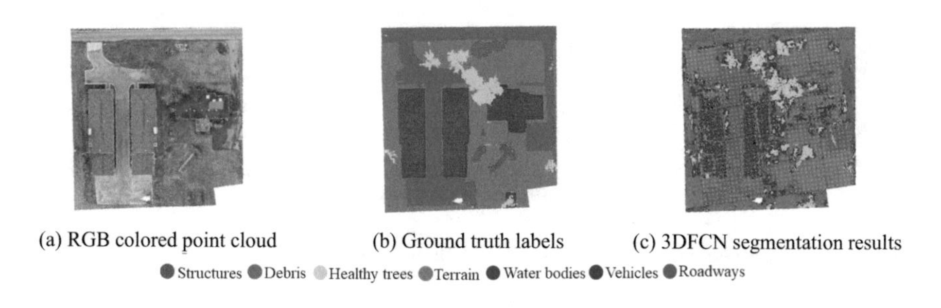

(a) RGB colored point cloud (b) Ground truth labels (c) 3DFCN segmentation results

Fig. 8 Demonstration of the developed 3DFCN on the part of the dataset

the structures were classified correctly and instead were classified as terrain, which matches the test results of the confusion matrix presented in Fig. 6. Misclassification of the points representing the structure class with the terrain class was observed in both training and testing data. As the training process demonstrates that the model can learn to distinguish the points within these two classifications, it is expected that further training improves the classification points that represent these classes. Once the 3DFCN model segmented the point cloud into seven classes, the workflow then transforms the voxel representation into the raw point cloud representation by applying the inverse process described in Sect. 4.1. Afterward, the workflow runs a query to collect the points with structure classification, as shown in Fig. 9a. Then, the workflow utilizes the SOR filter to eliminate the sparse points and uses a k-means clustering algorithm to segment the structures into individual segments. As shown in Fig. 9b, the algorithm eliminated most of the sparse points using a neighboring number of 31 and standard deviation factors of 0.25. The clustering algorithm uses six clusters (double the number of observed structures in the selected ROI) and a threshold of 0.20 m to segment the structures. The final result of k-means clustering segmentation results is shown in Fig. 9b. Once the segments are identified, the workflow converts these segments into the voxel representation, which the trained 3DCNN model then analyzes. Figure 10 demonstrates the results of the 3DCNN model, where 50% of segments were classified correctly. Figure 11 depicts the results of the classification.

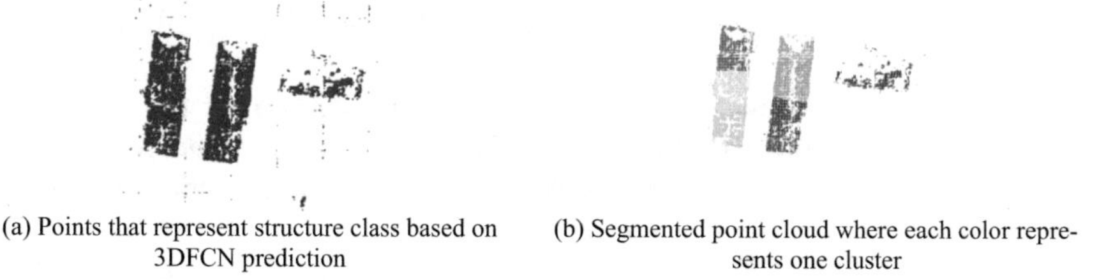

(a) Points that represent structure class based on 3DFCN prediction (b) Segmented point cloud where each color represents one cluster

Fig. 9 Demonstration of the k-means clustering algorithm

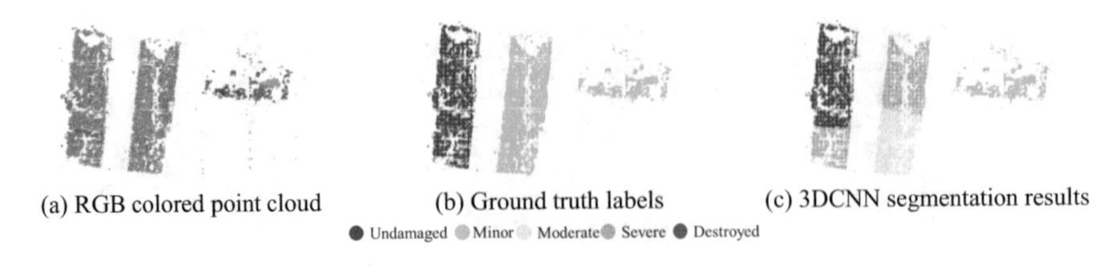

Fig. 10 Demonstration of the developed 3DCNN on the part of the dataset

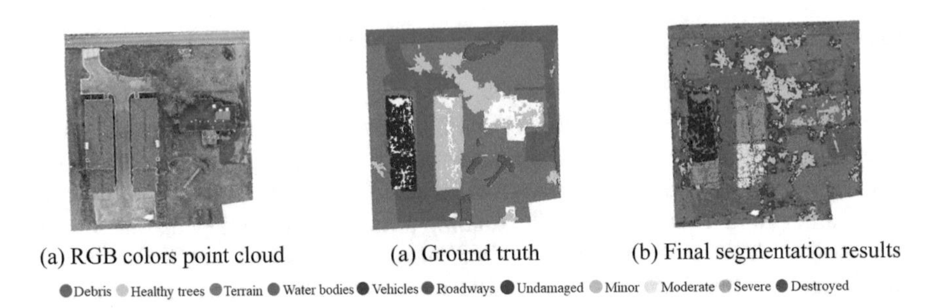

Fig. 11 Demonstration of the overall prediction for the developed workflow

6 Conclusions

This manuscript presented a new workflow that combines three distinct machine learning algorithms, including 3DFCN, k-means clustering, and 3DCNN. The developed workflow was developed based on UAS-SfM-derived point clouds to classify post-tornado scenes for damage assessment semantically. The UAS-SfM point cloud of built-up areas sustained damage following the March 2020 tornadic windstorm event in northern Tennessee was used to train and test the workflow. Each point within this dataset was segmented manually to have a label to match the door-to-door assessments performed by a ground team of investigators. Afterward, the point cloud data were divided into two parts for testing and training segments. Initially, the labeled point cloud was segmented into 10 m × 10 m (32 ft × 32 ft) parts to create training and testing instances. Then, the 64^3 voxel representation of the segments was created, which resulted in a sampling rate of 16 cm (6.2 in). The 3DFCN was trained based on these voxel models, and the 3DCNN was trained based on structure voxels.

The developed models were evaluated based on precision, recall, and IOU that are computed from the confusion matrix. As illustrated by the performance measures presented for training results, the developed models are optimized to learn the features; however, the convergence was shown to be slower in the 3DFCN model, primarily due to the number of learnable parameters. The 3DFCN model learned and predicted the correct labels of the neutral, terrain, and structure classes that are essential to the workflow's success. However, it demonstrated a lower precision and recall for objects with much smaller frequency within the dataset (e.g., vehicle). The 3DCNN model learned all structure classes and demonstrated significant improvement over the 3DFCN model in detecting damaged structures [26].

Acknowledgements This work was completed utilizing the Holland Computing Center of the University of Nebraska, which receives support from the Nebraska Research Initiative. Reconnaissance data were collected with the administrative and financial support of NSF-StEER, which included travel funds for the second author. The authors would like to thank Dr. Yijun Liao for her help and assistance in the field as well as Mr. Awang Ahmad Hashim, Mr. Mario Esquivel, and Ms. Pooja Rajeev for their assistance in the data preparation and processing.

References

1. He X, Cha EJ (2018) Modeling the damage and recovery of interdependent critical infrastructure systems from natural hazards. Reliab Eng Syst Safe 177:162–175
2. Wood RL, Roueche D, Cullum K, Davis B, Gutierrez Soto M, Javadinasab Hormozabad S, Liao Y, Lombardo F, Moravej M, Pilkington S, Prevatt S, Kijewski-Correa D, Djima S, W. Robertson I (2020) Early access reconnaissance report (EARR). In: StEER - 3 March 2020 Nashville Tornadoes. DesignSafe-CI. https://doi.org/10.17603/ds2-2zs2-r990
3. Vetrivel A, Gerke M, Kerle N, Vosselman G (2015) Identification of damage in buildings based on gaps in 3D point clouds from very high resolution oblique airborne images. ISPRS J Photogram 105:61–78
4. Duarte D, Nex F, Kerle N, Vosselman G (2018) Satellite image classification of building damages using airborne and satellite image samples in a deep learning approach. ISPRS Ann Photogramm Remote Sens Spatial Inf Sci 2:89–96. https://doi.org/10.5194/isprs-annals-IV-2-89-2018
5. Spencer Jr BF, Hoskere V, Narazaki Y (2019) Advances in computer vision-based civil infrastructure inspection and monitoring. Engineering. 5(2):199–222. https://doi.org/10.1016/j.eng.2018.11.030
6. Gao Y, Mosalam KM (2018) Deep transfer learning for image-based structural damage recognition. Comput Aided Civ Infrastruct Eng 33:748–768
7. Simonyan K, Zisserman A (2014) Very deep convolutional networks for large-scale image recognition. arXiv preprint at https://arxiv.org/pdf/1409.1556.pdf
8. Tilon S, Nex F, Kerle N, Vosselman G (2020) Post-disaster building damage detection from Earth observation imagery using unsupervised and transferable anomaly detecting generative adversarial networks. Remote Sens 12(24):4193. https://doi.org/10.3390/rs12244193
9. Ronneberger O, Fischer P, Brox T (2015) U-net: convolutional networks for biomedical image segmentation. In: International conference on medical image computing and computer-assisted intervention, Boston, MA, pp 234–241
10. Olsen MJ (2015) In situ change analysis and monitoring through terrestrial laser scanning. J Comput Civ Eng 29:04014040. https://doi.org/10.1061/(ASCE)CP.1943-5487.0000328
11. Tran THG, Ressl C, Pfeifer N (2018) Integrated change detection and classification in urban areas based on airborne laser scanning point clouds. Sensors 18(2):448. https://doi.org/10.3390/s18020448
12. Aixia D, Zongjin M, Shusong H, Xiaoqing W (2016) Building damage extraction from post-earthquake airborne LiDAR data. Acta Geologica Sinica 90(4):1481–1489. https://doi.org/10.1111/1755-6724.12781

13. He M, Zhu Q, Du Z, Hu H, Ding Y, Chen M (2016) A 3D shape descriptor based on contour clusters for damaged roof detection using airborne LiDAR point clouds. Remote Sens 8(3):189. https://doi.org/10.3390/rs8030189
14. Axel C, van Aardt JA (2017) Building damage assessment using airborne lidar. J Appl Remote Sens 11(4):046024. https://doi.org/10.1117/1.JRS.11.046024
15. Hackel T, Wegner JD, Schindler K (2016) Fast semantic segmentation of 3D point clouds with strongly varying density. ISPRS Ann Photogramm Remote Sens Spatial Inf Sci 3:177–184. https://doi.org/10.5194/isprs-annals-III-3-177-2016
16. Prokhorov D (2010) A convolutional learning system for object classification in 3-D lidar data. IEEE Trans Neural Netw 21:858–863. https://doi.org/10.1109/TNN.2010.2044802
17. Maturana D, Scherer S (2015) Voxnet: a 3d convolutional neural network for real-time object recognition. In: IEEE/RSJ international conference on intelligent robots and systems, Hamburg, Germany, pp 922–928. https://doi.org/10.1109/IROS.2015.7353481
18. Qi CR, Su H, Mo K, Guibas LJ (2017) Pointnet: deep learning on point sets for 3d classification and segmentation. In: IEEE conference on computer vision and pattern recognition, Honolulu, HI, pp 652–660
19. Qi CR, Yi L, Su H, Guibas LJ (2017) Pointnet++: Deep hierarchical feature learning on point sets in a metric space. arXiv preprint at https://arxiv.org/pdf/1706.02413.pdf
20. Can G, Mantegazza D, Abbate G, Chappuis S, Giusti, A (2020) Semantic segmentation on Swiss3DCities: a benchmark study on aerial photogrammetric 3D Point cloud dataset. arXiv preprint at https://arxiv.org/pdf/2012.12996.pdf
21. Klokov R, Lempitsky V (2017) Escape from cells: deep kd-networks for the recognition of 3d point cloud models. In: IEEE international conference on computer vision, Honolulu, HI, pp 863–872
22. Riegler G, Osman Ulusoy A, Geiger A (2017) Octnet: learning deep 3d representations at high resolutions. In: IEEE conference on computer vision and pattern recognition, Honolulu, HI, pp 3577–3586
23. Wen C, Li X, Yao X, Peng L, Chi T (2021) Airborne LiDAR point cloud classification with global-local graph attention convolution neural network. ISPRS J Photogramm Remote Sens 173:181–194. https://doi.org/10.1016/j.isprsjprs.2021.01.007
24. National Weather Services (NWS) (2020) March 2–3, 2020 Tornadoes and Severe Weather. https://www.weather.gov/ohx/20200303. Accessed 27 Feb 2020
25. Mohammadi ME, Watson DP, Wood RL (2019) Deep learning-based damage detection from aerial SfM point clouds. Drones 3(3):68. https://doi.org/10.3390/drones3030068
26. Liao Y, Mohammadi ME, Wood, RL (2020) Deep learning classification of 2D orthomosaic images and 3D point clouds for post-event structural damage assessment. Drones 4(4):24. https://doi.org/10.3390/drones4020024
27. Womble JA, Wood RL, Mohammadi ME (2018) Multi-scale remote sensing of tornado effects. Front Built Environ 4:66. https://doi.org/10.3389/fbuil.2018.00066
28. Long J, Shelhamer E, Darrell T (2015) Fully convolutional networks for semantic segmentation. In: IEEE conference on computer vision and pattern recognition, Boston, MA, pp 3431–3440
29. Rusu RB, Marton ZC, Blodow N, Dolha M, Beetz M (2008) Towards 3D point cloud based object maps for household environments. Robot Auton Syst 56(11):927–941. https://doi.org/10.1016/j.robot.2008.08.005

New Sensor Nodes, Cloud, and Data Analytics: Case Studies on Large Scale SHM Systems

Isabella Alovisi, Dario La Mazza, Monica Longo, Francescantonio Lucà, Marzia Malavisi, Stefano Manzoni, Diego Melpignano, Alfredo Cigada, Paola Darò, and Giuseppe Mancini

Abstract The progress in the world of new sensors is running fast, offering good performances, and reliable solutions with initial costs orders of magnitude lower than those faced only a few years ago. The spread of new electronic devices, like microcontrollers, the increasing power of the networks for data transmission and management and in the end the availability of the new data-driven approaches have created a revolution in SHM approaches, not yet fully mastered. The way to design a SHM system is going to be deeply revised in an industrial perspective, within a complex framework in which everything has to be planned into details since the beginning, including the development of a metrological culture, the personnel education, the need of spare parts, re-calibration, ... This also means a revolution in data management: huge data flows not only create hardware problems related to their transfer; the software too requires a great deal of effort to compress data, also due to the actual cost of cloud resources. All these facts, accounting for the real metrological performances of the best MEMS sensors available at present, also require simplified data analyses, as software complexity is now mainly transferred to the network management. A trade-off must be looked for between the big redundancy offered by the actual networks and the need of a simple and prompt information, granting the structure safety: That is why as the data rates increase, the algorithms to be adopted must be simple, reliable, eventually adapted to edge computing at the sensor level, where hardware power is now present though at a reduced scale. The chapter shows such an approach in a real case from the system design, its birth, and its proper use for damage detection, up to the detection of a structural failure.

Keywords Structural health monitoring · Mems sensors · Sensor network · Data analytics · Damage detection

---

I. Alovisi · D. La Mazza · M. Longo · M. Malavisi · P. Darò · G. Mancini
Sacertis Ingegneria S.r.l., Roma, Italy

F. Lucà · S. Manzoni · A. Cigada (✉)
Dipartimento di Meccanica, Politecnico di Milano, Milano, Italy
e-mail: alfredo.cigada@polimi.it

D. Melpignano
STMicroelectronics, Agrate, Italy

A. Cury et al. (eds.), *Structural Health Monitoring Based on Data Science Techniques*, Structural Integrity 21, https://doi.org/10.1007/978-3-030-81716-9_22

1 Introduction

The world of structural health monitoring (SHM) is going through a sort of destructuring, due to the completely new approaches to structural sensing. The new Internet of Things tools have created a revolution in the way measurements are being carried out, not just for machines, but also for structures, pushing hard toward the idea of dense sensing, often translated into that of smart structures.

As often discussed, the general idea of SHM is to mimic the human behavior [1]: This does not just mean having distributed sensors, but also an efficient network and an improved data management capability, to provide fast and reliable evaluations. A measurement system, pretending to act as a human, must be designed with a clear idea about its final aim, so it has to be deeply understood which functions can be transferred to the computer through models or data science approaches and which are still under the specialist's supervision. In addition, until recent days, the huge effort on SHM has been mainly carried out on the research side, yielding to a relatively poor number of routine industrial approaches [2].

The mentioned destructuring of SHM approaches is facing new problems, moving the most critical implementation phases toward new aspects, different from the past.

If new sensors are considered, they have different performances with respect to those commonly adopted today for monitoring applications, mainly consisting in laboratory instrumentation for short-term dynamic testing. In general, the new cheaper devices have slightly lower output quality, though this gap is narrowing; a huge difference remains if costs are addressed. The new sensors are often less expensive than those commonly adopted for SHM: this also means that a trade-off must be looked for between the available budget and the need for quality in measurements. Given a fixed budget, the higher number of allowed sensors and the availability of some redundancy, can help recovering a more than acceptable information content; at the same time, the new solution allows for a better hope to be closer to an eventual damage onset, for prompt operation.

Another revolution is offered by the possibility to couple each sensor to a micro-controller: this unit is a sort of tiny and cheap computer, programmed to work on simple software-defined operations. Thus, they allow to manage essential functions like data acquisition or communication to a supervising unit, but they are also capable of performing simple data analysis like the fast Fourier transform (FFT) or a running root mean square (RMS) at the sensor level, making the system cleverness distributed and, therefore, getting closer to the idea of a smart system: however, this implies the risk of an information jam, which has to be properly managed.

The mentioned destructuring also brings in new problems: in such monitoring systems with distributed intelligence, each sub-system lives a life on its own within its time framework, given by its internal clock, unless complex synchronization tools are applied: This does not just mean abandoning those approaches to damage detection requiring a single clock shared among all the sensors in the same network (for example, a loss of synchronization makes traditional modal analysis a harder task); rather this means changing the strategy to achieve the final result.

Even if sensors are now less expensive, it is not possible to assess the same for the complete network. In case of a wired solution, a more complex layout comes from the higher number of connections among sensors and supervising units, and this also implies the need to both provide power supply to all nodes and bring the data back to a gateway or more generally to a network.

It has to be considered that, in view of a complete description of what's going on to get a structure health check, data fusion approaches are often adopted, merging quite different data kinds, with diverse aims, generated by a large variety of sensors, each having its own output impedance, sampling rate, output range, etc. Wireless, often offered as the solution to every kind of problems, is not yet reliable enough when big data streams have to be managed, or when safety comes into play. The reverse applies to static data, where wireless can be an interesting option, as the reduced data flow allows to send a data packet again in case it gets lost over the network. At this point, the main bottlenecks appear to be those related to the data flux over the sensor network: this stream is getting bigger and bigger, harder to manage, especially if in the end a single synthetic piece of information has to be given to the structure manager, or if an alarm has to be promptly generated.

The new IoT tools for sensor networks are sometimes not really fit for SHM applications: a powerful protocol like LoRaWAN [3], widely adopted for smart city projects, due to its power to transmit over long distances, gives each sensor a very short time to transmit data. This solution is, therefore, not suitable for continuous data streaming, as in the case of accelerometers, having a higher sampling rate: new solutions like the narrowband IoT [4, 5] are helping to mitigate these problems. A high innovation rate is also seen for cabled solution, remaining the majority among SHM systems: interesting solutions point at the use of industrial protocols for data transmission, as these have proven to be robust and allow to transfer medium to high data rates even over complex and busy networks [6].

The preliminary design of the monitoring system becomes the main challenge [7]; once the aims are properly defined, density of measurements in both time and space are key issues. The designer has to define beforehand how many sensors are to be used, their location on the structure, and for how long measurements have to be taken; this resolves the risks of losing meaningful events in a trade-off against the need to keep the network not too busy and to store a reasonable though meaningful amount of data.

In the end, some issues are seldom addressed, and again they belong to the already mentioned industrial management of SHM. Maintenance of monitoring systems has costs which are a substantial part of the initial investment; sensors need periodical checks on their calibration, and this can be a difficult task, especially if they are not easily reached; a stock of spare parts has to be kept, as electronics systems undergo a very fast aging, not comparable to the life of the structure. In the end, the technical staff, taking care of the measurement systems, is asked to follow a proper training, which has to be strongly multi-disciplinary, also including a plan to take decisions over the different scenarios which can be depicted by the monitoring system.

All these aspects make clear that a monitoring system is not just installing sensors on a structure and getting data, this is only a part in a more complex framework, a process which requires proper design and awareness [8–10].

If on one hand the recent advances in electronics make new monitoring system hardware more accessible to everyone, on the other hand it must not be forgotten how many factors still constitute a problem for these approaches: one above the others is the required multi-disciplinary approach, needing experts in many different areas, from civil engineering to data sciences, to electronics, to metrology, to telecom, mathematics, and so on: all these skills are hardly found in a single professional and require a trained staff.

Under these principles, the present chapter will go through a list of specifications to develop a complete industrial project; a path among the others is laid, relying on the up to date innovation in electronics, networks, communication, data management, and above all the possibility to give back a prompt and synthetic evaluation about the structure health.

2 System Design

The approach developed by the authors of this paper aims at a pervasive sensing of structures. Providing a dense sensing in both time and space means managing impressive data flows, but it also means designing and realizing a new sensing infrastructure.

The new approach is shortly summarized in Fig. 1: a traditional SHM system consists in a number of sensors placed over the structure: their cost usually allows

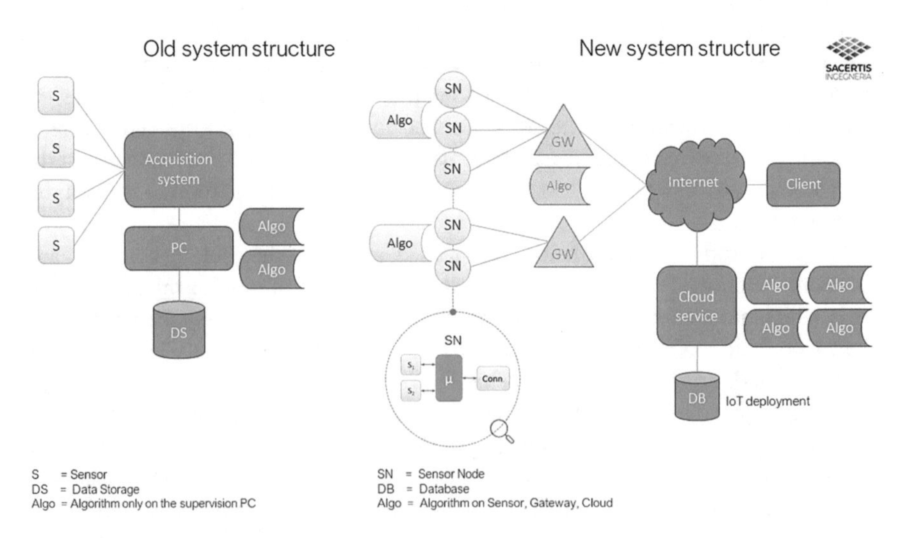

Fig. 1 Traditional (left) and the new (right) structure of SHM systems

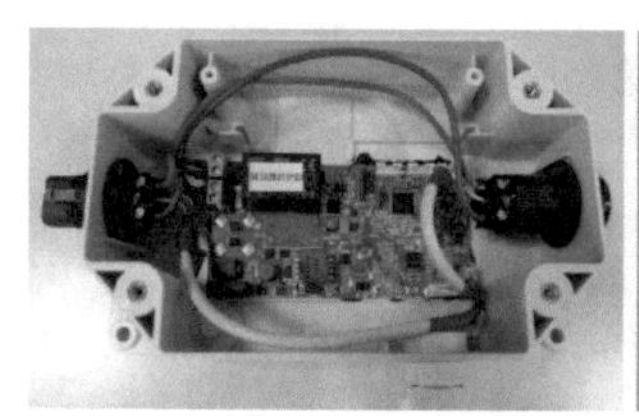

Fig. 2 Sensor node (left) and two examples of sensor nodes for measurements in tunnels (centre) and on bridges (right)

for a limited number of sensing points; the measurement chain is completed by cables sending data to a single central unit (or a number of data acquisition units, governed by a supervisor). This unit has the tasks of checking the system fundamental functions and data storage, managing issues like the needed redundancies to avoid any data loss. Any algorithm working on the data can start from this point, as this architecture does not offer any other chance to have "distributed cleverness." For most applications, a remotely controlled supervision computer can be enough to manage the data volumes.

The new SHM structure has to manage huge data flows, as cheaper and more numerous sensors are part of the same network: due to this reason, the mentioned "cleverness" has to be distributed, and the data streaming has to be compressed as much as possible, also in spite of helping a final evaluation which has to be synthetic, effective and fast.

Instead of having just the sensors, the new system has "sensor nodes" (Fig. 2), units made up of a board carrying some sensors (in spite of standardization, the suite is always the same and every application works on a chosen subset). A stabilization device for power supply is present, together with a microcontroller, a tag to identify all the on-board features, a NFC radio for an immediate check of the sensor output, and a connection with the rest of the network. For purpose of generality, the node can also accept inputs from external sensors, needed to complete the specific setup, if not already present on the board. The microcontroller, apart from supervising the node activities, can already perform some basic operation and send only the fundamental ones in case the network is too busy. Then, as the important data generation cannot be managed locally by a computer or even by a server, the power of a cloud infrastructure is often required. This means some device to locally manage the sensor network, collect data, and send them to the cloud. This task is performed by the gateways, gaining higher attention by the semiconductor industry, due to their growing importance. As technology develops, these devices have also added storage capability and also the chance to perform tasks similar to those working on the cloud, of course with a more limited power and memory. They further reduce, if possible, the amount of data traveling over the networks toward the cloud. Therefore, the real revolution consists in the possibility to distribute the system intelligence, increasing its smartness.

To develop a standardized and efficient monitoring system applicable to various structures and to different diagnostics aims, the SHM design process should consider some key aspects grouped in four main clusters:

- Hardware
  - Sensor node (metrology issues)
  - Network (node synchronization, data transfer, gateway tasks)
- Software
  - Sensor node (microcontroller functions)
  - Gateway (pre-processing routines)
- Cloud
  - Architecture
  - Data analytics (post-processing)
- Structural diagnostics and health evaluation
  - Sensor layout optimization
  - Health evaluation: alerts, thresholds, and communication systems.

In the following paragraphs, each feature is presented into details.

2.1 Hardware

The hardware design turns around the kind of sensors to be adopted for monitoring. If dense sensing is the main requirement, a deep study about the target performances and those made available by the chosen sensors is a needed prerequisite.

Micro-Electro-Mechanical-System (MEMS) sensors are considered the most popular solution when the chosen solution aims at reconstructing a structure kinematics due to some external excitation. MEMS sensors are cheap; therefore, we must assess the proper trade-off between the available budget for measurements and the desired information. As these systems belong to the world of traditional electronics and semiconductor industry, they are easily scalable, to include different sensor kinds and possible network integration.

Fiber optic sensing, the other widely adopted solution today for SHM, has not been chosen for the monitoring systems described in this chapter, having different requirements and analysis procedures, and in addition a specific network design and a completely different economic impact [11, 12].

The hardware analysis presented below is focused on both the sensor level and the network level.

Sensor-Node: Metrology Issues

A first point is about metrology issues: very often MEMS sensors already provide a digital output. This makes harder to qualify the sensor performances alone, as the output comes from a system already including the analog-to-digital conversion. Some main problems faced in the past are not always solved in a satisfactory way, even today, like the lack of a sufficiently constant sampling rate, sometimes different from the nominal one [13]. The experimenter is, therefore, forced to adopt solutions like resampling or some hardware periodic re-alignment of all the clocks belonging to the same network: sending the timestamp together with every sample can be an option not always feasible when the network is already busy due to the big amount of transmitted data, and it is a waste of resources.

The same clock problems can be experienced at the microcontroller level. Provided the clock rate is at least constant in time, some techniques for damage detection can anyway be adopted, as for example, the detection of a local behavior change with respect to the past.

Remaining in the area of metrological qualification, another issue is sensitivity to temperature or any other environmental parameter affecting the output. If temperature is considered, the clock frequency depends upon its changes: that is why the capability to properly filter out its effect is one of the main concerns in proper data management [14]. Changes in temperature do not just alter the integrated circuit behavior, and they also have effects on all the mechanical interfaces among the sensor itself and the structure surface to be monitored, including the fixing clamps. A hard task is separating the parasitic effects, like those listed above, from some temperature effects which have to be preserved, namely those directly acting on the structure, attaining to its real behavior and providing useful information also for damage detection.

Resolution for MEMS sensors has to be properly tuned to the sensor noise floor and to the performances of the A/D converter: if this resolution is too fine, both the network and the memory are going to be overloaded, to store what most probably is meaningless electronic noise; for battery powered devices, this means higher energy consumption for both data storage and wireless transmission, if present. On the other side, it might happen that the sensor performances are not well exploited due to a too coarse A/D resolution and to the desire to spare memory energy and money.

Reading a number does not necessarily means that the number really makes sense: only the meaningful ones have to be managed. This is another difference among industrial continuous monitoring approaches and a simple dynamic test, when a limited number of sensors and a shorter testing time allow to possibly waste some resources. In case of dynamic tests, with a limited number of sensors working for a limited time, even wasting memory space, by oversampling or by the use of a high A/D converter bit number, is not considered a big issue.

One among the recognized problems in the use of MEMS sensors is their higher noise floor with respect to laboratory instrumentation: this means a harder detectability of a structure dynamic performances, in case of low amplitude signals, as these can remain buried below the noise threshold (a clear example is given in Fig. 3).

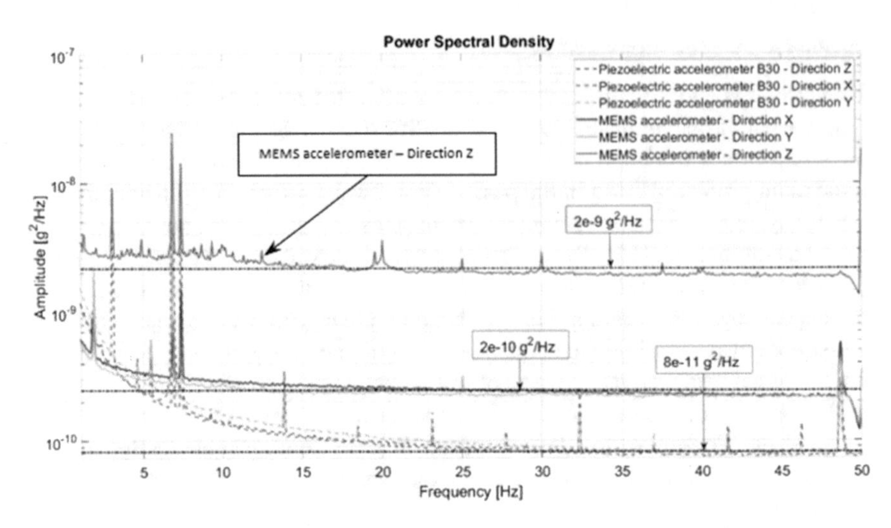

Fig. 3 Noise floor of MEMS accelerometers in comparison to piezo accelerometers

A positive aspect related to the use of cheap sensors is the possibility to implement new strategies oriented at reducing the effects of noise, such as using redundant or repeated measurements. Some preliminary attempts in the design phase of the new units have tried to use four inertial sensors on the same boards, both clinometers and accelerometers, with different aims. If one considers the same acceleration sensed by four sensors, then uncorrelated noise gets halved by a classical uncertainty analysis in case of "in band uncorrelated noise" [15]. In case of clinometers, the difference of the outputs from a twin configuration in which the two sensors are put close to each other with opposite sensing axis provides an amplification of the rotation output, at the same time deleting the effects of temperature; this only works if the sensitivity to this parameter on each MEMS component is the same. As pointed out in the introduction, this new strategy is allowed by the low overall cost increment, due to the increased number of bare sensors, but it has to be framed into an industrial approach, in which reliability of a new and complex system comes into play as the main need. A preliminary testing campaign has proven that the sensitivity of most MEMS sensors to temperature varies between each device, also for integrated circuits obtained from the same slice; that is why the clinometer temperature compensation is not always guaranteed or at least it is not always really achieved. At the same time, halving the accelerometer noise floor, by combining the output of four sensors, measuring the same quantity, has not been considered enough most structures provide a good enough S/N ratio, overcoming this problem; all the same some redundancy on the board design has been preserved. As in an industrial perspective, many devices have to be deployed at the same time, and an additional issue has been the need for proper calibration of a meaningful number of sensors together. The possibility to use static calibration also for accelerometers, ruled by international standards [16, 17], has been the adopted solution, provided a preliminary check against dynamic

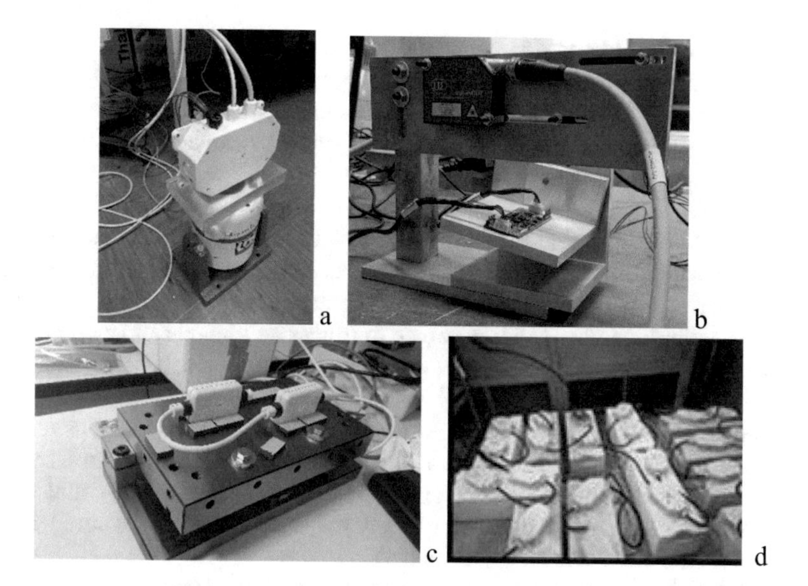

Fig. 4 Calibration of MEMS accelerometers (**a**), of MEMS clinometers (**b**, **c**), sensitivity to temperature in a temperature controlled room (**d**)

tests have been carried out too (see Fig. 4). Of course, a different strategy can also be implemented, with respect to traditional approaches, as continuous cross checks among redundant close sensors is a new possible approach to be developed in order to easily detect any device bad functioning or drift from good operation.

The clinometer calibration proved to be an easier task, as standard back-to-back approaches against a calibrated reference are an affordable and fast procedure [18]. In this case, careful attention has been paid to define a meaningful resolution limit, as the number of layers between the structure surface and the sensor has required a long and burdensome evaluation of any parasitic effect due to temperature. This has allowed to detect and account for any possible interference on measurements given by the behavior of glues, resins, plastic, and mechanical fixtures, used to transfer the rotation information through the mentioned intermediate layers.

Network: Node Synchronization

Each sensor node is also expected to manage the connections with the adjacent nodes and with all the rest of the network. A first issue has already been mentioned about the synchronization of nodes, allowing one to define different damage detection scenarios, if working on a single node or on the comparison among nodes.

Tools are available for at least a rough node synchronization, ranging from GPS to NTP protocols. These can be considered effective especially for some civil engineering applications, since the dynamics of interest for structures is usually confined in the low frequency range: even a not perfect synchronization does not alter the vibration mode detection in a meaningful way, as the percentage phase error is limited. By now, it has been stated not to implement such protocols, to avoid added complexity to the existing firmware governing the microcontroller: the adopted solutions have been considered in relation to the record lengths to be analyzed, with the possibility to periodically restart the recording and so realign the clocks over the network.

Network: Data Transfer

Another issue is the meaningful amount of data to be transferred, due to the presence of many sensors in the network: many solutions have been explored to make this traffic lighter. A first option is to implement the data elaboration at the sensor level (edge computing); another solution can be to trigger an immediate data transfer, in case of significant events, or possibly to store data locally and send them later to the cloud when the network is not so busy.

In our real case studies, we also tried new approaches. When the data streaming came mainly from static transducers like clinometers or temperature sensors, requiring a limited transfer rate, we opted for the use of power line solutions, so that the same cable bringing the power supply to the peripheral nodes could be also used to take measurements back (see Fig. 4).

In case of higher data flow rates, the typical case of acceleration measurements, the need to have robust, and reliable systems has led again to look for industrial solutions. Among the others, the use of field bus solutions, like the CAN bus, typical of the automotive industry, has been considered a winning asset, as protocols are already available, at the same time preventing from the risk of losing data (Fig. 5). As already pointed out, for such measurements, wireless solutions, though tempted several times in smaller networks, have not yet been considered reliable enough, especially when safety issues come into play.

Network: Gateway Tasks

The continuous improvement of electronics is rapidly making available to the market new devices to collect data in a smart way: Gateways are the heart of the new

Fig. 5 Cabled network installed on a bridge: a CAN bus connects the sensor nodes

monitoring systems, being capable at the same time to collect the big data streaming rates from the sensor nodes, to order and organize them, in the end to send them to a cloud server for any further processing. Gateways need to manage both cabled and wireless connections: in most monitoring systems, we are managing right now, i.e., bridges along highways crossing mountain areas, a stable and reliable connection to send data from the gateway to the cloud can be a challenging hazard. Problems here are mainly transferred to telecom networks, also requiring a detailed design in terms of the number of installed gateways to ensure that all data are properly collected, packed, and sent.

Another very important function, which has been intensively studied with the progress offered by new gateways, has been the development of "fog computing" approaches. The need to face SHM as an industrial process has forced to look for economically sustainable solutions. The cost of cloud resources is still high, especially if a wide number of sensors, having high data rates, is considered: that is why, the possibility to store data, at least temporarily, at the gateway level, seems to be more than attractive, also to guarantee some redundancy at least on the short term. The gateway can further perform some pre-analysis reducing the amount of data to be stored, also offering the chance to have a prompt response in case of anomalies: This is one of the most important trends for the close future.

Data processing at the gateways is an ongoing major technological trend, and cloud providers are offering solutions to load data processing SW components into such edge devices. This customer and application driven business logic often comes in the form of software "containers" that run in an isolated and secured environment. The gateway software architecture is such that important data are cached locally so that no information is lost even when Internet connectivity is temporarily unavailable (Fig. 5).

2.2 *Software*

Sensor Node: Microcontroller Functions

The selection of the functions to be transferred to the microcontroller has been the object of a long study.

As a conclusion of the optimization process, only some basic operations were designed to be performed at the sensor node level, the main function remaining the management of the data stream towards the gateways.

An accurate analysis must be carried out before any new software development. A reasonable compromise has to be reached between sensor performances and information required for a proper evaluation of the structure health status. In this respect, parameters have to be optimized considering the type of structure to be monitored and the physical phenomena under analysis. Sampling frequency, for example, should be customized based on the structure being monitored.

In case of structures like bridges, both sudden/brittle and slow/ductile events could happen, resulting in different requirements in terms of static and dynamic data processing. Moreover, modal parameter estimation is a key aspect of the structural identification that highly influences the data rate. For this reason, a relatively high sampling frequency is required, and the optimal value must be chosen in agreement with the structure eigenfrequencies. If tunnels are considered, these are subjected to deformation phenomena evolving slowly, and the interest is on the static rather than the dynamic behavior. In this case, a large amount of data is not necessary, and sampling frequency can be set to lower values: Specifically, the sensor output is sampled at 100 Hz, up to filling the available microcontroller memory. At this point average, standard deviation, maxima, and minima are evaluated and sent to the supervising unit.

Concerning acceleration measurements, the sensor node system is not provided with anti-aliasing filters; the risk of such phenomenon, due to undesired local effects, is always present. The microcontroller does not have problems in sampling at very fast rates; however, all these data cannot be managed in the network: that's why after sampling at 25.6 kHz at each sensor node, for sure preventing from aliasing, a filtering+down-sampling task is implemented on the sensor node, bringing the sampling frequency down to 100 Hz, considered suitable for most civil engineering applications and allowing for a continuous stream from three-axis accelerometers over industrial buses like CAN bus, provided a careful planning has been made. In fact, the aim is to find the best trade-off between the cable lengths and the required sensor data throughput.

In any case, if necessary, the network connection to the cloud allows one to remotely reprogram the sensor node firmware without the need to be on site.

This approach enables transferring data from all sensors to one or more gateways, where suitable processing can take place. The decision about which information needs to be sent to the cloud must often be taken on a case-by-case basis.

The basic info, like the peak values, the mean or the RMS, all of them calculated every second, have the value of a synthetic information content to be immediately evaluated at the sensor node, also providing some important diagnostic tools to assess good working conditions of the sensors and the electronics.

Gateway: Pre-processing Routines

The choice of which data need to be transmitted across the layers of our architecture and where they need to be processed is key success factors for the effectiveness and cost efficiency of the overall monitoring system. One peculiarity of such IoT systems is the need to operate on vectorial data (e.g., acceleration) on top of scalar ones (tilt, temperature), also requiring different data rates. As processing power at the edge (i.e., at the sensor level) is poised to increase exponentially, both microcontrollers in the sensor nodes and microprocessors in the gateways microcontrollers contribute to the transformation of raw data into meaningful information. Machine learning techniques are often used at the edge to perform inference tasks, such as recognizing anomalies or classifying different types of structure behavior. The accuracy of these tasks must kept under control even after system installation, and it is critical to be able to retrain a neural network or a machine learning task in field.

Another important aspect of these edge processing tasks is that they must be documented, replicable, and auditable because they act as the first data filtering stage, and any misbehavior may result in important or even critical information to be lost. This is why caution must be taken, and, as soon as an anomaly is suspected, raw data should always be sent upstream in the architecture up to where they can be analyzed more carefully and deeply.

As the sampling frequency may differ for each structure, the collected data quantity is variable. In any case, to guarantee a minimum of system redundancy, a larger amount of data are usually acquired, leading to problems in transmission, storage, and post-processing analysis, increasing costs and decreasing processing speed. Again, the challenge is finding the right balance between system capacity and information needed for structural characterization and health assessment. One possible solution could be to limit sensors acquisition only to time intervals in which the structure is really loaded, avoiding periods with no or too low external excitation (accounting for the sensors noise floor). In this way, selected datasets will always have a sufficient energy content, allowing for the evaluation of both long-term and short-term processes, the correct estimation of modal parameters and ensuring data acquisition during extreme events.

To implement this kind of smart acquisition, several issues have to be considered and fixed. Firstly, features able to drive data selection have to be identified, and these features should, at the same time, consist in simple functions, so that they can be performed by the gateway or even by the just mentioned sensor microcontroller. Then, thresholds have to be calculated for the chosen parameters: the idea is that when a certain number of sensors simultaneously exceed these values, data acquisition starts.

Groups of sensors designed to be responsible of starting the acquisition are another key point to be carefully considered. They are essential to give robustness and reliability to the system, avoiding the storage of redundant information, and guaranteeing data collection for all the significant scenarios.

For this reason, groups are various; their size is different and function of the expected phenomenon: in case of global events, such as an earthquake, for example, all the structure will be involved, and almost all sensors will detect it; conversely, when a local damage occurs, only sensors installed on the interested structural element or in its proximity will show anomalous values, so allowing, by means of procedures of modal shape identification, to localize with an acceptable approximation the region in which the local damage intervened. Furthermore, a robust cloud architecture should support an effective and reliable alert system (see Sect. 2.3). Once different alerts statuses have been defined, with related control parameters and threshold values, false alarms must be prevented, limiting the alerts in case of extreme events, when groups of sensors exceed the limit values.

2.3 Cloud

Architecture

The design of a cloud architecture that can scale and support large amounts of heterogeneous SHM systems has been based on the following principles: (1) data of different customers need to be kept isolated at all times, (2) real-time alerts should be handled in the minimum amount of time and clearly notified to interested staff, (3) cloud should support all the supply chain involved in the design, provisioning, production, testing, calibration, and installation of full SHM systems, (4) computing resources necessary to calculate slowly evolving structure trends on massive amounts of data are allocated dynamically and take advantage of the most advanced algorithm parallelization techniques, to minimize processing time and associated costs, (5) dependencies to cloud provider specific services has to be minimized.

For each SHM installation, gateway and device provisioning are part of the SHM design process; their configuration parameters are stored in a database to drive the production and testing process in the fab. After installation, sensor data are ingested by using an IoT platform and saved into S3 buckets as parquet files with Hive partitioning to ease later retrieval [19]. Real-time anomalies are detected during data ingestion, while slowly evolving phenomena and alarm thresholds are checked and adapted by running parallel Spark jobs in a Kubernetes cluster. This cluster runs pods for remote gateway control, a LoRaWAN network server, and other housekeeping tasks that are active 24/7. Once a day, the dimension of the cluster is increased to launch parallel jobs that query S3 buckets for each structure, run structure dependent algorithms, and store analysis results that can be visualized for the engineering staff.

Since an important part of sensor data processing needs to take place in the gateway, a solution derived from OpenStack has been developed; it enables its full remote control through a combination of web-sockets and MQTT, including injecting signed Python code, remapping/exporting service ports, and storing one week of full sensor data locally. If an anomaly is suspected after analysis in the cloud and more sensor data are needed, it can be obtained by querying the gateway within the one-week time buffer. At the present stage, no data cryptography has been implemented yet, though it is considered a fundamental next step.

Data Analytics (Post-processing)

As civil infrastructures are naturally subject to a gradual aging process, induced by the progressive physical deterioration or accidental actions, there is a specific need to make the damage evolution controllable and predictable over time through an effective diagnostic system that allows an adequate asset management and scheduled maintenance. Thus, the data collection, selection, and pre-processing must be followed by a thorough, more detailed, and comprehensive analysis process, in which the engineering judgment plays a key role in assessing the structural state of health and conservation. Model updating procedures based on sensors acquisitions, detailed modal analysis, and reliability evaluation are needed to explore critical scenarios, identify threshold limits, and develop algorithms to perform long-term data interpretation to support subsequent decision-making processes. The SHM system should be generally designed to display data that have been processed and synthesized to a simple, meaningful format, to guide the structural engineer and/or the asset owner to clearly understand and handle potential risks (ref. Sect. 2.4).

2.4 *Structural Diagnostics and Health Evaluation*

Sensor Layout Optimization

The idea of dense sensing creates a dramatic revolution in the approaches to SHM. The desire to standardize a product, relying on a strong information redundancy, has not yet forced the system designer to find algorithms for the best optimization in the sensor positioning and in their number, although this is a common practice widely addressed in literature [20]. The best preferred solution is to have a single type of sensing device, the already mentioned sensor node, with all the needed sensors on-board, then choosing only those sensors needed for the specific application, according to the specifications and requirements provided by the expert in charge of the structure evaluation (the other sensors on the board are shut down). In the end, as the sensor node chains should be as standard as possible to reduce costs; it is preferred to have redundant nodes, rather than tailoring a different system for any new structure, if possible. All the same, preliminary design is needed for an effective synthesis about the structure health; this is a very sensitive aspect, as sensors are to be placed according to the evaluation plan designed in advance and not the opposite. Again, an industrial approach to SHM leads to specific requirements: the system needs some flexibility and scalability, but an effort towards? a standard which can be easily replicated on different structures is a very strong link. Sometimes the ease of installation and maintenance is privileged with respect to the best position, and this is quite important for systems needing to work for years.

Sensors are typically grouped according to the overall monitoring objective first, and then according to structural or technological constraints. The grouping and placement of sensors are a design activity that takes place early in the monitoring planning and are subjected to optimization in terms of redundancy and cost tradeoffs. Sensors that belong to the same group typically share key parameters such as sampling frequencies, alert thresholds, and processing tasks to be applied to raw data.

Health Evaluation: Alerts, Thresholds, and Communication Systems

Anomaly conditions may be detected by individual sensors and reported to the gateway. Here, anomalous events coming from different sensors may be correlated (in space and time) to increase reliability of this type of inference. A trivial remark, at this point, is that the capability to distinguish among anomalies that may be due to sensor mechanical degradation, and real structural changes are the key in avoiding false positives. Alerts are propagated to the cloud with an estimated severity metadata attached, so that they can be handled with the appropriate urgency. As stated before, an alert should always come with raw data attached, so that an in-depth analysis can be done to confirm or deny the event. Similarly, the frequency of alerts and their correlation with other critical factors (weather conditions, time, season) should be kept under control and subject to periodical audit.

The complexity of such monitoring systems coupled to the need to provide a fast and correct assessment has pushed toward a "back to basics" evaluation of simple and robust approaches. Given the performances of the new sensors and networks, cheaper and denser, a careful review of the commonly adopted feature detection strategies is still running to explore their real effectiveness and capability to provide satisfactory health evaluations, accounting for the new process uncertainties.

The need of standardization required by the industrial approach is often winning over the requirements of optimization, also due to the need to guarantee a high redundancy level in case of sensor failures.

3 A Case Study on a Large Scale SHM System

The SHM system developed by Sacertis, including new sensor nodes, cloud, data analytics, and diagnostics, as outlined in the previous paragraphs, has been designed, physically realized, and it is currently active on more than 40 bridges and tunnels within the Italian road network. These infrastructures are equipped to provide continuous assessment of structural integrity, essential to detect the occurrence of structural changes, or the evolution of damage that could affect the performance and safety of a structure. In the present chapter, a case study on large-scale SHM system is presented, to highlight the effectiveness of an innovative, affordable, and minimally invasive monitoring system in the perspective of damage detection, structural diagnostics, and proactive asset maintenance.

Data analysis has been carried out following two quite different methods: a data processing technique related to the industrial approach to SHM, based on simple features and a complex cross check on the data reliability, which must provide simple and prompt information, helping the bridge manager with safety issues (Sect. 3.2); a second, more scientific approach, implements unsupervised learning techniques to track any eventual damage (Sect. 3.3). Having recognized a real damage during monitoring, a supervised approach could be implemented too; but, as a single failure was measured, this event was considered without real statistical relevance, therefore not suitable for classical classification tools. The two methods have been considered in a benchmark approach, one as the reference for the other. The first proved to be more robust and reliable to support an automatic alert and notification system; the second provided promising results, however, not yet at a technology readyness level (TRL) allowing it to be delivered as a final product. As mentioned earlier, the sensor output quality and the will to work with edge computing, to compress the data flow, has forced to favor the implementation of simplified approaches.

3.1 Long-Term Damage Detection Strategies on Bridge External Tendons

This section describes the application of MEMS accelerometers in a high performance, cost-effective, and fully automated SHM system installed on a highway concrete bridge located in a central region of Italy. The need to set-up a monitoring system was originated after the failure of one of the external tendons located in the hollow section of the box girder of the concrete bridge, due to an incorrect grout composition and consequent tendons corrosion, only two years after the construction completion. The final aim of the SHM system was to develop an effective long-term damage detection process to early identify further signs of the corrosion progression in time and check the behavior of the other tendons as well as the overall bridge. The use of external pre-stressing is increasing in motorway bridge structures due to the considerable construction time reduction. In this specific case, internal and external steel tendons are one of the main load-carrying components; as such, the integrity of these elements has to be controlled and guaranteed through the monitoring system, to avoid damage that could lead to catastrophic consequences for the entire structure [21–23]. The progressive collapse of each of the strands composing the cable proceeds until the tendon break is reached: The monitoring system has the aim to promptly detect the evolution of the phenomenon as corrosion proceeds. The installation of a real-time monitoring system also responds well to the arduous accessibility of the structure to carry out visual inspections, limiting any closer site observations and human interventions in case of any change of the monitored key structural performance indicators.

Description of the Structure and the Monitoring System

The monitored bridge has a total length of about 624 m and is characterized by a counterweight span of 30 m, four hyperstatic spans of 120 m, a hyperstatic span of 71 m, and an isostatic one of 43 m (see Fig. 6). The structural box girder cross-section height varies from 6.0 m (at the bearings) to 3.0 m (at the span centerline). The usual monitoring approach is to sense the deck with a number of devices capable of describing the bridge dynamics; if cables are the elements eventually damaged, the possibility to quickly and effectively sense a problem is given with some delay and uncertainty, being far from the damage onset. The new monitoring structure has allowed for a completely different plan: once recognized that the problem was the external tendons, each one has been instrumented with 2 three-axis accelerometer, also including additional sensors to get the environmental conditions inside the bridge deck, giving the occasion to practice on all the aspects mentioned earlier, up to the big data analysis [24–26].

The external bridge tendons have been instrumented with a total of 88 tri-axial accelerometers, 2 for each monitored tendon, installed at the fixed ends. Each tendon is made of 27 strands, and it is protected by an external polyethylene duct filled with injected grout. Due to a problem in the grout composition, the strands experienced a quick and extensive corrosion process that resulted in some tendon failures: luckily,

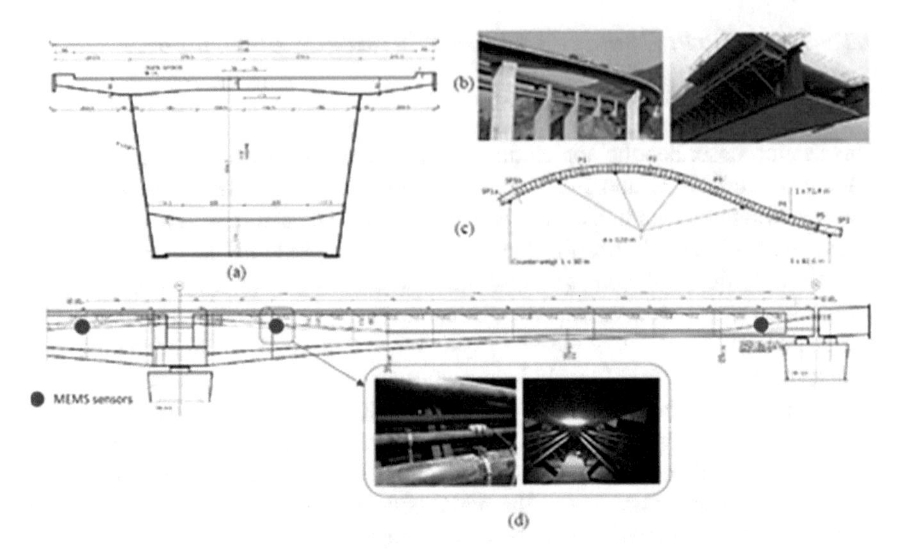

Fig. 6 Structural drawings of the bridge: (**a**) bridge cross-section; (**b**) bridge images after and during construction; (**c**) plan view of the highway bridge; (**d**) longitudinal section of a span, with sensor positioning

the bridge could survive even without some tendons, but these had to be substituted. The application of MEMS accelerometers on the ducts, so as close as possible to any possible damage, gave the structure owner a prompt and detailed warning about the real damage position, providing a measure of the fundamental properties necessary to predict the long-term performance of the bridge and foresee any potential further damages induced by the corrosion of the strands. Of course the price was a huge data flow, an important test bed for automatic early warning to detect damage.

The monitoring system includes (Fig. 7):

- the sensor nodes, comprehensive of tri-axial MEMS accelerometers (full scale $\pm$ 2.5 g, bandwidth of 50, 100 Hz sampling rate), temperature and humidity sensors, and a microcontroller used for data sampling. This choice complies with the need to save space, at the same time keeping some basic information: the band 0–50 Hz already includes some cable natural frequencies and, thus, fits for the application of several damage detection approaches.
- The CAN bus network to transfer the data from the end nodes to the local gateways (as the bridge is long and both the network and the data streaming could not be managed by a single unit, in this case, we had two at the bridge ends).
- The IoT gateway, where the data are collected, pre-preprocessed, and filtered to send a selection of significant information to the cloud for further analyses.
- The cloud, where data are stored and more complex data analytics can be performed.

The SHM system is active since September 20, 2017, and therefore, several years of data have been stored and analyzed, and the developed damage detection strategy

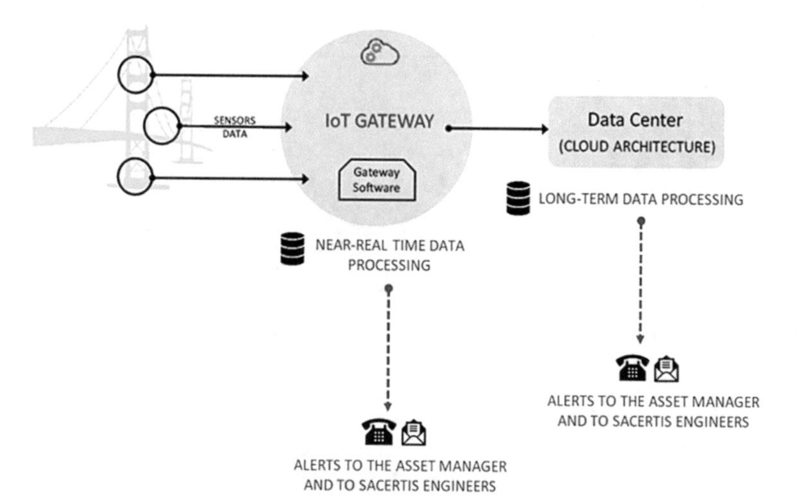

Fig. 7 Monitoring system overview

has been validated through a continuous long-term study, proving to be an effective tool for an efficient structural proactive maintenance (Fig. 7).

3.2 *Approach 1: Damage Detection Strategies and Results*

This monitoring strategy reflects the network structure and responds to need of providing both a quick and reliable near-real-time analysis to detect fragile and sudden anomalies (local abrupt structural damages, earthquakes vibrations,…) as well as the identification of long-term damage developed in a longer time-frame (as corrosion effects).

The choice has been directed toward a system based on two alarm levels:

- Level 1: alarm characterized by a check over the RMS (over 60 s), immediately estimated and verified at the IoT gateway level, with the aim to highlight the presence of anomalous vibration levels in the tendon dynamic responses under external loads.
- Level 2: alarm characterized by a check on a physical parameter, the natural frequencies of the cables, performed on the cloud platform and activated only if the first one is exceeded. This approach, among the many possible, has been considered for its ease of implementation, also due to the general high innovation and complexity of the whole system.

The Level 1 type of alert highlights the occurrence of anomalous vibrations of the tendons through the evaluation of the RMS level of each sensor over a sampling window of 60 s. The release of high energy levels, corresponding to impulses detected by the accelerometers, may be induced by random actions or to the brittle failure of

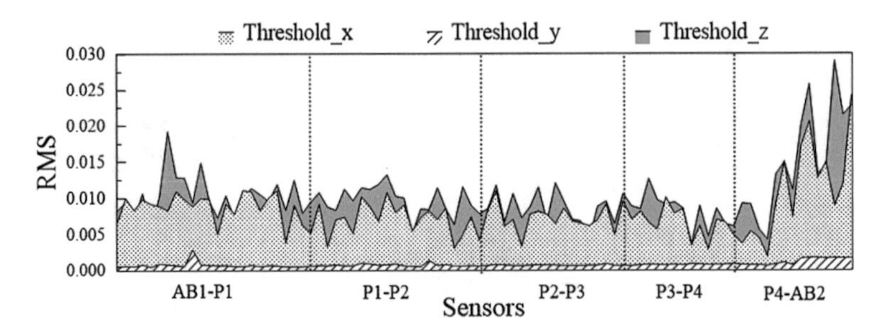

Fig. 8 Threshold values for all the 88 sensors—directions x, y, z

single strands. The RMS is compared to a threshold limit, in real-time, at the gateway level. The thresholds have been estimated for each sensor over a training dataset representative of the standard behavior of the tendon under external loads (traffic, wind, thermal loads, …). With the increase of the monitoring period, the thresholds have been recalculated over a longer training set until reaching a convergence (stabilization of statistical indicators), obtaining a robust alert level. The thresholds levels for each sensor and for each of the three measurement directions x, y, and z are shown in Fig. 8.

In case Level 1 thresholds are exceeded in a group of nearby sensors, a more complex data processing in the frequency domain is performed at the cloud level, based on a physical interpretation of the vibration data. The power spectral density is evaluated by averaging over a window duration of 200 s, (66% of overlap, Hanning window). If a strand collapses, the cross-section of the tendon is suddenly reduced, causing a series of complex phenomena in which the overall load is differently distributed inside each single tendon and among nearby tendons, anyway producing a natural frequencies drop; to identify a shift in the natural frequency of the tendons, the frequency domain decomposition (FDD) has been adopted. In case a change in the natural frequencies is detected (in this case detected over short times, so hardly depending upon environmental parameters), the second-level alert alarm is produced. In this section, two main events are described, occurred over the 4-year monitoring period, respectively, during November and December 2017, showing the effectiveness of the two-level alert approach. As shown in Fig. 9, on 11/19/2017, the RMS threshold was exceeded by all the accelerometers of the bridge, pointing at an anomaly occurring to the entire structure simultaneously. As per the alert procedure, the second-level analysis was triggered, confirming that the natural frequency of the tendons did not experience any variation. The sudden increase of the vibration levels was caused by an earthquake. The SHM system correctly pointed out the anomalous condition, proving its reliability and robustness, and confirmed the structural integrity with the second-level check.

In December 2017, instead, the first level threshold was exceeded by a limited number of sensors, all of them installed in the first span of the bridge. Again, the second-level analysis was automatically activated for all the sensors that exceeded the first threshold. As shown in Fig. 10, one of the tendons installed in the first span

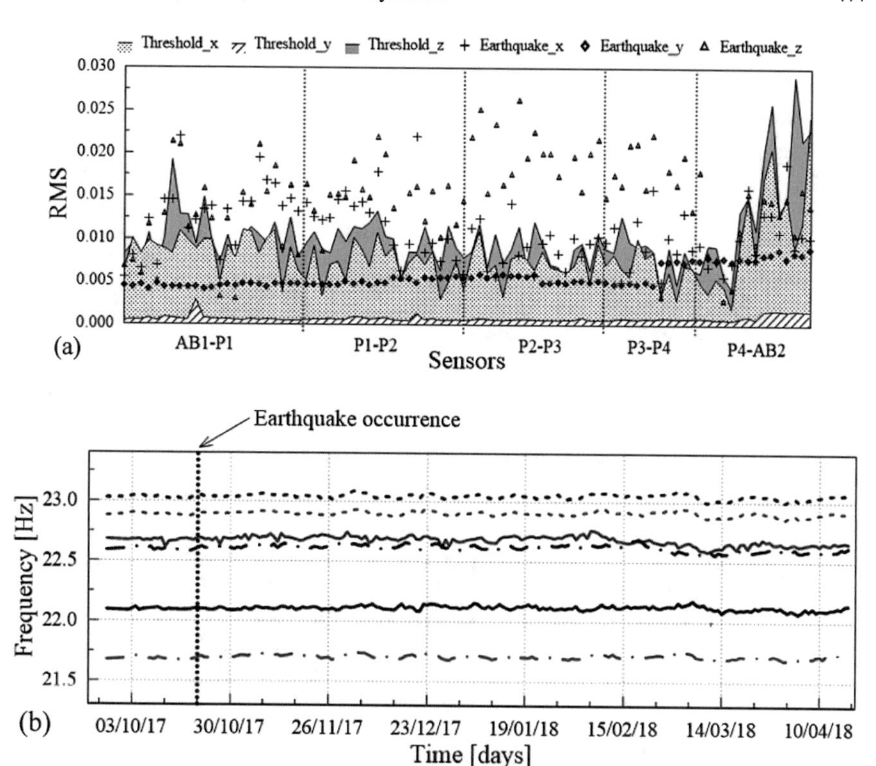

Fig. 9 (**a**) Level 1 and (**b**) Level 2 check on the earthquake excitation on the 19/11/2017

of the monitored bridge also exceeded the second threshold level, experiencing a frequency reduction caused by the failure of one of the strands, and a partial loss of prestress force in the structural element. This reduction in the stress distribution is a function of the severity of the tendon damage (the number of broken strands) and of the damage location. The energy released when the strand broke was recorded as a dynamic input also by the nearby sensors.

During the 4 years monitoring period, Sacertis system automatically detected the failure of strands on other 3 tendons, providing a clear overview of the pre stressing status of the structure, and significantly improving the maintenance schedule of the bridge.

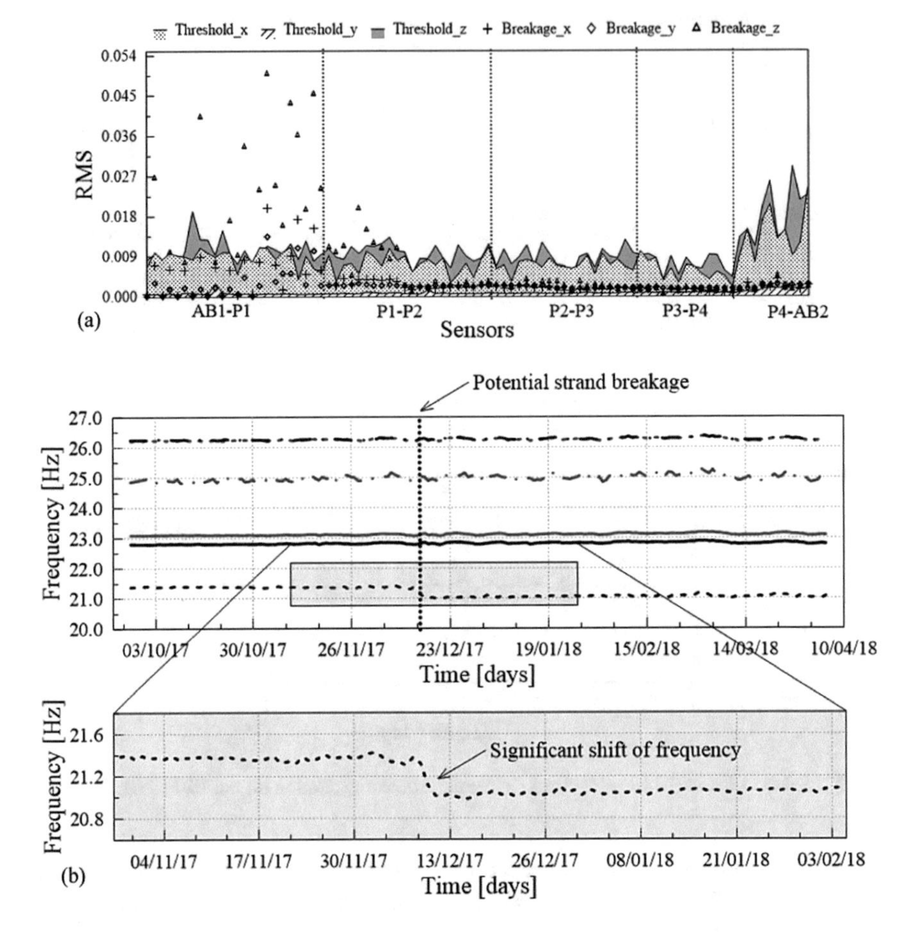

Fig. 10 (**a**) Level 1 and (**b**) Level 2 check on the 12/21/2017 event

3.3 Approach 2: An Unsupervised Method with Auto Regressive Models and PCA

The choice adopted for a data-driven approach, to detect damage through synthetic features, has relied on an unsupervised strategy consisting in a proper mix of autoregressive models AR [27] and principal component analysis PCA [28]. The underlying idea was to separate the effect of environmental quantities out of the dynamic features pointing at damage.

Given X_t, the value of the time record X at time t and X_{t-j}, at time $t - j$, it is possible to define the autoregressive model of order p, namely $AR(p)$, thanks to the well-known expression:

$$X_t = \sum_{j=1}^{p} \phi_j X_{t-j} + a_t \tag{1}$$

Specific interest is devoted to the weights ϕ_j for the p preceding steps used to get the actual value in the time series and to the residual a_t, expected to belong to a random Gaussian distribution with zero mean $\mu_a = 0$ and variance σ_a^2.

A first step in the effort to compress information has consisted in the optimization for the p value: the adopted tool has been the Bayesian Information Criterion (BIC) [29], consisting in a likelihood method including a penalty as the number of autoregressive coefficients increases, avoiding overfitting.

The founding steps of both AR and PCA are well known in damage detection: the novelty here is their combined use. We chose to focus on implementation aspects and on the main results. In the following, attention is devoted to a single sensor (1S33) over a period between October 2017 and April 2018; this is considered representative of the whole sensor set: the cloud resources had to be shared between ordinary analysis, routinely performed on all sensors to provide real-time alarms, and the present application.

Each record stored on the cloud consists of 200 s sampled at 100 Hz: the bridge is always very busy, subject to a continuous random and high-level input excitation; the AR approach has been applied to records of the same length as the file length. As the accelerometers are g-sensitive, the average value has been high pass filtered. A data normalization is considered a good practice, as the AR coefficients are representative of the poles and not of the zeroes: the input data $\tilde{X}$ to AR models is:

$$\tilde{X} = \frac{X_{\text{orig}} - \mu_{X_{\text{orig}}}}{\sigma_{X_{\text{orig}}}} \tag{2}$$

being $\mu_{X_{\text{orig}}}$ e $\sigma_{X_{\text{orig}}}$ the mean and standard deviation of the original data X_{orig}.

For the considered data, we have precious information about the ongoing damage, occurring at subsequent steps; although the monitoring system has been installed on a bridge already in service, we have considered the first dataset, up to the first known damage event, as a healthy baseline.

The Bayesian Information Criterion (BIC) has been applied to a relevant quantity of data: an example is given in Fig. 11, showing the estimated maximum order of the AR model evaluated over 200 records belonging to the training set, in which the bridge is assumed under reference undamaged conditions.

Although some scattering, the mean has been considered a reasonable compromise.

To prove the quality of the adopted approach, Fig. 12 provides an original record and its reconstruction given by the AR (34) model. A further confirmation about the goodness of fit is proved by the Gaussian distribution of residuals. The next step has then consisted in modeling every record of the training set by means of an AR (34) model. Figure 13 provides the 34 ϕ coefficients evaluated over the period between 10/1/2017 and 4/7/2018 for the selected accelerometer, vertical axis.

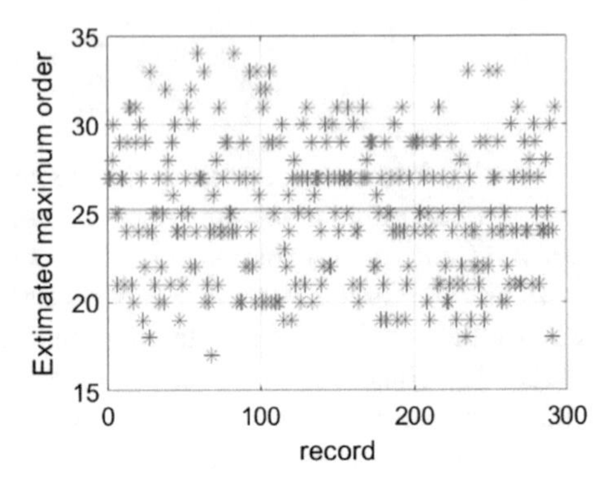

Fig. 11 BIC evaluation over 200 baseline undamaged records

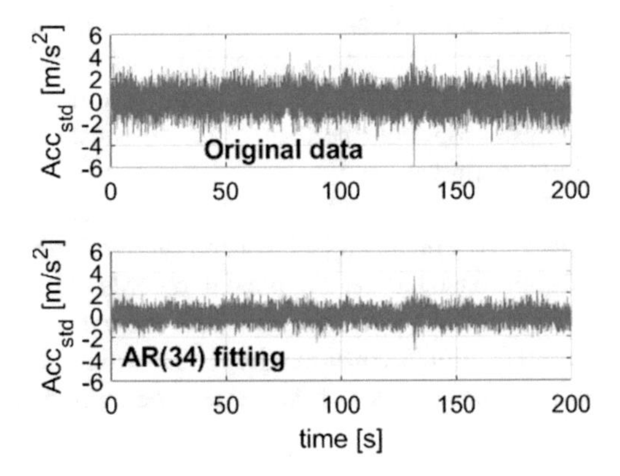

Fig. 12 Original (top) and reconstructed (bottom) time record

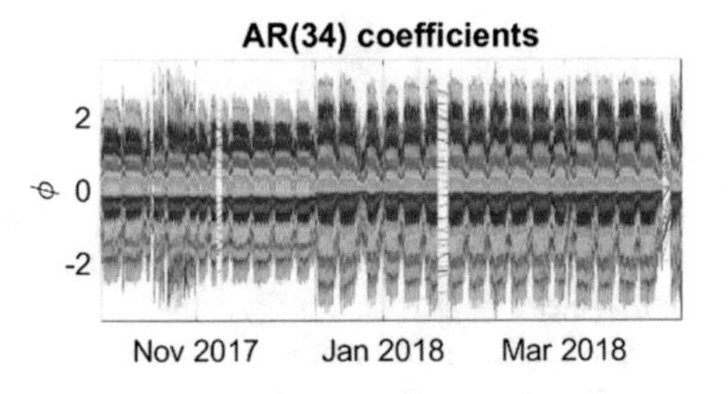

Fig. 13 Trend over time of 34 AR coefficients

The 34 coefficients are not yet a synthetic damage feature; that is why a further step consisted in compressing the available information in a lower number of coefficients $m \ll p$. Once evaluated the PCA over the whole AR coefficient set, the comparison of each eigenvalue of the covariance matrix against their sum (the variance of the whole set) provides the histogram in Fig. 14: the first three principal component values provide around 97% of the total data variability. Instead of the 34 AR coefficients; these three components are a much more compressed information set, allowing one to preserve most of the desired information content (Fig. 15).

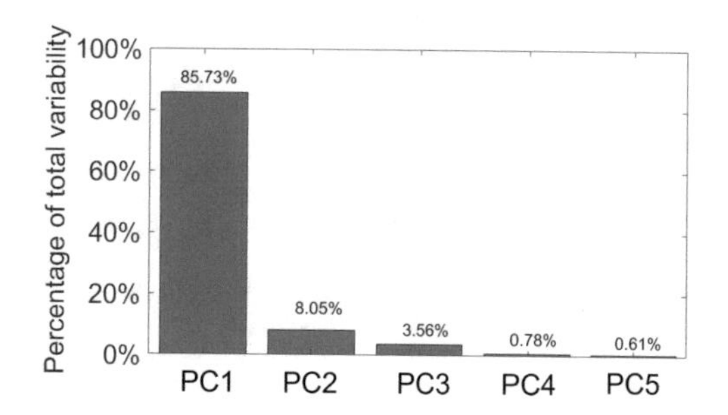

Fig. 14 Percentage of total variability accounted for by each of the first five principal components of the AR coefficients

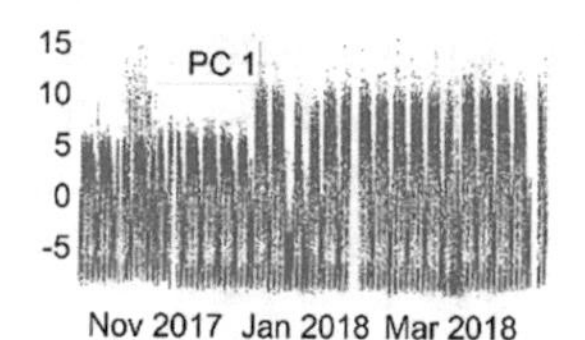

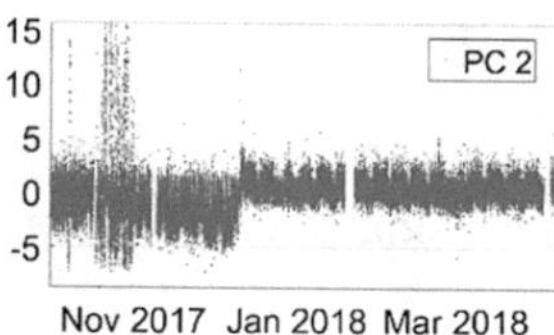

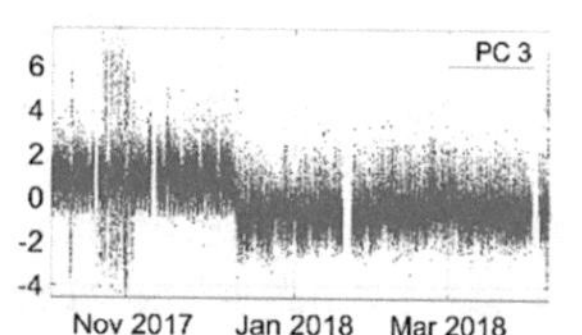

Fig. 15 Trend over time of the first three principal components of the autoregressive coefficients

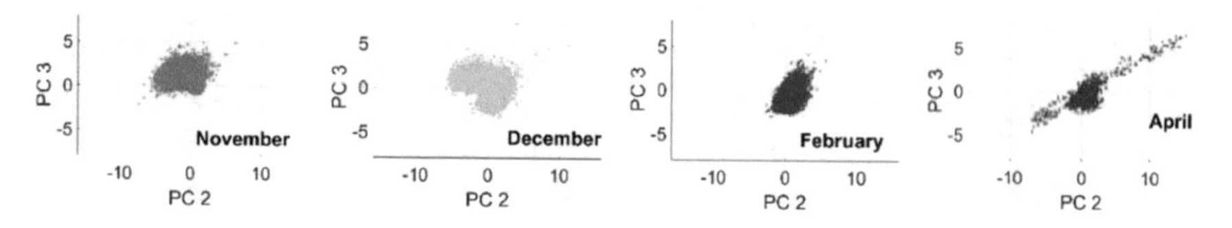

Fig. 16 Score plot of the second and third principal components of the autoregressive coefficients (November 2017, December 2017, February 2018, April 2018)

As an example, the first three principal component coefficients are represented in Fig. 16: they refer to the already mentioned accelerometer, spanning a period from 1/10/2017 to the cable complete failure, in April 2018.

All the three principal components show a marked discontinuity on December 11, 2017, when the analyses of Sect. 3.2 detected an anomaly. A similar approach carried out for a nearby tendon, considered healthy and close to the damaged one, does not show any discontinuity.

As the process is mainly described by three PCA coefficients, their representation in a 3D space or the projection over a plane can have an immediate and effective visual meaning. It is interesting to plot the projection on the Component 2 / Component 3 plane, during the considered period (Fig. 16). The point cloud changes its shape moving from a circular contour toward an elliptical one, denoting a change in one of the features describing the tendon behavior: December appears to be the month in which the transition occurs.

A further step consisted in trying to filter the environmental effects (mainly temperature) out of the considered dataset. To better describe the situation, a zoom over a shorter time interval has been carried out in Fig. 17 for the three considered

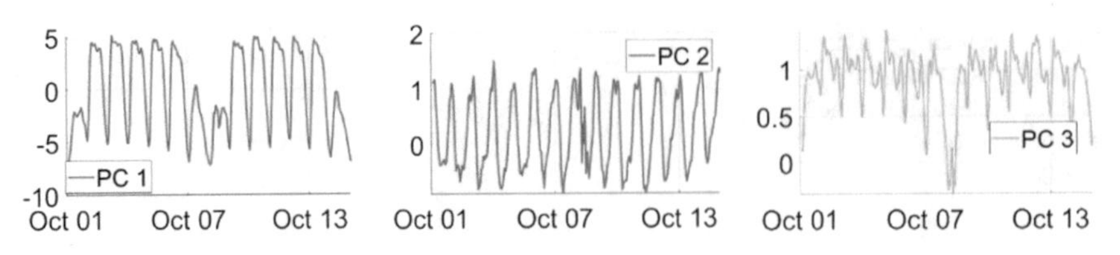

Fig. 17 Trend over time of the first three principal components of the autoregressive coefficients (from 10/01/2017 to 10/15/2017)

PCA coefficients, respectively (data collected from 10/01/2017 to 10/15/2017): data have been averaged every hour. All coefficients show a clear daily trend; the first principal component also shows a weekly trend: Oct 1st, Oct 8th, and Oct 14th are Sundays, and the corresponding peaks are lower. The same is not happening for the other two components, having similar peak-to-peak value, independent from the considered day of the week.

This suggests linking the first principal component to traffic, lowering during the night and also during the weekend, while the second PCA coefficient, having a similar daily trend, but independent from the day of the week, can be linked to temperature. As the principal components are by definition uncorrelated, this fact offers a way to filter out the effects of any environmental quantity. This study offers a challenging perspective, and it is still being carried out; as the database gets richer and richer, at the moment, an extensive benchmarking is being carried out to assess the bridge health status without the support and results confirmation offered by other methods.

4 Conclusions

This chapter aims at demonstrating how an industrial approach to SHM implies a completely new design of SHM systems, requiring contemporary innovation in several fields. Data analytics is a fundamental part within a complex framework in which new sensors, new network designs, digital twins, and final evaluation criteria have to interact since the system preliminary design, to create a sustainable and effective process for structural health control. In the end, a real application shows the feasibility of the proposed approaches. Further methods and systems improvements are yet to come, as the monitoring field develops over time. The uncertainties related to the new approaches, richer in information content, cheaper but with performances still to be fully understood, should be progressively identified and dealt with, as well as checking the applicability of traditional damage features to the new networks. The final aim is to review the whole process under the light of risk-based and value of information analyses, to get to the best design of the new monitoring systems.

References

1. Farrar CR, Worden K (2012) Structural health monitoring: a machine learning perspective. Wiley. ISBN: 978-1-119-99433-6
2. Cawley P (2018) Structural health monitoring: Closing the gap between research and industrial deployment. Struct Health Monit 17(5):1225–1244. https://doi.org/10.1177/1475921717750047
3. Augustin A, Yi J, Clausen T, Townsley WM (2016) A study of LoRa: long range & low power networks for the internet of things. Sensors 16:1466. https://doi.org/10.3390/s16091466
4. Alonso L, Barbarán J, Chen J, Díaz M, Llopis L, Rubio B (2018) Middleware and communication technologies for structural health monitoring of critical infrastructures: a survey. Comput Standards & Interf 56: 83–100, ISSN 0920-5489. https://doi.org/10.1016/j.csi.2017.09.007
5. Arcadius Tokognon C, Gao B, Tian GY, Yan Y (2017) Structural health monitoring framework based on internet of things: a survey. IEEE Internet of Things J 4(3): 619–635. https://doi.org/10.1109/JIOT.2017.2664072
6. Valeske B, Osman A, Römer F, Tschuncky R (2020) Next generation NDE sensor systems as IIoT elements of Industry 4.0. Res Nondestruct Evaluat 31(5–6):340–369. https://doi.org/10.1080/09349847.2020.1841862
7. Farrar CR, Park G, Allen DW, Todd MD (2006) Sensor network paradigms for structural health monitoring. Struct Control Health Monit 13:210–225. https://doi.org/10.1002/stc.125
8. Sousa H, Wenzel H, Thöns S (2019) Quantifying the value of structural health information for decision support guide for operators. COST Action TU 1402
9. Diamantidis D, Sykora M, Sousa H (2019) Quantifying the value of structural health information for decision support guide for practicing engineers. COST Action TU 1402
10. Thöns S (2019) Quantifying the value of structural health information for decision support guide for scientists. COST Action TU 1402
11. Ansari F (2007) Practical implementation of optical fiber sensors in civil structural health monitoring. J Intell Mater Syst Struct 18(8):879–889. https://doi.org/10.1177/1045389X06075760
12. Wu T, Liu G, Fu S, Xing F (2020) Recent progress of fiber-optic sensors for the structural health monitoring of civil infrastructure. Sensors 20(16):4517. https://doi.org/10.3390/s20164517
13. Cheung W-S (2020) Effects of the sample clock of digital-output MEMS accelerometers on vibration amplitude and phase measurements. Metrologia 57(1)
14. Cigada A, Lurati M, Redaelli M, Vanali M (2007) Mechanical performance and metrolocigal characterization of mems accelerometers and application in modal analysis. In: IMAC XXV (International modal analysis conference), Orlando, 19–22 Feb 2007
15. JCGM 100 series—Guides to the expression of uncertainty in measurement (GUM series) ISO
16. ISO16063-16: Methods for the calibration of vibration and shock transducers—Part 16: Calibration by Earth's gravitation
17. ISO16063-21 Methods for the calibration of vibration and shock transducers—Part 21 Vibration calibration by comparison to a reference transducer
18. Zhu J, Wang W, Huang S, Din W (2020) An improved calibration technique for MEMS accelerometer based inclinometers. Sensors 20:452. https://doi.org/10.3390/s20020452
19. Levin A, Garion S, Kolodner EK, Lorenz DH, Barabash K, Kugler M, McShane N (2019) AIOps for a cloud object storage service. In: 2019 IEEE international congress on big data (BigData Congress). IEEE. https://doi.org/10.1109/BigDataCongress.2019.00036
20. Flynn EB, Todd MD (2010) A Bayesian approach to optimal sensor placement for structural health monitoring with application to active sensing. Mechan Syst Signal Proces 24:891–903
21. Anania L, Badalà A, D'Agata G (2018) Damage and collapse mode of existing post tensioned precast concrete bridge: the case of Petrulla viaduct. Eng Struct 162:226–244
22. Colajanni P, Recupero A, Ricciardi G, Spinella N (2016) Failure by corrosion in PC bridges: a case history of a viaduct in Italy. Int J Struct Integr 7(2):181–193

23. Minh H, Mutsuyoshi H, Niitani K (2007) Influence of grouting condition on crack and load-carrying capacity of post-tensioned concrete beam due to chloride-induced corrosion. Constr Build Mater 21(7):1568–1575
24. Levin A, Garion S, Kolodner EK, Lorenz DH, Barabash K (2019) IBM Research—Haifa AIOps for a cloud object storage service. 2019 IEEE international congress on big data
25. Cigada A, Lucà F, Malavisi M, Mancini G (2021) A damage detection strategy on bridge external tendons through long-time monitoring. In: Conference proceedings of the society for experimental mechanics series, pp 159–168
26. Bertagnoli G, Lucà F, Malavisi M, Melpignano D, Cigada A (2020) A large scale SHM system: a case study on pre-stressed bridge and cloud architecture. In: Conference proceedings of the society for experimental mechanics series, pp 75–83
27. Pandit SM, Wu S-M (2000) Time series and system analysis with applications. Krieger Publishing Company
28. Jolliffe IT (2002) Principal component analysis. Springer, Berlin
29. Ljung L (1999) System identification: theory for the user. Prentice-Hall, Englewood Cliffs

Printed in the United States
by Baker & Taylor Publisher Services